D1700155

Catalysis for a Sustainable Environment

Catalysis for a Sustainable Environment

Reactions, Processes and Applied Technologies

Volume 2

Edited by

Professor Armando J. L. Pombeiro
Instituto Superior técnico
Lisboa, Portugal

Dr. Manas Sutradhar
Universidade Lusófona de Humanidades e Tecnologias
Faculdade de Engenharia
Lisboa, Portugal

Professor Elisabete C. B. A. Alegria
Instituto Politécnico de Lisboa
Departamento de Engenharia Química
Lisboa, Portugal

This edition first published 2024
© 2024 John Wiley and Sons Ltd

All rights reserved. No part of this publication may be reproduced, stored in a retrieval system, or transmitted, in any form or by any means, electronic, mechanical, photocopying, recording or otherwise, except as permitted by law. Advice on how to obtain permission to reuse material from this title is available at http://www.wiley.com/go/permissions.

The right of Armando J.L. Pombeiro, Manas Sutradhar, and Elisabete C.B.A. Alegria to be identified as the author of the editorial material in this work has been asserted in accordance with law.

Registered Offices
John Wiley & Sons, Inc., 111 River Street, Hoboken, NJ 07030, USA
John Wiley & Sons Ltd, The Atrium, Southern Gate, Chichester, West Sussex, PO19 8SQ, UK

Editorial Office
The Atrium, Southern Gate, Chichester, West Sussex, PO19 8SQ, UK

For details of our global editorial offices, customer services, and more information about Wiley products visit us at www.wiley.com.

Wiley also publishes its books in a variety of electronic formats and by print-on-demand. Some content that appears in standard print versions of this book may not be available in other formats.

Trademarks: Wiley and the Wiley logo are trademarks or registered trademarks of John Wiley & Sons, Inc. and/or its affiliates in the United States and other countries and may not be used without written permission. All other trademarks are the property of their respective owners. John Wiley & Sons, Inc. is not associated with any product or vendor mentioned in this book.

Limit of Liability/Disclaimer of Warranty
In view of ongoing research, equipment modifications, changes in governmental regulations, and the constant flow of information relating to the use of experimental reagents, equipment, and devices, the reader is urged to review and evaluate the information provided in the package insert or instructions for each chemical, piece of equipment, reagent, or device for, among other things, any changes in the instructions or indication of usage and for added warnings and precautions. While the publisher and authors have used their best efforts in preparing this work, they make no representations or warranties with respect to the accuracy or completeness of the contents of this work and specifically disclaim all warranties, including without limitation any implied warranties of merchantability or fitness for a particular purpose. No warranty may be created or extended by sales representatives, written sales materials or promotional statements for this work. The fact that an organization, website, or product is referred to in this work as a citation and/or potential source of further information does not mean that the publisher and authors endorse the information or services the organization, website, or product may provide or recommendations it may make. This work is sold with the understanding that the publisher is not engaged in rendering professional services. The advice and strategies contained herein may not be suitable for your situation. You should consult with a specialist where appropriate. Further, readers should be aware that websites listed in this work may have changed or disappeared between when this work was written and when it is read. Neither the publisher nor authors shall be liable for any loss of profit or any other commercial damages, including but not limited to special, incidental, consequential, or other damages.

A catalogue record for this book is available from the Library of Congress

Hardback ISBN: 9781119870524; ePub ISBN: 9781119870630; ePDF ISBN: 9781119870623;
oBook ISBN: 9781119870647

Cover image: © Sasha Fenix/Shutterstock
Cover design by Wiley

Set in 9.5/12.5pt STIXTwoText by Integra Software Services Pvt. Ltd, Pondicherry, India
Printed and bound by CPI Group (UK) Ltd, Croydon, CR0 4YY

Contents

VOLUME 1

About the Editors *xiii*
Preface *xv*

1 **Introduction** *1*
   Armando J.L. Pombeiro, Manas Sutradhar, and Elisabete C.B.A. Alegria

   Part I Carbon Dioxide Utilization *5*

2 **Transition from Fossil-C to Renewable-C (Biomass and CO_2) Driven by Hybrid Catalysis** *7*
   Michele Aresta and Angela Dibenedetto

3 **Synthesis of Acetic Acid Using Carbon Dioxide** *25*
   Philippe Kalck

4 **New Sustainable Chemicals and Materials Derived from CO_2 and Bio-based Resources: A New Catalytic Challenge** *35*
   Ana B. Paninho, Malgorzata E. Zakrzewska, Leticia R.C. Correa, Fátima Guedes da Silva, Luís C. Branco, and Ana V.M. Nunes

5 **Sustainable Technologies in CO_2 Utilization: The Production of Synthetic Natural Gas** *55*
   M. Carmen Bacariza, José M. Lopes, and Carlos Henriques

6 **Catalysis for Sustainable Aviation Fuels: Focus on Fischer-Tropsch Catalysis** *73*
   Denzil Moodley, Thys Botha, Renier Crous, Jana Potgieter, Jacobus Visagie, Ryan Walmsley, and Cathy Dwyer

7 **Sustainable Catalytic Conversion of CO_2 into Urea and Its Derivatives** *117*
   Maurizio Peruzzini, Fabrizio Mani, and Francesco Barzagli

Part II Transformation of Volatile Organic Compounds (VOCs) *139*

8 Catalysis Abatement of NO$_x$/VOCs Assisted by Ozone *141*
Zhihua Wang and Fawei Lin

9 Catalytic Oxidation of VOCs to Value-added Compounds Under Mild Conditions *161*
Elisabete C.B.A. Alegria, Manas Sutradhar, and Tannistha R. Barman

10 Catalytic Cyclohexane Oxyfunctionalization *181*
Manas Sutradhar, Elisabete C.B.A. Alegria, M. Fátima C. Guedes da Silva, and Armando J.L. Pombeiro

Part III Carbon-based Catalysis *207*

11 Carbon-based Catalysts for Sustainable Chemical Processes *209*
Katarzyna Morawa Eblagon, Raquel P. Rocha, M. Fernando R. Pereira, and José Luís Figueiredo

12 Carbon-based Catalysts as a Sustainable and Metal-free Tool for Gas-phase Industrial Oxidation Processes *225*
Giulia Tuci, Andrea Rossin, Matteo Pugliesi, Housseinou Ba, Cuong Duong-Viet, Yuefeng Liu, Cuong Pham-Huu, and Giuliano Giambastiani

13 Hybrid Carbon-Metal Oxide Catalysts for Electrocatalysis, Biomass Valorization and, Wastewater Treatment: Cutting-Edge Solutions for a Sustainable World *247*
Clara Pereira, Diana M. Fernandes, Andreia F. Peixoto, Marta Nunes, Bruno Jarrais, Iwona Kuźniarska-Biernacka, and Cristina Freire

VOLUME 2

About the Editors *xiii*
Preface *xv*

Part IV Coordination, Inorganic, and Bioinspired Catalysis *299*

14 Hydroformylation Catalysts for the Synthesis of Fine Chemicals *301*
Mariette M. Pereira, Rui M.B. Carrilho, Fábio M.S. Rodrigues, Lucas D. Dias, and Mário J.F. Calvete
14.1 Introduction *301*
14.2 Homogeneous Catalytic Systems *303*
14.2.1 Development of Phosphorus Ligands *303*
14.2.2 Hydroformylation of Biologically Relevant Substrates *307*
14.2.3 Hydroformylation-based Sequential Reactions *309*
14.3 Heterogeneized Catalytic Systems *313*

14.4	Conclusions *320*	
	References *320*	

15	**Synthesis of New Polyolefins by Incorporation of New Comonomers** *323*	
	Kotohiro Nomura and Suphitchaya Kitphaitun	
15.1	Introduction *323*	
15.2	Synthesis of New Ethylene Copolymers by Incorporation of Sterically Encumbered Olefins, Cyclic Olefins *325*	
15.2.1	Ethylene Copolymerization with Sterically Encumbered Olefins *325*	
15.2.2	Ethylene Copolymerization with Cyclic Olefins *327*	
15.3	Ethylene Copolymerization with Alken-1-ol for Introduction of Hydroxy Groups into Polyefins *330*	
15.4	Synthesis of Biobased Ethylene Copolymers by the Incorporation of Linear and Cyclic Terpenes *332*	
15.5	Concluding Remarks and Outlook *334*	
	Acknowledgements *335*	
	References *335*	

16	**Catalytic Depolymerization of Plastic Waste** *339*	
	Noel Angel Espinosa-Jalapa and Amit Kumar	
16.1	Introduction *339*	
16.2	Pyrolysis *340*	
16.3	Gasification *340*	
16.4	Solvolysis *342*	
16.4.1	Hydrolysis *342*	
16.4.2	Glycolysis and Methanolysis *344*	
16.4.3	Aminolysis *344*	
16.5	Hydrogenation *345*	
16.5.1	Hydrogenative Depolymerisation of Polyesters *345*	
16.5.2	Hydrogenative Depolymerisation of Polycarbonates *348*	
16.5.3	Hydrogenative Depolymerisation of Nylons and Polyurethanes *350*	
16.5.4	Hydrogenative Depolymerisation of Polyureas *354*	
16.6	Depolymerisation of Polyethylene Using Alkane Metathesis *356*	
16.7	Hydrogenolysis *357*	
16.8	Hydrosilylation and Hydroboration *359*	
16.9	Summary *363*	
	References *363*	

17	**Bioinspired Selective Catalytic C-H Oxygenation, Halogenation, and Azidation of Steroids** *369*	
	Konstantin P. Bryliakov	
17.1	Introduction *369*	
17.2	The Mechanistic Frame of Bioinspired Oxidative C-H Functionalizations *370*	
17.3	Transition Metal Catalyzed C-H Oxofunctionalizations: Early Examples *372*	
17.4	Transition Metal Catalyzed C-H Oxofunctionalizations with H_2O_2 *375*	
17.5	Transition Metal Catalyzed Halogenations and Azidations *381*	
17.6	Conclusions and Outlook *383*	
	Acknowldegment *384*	
	References *384*	

18	**Catalysis by Pincer Compounds and Their Contribution to Environmental and Sustainable Processes** *389*
	Hugo Valdés and David Morales-Morales
18.1	Introduction *389*
18.2	Hydrogenation of CO_2 to Methanol and Related Processes *390*
18.3	Hydrogen Production from Methanol and Water *399*
18.4	Biomass Transformation *402*
18.5	Conclusions *403*
	Acknowledgments *404*
	References *404*
19	**Heterometallic Complexes: Novel Catalysts for Sophisticated Chemical Synthesis** *409*
	Franco Scalambra, Ismael Francisco Díaz-Ortega, and Antonio Romerosa
19.1	Introduction *409*
19.2	C-X Formation (X = C, H, N, O, Metal) *410*
19.3	Oxidation Processes *414*
19.4	CO_2 Activation *416*
19.5	O_2 and H_2 Generation from Water *420*
19.6	Conclusions *424*
	Note *424*
	Acknowledgments *424*
	References *425*
20	**Metal-Organic Frameworks in Tandem Catalysis** *429*
	Anirban Karmakar and Armando J.L. Pombeiro
20.1	Introduction *429*
20.2	MOFs as Catalysts for Tandem Reactions *430*
20.2.1	Active Sites in MOFs *430*
20.2.2	Advantages and Limitations of MOFs *431*
20.3	Examples of MOFs Used as Catalysts for Tandem Reactions *431*
20.3.1	Deacetalization-Knoevenagel Condensation *431*
20.3.2	Deacetalization-Henry Reaction *433*
20.3.3	Meinwald rearrangement-Knoevenagel Condensation *434*
20.3.4	Reductive Amination of Aldehydes with Nitroarenes *435*
20.3.5	Epoxidation–Ring Opening of Epoxide *436*
20.3.6	Oxidation-Esterification *437*
20.3.7	Oxidation-Hemiacetal Reaction *438*
20.3.8	Asymmetric Tandem Reactions *439*
20.3.9	Photocatalytic Tandem Reactions *440*
20.4	Conclusions *441*
	Acknowledgements *442*
	References *442*
21	**(Tetracarboxylate)bridged-di-transition Metal Complexes and Factors Impacting Their Carbene Transfer Reactivity** *445*
	LiPing Xu, Adrian Varela-Alvarez, and Djamaladdin G. Musaev
21.1	Introduction *445*

21.2	Computational Procedure *447*	
21.3	Results and Discussion *447*	
21.3.1	Geometrical and Electronic Structures of the (RCOO)$_4$-[M$_2$] Complexes *447*	
21.3.2	Diazocarbene Decomposition by the Reported [RCOO]$_4$–[M$_2$] Complexes *455*	
	Acknowledgement *457*	
	Note *457*	
	References *457*	

22 **Sustainable Cu-based Methods for Valuable Organic Scaffolds** *461*
Argyro Dolla, Dimitrios Andreou, Ethan Essenfeld, Jonathan Farhi, Ioannis N. Lykakis, and George E. Kostakis

22.1	Introduction *461*
22.2	Synthetic Aspects *462*
22.2.1	Propargylamines *462*
22.2.1.1	Seminal Work *462*
22.2.1.2	Propargylamines from in Situ Generated Cu Catalytic Species *463*
22.2.1.3	Propargylamines from Well-Characterised Cu Catalytic Species *464*
22.2.1.4	Propargylamines as Intermediates *465*
22.2.2	Pyrroles *466*
22.2.2.1	General Synthetic Aspects of Pyrroles *466*
22.2.2.2	Cyclization/annulation Reactions of Enamino Compounds *467*
22.2.2.3	Condensation/Cyclization with Diazo Compounds or Oximes *468*
22.2.2.4	Condensation/cyclization Reactions with Alkynes or Isonitriles *468*
22.2.2.5	Reaction of Nitroalkenes *469*
22.2.2.6	Reaction of Dicarbonyl Compounds *470*
22.2.2.7	Miscellaneous Reactions *470*
22.2.3	Dihydropyridines *471*
22.2.3.1	Synthesis of 1,4-DHPs with Copper Catalysts *472*
22.2.3.1.1	Hantzsch-Type Reactions *472*
22.2.3.1.2	Nucleophilic Addition to Iminium Salts *473*
22.2.3.1.3	Cycloaddition and C–C/C–N Coupling *474*
22.2.3.2	Synthesis of 1,2-DHPs with Copper Catalysts *475*
22.3	Conclusions *476*
	References *476*

23 **Environmental Catalysis by Gold Nanoparticles** *481*
Sónia Alexandra Correia Carabineiro

23.1	Introduction *481*
23.2	Preparation Methods *482*
23.2.1	Sol-immobilisation (COL) *482*
23.2.2	Impregnation (IMP) and Double Impregnation (DIM) *483*
23.2.3	Co-Precipitation (CP) *483*
23.2.4	Deposition Precipitation (DP) *484*
23.2.5	Liquid-phase Reductive Deposition (LPRD) *484*
23.2.6	Ion-Exchange *485*
23.2.7	Photochemical Deposition (PD) *485*
23.2.8	Ultrasonication (US) *486*

23.2.9	Vapor-Phase and Grafting 486
23.2.10	Bi- and Tri-metallic Au Catalysts 487
23.2.11	Post-treatment and Storage 487
23.3	Properties of Gold Nanoparticles 487
23.3.1	Activity 487
23.3.2	Selectivity 488
23.3.3	Durability 489
23.3.4	Poison Resistance 490
23.4	Reactions Efficiently Catalyzed by Gold Nanoparticles 490
23.4.1	CO Oxidation 490
23.4.2	Preferential Oxidation of CO in the Presence of H_2 (PROX) 496
23.4.3	Water-gas Shift 499
23.4.4	Total Oxidation of Volatile Organic Compounds (VOCs) 502
23.5	Conclusions and Outlook 505
	Acknowledgements 505
	References 505

24	**Platinum Complexes for Selective Oxidations in Water** 515
	Alessandro Scarso, Paolo Sgarbossa, Roberta Bertani, and Giorgio Strukul
24.1	Hydrogen Peroxide and Its Activation 515
24.2	Platinum Complexes 516
24.3	Enantioselective Oxidations 520
24.4	Water as the Reaction Medium 522
24.5	Catalytic Oxidation Reactions in Water 524
24.6	The Catalyst/micelle Interaction 527
24.7	Environmental Acceptability Evaluation and Possible Industrial Applications 530
24.8	Conclusions 532
	References 532

25	**The Role of Water in Reactions Catalyzed by Transition Metals** 537
	A.W. Augustyniak and A.M. Trzeciak
25.1	Water as a Solvent in Organic Reactions 537
25.2	The Role of Water in Heterogeneous Catalytic Systems 539
25.2.1	The Transformations of Furfuryl Derivatives 539
25.2.2	Oxidation and Deoxygenation 540
25.2.3	Arylcyanation and C–C Cross-Coupling 542
25.2.4	Hydrogenation 544
25.2.5	Hydroformylation 545
25.2.6	Catalytic Reactions with MOF-based Catalysts in an Aqueous Medium 546
25.2.6.1	Cross-Coupling Reactions 546
25.2.6.2	Hydrogenation Reactions 548
25.2.6.3	Hydroamination 549
25.3	The Contribution of Water in Homogeneous Catalytic Systems 550
25.3.1	Oxidation and Epoxidation 550
25.3.2	The Hydrogenation of Carbonyl Compounds and CO_2 550
25.3.3	The Cyclotrimerization of Alkynes 552
25.3.4	The Isomerization of Allylic Alcohols 553
25.3.5	Hydroarylation with Boron Compounds 554

25.3.6	Hydroformylation *555*
25.4	Conclusions *556*
	References *556*

26	**Using Speciation to Gain Insight into Sustainable Coupling Reactions and Their Catalysts** *559*
	Skyler Markham, Debbie C. Crans, and Bruce Atwater
26.1	Introduction *559*
26.2	The First Cross-coupling Reaction *559*
26.3	Phosphine Ligands for Catalysts of Cross-Coupling Reactions *560*
26.4	Speciation *564*
26.5	Palladium Nanoparticle Catalysts *564*
26.6	Speciation of Palladium (Pd) Catalysts *567*
26.7	Alternative Metal Catalysts *570*
26.7.1	Nickel *570*
26.7.2	Cobalt *570*
26.8	Speciation of Nickel and Cobalt Catalysts *572*
26.9	Cross-coupling Reactions and Sustainability: Summary and the Future *574*
	References *574*

27	**Hierarchical Zeolites for Environmentally Friendly Friedel Crafts Acylation Reactions** *577*
	Ana P. Carvalho, Angela Martins, Filomena Martins, Nelson Nunes, and Rúben Elvas-Leitão
27.1	Introduction *577*
27.2	Zeolites and Hierarchical Zeolites *580*
27.3	Zeolites and Hierarchical Zeolites as Catalysts for Friedel Crafts Acylation Reactions *585*
27.4	Understanding Friedel-Crafts Acylation Reactions through Quantitative Structure-property Relationships *597*
27.5	Final Remarks *604*
	Acknowledgements *605*
	References *605*

VOLUME 3

About the Editors *xiii*
Preface *xv*

Part V Organocatalysis *609*

28	**Sustainable Drug Substance Processes Enabled by Catalysis: Case Studies from the Roche Pipeline** *611*
	Kurt Püntener, Stefan Hildbrand, Helmut Stahr, Andreas Schuster, Hans Iding, and Stephan Bachmann

29	**Supported Chiral Organocatalysts for Accessing Fine Chemicals** *639*
	Ana C. Amorim and Anthony J. Burke

30	**Synthesis of Bio-based Aliphatic Polyesters from Plant Oils by Efficient Molecular Catalysis** *659* *Kotohiro Nomura and Nor Wahida Binti Awang*
31	**Modern Strategies for Electron Injection by Means of Organic Photocatalysts: Beyond Metallic Reagents** *675* *Takashi Koike*
32	**Visible Light as an Alternative Energy Source in Enantioselective Catalysis** *687* *Ana Maria Faisca Phillips and Armando J.L. Pombeiro*

Part VI   Catalysis for the Purification of Water and Liquid Fuels *717*

33	**Heterogeneous Photocatalysis for Wastewater Treatment: A Major Step Towards Environmental Sustainability** *719* *Shima Rahim Pouran and Aziz Habibi-Yangjeh*
34	**Sustainable Homogeneous Catalytic Oxidative Processes for the Desulfurization of Fuels** *743* *Federica Sabuzi, Giuseppe Pomarico, Pierluca Galloni, and Valeria Conte*
35	**Heterogeneous Catalytic Desulfurization of Liquid Fuels: The Present and the Future** *757* *Rui G. Faria, Alexandre Viana, Carlos M. Granadeiro, Luís Cunha-Silva, and Salete S. Balula*

Part VII   Hydrogen Formation, Storage, and Utilization *783*

36	**Paraformaldehyde: Opportunities as a C1-Building Block and H_2 Source for Sustainable Organic Synthesis** *785* *Ana Maria Faísca Phillips, Maximilian N. Kopylovich, Leandro Helgueira de Andrade, and Martin H.G. Prechtl*
37	**Hydrogen Storage and Recovery with the Use of Chemical Batteries** *819* *Henrietta Horváth, Gábor Papp, Ágnes Kathó, and Ferenc Joó*
38	**Low-cost Co and Ni MOFs/CPs as Electrocatalysts for Water Splitting Toward Clean Energy-Technology** *847* *Anup Paul, Biljana Šljukić, and Armando J.L. Pombeiro*

Index *871*

About the Editors

Armando Pombeiro is a Full Professor Jubilado at Instituto Superior Técnico, Universidade de Lisboa (ULisboa), former Distant Director at the People's Friendship University of Russia (RUDN University), a Full Member of the Academy of Sciences of Lisbon (ASL), the President of the Scientific Council of the ASL, a Fellow of the European Academy of Sciences (EURASC), a Member of the Academia Europaea, founding President of the College of Chemistry of ULisboa, a former Coordinator of the Centro de Química Estrutural at ULisboa, Coordinator of the Coordination Chemistry and Catalysis group at ULisboa, and the founding Director of the doctoral Program in Catalysis and Sustainability at ULisboa. He has chaired major international conferences. His research addresses the activation of small molecules with industrial, environmental, or biological significance (including alkane functionalization, oxidation catalysis, and catalysis in unconventional conditions) as well as crystal engineering of coordination compounds, polynuclear and supramolecular structures (including MOFs), non-covalent interactions in synthesis, coordination compounds with bioactivity, molecular electrochemistry, and theoretical studies.

He has authored or edited 10 books, (co-)authored *ca.* 1000 research publications, and registered *ca.* 40 patents. His work received *over.* 30,000 citations (over 12,000 citing articles), h-index *ca.* 80 (Web of Science).

Among his honors, he was awarded an Honorary Professorship by St. Petersburg State University (Institute of Chemistry), an Invited Chair Professorship by National Taiwan University of Science & Technology, the inaugural SCF French-Portuguese Prize by the French Chemical Society, the Madinabeitia-Lourenço Prize by the Spanish Royal Chemical Society, and the Prizes of the Portuguese Chemical and Electrochemical Societies, the Scientific Prizes of ULisboa and Technical ULisboa, and the Vanadis Prize. Special issues of Coordination Chemistry Reviews and the Journal of Organometallic Chemistry were published in his honor.

https://fenix.tecnico.ulisboa.pt/homepage/ist10897

Manas Sutradhar is an Assistant Professor at the Universidade Lusófona, Lisbon, Portugal and an integrated member at the Centro de Química Estrutural, Instituto Superior Técnico, Universidade de Lisboa, Portugal. He was a post-doctoral fellow at the Institute of Inorganic and Analytical Chemistry of Johannes Gutenberg University of Mainz, Germany and a researcher at the Centro de Química Estrutural, Instituto Superior Técnico, Universidade de Lisboa. He has published 72 papers in international peer review journals (including three reviews + 1 reference module), giving him an h-index 28 (ISI Web of Knowledge) and more than 2250 citations. In addition, he has 11 book chapters in books with international circulation and one patent. He is one of the editors of the book *Vanadium Catalysis*, published by the Royal Society of Chemistry. His main areas of work include metal complexes with aroylhydrazones, oxidation catalysis of industrial importance and sustainable environmental significance, magnetic properties of metal complexes, and bio-active molecules. The major contributions of his research work are in the areas of vanadium chemistry and oxidation catalysis. He received the 2006 Young Scientist Award from the Indian Chemical Society, India and the Sir P. C. Ray Research Award (2006) from the University of Calcutta, India.

https://orcid.org/0000-0003-3349-9154

Elisabete C.B.A. Alegria is an Adjunct Professor at the Chemical Engineering Department of the Instituto Superior de Engenharia de Lisboa (ISEL) of the Polytechnic Institute of Lisbon, Portugal. She is a researcher (Core Member) at the Centro de Química Estrutural (Coordination Chemistry and Catalysis Group). She has authored 86 papers in international peered review journals and has an h-index of 23 with over 1600 citations, four patents, five book chapters, and over 180 presentations at national and international scientific meetings. She was awarded an Honorary Distinction (2017–2020) for the Areas of Technology and Engineering (Scientific Prize IPL-CGD). She is an editorial board member, and has acted as a guest editor and reviewer for several scientific journals. Her main research interests include coordination and sustainable chemistry, homogeneous and supported catalysis, stimuli-responsive catalytic systems, green synthesis of metallic nanoparticles for catalysis, and biomedical applications. She is also interested in mechanochemistry (synthesis and catalysis) and molecular electrochemistry.

https://orcid.org/0000-0003-4060-1057

Preface

Aiming to change the world for the better, 17 Sustainable Development Goals (SDGs) were adopted by the United Nations (UN) Member States in 2015, as part of the UN 2030 Agenda for Sustainable Development that concerns social, economic, and *environmental sustainability*. Hence, a 15-year plan was set up to achieve these Goals and it is already into its second half.

However, the world does not seem to be on a good track to reach those aims as it is immersed in the Covid-19 pandemic crisis and climate emergency, as well as economic and political uncertainties. Enormous efforts must be pursued to overcome these obstacles and chemical sciences should play a pivotal role. *Catalysis* is of particular importance as it constitutes the most relevant contribution of chemistry towards sustainable development. This is true even though the SDGs are integrated and action in one can affect others.

For example, the importance of chemistry and particularly catalysis is evident in several SDGs. Goal 12, addresses "Responsible Consumption and Production Patterns" and is aligned with the circularity concept with sustainable loops or cycles (e.g., in recycle and reuse processes that are relevant within the UN Environmental Program). Goal 7 addresses "Affordable and Clean Energy" and relates to efforts to improve energy conversion processes, such as hydrogen evolution and oxygen evolution from water, that have a high environmental impact. Other SDGs in which chemistry and catalysis play an evident role with environmental significance include Goal 6 ("Clean Water and Sanitation"), Goal 9 ("Industry, Innovation and Infrastructure" 13 ("Climate Action"), Goal 14 ("Life Below Water"), and Goal 15 ("Life on Land").

The book is aligned with these SDGs by covering recent developments in various *catalytic processes* that are designed for a *sustainable environment*. It gathers skilful researchers from around the world to address the use of catalysis in various approaches, including homogeneous, supported, and heterogeneous catalyses as well as photo- and electrocatalysis by searching for innovative green chemistry routes from a sustainable environmental angle. It illustrates, in an authoritative way, state-of-the-art knowledge in relevant areas, presented from modern perspectives and viewpoints topics in coordination, inorganic, organic, organometallic, bioinorganic, pharmacological, and analytical chemistries as well as chemical engineering and materials science.

The chapters are spread over seven main sections focused on Carbon Dioxide Utilization, Transformation of Volatile Organic Compound (VOCs), Carbon-based Catalysts, Coordination, Inorganic, and Bioinspired Catalysis, Organocatalysis, Catalysis for the Purification of Water and Liquid Fuels,and Hydrogen Formation, Storage, and Utilization. These sections are gathered together as a contribution towards the development of the challenging topic.

The book addresses topics in (i) activation of relevant small molecules with strong environmental impacts, (ii) catalytic synthesis of important added value organic compounds, and (iii) development of systems operating under environmentally benign and mild conditions toward the establishment of sustainable energy processes.

This work is expected to be a reference for academic and research staff of universities and research institutions, including industrial laboratories. It is also addressed to post-doctoral, postgraduate, and undergraduate students (in the latter case as a supplemental text) working in chemical, chemical engineering, and related sciences. It should also provide inspiration for research topics for PhD and MSc theses, projects, and research lines, in addition to acting as an encouragement for the development of the overall field.

The topic Catalysis for Sustainable Environment is very relevant in the context of modern research and is often implicit, although in a non-systematic and disconnected way, in many publications and in a number of initiatives such as international conferences. These include the XXII International Symposium on Homogeneous Catalysis (ISHC) that we organized (Lisbon, 2022) and that to some extent inspired some parts of this book.

In contrast to the usual random inclusion of the topic in the literature and scientific events, the applications of catalytic reactions focused on a sustainable environment in a diversity of approaches are addressed in this book.

The topic has also contributed to the significance of work that led to recent Nobel Prizes of Chemistry. In 2022, the Nobel Prize was awarded to Barry Sharpless, Morten Meldal, and Carolyn Bertozzi for the development of click chemistry and bioorthogonal chemistry. The set of criteria for a reaction or a process to meet in the context of click chemistry includes, among others, the operation under benign conditions such as those that are environmentally friendly (e.g., preferably under air and in water medium). In 2021, the Nobel Prize was awarded to Benjamin List and David W.C. MacMillan for the development of asymmetric organocatalysis, which relies on environmentally friendly organocatalysts.

The book illustrates the connections of catalysis with a sustainable environment, as well as the richness and potential of modern catalysis and its relationships with other sciences (thus fostering interdisciplinarity) in pursuit of sustainability.

At last, but not least, we should acknowledge the authors of the chapters for their relevant contributions, prepared during a particularly difficult pandemic period, as well as the publisher, John Wiley, for the support, patience, and understanding of the difficulties caused by the adverse circumstances we are experiencing nowadays and that constituted a high activation energy barrier that had to be overcome by all of us… a task that required rather active catalysts.

We hope the readers will enjoy reading its chapters as much as we enjoyed editing this book.

Armando Pombeiro
Manas Sutradhar
Elisabete Alegria

Part IV

Coordination, Inorganic, and Bioinspired Catalysis

14

Hydroformylation Catalysts for the Synthesis of Fine Chemicals

Mariette M. Pereira[1], Rui M.B. Carrilho[1], Fábio M.S. Rodrigues[1], Lucas D. Dias[2], and Mário J.F. Calvete[1]

[1] Coimbra Chemistry Centre, Department of Chemistry, University of Coimbra, Rua Larga Coimbra, Portugal
[2] Laboratório de Novos Materiais, Universidade Evangélica de Goiás, Anápolis GO, Brazil

14.1 Introduction

The hydroformylation reaction consists of the addition of CO and H_2, also known as synthesis gas or *syngas*, across the π system of a C=C double bond, in the presence of a catalyst, to form aldehydes (Figure 14.1). As a pure addition reaction, in which all raw materials are incorporated in the aldehyde products, the catalytic hydroformylation is a paradigmatic example of a 100% atom economy chemical process [1–5].

The reaction was discovered accidentally by Otto Roellen (1897–1993) in 1934, when he worked in the German Ruhrchemie industry, aiming the optimization of the Fischer-Tropsch reaction for the preparation of fuels [6, 7]. During an experiment to prepare ethylene, Roelen surprisingly found that, in the presence of ammonia, the imine of propionaldehyde was obtained as a white solid when a mixture of cobalt, thorium, and magnesium oxides were used as catalysts [4]. In the following years, he investigated the potentialities and limitations of the new reaction and discovered that, indeed, homogeneous cobalt salts were the catalytic species responsible for the synthesis of aldehydes from olefins and *syngas*. Roelen recognized the high synthetic and economic potential of the *oxo-synthesis* and, in 1938, he submitted a patent application [6]. Soon after its discovery, the scientific community quickly found that, to turn the hydroformylation reaction into a viable synthetic alternative for the industrial production of fine chemicals, the simultaneous control of the catalyst activity and chemo-, regio-, and enantioselectivity or diastereoselectivity would be the greatest challenges to achieve (Figure 14.2).

The first challenging goal was to obtain high chemoselectivity for the desired aldehydes. This was almost achieved through replacement of cobalt-carbonyl catalysts by rhodium-carbonyl complexes coordinated with phosphorus ligands. With respect to the regio-, enantio-, and diastereoselectivity issues, scientists and industrials have been dedicated to the design and synthesis of elaborate phosphorus ligands over the years, which led to the development of both homogenous and immobilized rhodium catalysts with high activity and selectivity, allowing the implementation of hydroformylation in fine chemical industries [8]. Since then, the hydroformylation has turned into one of the largest scale industrial homogeneous catalytic processes for production of

---

Catalysis for a Sustainable Environment: Reactions, Processes and Applied Technologies Volume 2, First Edition. Edited by Armando J. L. Pombeiro, Manas Sutradhar, and Elisabete C. B. A. Alegria.
© 2024 John Wiley & Sons Ltd. Published 2024 by John Wiley & Sons Ltd.

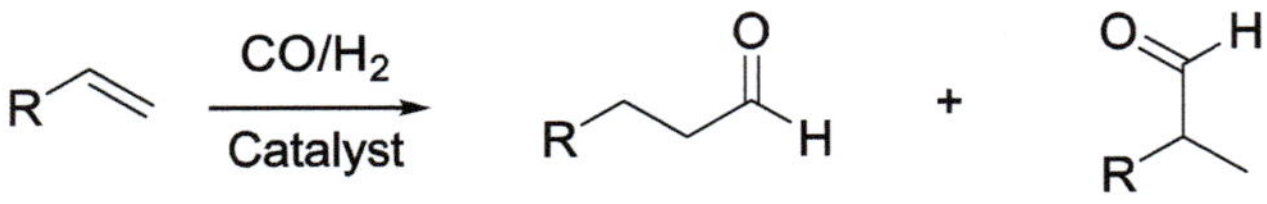

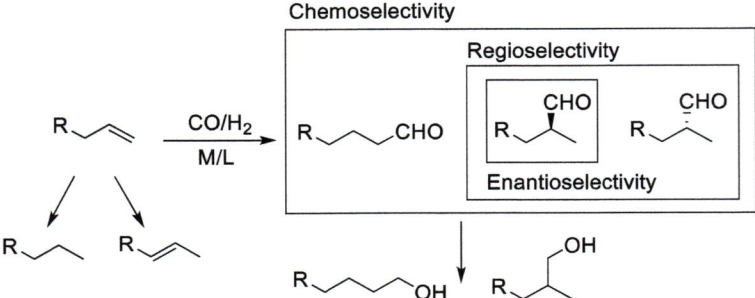

Figure 14.1 General scheme of the hydroformylation reaction.

Figure 14.2 Control of chemo-, regio-, and enantioselectivity in catalytic olefin hydroformylation.

aldehydes and their derivatives (nearly 10 million tons/year) [4]. Furthermore, the aldehydes formed are key intermediates for the preparation of added-value fine chemical products such as alcohols, amines, acetals, and carboxylic acid derivatives (Figure 14.3). Thus, the catalytic hydroformylation is currently considered a central reaction for sequential synthetic processes to transform aldehydes into other functional groups under sustainable reaction conditions [9–11].

In this chapter, we present the most paradigmatic examples of homogeneous and immobilized hydroformylation catalysts including a brief description of their use in hydroformylation reactions and hydroformylation centered sequential reactions, particularly for the synthesis of fine chemicals. A special emphasis will be given to the contribution of Portuguese researchers to the development of this topic.

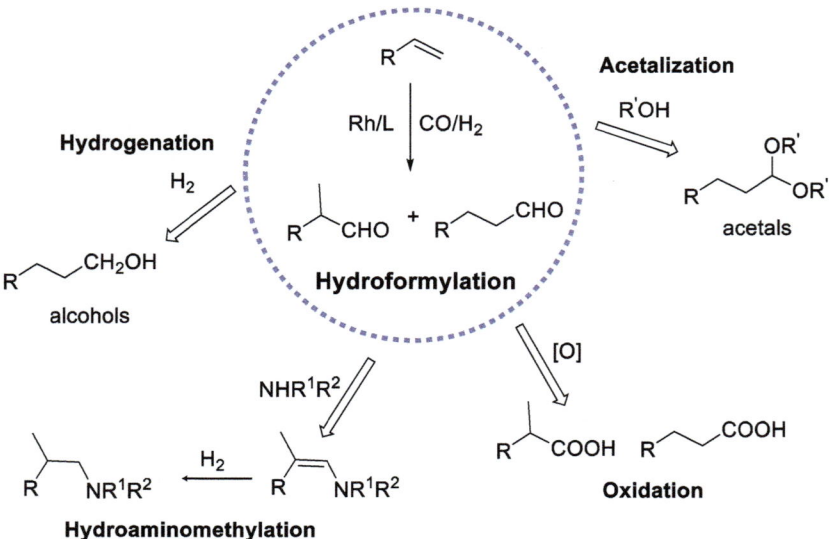

Figure 14.3 Examples of hydroformylation-based sequential reactions for fine-chemical synthesis.

14.2 Homogeneous Catalytic Systems

14.2.1 Development of Phosphorus Ligands

As mentioned, rhodium complexes are considered the catalysts of choice for hydroformylation reactions. To increase their activity and selectivity, over the years, a plethora of phosphorus ligands have been developed. Wilkinson et al. [12–14] reported the use of rhodium complexes coordinated with triphenylphosphine **1**, providing the first highly chemoselective systems. Soon after Wilkinson's work, Pruett and Smith [15, 16] introduced the use of rhodium phosphite complexes (Figure 14.4). In the 80s, van Leeuwen et al. [17–19] discovered a peculiar effect of bulky aryl monophosphites, such as tris(o-tertbutylphenyl)phosphite **2** (Figure 14.4), with a large cone angle (c. 175°), whose rhodium complexes led to highly active chemo- and regioselective catalysts in the hydroformylation of substituted olefins under mild reaction conditions.

Regarding the development of highly active rhodium-phosphite catalysts, Pereira et al. and Bayón et al. designed and prepared a family of chiral bulky monophosphites with C_3-symmetry [20–22]. Their synthesis involved the mono-etherification of (R) or (S)-BINOL with primary, secondary or tertiary alcohols, via Mitsunobu reaction, followed by reaction of the BINOL monoethers with 1/3 equivalent of PCl_3 in triethylamine, used simultaneously as base and reaction solvent (Figure 14.5).

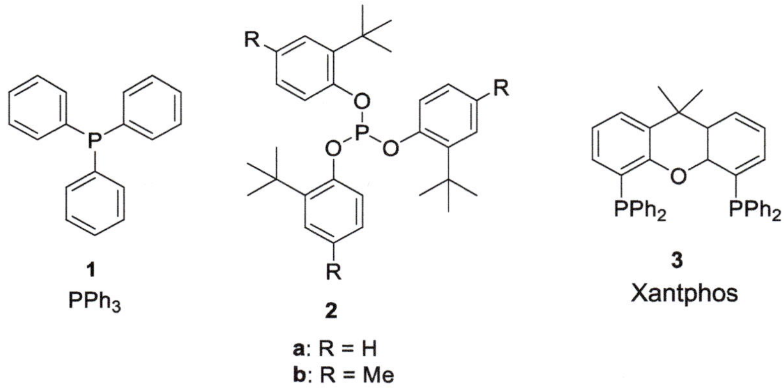

Figure 14.4 Some of the first phosphorus ligands used in Rh-catalyzed hydroformylation.

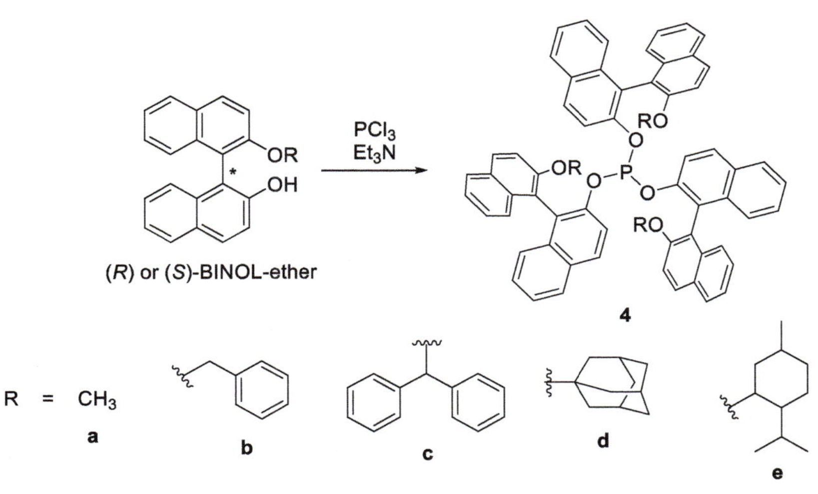

Figure 14.5 Synthesis of tris-BINOL based monophosphites.

The cone angles of **4a-d** were calculated by semi-empirical computational methods, using the PM6 Hamiltonian, and the values obtained were within the range of 240 – 270º. In addition, their rhodium complexes, formed in situ, showed remarkable catalytic activity in the hydroformylation of disubstituted styrene derivatives under relatively mild reaction conditions (80 °C, 30 bar CO/H_2), achieving 100% chemoselectivity for aldehydes and regioselectivities between 90–99%; however, low enantioselectivity was obtained (up to 20% ee). More recently [23], the family of tris-binaphthyl phosphites was expanded, with the preparation of four stereoisomeric monophosphites **4e**, based on axially chiral (R)- or (S)-BINOL bearing a chiral (+)- or (−)-neomenthyloxy group. These new tris-binaphthyl phosphite ligands were characterized by DFT computational methods, which allowed calculatation of an extremely large cone angle of 345°. Furthermore, tris-BINOL neomenthol mononophosphites were applied in rhodium-catalyzed hydroformylation of styrene, leading to complete conversions in 4h with 100% chemoselectivity for aldehydes and up to 98% iso-regioselectivity. The Rh(I)/phosphite catalytic system was also highly active and selective in the hydroformylation of disubstituted olefins, including (E)-prop-1-en-1-ylbenzene and prop-1-en-2-ylbenzene, yielding the corresponding branched aldehyde in 80% regioselectivity and the linear aldehyde in 99% regioselectivity, respectively.

The effect of the phosphine ligands bite angles on the reaction's regioselectivity was reported by several research groups [24], including van Leeuwen et al., who reported the use of Xantphos ligand **3**, with a bite angle of 111°, which led to exclusive formation of linear aldehydes from alkyl terminal olefins [25–27]. The most remarkable enantioselectivities have been accomplished using rhodium complexes coordinated with bidentate phosphorus ligands. Babin and Whiteker described the use of bulky diphosphites, derived from homochiral (2R,4R)-pentane-2,4-diol, including Chiraphite **5**, in the Rh-catalyzed hydroformylation of styrene (Figure 14.6) [28]. This led to 99% regioselectivity for the branched aldehyde and up to 90% ee, which was attributed to formation of eight-membered chelates with rhodium(I), through a bis-equatorial coordination mode. Another family of chiral diphosphite ligands, based on carbohydrate backbones, was developed by Claver et al. [29, 30]. Among them, the rhodium(I) complex of ligand **6**, containing a three-carbon bridge, achieved 93% ee, along with 98% of iso-regioselectivity, in styrene hydroformylation (Figure 14.6). Klosin et al. reported the use of bis-diazaphospholane ligand **7** in the asymmetric hydroformylation of styrene, vinyl acetate and allyl cyanide, achieving for these substrates with 89%, 96%, and 87% ee, respectively (Figure 14.6) [31, 32].

The major breakthrough on enantioselectivity came from Takaya's laboratory with the synthesis of (R,S)-BINAPHOS **8** [33, 34], which led to up to 95% ee with 91% regioselectivity toward the branched product in the hydroformylation of styrene, and more than 90% ee for a range of other functionalized substrates, including internal alkenes [35–37]. HP-NMR studies showed that the C_1-symmetric ligand **8** coordinate to Rh(I) in equatorial-axial fashion, in which the more σ-donor phosphine P atom sits in the plane with the CO ligands, while the more π-accepting phosphite P atom binds apical to the hydride (complex **9**). This unique dissymmetric environment in a single catalytically active species was found to be a decisive factor for the high enantiodiscrimination. Later, Zhang et al. reported the synthesis of the mixed phosphine-phosphoramidite ligand (R,S)-YANPHOS **10** [38], derived from 2-amino-2′-hydroxy-1,1′-binaphthyl (NOBIN). The N-alkyl substituent was intended to supply a rigid conformation able to provide a more closed chiral pocket than the BINAPHOS complex and, as expected, it led to a significant improvement on asymmetric induction. Indeed, the Rh/(R,S)-YANPHOS complex provided up to 99% ee along with 90% iso-regioselectivity, in the Rh-catalyzed hydroformylation of styrene and a number of other olefins [39].

Freixa and Bayón reported a BINOL-based diphosphite ligand **11** (Figure 14.8), linked through an isophthalate bridge, forming a large chelating ring [40]. Their rhodium complexes provided

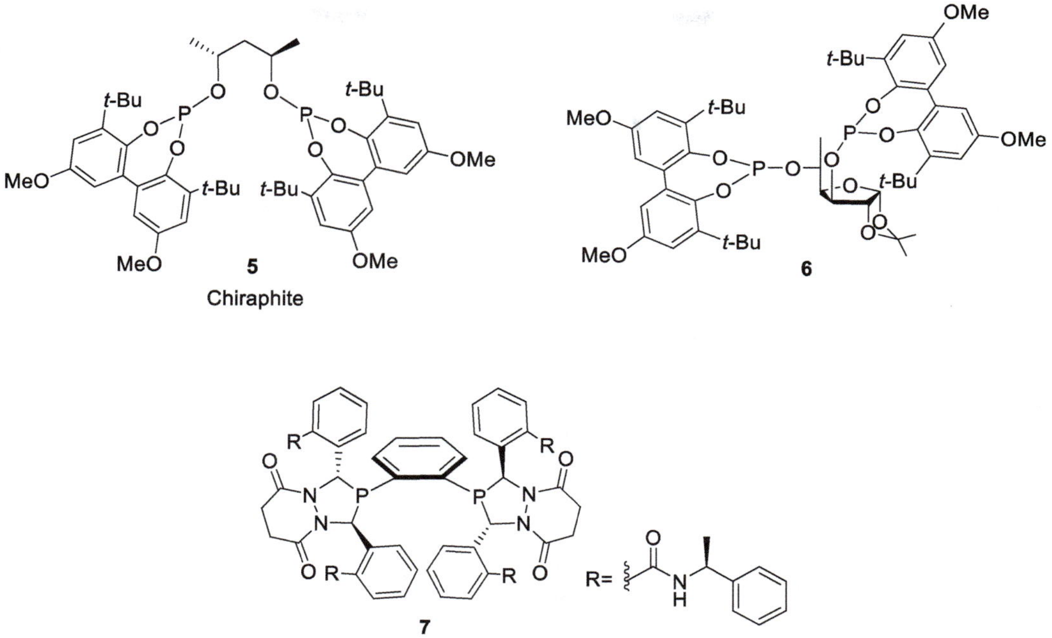

Figure 14.6 Examples of bidentate phosphites and diazaphospholane ligands.

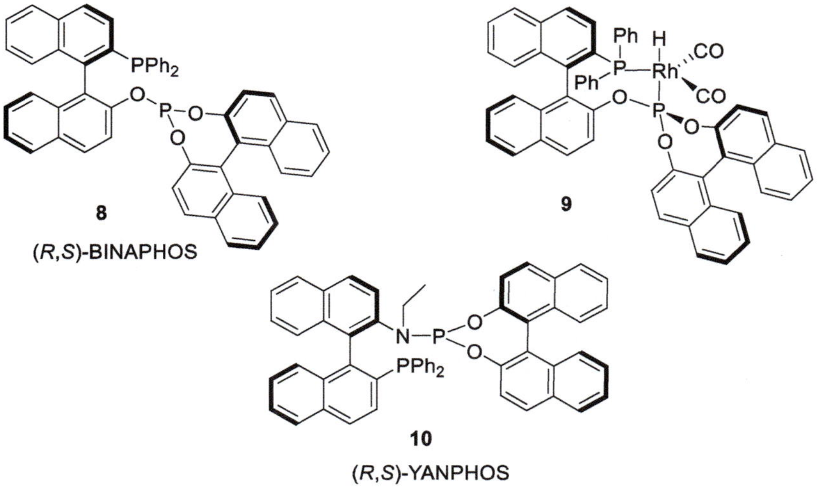

Figure 14.7 Mixed phosphine-phosphite (**8**) and phosphine-phosphoramidite ligands (**10**).

active and regioselective catalysts in the hydroformylation of vinyl arenes and the appropriate combination of stereogenic centers provided by chiral binaphthyl and propan-1,2-diol backbones allowed to achieve up to 76% enantiomeric excess. Following these studies, Pereira et al. [41] developed a family of related ditopic (*R*)-BINOL-based phosphites **12–14**, containing pyridine bridges linking two binaphthyl fragments (Figure 14.8). The same authors developed the synthesis of diphosphites **15**, containing alkyl ether spacers linking the two binaphthyl fragments. These were synthesized by reaction of (1*R*,1″*R*)-2′,2″-[propane-1,3-diyl-bis(oxy)]di-1,1′-binaphthyl-2-ol or

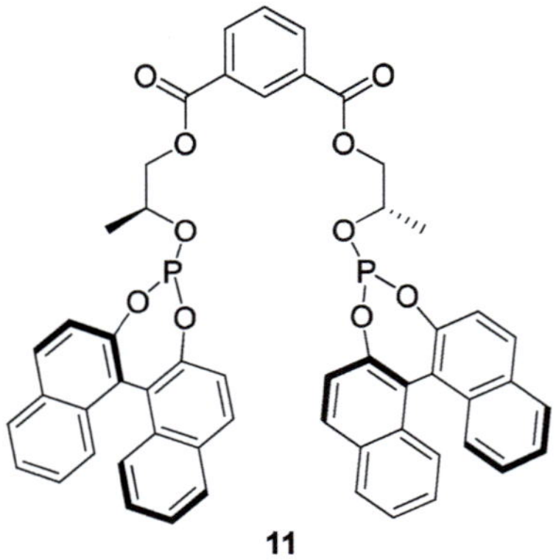

11

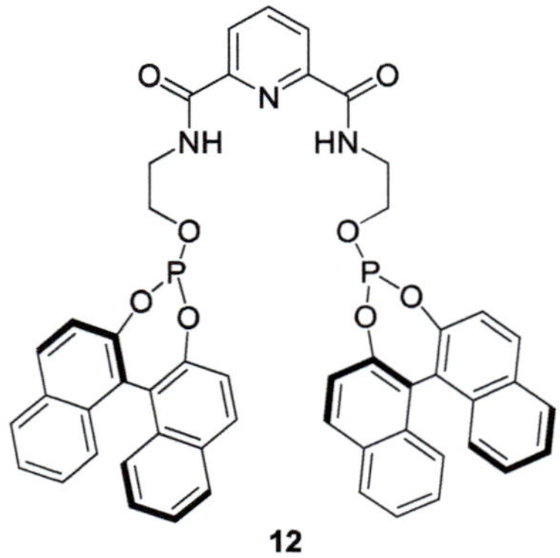

12

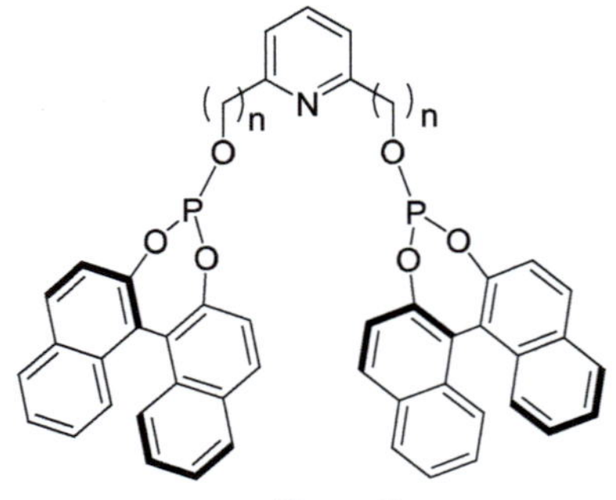

13: n = 0
14: n = 1

Figure 14.8 Examples of large chelating ring diphosphites developed by Bayón et al. (**11**) and Pereira et al. (**12–14**).

(1*R*,1''*R*)-2',2''-[2,2-dimethylpropane-1,3-diyl-bis(oxy)]di-1,1'-binaphthyl-2-ol, with pyrocathecol phosphochloridite, using triethylamine as base (Figure 14.9). Their rhodium complexes achieved up to 95% regioselectivity for the formation of branched aldehydes, but enantiomeric excesses were only ca. 10% in all cases. For the Rh/**14** catalytic system, the authors further studied the effect of different reaction variables (pressure, temperature, and the addition of a lithium salt) on the regioselectivity by 2^3 factorial design computational methods, which revealed that temperature was the most decisive reaction parameter[41].

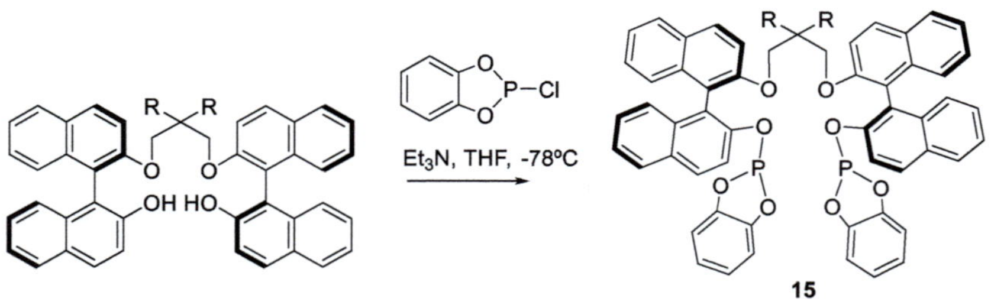

15
a: R = H
b: R = Me

Figure 14.9 Synthesis of bis-BINOL-pyrocathecol disphosphite ligands.

14.2.2 Hydroformylation of Biologically Relevant Substrates

Steroid and terpene molecules are important classes of biologically relevant compounds that are extensively applied in the synthesis of fine chemicals in medicinal chemistry [42]. However, the hydroformylation of these substrates is usually very difficult to perform because they contain highly substituted and internal double C=C bonds, which are less reactive due to steric hindrance that prevents the approach of the metal catalyst. The only exceptions are the use of Rh/bulky aryl phosphite catalytic systems, whose exceptional activity results from both electronic and stereo effects. On one hand, the π-acidic character of the phosphite weakens the metal-CO bond, thereby allowing a faster CO dissociation; on the other hand, the phosphite ligand's large cone angle allows the coordination of only one phosphite to the metal center, even when used in large excess, which results in a low global steric hindrance around the metal center [19]. Therefore, Rh-phosphite catalysts have opened new perspectives for application of hydroformylation as key tool for functionalization of natural products such as steroids and terpene molecules.

In this regard, the use of Rh/tris(o-tert-butylphenyl)phosphite catalyst (Rh(I)/**2a**) in the hydroformylation of highly substituted steroid olefins was also reported, namely for 17β-acetoxyandrost-4-ene and 3β,17β-diacetoxyandrost-4-ene (Figure 14.10). In both cases, the major reaction product (68%) was 4β-formyl-17β-acetoxy-5β-androstane. The authors concluded that the hydroformylation reaction occurred preferentially at the β face of the steroid nucleus. This reaction was the first example of a catalytic carbonylation to the β face of a steroid backbone [43]. The studies were expanded further, to the hydroformylation of cholest-4-ene and 3β-acetoxycholest-4-ene, using the same catalytic system. Under the reaction conditions assayed, these steroids were hydroformylated, producing the corresponding 4-formyl derivatives with 100% regioselectivity and up to 70% β-diastereoselectivity. Both formyl steroids were isolated in their acetal forms [44]. More recently, the application of Rh(I)/tris-binaphthyl monophosphite catalyst Rh(I)/(R)-**4b** was also reported in the hydroformylation of 17β-acetoxyandrost-4-ene, leading to 95% conversion in 48 h, with 86% chemoselectivity for aldehydes and 70% diastereoselectivity for the β-formyl aldehyde [45].

Terpenes are another relevant example of natural molecules that have been extensively used as substrate for hydroformylation reactions aiming the preparation of value-added fine chemicals [46, 47]. In this context, Pereira et al. reported the hydroformylation of monoterpenes, such as (1R)-(−)-myrtenyl acetate using and (−)-isopulegol benzyl ether, using highly active Rh(I)/**2a** or Rh(I)/**4b** bulky aryl monophosphite catalytic systems (Figure 14.11).

The hydroformylation of diterpenes with exocyclic methylenic double bonds was also reported [48], namely the methyl esters of kaurenic and grandiflorenic acids, and the trimethylsilylether of kaurenol. In this reactions, the authors used Rh/triphenulphosphine (**1**) and Rh/tris-(o-

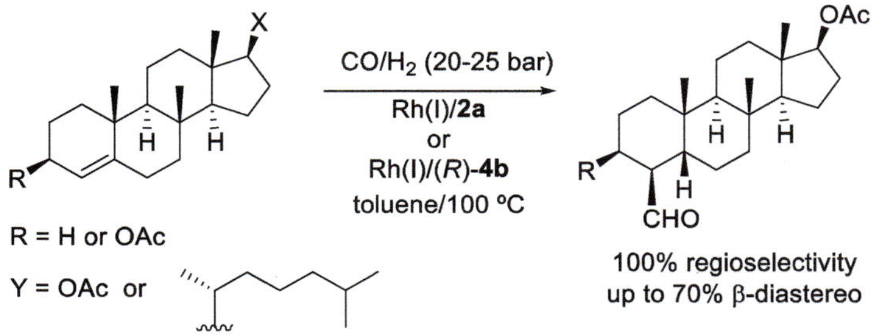

Figure 14.10 Hydroformylation of steroid derivatives using Rh(I)/bulky aryl monophosphite catalytic systems.

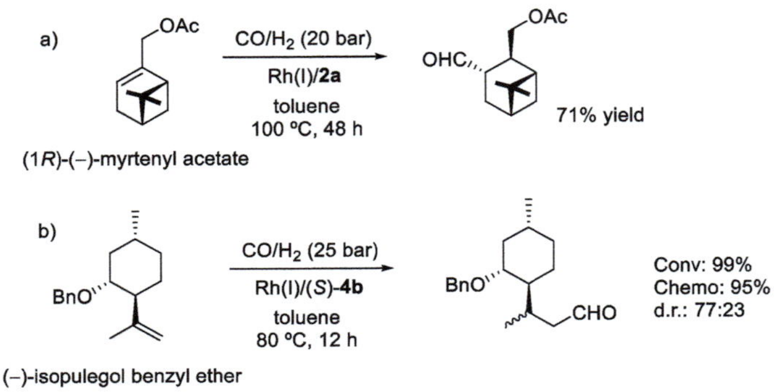

Figure 14.11 Hydroformylation of monoterpenes: a) (1R)-(−)-myrtenyl acetate; and b) (−)-isopulegol benzyl ether.

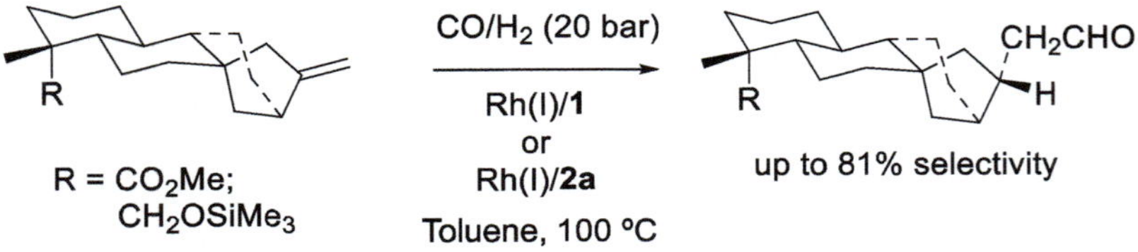

Figure 14.12 Hydroformylation of kaurenic acid methyl ester and kaurenol trimethylsilylether.

t-butylphenyl)phosphite (**2a**) as catalysts, with P(CO:H$_2$) = 20 bar and T = 100 °C (Figure 14.12). High yields, chemo- and regioselectivity toward the formation of the linear aldehydes were achieved with both catalytic systems. The major isolated linear aldehyde diastereoisomer resulted from the coordination of rhodium and consequent introduction of CO/H$_2$ in the less hindered face of the exocyclic double bond. Additionally, the authors observed that the isomerization to the internal ring position is a competitive process, which is strongly dependent of the catalyst used (% of isomerization: PPh$_3$-modified system < P(o-^{t}BuC$_6$H$_4$)$_3$-modified system < unmodified system).

Another relevant example for the use of hydroformylation as a key tool for preparation of fine chemicals for medicinal chemistry, carried out in Pereira's laboratory, was the transformation of β-substituted vinyl porphyrins into the corresponding aldehydes [49, 50]. The authors reported the hydroformylation of the Zn(II) and Ni(II) complexes of protoporphyrin-IX dimethyl ester (Figure 14.13), using the rhodium catalytic systems modified with triphenylphosphine (**1**) and Xantphos (**3**). A remarkable influence of the central metal at the porphyrin ring was observed, with the zinc(II) protoporphyrin-IX dimethyl ester inducing the formation of the branched aldehyde (99%), whereas, in the nickel(II) metalloporphyrin, the regioselectivity strongly depended on the reaction conditions (temperature and *syngas* pressure). This behavior was explained by PM3 calculations, which showed that the different dihedral angles of vinyl double bond atoms observed have a crucial role on the stabilization of branched-alkyl complexes, favoring the preferential formation of branched aldehydes in the case of Zn(II) complexes[49].

The same authors reported the hydroformylation of the *methyl pheophorbide-a* allomer (Figure 14.14), using the rhodium catalytic systems modified with triphenylphosphine (**1**). Even at nonfavorable reaction conditions, the branched aldehyde (98%) was obtained, which pointed out the relevance of the phosphorus ligand and the substrate macrocyclic structure on the regioselectivity of the catalytic system [50].

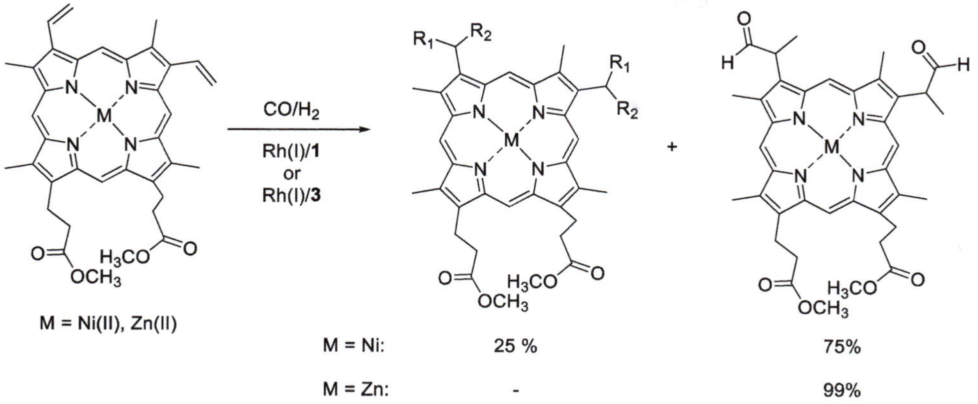

Figure 14.13 Hydroformylation of Ni(II) and Zn(II) metal complexes of protoporphyrin-IX dimethyl ester, using [Rh$_2$(-OMe)$_2$(cod)$_2$]/PPh$_3$ (**1**) or [Rh$_2$(-OMe)$_2$(cod)$_2$]/Xantphos (**3**) as catalytic systems.

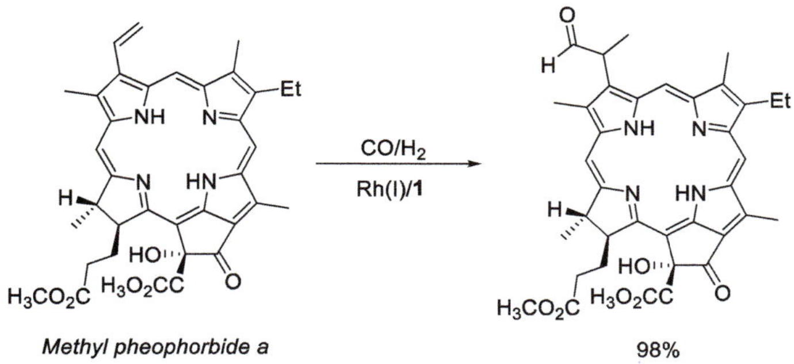

Figure 14.14 Hydroformylation of methyl pheophorbide-a allomer, using [Rh$_2$(-OMe)$_2$(cod)$_2$]/PPh$_3$ (**1**) as catalytic system.

14.2.3 Hydroformylation-based Sequential Reactions

As mentioned, the hydroformylation products (aldehydes) are versatile synthetic intermediates for preparation of fine chemicals through a vast number of sequential processes [10, 11, 51]. In particular, the strategy of operating different catalytic steps in a single vial, using one or several catalysts, is a useful approach to increment the molecular complexity and minimize the intermediate work-up and purification procedures. In this context, Pereira et al. described an efficient multicatalytic process involving a single rhodium/phosphine catalyst to promote tandem hydroformylation/arylation reactions, to prepare secondary alcohols directly from olefins (Figure 14.15) [52].

After optimization of the reaction conditions, the authors found that Rh/PPh$_3$ is the best catalyst for both hydroformylation and arylation reactions. This system gave up to 99% conversion and up to 98% regioselectivity for the branched aldehyde, in the hydroformylation step of aryl olefins, combined with an efficient arylation step, which allowed to achieve secondary alcohols in high combined yields (up to 87%). The synthetic potential of this sequential methodology for preparation of fine chemicals at multigram-scale, was well demonstrated by the large scope of the olefins and arylboronic acids used (Figure 14.16) [52].

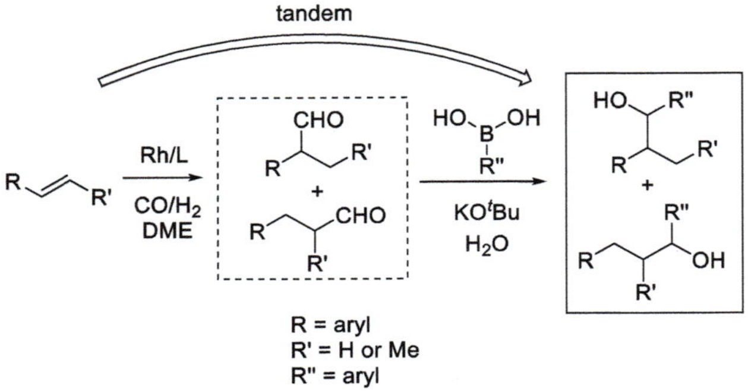

Figure 14.15 Tandem hydroformylation/arylation process to transform olefins into secondary alcohols.

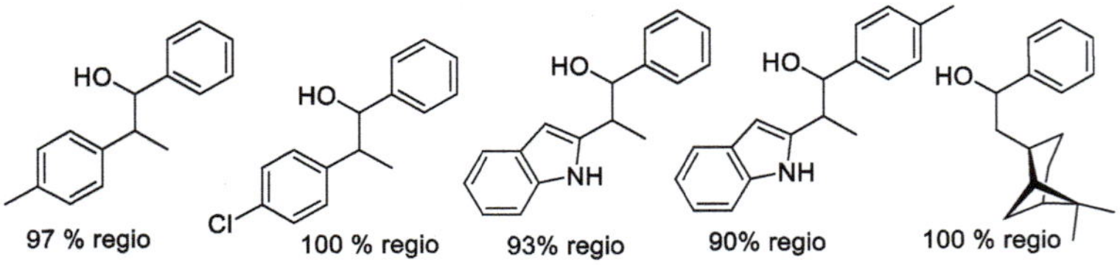

Figure 14.16 Examples of secondary aryl alcohols products obtained by tandem hydroformylation/arylation reactions.

Another relevant example of the application of hydroformylation as central reaction is the synthesis of aminonitriles and amino acids via catalytic sequential hydroformylation/Strecker process [53, 54]. In this context, a family of non-natural amino acids, starting from the olefins, was reported (Figure 14.17) [55]. Aiming at the preparation of linear aminoacids, the hydroformylation step was carried out using Rh/Xantphos as catalyst at 20 bar H_2:CO pressure and T = 80 °C. Under these conditions, conversions up to 99% and regioselectivity for linear aldehyde up to 98% were achieved. Then, without isolating the aldehyde, an aqueous solution of NaCN and NH_4Cl was added to the reaction's crude, to afford a set of alkyl amino nitriles with yields up to 98%.

The authors expanded the reaction scope to the preparation of potentially bioactive compounds, particularly aminonitriles derived from cholest-4-ene and 3-vinyl-1H-indole (Figure 14.18) [55]. In these reactions, the authors used Rh/tris-binaphthyl phosphite (**4b**) as catalyst to obtain the desired 4-formyl-cholestane derivative with 100% regioselectivity and 63% β-diastereoselectivity. The subsequent addition of stoichiometric amounts of NaCN and NH_4Cl allowed the production

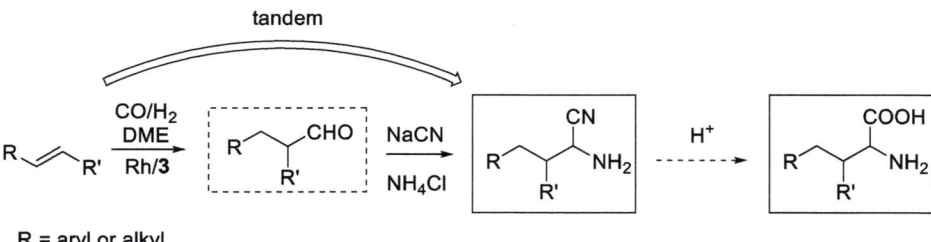

Figure 14.17 Hydroformylation-Strecker sequential reaction for preparation of aminonitriles and amino acids.

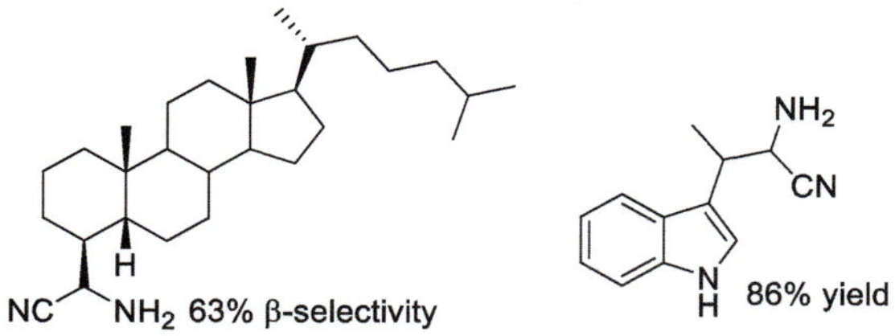

Figure 14.18 Potentially bioactive α-aminonitriles obtained via one-pot hydroformylation/Strecker reaction.

of the desired aminonitriles, which upon acid hydrolysis afforded the desired amino acids, isolated in high yields by column chromatography. Additionally, through the same strategy, the authors also prepared an indole-derived aminonitrile and the corresponding amino acid, using Rh/PPh$_3$ as catalyst.

Another relevant example of the synthetic potential of hydroformylation-based sequential reactions is the one-pot synthesis of tertiary amines directly from olefins *via* hydroaminomethylation [56, 57] (Figure 14.19). The hydroformylation step was performed using the rhodium/phosphite catalyst Rh(I)/**4b**, at 100 °C and 20 bar CO/H$_2$, which afforded the 4-formyl-cholestane derivative, with 70% β-diastereoselectivity. The subsequent reductive amination was performed in situ through addition of piperidine or morpholine, along with the change to a H$_2$ atmosphere, which allowed production of the corresponding amine products in 59% and 51% yields, respectively (Figure 14.19).

A dual catalytic system was developed by Pereira et al. and Beller et al. [58] to promote the sequential isomerization/hydroformylation/hydrogenation of olefins (Figure 14.20). The catalytic system Rh/**4a** was found to be efficient for the hydroformylation of 2,3-dimethylbut-2-ene, under mild reaction conditions (120 °C, 40 bar, 1:3 CO/H$_2$), upon fast internal olefin isomerization to the terminal position. In addition, in the presence of the highly active hydrogenation Shvo's catalyst, the **in situ** aldehyde reduction allowed to obtain the corresponding alcohol in up to 97% yield.

The reaction scope was expanded to various highly substituted olefins, containing internal C=C double bonds (16 examples), and the combination of the Rh/bulky monophosphite catalysts (Rh(I)/**4b**) with Ru-based Shvo's complex allowed the direct synthesis of a set of aromatic and aliphatic *oxo-alcohols* in good to excellent yields (Figure 14.21)

The direct synthesis of a set of formylcarboxamides with biological relevance was reported through a sequential aminocarbonylation/hydroformylation methodology (Figure 14.22) [59]. The first step was the microwave-assisted palladium-catalyzed aminocarbonylation of iodoaromatic compounds,

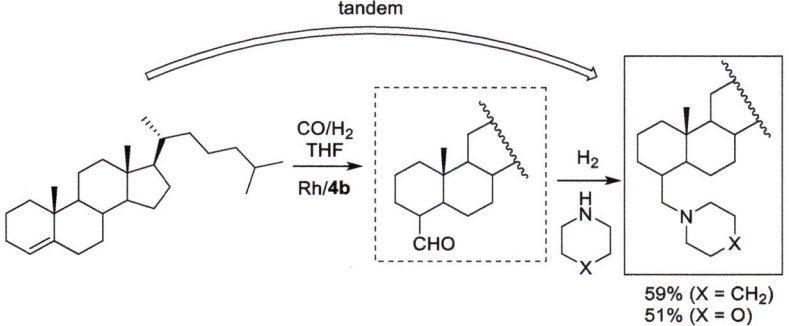

Figure 14.19 Synthesis of steroid-based tertiary amines via one-pot hydroaminomethylation reaction.

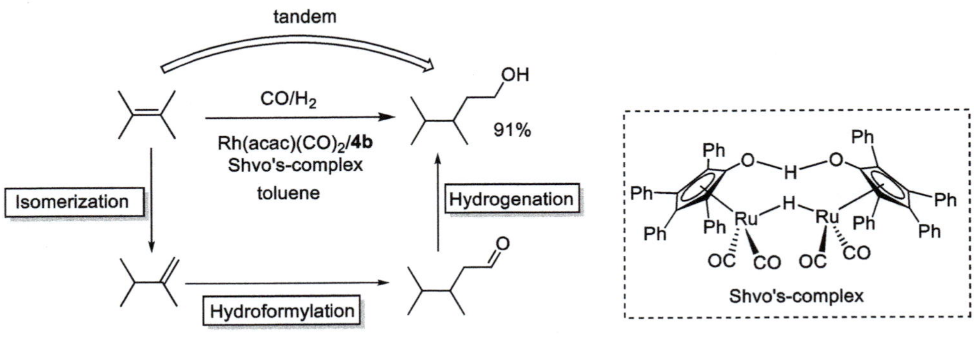

Figure 14.20 Sequential isomerization/hydroformylation/hydrogenation of 2,3-dimethylbut-2-ene.

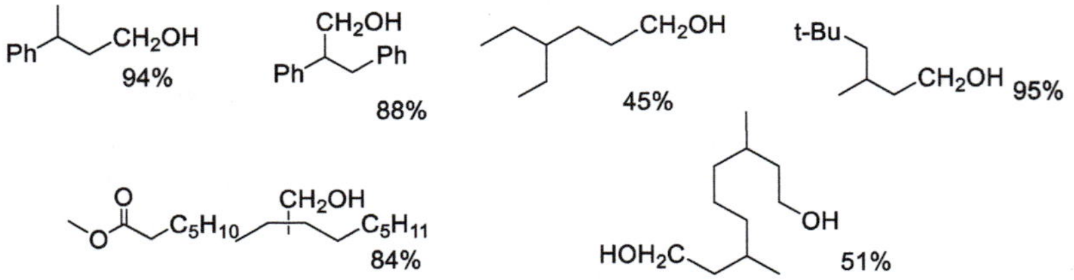

Figure 14.21 Examples of alcohols prepared through sequential isomerization/hydroformylation/hydrogenation of highly substituted/internal olefins.

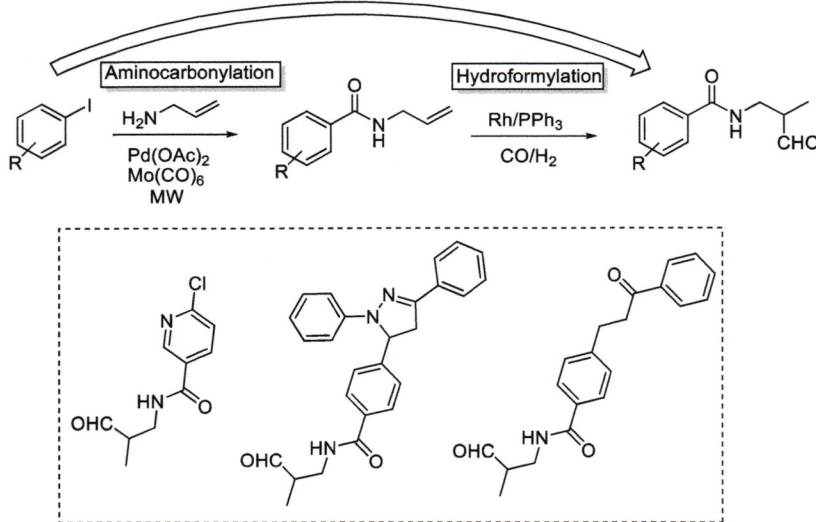

Figure 14.22 Synthesis of biologically relevant formylcarboxamides by sequential aminocarbonylation/hydroformylation reaction.

using allylamine as nucleophile and molybdenum hexacarbonyl as an alternative source of carbon monoxide, which provided an efficient method for the synthesis of allylcarboxamide derivatives. The subsequent hydroformylation of the allylcarboxamide intermediates, using Rh(I)/triphenylphosphine (**1**) catalyst, led to the preparation of biologically relevant new pyridine, pyrazoline and chalcone derivatives, containing both carboxamide and formyl moieties (Figure 14.22).

To turn the hydroformylation-based reactions into viable processes with industrial applications for fine chemistry, it is necessary to develop reusable catalysts. In this context, the strategies for immobilizing homogeneous ligands and catalysts on solid supports are presented in the following section, with special emphasis on the most relevant contribution from Portuguese research groups.

14.3 Heterogeneized Catalytic Systems

The immobilization of ligands and their metal-based complexes onto solid supports presents several advantages, because it allows preservation of the activity and selectivity of homogeneous catalytic systems, but also allows their easy recovery and reutilization, which significantly increases the sustainability of the hydroformylation process. This is particularly relevant when handling expensive catalysts, such as those typically used in hydroformylation-based reactions (mainly rhodium complexes) [5, 60–62]. In that respect, numerous solid supports, both inorganic and organic, have been designed to immobilize typical hydroformylation catalysts [63, 64]. The strategies adopted for immobilization can be classified into three types: non-covalent interactions, encapsulation, and covalent bonding, as depicted in Figure 14.23. In non-covalent interactions strategies, which are often based on adsorption phenomena, the ligands/catalysts are adsorbed on support's surface through weak intermolecular interactions such as van der Waals or electrostatic interactions (Figure 14.23A), whereas encapsulation denotes the physical immobilization of a ligand/catalyst inside the cavities or pores of a support (Figure 14.23,B). On the other hand, in the covalent bonding strategies, there are covalent tethering between the ligands/catalysts and the supports (Figure 14.23C). The latter offers additional protection against catalyst leaching, avoiding deactivation of the catalytic system upon each reutilization cycle [5].

Among the solid supports commonly used for ligand/metal complex immobilization, some possess hydroxyl groups (Figure 14.24A) at their surfaces, which can react directly with the appropriately functionalized ligands or be transformed into the selected functional group (Cl, NH_2, OH) via reaction with alkyl substituted silicon alkoxides (Figure 14.24B) [65]. Then, the appropriately functionalized ligand/metal complex immobilization occurs through covalent reaction with the solid supports (Figure 14.24C). The reaction of the functionalized supports occurs according to coupling reactivity pairs. For example, when X in the solid support is a halogen, the functionalized ligand/metal complex should have nucleophilic Y groups, such as NH_2 or OH, to react through an S_N mode. On the other hand, when X is a NH_2 group, Y could be a halogen (SN reaction) or a CO_2H group (carbodiimide-mediated coupling reaction). Occasionally, the ligand/metal complex might be functionalized

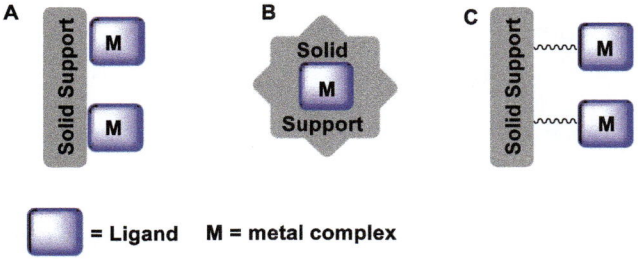

☐ = Ligand   M = metal complex

Figure 14.23 General strategies for catalysts' immobilization onto solid supports.

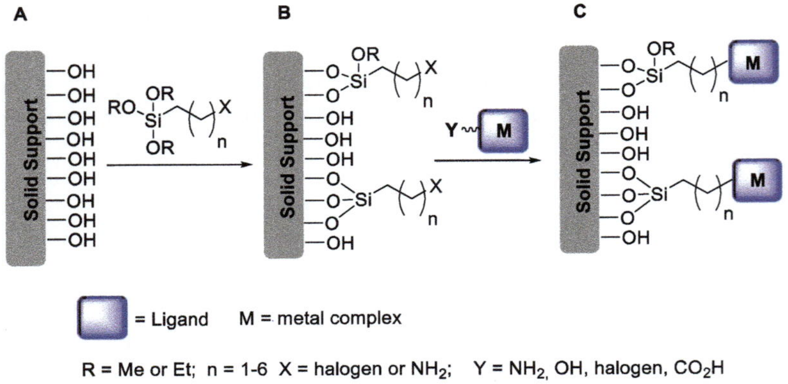
A
B
C
Solid Support
= Ligand
M = metal complex
R = Me or Et; n = 1-6 X = halogen or NH2; Y = NH2, OH, halogen, CO2H

Figure 14.24 Covalent immobilization of ligands and metal complexes onto functionalized supports.

with the alkyl substituted silicon alkoxide, allowing the direct reaction of the solid support with the ligand/metal complex, promoting the formation of the heterogenized ligand/catalyst.

Ding et al. [66] provided one of the first reports on the immobilization of a phosphorus ligand onto an inorganic solid support (Figure 14.25), by locking a phosphine derivative ligand and rhodium complex directly on the surface of the support, through mixing 2-(diphenylphosphino) ethyltriethoxysilane with silica (SiO_2), freshly impregnated with rhodium (Rh@SiO_2). This strategy was intended to avoid metal leaching and promote the in situ formation of the active species, by coordination of the flexible ligand to the metal on the surface of SiO_2. The material **PPh_2-Rh@ SiO_2** (Figure 14.25) was fully characterized by standard spectroscopic means, including thermogravimetry (TG), solid-state nuclear magnetic resonance (NMR), and Fourier transformed infrared (FTIR). This material was then applied in ethylene hydroformylation, showing quite high activity, increased over reaction time, and displaying improved TOF = 18 to 48 h^{-1} when reaction time increased from 10 to 155 hours. The authors attributed the high activity of the immobilized catalyst to the homogeneous active species formed in situ during the reaction. Based on carbon elemental analysis, no phosphine leached over the heterogeneous catalyst after a run of 1,000 hours, which was attributed to the covalent immobilization of the ligand.

Pereira et al. [67] reported an innovative strategy to promote the covalent immobilization of triphenylphosphine onto both amino-functionalized carbon nanotubes (NH_2@MWCNT) and iron oxide magnetic nanoparticles (NH_2@MNP) *via* catalytic hydroaminomethylation reaction (Figure 14.26). Vinyl-triphenylphosphine **16** was used simultaneously as substrate and as phosphorus

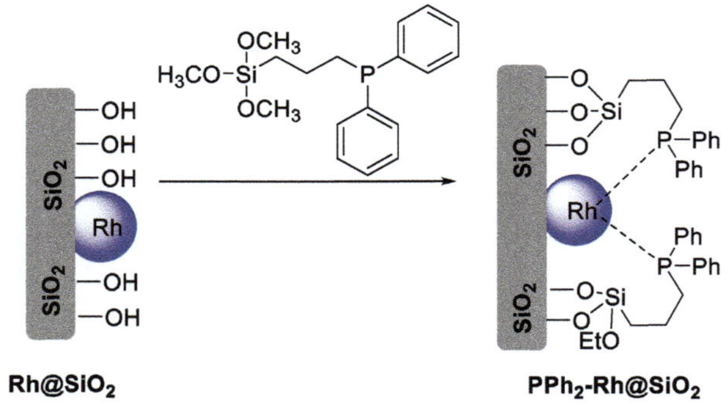
Rh@SiO2
PPh2-Rh@SiO2

Figure 14.25 Immobilization of 2-(diphenylphosphino)ethyltriethoxysilane onto silica (SiO_2) surface.

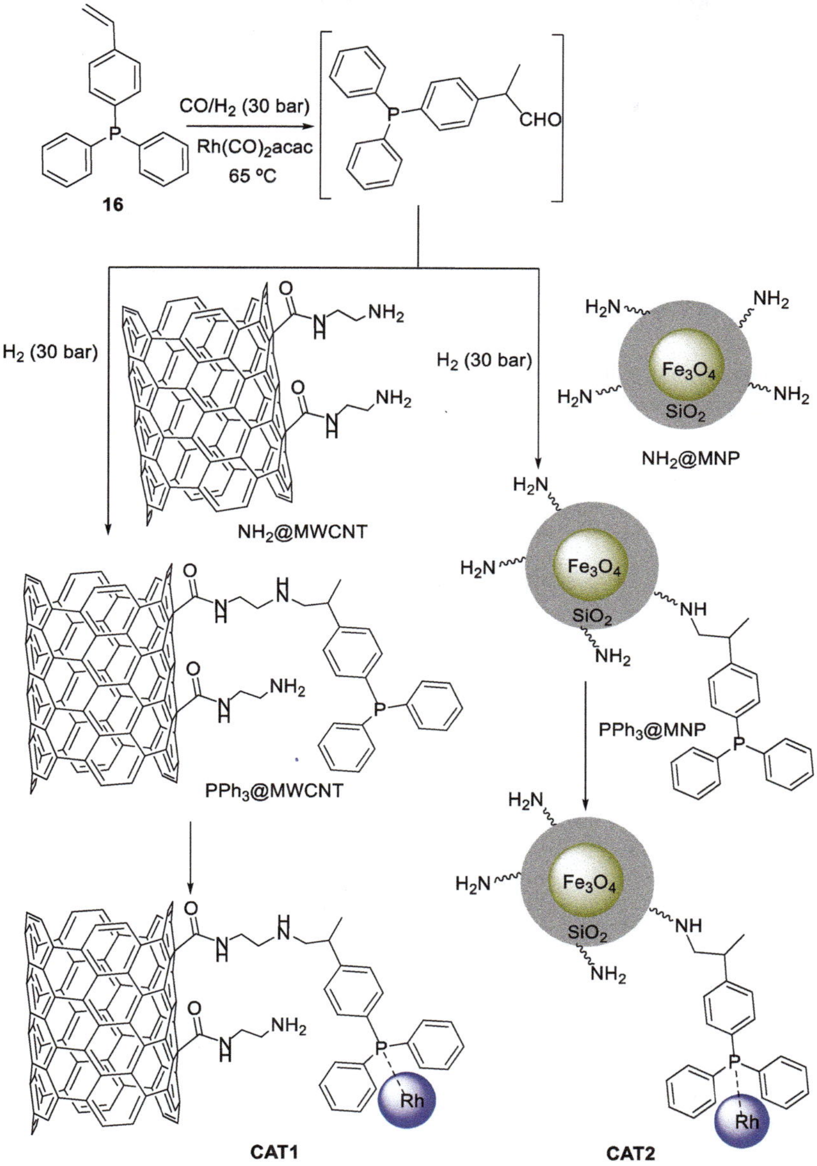

Figure 14.26 Immobilization of vinyl-triphenylphosphine onto amino-functionalized carbon nanotubes (NH$_2$@MWCNT) and iron oxide magnetic nanoparticles (NH$_2$@MNP) *via* catalytic hydroaminomethylation reaction.

ligand, being subjected to hydroformylation conditions, followed by addition of the amine-functionalized supports, NH$_2$@MWCNT or NH$_2$@MNP, to produce the corresponding supported ligands and their rhodium complexes **CAT1** and **CAT2**, respectively. The catalysts were characterized by TG, FTIR, transmission electronic microscopy (TEM), X-ray photoelectron spectroscopy (XPS) and inductively coupled plasma atomic emission spectroscopy (ICP-OES), which allowed to determine the rhodium content for each material, displaying values of 0.16 mmol/g for **CAT1** and 0.11 mmol/g for **CAT2**. Both catalysts were evaluated in the hydroformylation of styrene (T = 80 °C and P = 30 bar CO/H$_2$). Remarkably, **CAT2** was revealed to be stable under hydroformylation conditions, being active over three reutilization cycles without significant loss of activity or regioselectivity toward the branched isomer (ca. 70%). The ICP analysis of **CAT2**, after the reutilization studies, confirmed a rhodium content of 0.11 mmol/g, demonstrating that no Rh leaching occurred. However, under the same reaction conditions, when using **CAT1**, Rh leached from the carbon nanotubes support, leading to complete loss of activity and selectivity over the reutilization tests. The authors attributed this result to the occurrence of catalyst adsorption rather than immobilization when using carbon nanotubes as supporting material.

Pereira et al. synthesized phosphoramidite ligand **18** [68] from reaction of the BINOL chlorophosphite **17** with 3-aminopropylmethoxysilane in toluene, using triethylamine as base (Figure 14.27). Subsequently, this ligand was reacted with MCM-41, affording BINOL@MCM-41 supported ligand which, upon reaction with the rhodium precursor Rh(CO)$_2$(acac), yielded the immobilized catalyst **CAT 3**. Both the immobilized ligand and catalyst were characterized by TG, FTIR, N$_2$-sorption and elemental analysis, which indicated a rhodium content of 0.28% (w/w) immobilized in **CAT3** (Figure 14.27). This hybrid material was applied as catalyst in styrene hydroformylation, under moderate reaction conditions (80 °C, 25 bar CO/H$_2$) achieving 90% conversion and 93% regioselectivity for the branched aldehyde. Furthermore, the catalyst **CAT3** was recycled and reused in four consecutive runs without loss of activity and regioselectivity.

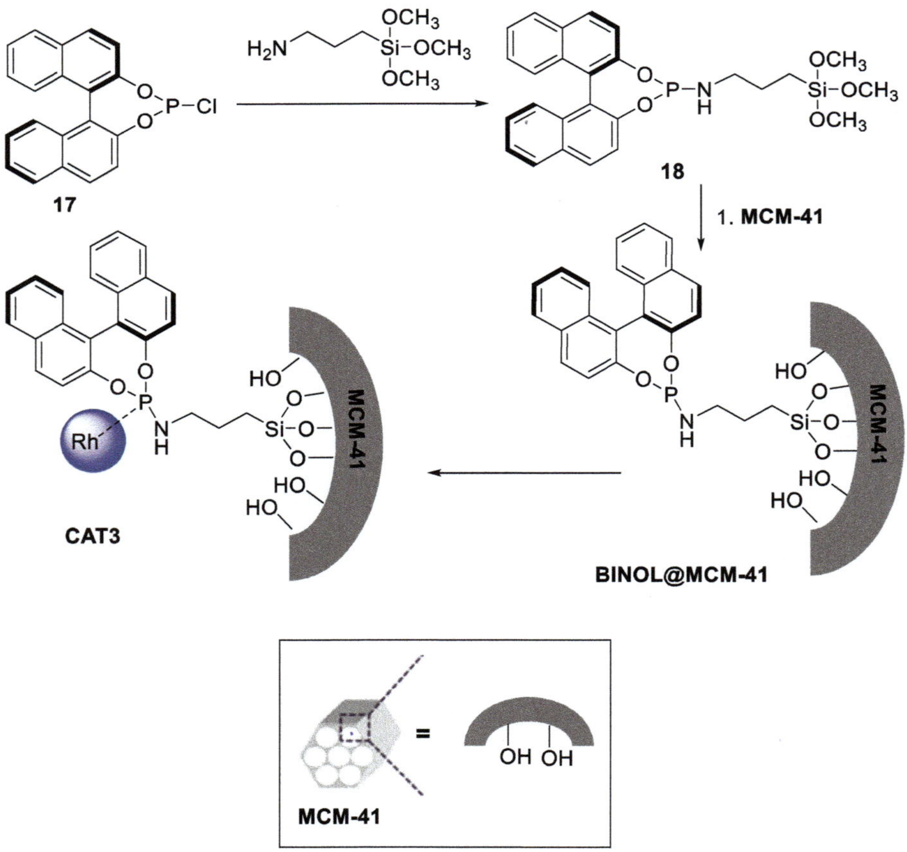

Figure 14.27 Covalent immobilization strategy for binaphthyl-based phosphoramidite onto MCM-41.

As previously mentioned, diphosphine ligands with large bite angles, such as Xantphos (*ca.* 110°), preferentially form bis-equatorial rhodium complexes, which favor the formation of linear aldehydes [27]. Therefore, the immobilization of Xantphos is a logical approach for heterogeneization of this emblematic ligand. Aiming to promote the immobilization of a rhodium/Xantphos complex for selective hydroformylation, Pereira et al. [69] reported the reaction of N-functionalized Xantphos analogue, *N*-Xantphos (**18**) onto a chloro-functionalized magnetic nanoparticle **Cl@MNP**, through a nucleophilic substitution reaction, in presence of a base to yield the immobilized ligand **18@MNP** (Figure 14.28). Subsequently, the addition of the rhodium precursor [Rh(acac)(CO)$_2$] resulted in the formation of **CAT4**. The resulting material was analyzed by ICP-OES analysis, which indicated a Rh content of 0.061 mmol/g of solid material.

Regarding the hydroformylation/acetalization sequential process, the previously developed hydroaminomethylation immobilization strategy [67], was used to promote the attachment of a C-scorpionate ligand onto amino-functionalized iron oxide magnetic nanoparticles **NH$_2$@MNP** (Figure 14.29). First, the allyl functionalized C-scorpionate ligand **19** was reacted with FeBr$_2$ to provide the corresponding Fe(II)/**19** complex. Then, this complex was subjected to hydroaminomethylation conditions, in presence of Rh(acac)(CO)$_2$/Xantphos (**3**) and the aminated support **NH$_2$@MNP** as nucleophile (first 30 bar with 1:1 CO/H$_2$, for 48 h and then 30 bar with H$_2$, for four hours), at 65 °C. Finally, the resulting **CAT5** was separated from the reaction medium by simple use of a magnet, isolated, and analyzed through ICP-OES and TG analysis, which indicated a content of 0.12 mmol of iron complex per gram of material [69].

Concerning the hydroformylation/cyanosilylation sequential process [70], the immobilization of BINOL **22** onto silica support was carried out (Figure 14.30). First, the methoxymethyl (MOM) protected 6,6′-dibromo BINOL derivative **20** was reacted, by Suzuki coupling, with 4-(methoxycarbonyl)phenylboronic acid to provide the BINOL derivative **21**. Then, MOM and methyl ester deprotection were performed in acidic and basic media, respectively to afford **22**. Finally, the carboxylate compound was coupled by amino acid coupling, using 1-hydroxybenzotriazole (HOBt)

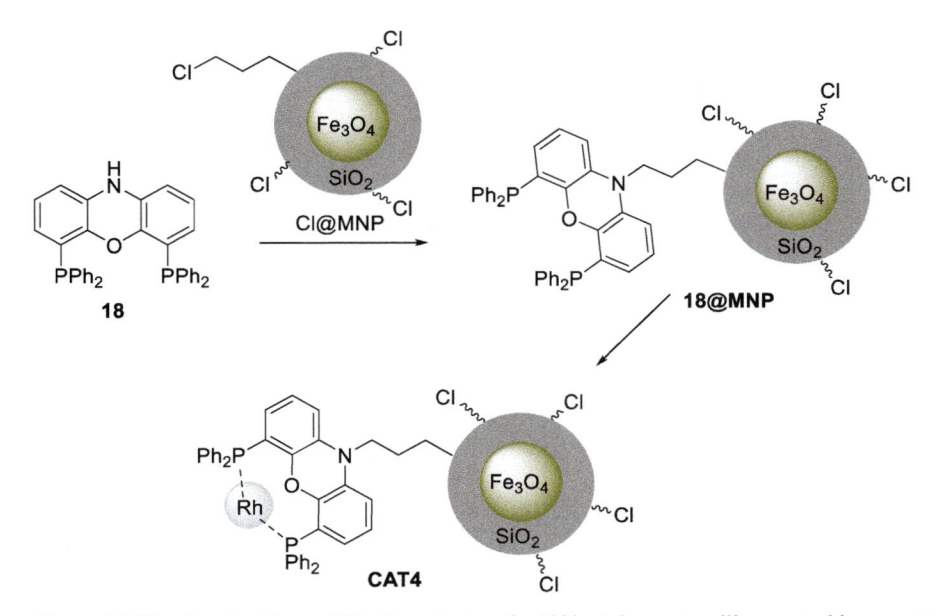

Figure 14.28 Covalent immobilization strategy for N-Xantphos onto silica-coated iron magnetic nanoparticle (MNP).

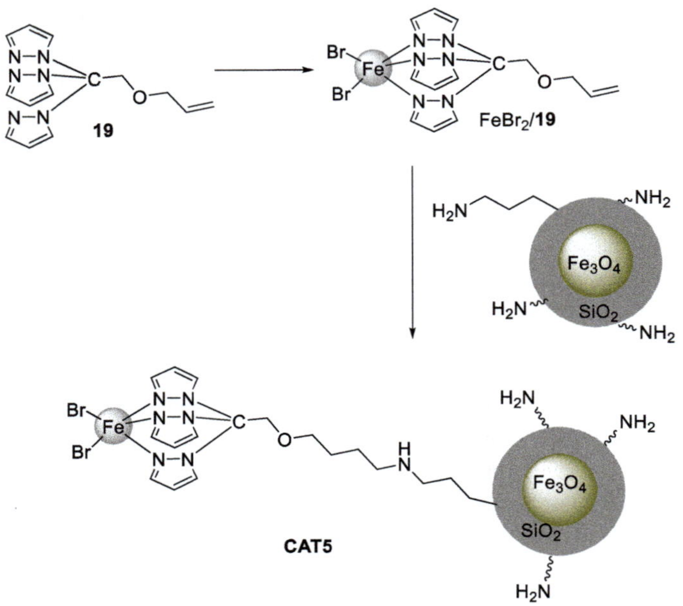

Figure 14.29 Preparation of C-scorpionate FeBr$_2$ (**CAT5**) immobilized onto silica coated magnetic nanoparticles *via* hydroaminomethylation reaction.

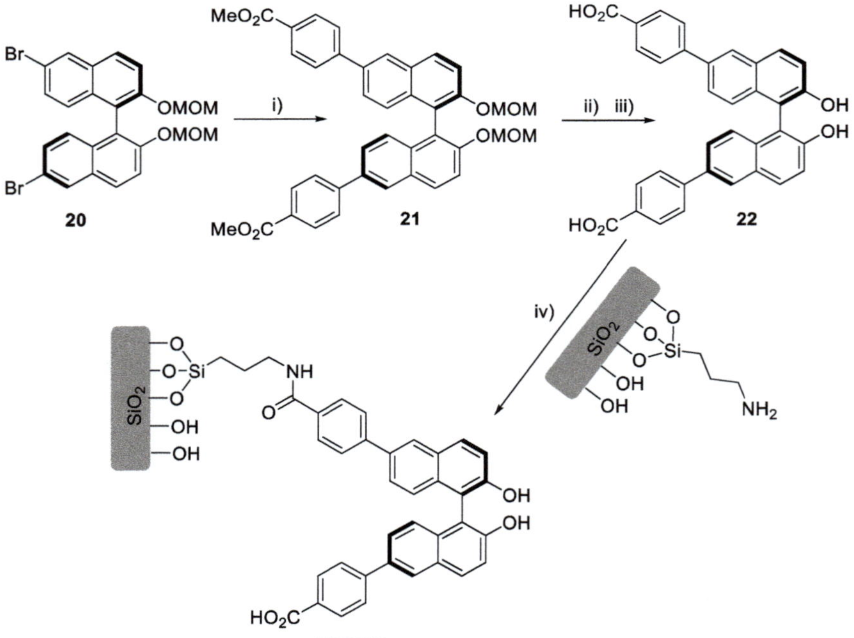

Figure 14.30 Synthetic route for the preparation of silica-immobilized BINOL derivative **22**@SiO$_2$. i) 4-(methoxycarbonyl)phenylboronic acid, Cs$_2$CO$_3$; Pd(PPh$_3$)$_4$; ii) TFA, CH$_2$Cl$_2$; iii) NaOH, THF; iv) HOBt, DCC.

and *N,N'*-dicyclohexylcarbodiimide (DCC) as the coupling agents to afford **22**@SiO$_2$. The resulting hybrid material **22**@SiO$_2$ was characterized by TG, FTIR and N$_2$ sorption, and the amount of immobilized BINOL derivative **22** onto the silica support was calculated in 0.8 mmol/g.

The immobilized Rh/Xantphos catalysts **CAT4** was applied in the hydroformylation of oct-1-ene, under moderate conditions (P = 20 bar CO/H$_2$, 1:1 and T = 80 °C), achieving close to full conversion, with 98% regioselectivity for the linear aldehyde (nonanal). This catalyst **CAT4** was retained in the reaction vessel through application of an external magnetic field and was used in five additional cycles without significant loss of activity and selectivity [69]. Then, the aldehydes obtained from the hydroformylation reaction were separated from **CAT4** and added without further purification, to a second reaction vessel, under an inert atmosphere, to proceed with the sequential process. In the case of the acetalization (Figure 14.31A), the aldehydes were added to a reaction vessel containing the iron C-scorpionate catalyst immobilized onto magnetic nanoparticles **CAT5** and ethanol. After 20 h at 80 °C, the correspondent acetals were obtained in up to 86% yield. This catalyst (**CAT5**) was further recovered and reused in five additional cycles and a small decrease in the reaction yield toward acetals was observed (75%) [69].

Similarly, the hydroformylation/cyanosilylation [70] was carried out using a two-vessel sequential strategy (Figure 14.31B). In this case, the aldehyde crude mixture, obtained from the hydroformylation reaction, was added to the second vessel containing the in situ generated **CAT6** (formed by addition of titanium(IV) tetra-isopropoxide (Ti(O-i-Pr)$_4$) to the supported BINOL ligand **22**@SiO$_2$) and trimethylsilyl cyanide (TMSCN). After 20 hours at 0 °C, full conversion was observed toward the hydroxydecanenitrile trimethylsilyl ether. The activity of the immobilized catalyst remained almost unchanged after three consecutive cycles. In sum, this two-vessel strategy was revealed to be an efficient methodology for the implementation of the described sequential processes already described. Additionally, the catalyst filtration/magnetic separation allowed a straightforward reutilization improving the sustainability of the catalytic systems.

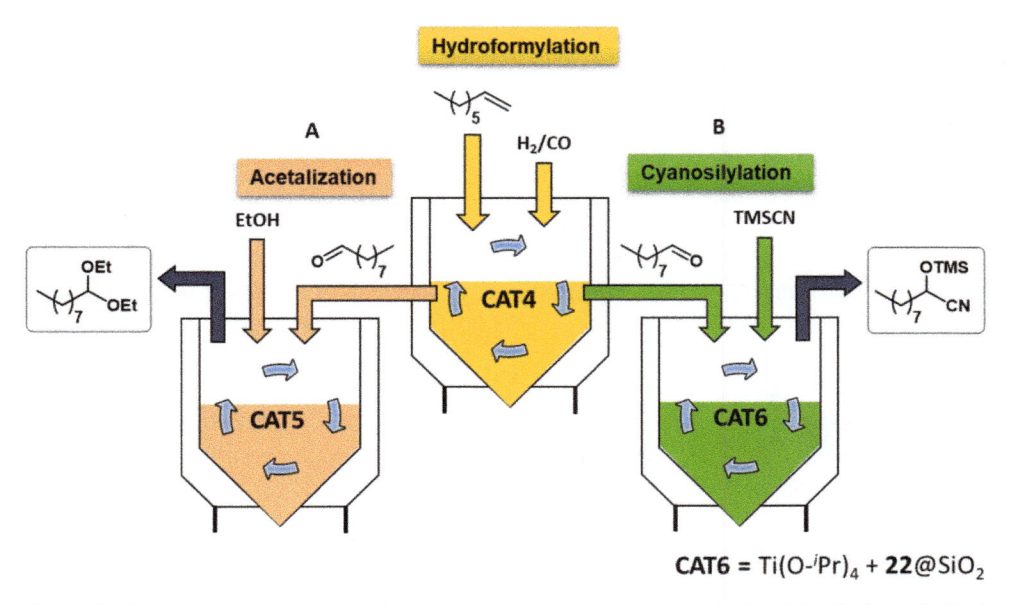

Figure 14.31 Two-vessel sequential processes using oct-1-ene as substrate: (A) hydroformylation/acetalization; (B) hydroformylation/cyanosilylation.

14.4 Conclusions

This chapter demonstrates the great interest in pursuing the development of highly active and selective immobilized catalysts for the synthesis of fine chemicals based on sustainable sequential processes. Throughout the years, emphasis has been put on the development of highly active and selective homogeneous hydroformylation catalysts, mostly based on rhodium transition metal, complexated within a plethora of ligands. More recently, these paradigmatic catalysts started being immobilized in a variety of supports, both organic and inorganic, with covalent linkage being the most effective strategy to immobilize ligands/catalysts. Since hydroformylation is considered a central reaction, these catalysts are suited for many hydroformylation centered sequential reactions, particularly for the synthesis of fine chemicals.

References

1 Brezny, A.C. and Landis, C.R. (2018). *Acc. Chem. Res.* 51 (9): 2344–2354. doi: 10.1021/acs.accounts.8b00335.
2 Nurttila, S.S., Linnebank, P.R., Krachko, T., and Reek, J.N.H. (2018). *ACS Catal.* 8 (4): 3469–3488. doi: 10.1021/acscatal.8b00288.
3 Chakrabortty, S., Almasalmaa, A.A., and de Vries, J.G. (2021). *Catal. Sci. Technol.* 11: 5388–5411. doi: 10.1039/D1CY00737H.
4 Zhang, B., Fuentes, D.P., and Börner, A. (2022). *ChemTexts* 8: 2. doi: 10.1007/s40828-021-00154-x.
5 Rodrigues, F.M.S., Carrilho, R.M.B., and Pereira, M.M. (2021). *Eur. J. Inorg. Chem.* 2021: 2294–2324. doi: 10.1002/ejic.202100032.
6 Roelen, O. (1938). *German Patent* DE 849548. (To Chemische Verwertungsgesellschaft Oberhausen mbH).
7 Cornils, B., Herrmann, W.A., and Rasch, M. (1994). *Angew. Chem. Int. Ed.* 33 (21): 2144–2163. doi: 10.1002/anie.199421441.
8 Gusevskaya, E.V., Jiménez-Pinto, J., and Börner, A. (2014). *ChemCatChem.* 6 (2): 382–411. doi: 10.1002/cctc.201300474.
9 Behr, A., Vorholt, A.J., Ostrowski, K.A., and Seidensticker, T. (2014). *Green Chem.* 16: 982–1006. doi: 10.1039/C3GC41960F.
10 Eilbracht, P., Barfacker, L., Buss, C. et al. (1999). *Chem. Rev.* 99: 3329–3366. doi: 10.1021/cr970413r.
11 Eilbracht, P. and Schmidt, A.M. (2006). Synthetic applications of tandem reaction sequences involving hydroformylation. In: *Catalytic Carbonylation Reactions*, 18 (ed. M. Beller), *Topics in Organometallic Chemistry*, 65–95. Berlin, Heidelberg: Springer. doi: 10.1007/3418_017.
12 Osborn, E.D. and Wilkinson, J.A. (1968). *J. Chem Soc. A.* 3133–3142. doi: 10.1039/J19680003133.
13 Yagupsky, E.D. and Wilkinson, G. (1968). *J. Chem. Soc. A.* 2660–2665. doi: 10.1039/J19680002660.
14 Brown, C.K. and Wilkinson, G. (1970). *J. Chem Soc. A.* 2753–2764. doi: 10.1039/J19700002753.
15 Pruett, R.L. and Smith, J.A. (1969). *J. Org. Chem.* 34 (2): 327–330. doi: 10.1021/jo01254a015.
16 Pruett, R.L. and Smith, J.A. (1968). *South African Patent 6804937*. (to Union Carbide Corporation).
17 van Leeuwen, P.W.N.M. and Roobeek, C.F. (1983). *J. Organomet. Chem.* 258 (3): 343–350. doi: 10.1016/S0022-328X(00)99279-9.
18 Jongsma, T., Challa, G., and van Leeuwen, P.W.N.M. (1991). *J. Organomet. Chem.* 421 (1): 121–128. doi: 10.1016/0022-328X(91)86436-T.
19 van Rooy, A., Orij, E.N., Kamer, P.C.J., and van Leeuwen, P.W.N.M. (1995). *Organometallics* 14 (1): 34–43. doi: 10.1021/om00001a010.
20 Carrilho, R.M.B., Abreu, A.R., Petöcz, G. et al. (2009). *Chem. Lett.* 38 (8): 844–845. doi: 10.1246/cl.2009.844.

21 Carrilho, R.M.B., Neves, A.C.B., Lourenço, M.A.O. et al. (2012). *J. Organomet. Chem.* 698: 28–34. doi: 10.1016/j.jorganchem.2011.10.007.
22 Damas, L., Costa, G.N., Ruas, J.C. et al. (2015). *Curr. Microwave Chem.* 2: 53–60. doi: 10.2174/2213335602011150212110805.
23 Felgueiras, A.P., Rodrigues, F.M.S., Carrilho, R.M.B. et al. (2022). *Molecules* 27 (6): 1989. doi: 10.3390/molecules27061989.
24 Diéguez, M., Pereira, M.M., Masdeu-Bultó, A.M. et al. (1999). *J. Mol. Catal. A: Chem.* 143 (1–3): 111–122. doi: 10.1016/S1381-1169(98)00373-2.
25 van Leeuwen, P.W.N.M., Kamer, P.C.J., and Reek, J.N.H. (1999). *Pure Appl. Chem.* 71 (8): 1443–1452. doi: 10.1351/pac199971081443.
26 Dierkes, P. and van Leeuwen, P.W.N.M. (1999). *J. Chem. Soc., Dalton Trans.* 10: 1519–1530. doi: 10.1039/A807799A.
27 Kamer, P.C.J., van Leeuwen, P.W.N.M., and Reek, J.N.H. (2001). *Acc. Chem. Res.* 34 (11): 895–904. doi: 10.1021/ar000060+.
28 Babin, J.E. and Whiteker, G.T. (1994). *United States Patent 5360938A*. (To Union Carbide Corporation).
29 Diéguez, M., Pàmies, O., Ruiz, A. et al. (2001). *Chem. Eur. J.* 7 (14): 3086–3094. doi: 10.1002/1521-3765(20010716)7:14<3086::AID-CHEM3086>3.0.CO;2-O.
30 Diéguez, M., Pàmies, O., Ruiz, A., and Claver, C. (2002). *New J. Chem.* 26: 827–833. doi: 10.1039/B200669C.
31 Clark, T.P., Landis, C.R., Freed, S.L. et al. (2005). *J. Am. Chem. Soc.* 127 (14): 5040–5042. doi: 10.1021/ja050148o.
32 Axtell, A.T., Klosin, J., and Abboud, K.A. (2006). *Organometallics* 25 (21): 5003–5009. doi: 10.1021/om060340e.
33 Sakai, N., Mano, S., Nozaki, K., and Takaya, H. (1993). *J. Am. Chem. Soc.* 115 (15): 7033–7034. doi: 10.1021/ja00068a095.
34 Nozaki, K., Sakai, N., Nanno, T. et al. (1997). *J. Am. Chem. Soc.* 119 (19): 4413–4423. doi: 10.1021/ja970049d.
35 Nozaki, K. (2005). *Chem. Rec.* 5 (6): 376–384. doi: 10.1002/tcr.20046.
36 Tanaka, R., Nakano, K., and Nozaki, K. (2007). *J. Org. Chem.* 72 (23): 8671–8676. doi: 10.1021/jo0712190.
37 Shibahara, F., Nozaki, K., and Hiyama, T. (2003). *J. Am. Chem. Soc.* 125 (28): 8555–8560. doi: 10.1021/ja034447u.
38 Yan, Y. and Zhang, X. (2006). *J. Am. Chem. Soc.* 128 (22): 7198–7202. doi: 10.1021/ja057065s.
39 Zhang, X., Cao, B., Yan, Y. et al. (2010). *Chem. Eur. J.* 16 (3): 871–877. doi: 10.1002/chem.200902238.
40 Freixa, Z. and Bayón, J.C. (2001). *J. Chem. Soc., Dalton Trans.* 2067–2068. doi: 10.1039/B105208J.
41 Peixoto, A.F., Abreu, A.R., Almeida, A.R. et al. (2014). *Curr. Org. Synth.* 11 (2): 301–309. doi: 10.2174/15701794113106660072.
42 Biellmann, J.-F. (2003). *Chem. Rev.* 103 (5): 2019–2034. doi: 10.1021/cr020071b.
43 Freixa, Z., Pereira, M.M., Bayon, J.C. et al. (2001). *Tetrahedron Asymmetry* 12 (7): 1083–1087. doi: 10.1016/S0957-4166(01)00191-4.
44 Peixoto, A.F., Pereira, M.M., Silva, A.M.S. et al. (2007). *J. Mol. Catal. A, Chem.* 275 (1–2): 121–129. doi: 10.1016/j.molcata.2007.05.035.
45 Costa, G.N., Carrilho, R.M.B., Dias, L.D. et al. (2016). *J. Mol. Catal. A, Chem.* 416: 73–80. doi: 10.1016/j.molcata.2016.02.016.
46 Vieira, C.G., Freitas, M.C., Oliveira, K.C.B. et al. (2015). *Catal. Sci. Technol.* 5: 960–966. doi: 10.1039/C4CY01020E.
47 Faria, A.C., Oliveira, K.C.B., Monteiro, A.C. et al. (2020). *Catal. Today* 344: 24–31. doi: 10.1016/j.cattod.2018.10.024.

48 Peixoto, A.F., Melo, D.S., Fernandes, T.F. et al. (2008). *Appl. Catal. A, Gen.* 340 (2): 212–219. doi: 10.1016/j.apcata.2008.02.015.

49 Peixoto, A.F., Pereira, M.M., Neves, M.G.P.M.S. et al. (2003). *Tetrahedron Lett.* 44 (30): 5593–5595. doi: 10.1016/S0040-4039(03)01379-0.

50 Peixoto, A.F., Pereira, M.M., Sousa, A.F. et al. (2005). *J. Mol. Catal. A, Chem.* 235 (1–2): 185–193. doi: 10.1016/j.molcata.2005.04.008.

51 Kalck, P. and Urrutigoity, M. (2017). Tandem rhodium carbonylation reactions. In: *Rhodium Catalysis*, 61 (ed. C. Claver), *Topics in Organometallic Chemistry*, 69–98. Cham: Springer. doi: 10.1007/3418_2016_170.

52 Almeida, A.R., Dias, R.D., Monteiro, C.J.P. et al. (2014). *Adv. Synth. Catal.* 356 (6): 1223–1228. doi: 10.1002/adsc.201300968.

53 Strecker, A. (1850). *Liebigs Ann. Chem.* 75 (1): 27–45. doi: 10.1002/jlac.18500750103.

54 Bondžić, B.P. (2015). *J. Mol. Catal. A, Chem.* 408: 310–334. doi: 10.1016/j.molcata.2015.07.026.

55 Almeida, A.R., Carrilho, R.M.B., Peixoto, A.F. et al. (2017). *Tetrahedron* 73 (17): 2389–2395. doi: 10.1016/j.tet.2017.03.023.

56 Kalck, P. and Urrutigoïty, M. (2018). *Chem. Rev.* 118 (7): 3833–3861. doi: 10.1021/acs.chemrev.7b00667.

57 Chen, C., Dong, X.-Q., and Zhang, X. (2016). *Org. Chem. Front.* 3: 1359–1370. doi: 10.1039/C6QO00233A.

58 Rodrigues, F.M.S., Kucmierczyk, P.K., Pineiro, M. et al. (2018). *ChemSusChem.* 11 (14): 2310–2314. doi: 10.1002/cssc.201800488.

59 Damas, L., Rodrigues, F.M.S., Gonzalez, A.C.S. et al. (2020). *J. Organomet. Chem.* 923: 121417. doi: 10.1016/j.jorganchem.2020.121417.

60 Benaglia, M. and Puglisi, A. (eds.) (2020). *Catalyst Immobilization: Methods and Applications*. Weinheim: Wiley VCH.

61 Li, C., Wang, W., Yan, L. et al. (2018). *Front. Chem. Sci. Eng.* 12 (1): 113–123. doi: 10.1007/s11705-017-1672-9.

62 Hübner, S., de Vries, J.G., and Farina, V. (2016). *Adv. Synth. Catal.* 358 (1): 3–25. doi: 10.1002/adsc.201500846.

63 Regalbuto, J. (ed.) (2007). *Catalyst Preparation: Science and Engineering*. Boca Raton, FL: CRC Press, Taylor & Francis Group.

64 Li, C. and Liu, Y. (eds.) (2014). *Bridging Heterogeneous and Homogeneous Catalysis: Concepts, Strategies, and Applications*. Weinheim: Wiley-VCH.

65 Brinker, C.J. and Scherer, G.W. (eds.) (1990). *Sol-Gel Science: The Physics and Chemistry of Sol-Gel Processing*. San Diego, CA: Elsevier Inc.

66 Li, X.M., Ding, Y.J., Jiao, G.P. et al. (2009). *Appl. Catal. A, Gen.* 353 (2): 266–270. doi: 10.1016/j.apcata.2008.10.052.

67 Rodrigues, F.M.S., Calvete, M.J.F., Monteiro, C.J.P. et al. (2020). *Catal. Today* 356: 456–463. doi: 10.1016/j.cattod.2019.05.045.

68 Henriques, C.A., Rodrigues, F.M.S., Carrilho, R.M.B. et al. (2016). *Curr Org Chem* 20: 1445.

69 Rodrigues, F.M.S., Dias, L.D., Calvete, M.J.F. et al. (2021). *Catalysts* 11: 608. doi: 10.3390/catal11050608.

70 Pereira, M.M., Neves, A.C.B., Calvete, M.J.F. et al. (2013). *Catal. Today* 218-219: 99–106. doi: 10.1016/j.cattod.2013.09.050.

15

Synthesis of New Polyolefins by Incorporation of New Comonomers

Kotohiro Nomura and Suphitchaya Kitphaitun

Department of Chemistry, Tokyo Metropolitan University, Hachioji, Tokyo, Japan

15.1 Introduction

Olefin polymerization by early transition metal catalysts has been the core technology for industrial production of polyolefins (such as high-density polyethylene [HDPE], linear low-density polyethylene [LLDPE], isotactic polypropylene [PP], etc.], widely used as commodity polymers in our daily life. Synthesis of the new functional polymers, especially copolymers by incorporation of monomers that have not been prepared using conventional catalysts (Ziegler-Natta, metallocene catalysts), such as sterically encumbered, cyclic olefins has been a long-term interest, because physical, mechanical, and electronic properties of the copolymers can be tuned by modification of individual components. Moreover, the approach should be important in terms of better materials and chemical recycling due to unification of monomers (olefins) and the others; polyolefins with specified functions should be more sustainable compared to the functional polymers from rather complicated monomers (many steps required from fossil oil). Development of the molecular catalysts, especially the catalysts exhibiting high activities with better comonomer incorporations, has thus been a promising subject for the successful synthesis [1–15].

Generally, structural features of the catalyst (such as basic geometry and electronic and steric bulk of the ligands) play a key role toward the reaction, and this is an apparent difference from conventional radical and ionic polymerizations. It has been known that bridged metallocene showed better α-olefin incorporation than the unbridged ones (assumed as due to providing more open coordination space, Scheme 15.1) [15]. Linked half-titanocenes exemplified as constrained geometry catalyst (CGC), [Me$_2$Si(C$_5$Me$_4$)(N^{t}Bu)]TiCl$_2$ [2], exhibit more efficient α-olefin incorporation in the ethylene copolymerization [2, 16]. The catalyst also demonstrated a capability of (rather) efficient styrene incorporation in the ethylene copolymerization [11], but showed invariably of the incorporation (<50 mol%) [2, 10, 13–15, 17]. Later, the bimetallic system (bimetallic CGC, Scheme 15.1) enabled synthesis of the copolymer with high styrene content (76 mol%) [10, 11]. Half-titanocenes modified with anionic ancillary donor ligands of type, Cp'TiX$_2$(Y) (Y = aryloxide, ketimide, phosphinimide, iminoimidazolide, amidinate, and others, Scheme 15.1), demonstrated synthesis of the new ethylene

---

Catalysis for a Sustainable Environment: Reactions, Processes and Applied Technologies Volume 2, First Edition. Edited by Armando J. L. Pombeiro, Manas Sutradhar, and Elisabete C. B. A. Alegria.
© 2024 John Wiley & Sons Ltd. Published 2024 by John Wiley & Sons Ltd.

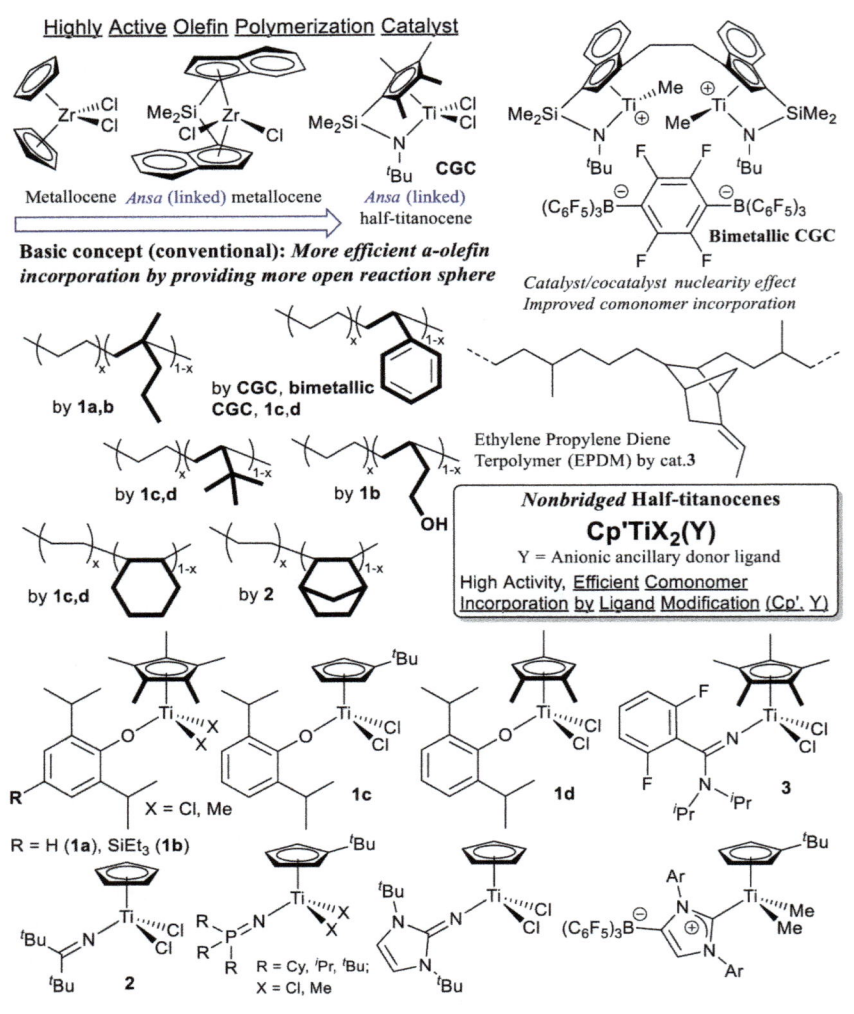

Scheme 15.1 Selected group 4 transition metal complex catalysts for olefin polymerisation.

copolymers by incorporations of various olefins (sterically encumbered olefins, cyclic olefins, aromatic vinyl monomers, and others) [13–15]; both the aryloxide (**1**) and the ketimide (**2**) analogues have been the known successful examples. Later, the η^1-amidinate analogue (**3**) demonstrated an industrial production of chlorine-free synthetic rubber (EPDM, ethylene propylene diene terpolymer) without deep cooling, commonly employed in the conventional (Ziegler type) catalyst systems in industry [18].

This chapter introduces reported examples for synthesis of (new) ethylene copolymers by incorporation of monomers, that were very difficult (considered to be impossible) with the conventional catalysts (metallocene, linked half-titanocenes), such as sterically encumbered monomers, cyclic olefins, α-olefins containing hydroxy group etc. Recently, synthesis of functional polymers from renewable feedstocks has been considered as an important subject (as alternatives from fossil oil, petroleum based chemicals) [19–22], the chapter thus also introduces recent results in synthesis of bio-based polyolefins in the ethylene copolymerization with terpenes [23, 24].

15.2 Synthesis of New Ethylene Copolymers by Incorporation of Sterically Encumbered Olefins, Cyclic Olefins

15.2.1 Ethylene Copolymerization with Sterically Encumbered Olefins

Ethylene copolymerization with isobutene (IB) using [Et(indenyl)$_2$]ZrCl$_2$–MAO catalysts afforded the copolymer with low IB content (<2.8 mol %) even under large excess IB conditions (IB:ethylene = 4000:1) [25]; therefore, these monomers (1,1-disubstituted α-olefins) were called as traditionally unreactive monomers in the metal catalyzed coordination polymerization [26]. The cyclododecylamide CGC analogue (CGC–CDD, Scheme 15.2) showed efficient incorporation of IB in the ethylene polymerization in the presence of MAO or AlEt$_3$–[Ph$_3$C][B(C$_6$F$_5$)$_4$], AlEt$_3$–[PhN(H)Me$_2$][B(C$_6$F$_5$)$_4$] cocatalyst (ethylene 4.4 atm, r.t.) [27], although the ordinary CGC showed negligible (low) IB incorporation under the same conditions [28–32]. The copolymerization with 2-methyl-1-pentene afforded the polymer with broad molecular weight distribution (M_w/M_n = 5.9) [27]. Later, the fluorenyl analogue (Flu–CGC–CDD, Scheme 15.2) also showed better IB incorporation, affording copolymers with alternating IB incorporations (ethylene 1 atm, 25 °C); the attempted IB polymerization failed [28]. The bimetallic CGC also exhibited improved IB incorporation compared with the mononuclear CGC in the ethylene copolymerization (IB 3.1 mol% vs 15.2 mol%, ethylene 1 atm, IB 1.3 M, 24 °C), although the activity decreased and the molecular weight distribution in the resultant polymer possessed was rather broad (M_w/M_n = 3.67) [29, 30]. The bimetallic system thus enabled synthesis of ethylene copolymers with methylenecyclopentene and methylenecyclohexene, whereas the attempted copolymerization with 2-methyl-2-butene gave the copolymer by incorporation of 2-methyl-1-butene after the isomerization [30].

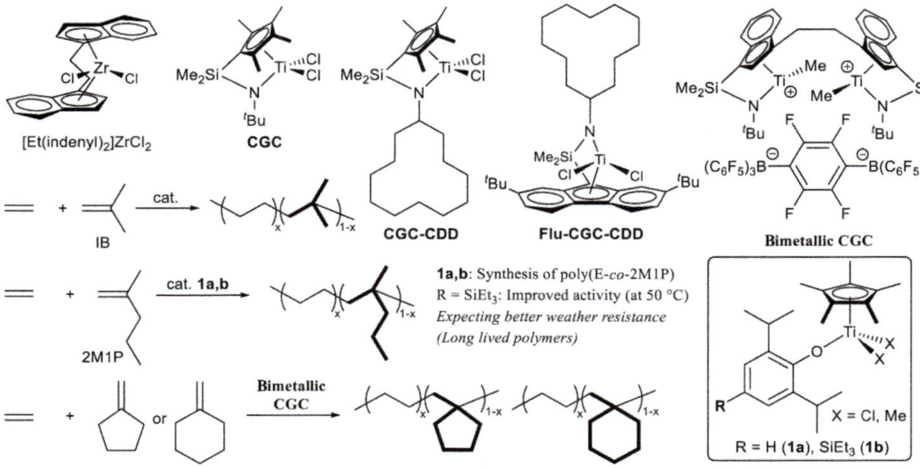

Scheme 15.2 Selected reports in ethylene copolymerization with disubstituted α-olefins. Adapted from [25, 27–34].

The phenoxide modified nonbridged half-titanocene (**1a**) demonstrated synthesis of ethylene copolymers by incorporation of 2-methyl-1-pentene in the presence of MAO cocatalyst. The observed activity (calculated on the basis of the polymer yield and catalyst charged) was initially higher than ethylene polymerization under low 2M1P concentration conditions but the activity decreased with increase in the 2M1P concentration and/ or decrease in ethylene pressure. The resultant polymers were poly(ethylene-co-2M1P)s confirmed by ^{13}C NMR spectra and DSC thermograms with uniform compositions; the copolymer possessed high molecular weights with unimodal molecular weight distributions (M_n = 3.3–13 × 10^4, M_w/M_n = 1.7–1.9). The activity at 50 °C increased (44,200 kg-polymer/mol-Ti·h, ethylene 4 atm, 2M1P 1.35 M) when the SiEt$_3$ analogue (**1b**) and the SiMe$_3$ analogue were employed as the catalyst [33]; the activity by **1b** further increased at 80 °C [34].

As shown in Scheme 15.3, the complex **1a** also incorporated vinylcyclohexane in an efficient manner, although as described below, the copolymerization with γ-disubstituted (branched) α-olefins by the conventional catalysts seemed very difficult [35]. The copolymerization with tert-butyl ethylene (TBE) was achieved by the tert-butyl cyclopentadienyl analogue (**1c**) and the 1,2,4-trimethylcyclopnetadienyl analogue (**1d**) [36]. Similarly, the copolymerization with vinyltrimethylsilane (vinyltriethylsilane) was also demonstrated by adopting the Cp-ketimide analogue (**2**) as the catalyst, which showed negligible TBE incorporation [37]. The copolymerization with allyltrialkylsilanes also proceeded by **2**, whereas both **1a** and **1c** showed better allyltrimethylsilane (ATMS) incorporations than **2** but the M_n values in the copolymers were rather low [38]. Therefore, **2** seemed to be more suited for synthesis of the copolymers with high molecular weights [37, 38].

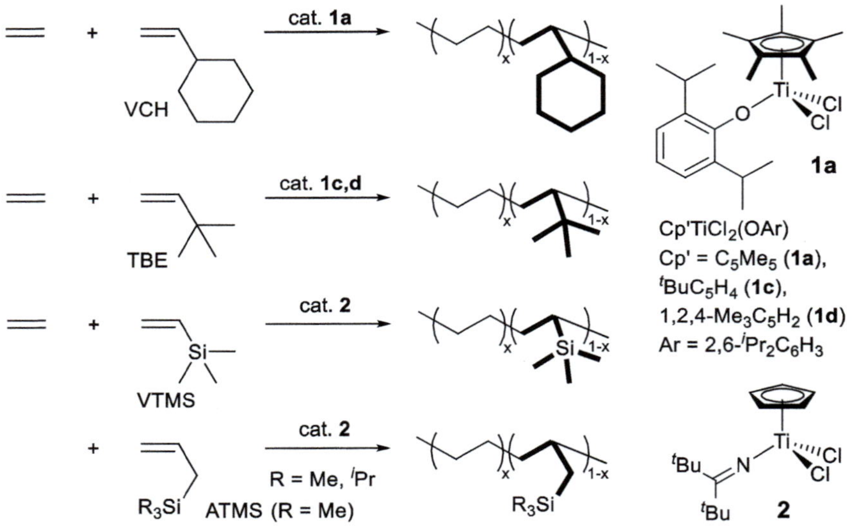

Scheme 15.3 Ethylene copolymerization with vinylcyclohexane, tert-butyl ethylene (TBE), vinyltrimethylsilane, and allyltrialkylsilanes. Adapted from [35–38].

Scheme 15.4 summarizes effect of monomer structure (the substituent in ethylene) toward monomer reactivity in the ethylene copolymerization [39]. The monomer reactivity ratio [r_E value, k_{EE}/k_{EC} (E: ethylene, C: comonomer)], placed in an explanation in the scheme, was used to compare the monomer reactivity with ethylene. It has been known, as demonstrated by **CGC** (r_E=>75), that the incorporation of γ-disubstituted (branched) α-olefins (exemplified as 3-methyl-1-pentene) is very difficult using conventional catalysts (especially metallocene) [1, 25, 26]. In contrast, the copolymerization by **1a** proceeded in an efficient manner (r_E = 8.47, 8.73), and **1a** also showed an efficient incorporation of 4,4-dimethyl-1-pentene (NHEP). Interestingly, the r_E value by **CGC** is close to that by **1a**, although the activity by **1a** was higher than for **CGC** [39]. The results indicate that **1a** should be suited for efficient ethylene copolymerization with branched α-olefins. Since **2** showed

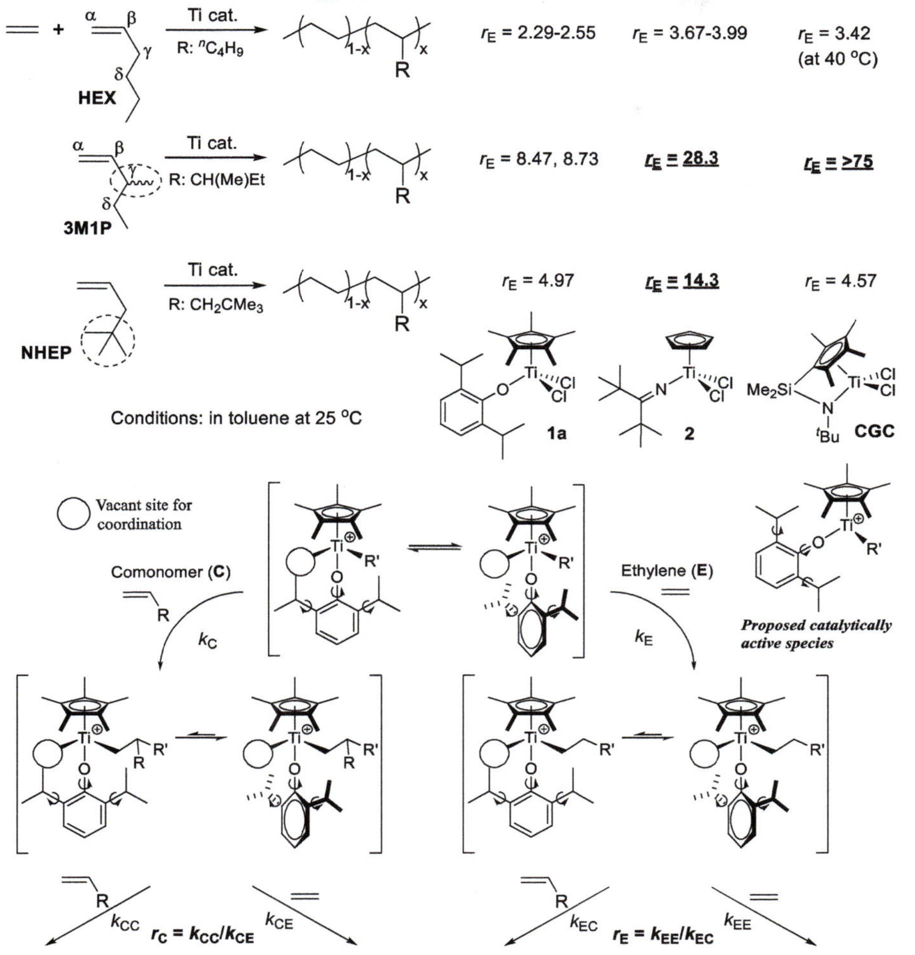

Scheme 15.4 Selected reports in ethylene copolymerization with disubstituted α-olefins [39] / American Chemical Society.

less efficient NHEP incorporation ($r_E = 6.77$), suggesting that the monomer reactivity (r_E value) by **2** was influenced by the substituent in the δ-position (in addition to the substituent in the γ-position); the substituent in the γ-position rather than the δ-position affected the monomer reactivity by **CGC**. One possible explanation why **1a** shows a high level of comonomer incorporation would be due to the flexible internal rotation of both cyclopentadienyl and phenoxide ligands (Scheme 15.4), which can change the optimized geometry in each stage by these rotations [39].

15.2.2 Ethylene Copolymerization with Cyclic Olefins

Amorphous cyclic olefin copolymers (COCs) are promising functional materials due to high transparency, thermal and humidity resistance, and dimensional stability. The copolymerization approach enables modification of their compositions (such as cyclic olefin contents) and microstructures (including tacticity) as commercialized (as TOPAS® and APEL®) as ultra-pure, crystal-clear, and high barrier materials. Many reports are known for the ethylene copolymerization with norbornene (NBE) using group 4 transition metal catalysts such as ordinary metallocenes,

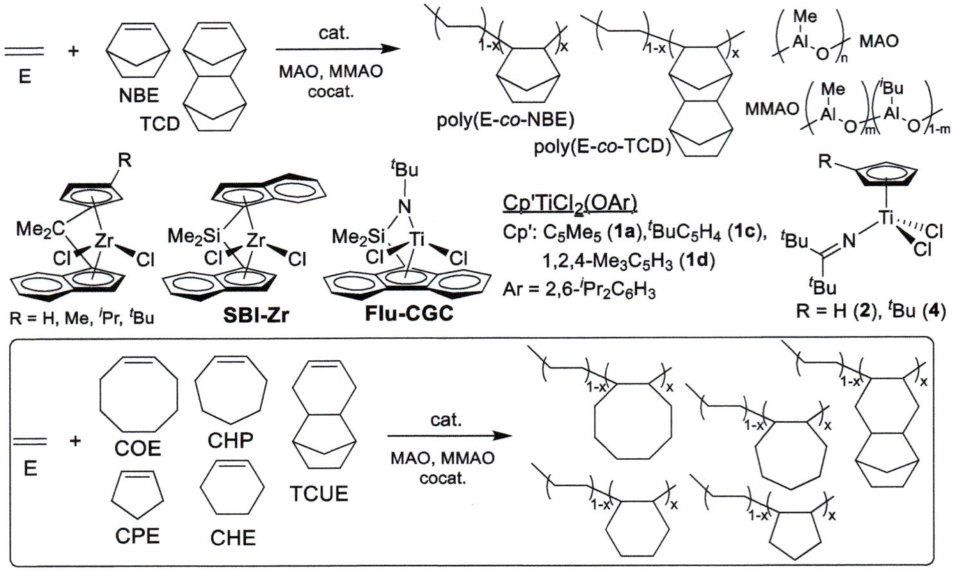

Scheme 15.5 Selected reports in ethylene copolymerization with cyclic olefins. Adapted from [40–52, 54, 64, 68, 70].

half-titanocenes, and the others (Scheme 15.5) [40–45]. However, the successful examples for the efficient synthesis of random, high molecular weight copolymers with high NBE contents (high glass transition temperature [T_g values]) have still been limited. Moreover, the successful examples in synthesis of amorphous copolymers by incorporation of low strain monomers (e.g. cyclopentene, cycloheptene, and cyclooctene, Scheme 15.5) have also still been limited.

It has been known both the activity and the M_n values in the resultant copolymers decreased when ordinary metallocene (exemplified as SBI-Zr, Scheme 15.5) and CGC were employed as catalysts in the ethylene/NBE copolymerization [40, 44]. In contrast, living polymerization of NBE proceeded by the fluorenyl analogue (Flu-CGC) in the presence of appropriate Al cocatalyst (dried MAO prepared by removing AlMe$_3$ from the commercially available MAO in toluene solution, or MMAO) [46, 47]. The catalyst showed efficient NBE incorporation than CGC in the ethylene copolymerization, affording the copolymers with high NBE contents [48]. The NBE copolymerizations with 1-alkene [49], as well as propylene [50], were demonstrated in the presence of dried MMAO (modified MAO, methyisobutylaluminoxane); [Me$_2$Si(C$_{29}$H$_{36}$)(N^{t}Bu)]TiMe$_2$ (C$_{29}$H$_{36}$ = octamethyloctahydrodibenzofluorenyl) showed the highest activity [50]. Synthesis of gradient copolymers of NBE and 1-alkene (1-octene, 1-decene, and 1-dodecene) was demonstrated by the Flu-CGC in the presence of borate and Al cocatalyst modified with 2,6-^{t}Bu-4-MeC$_6$H$_2$OH [51]. In this catalyst system, the effect of the cocatalyst strongly affects the activity and the copolymerization behavior.

The Cp-ketimide analogue (**2**) demonstrated efficient NBE incorporation in the ethylene copolymerization to afford the high molecular weight random copolymers with high NBE contents (e.g. NBE 73.5 mol%, M_n = 444000, M_w/M_n = 2.01; activity 31,500 kg-polymer/mol-Ti·h, ethylene 2 atm, NBE 10.0 M) [52]; the copolymerization behavior (increase in the activity upon addition of NBE) was apparently different from those by the conventional catalysts (metallocene, linked half-titanocenes such as CGC). The activity increased at 60 °C, and the significant decrease in the activity was not observed at 80 °C [52]. Moreover, **2** enabled highly efficient synthesis of COCs with high T_g values not only by copolymerization of NBE with α-olefins (1-hexene, 1-octene, 1-dodecene), but also by copolymerization of tetracyclododecene (TCD) with α-olefins [53]. Linear relationships

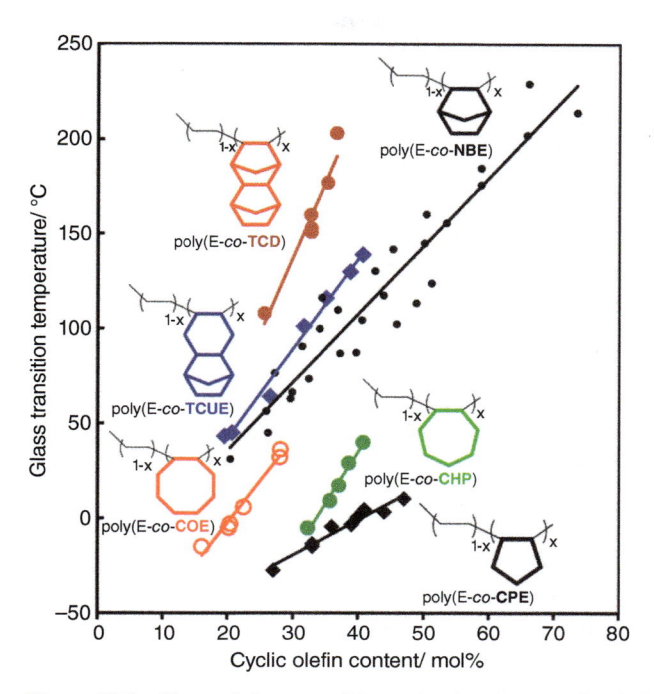

Figure 15.1 Plots of glass transition temperature *vs* cyclic olefin content. Effect of monomer structure on thermal property in ethylene cyclic olefin copolymers (COCs). Adapted from [70] Harakawa et.al., 2020.

between the T_g values and the NBE or TCD contents were observed, as observed in poly(ethylene-co-NBE)s (Figure 15.1) [44]. The copolymerization of NBE with 1-octene in the presence of 1,7-octadiene by the *tert*-BuC$_5$H$_4$ analogue (**4**) gave the polymer containing terminal olefinic double bond in the side chain [53].

Ethylene copolymers with TCD, that are produced by using classical Ziegler-type vanadium catalyst systems [VOCl$_3$, VO(OEt)Cl$_2$–EtAlCl$_2$·Et$_2$AlCl etc.] in industry [54], possess higher glass transition temperature (T_g) compared to poly(ethylene-*co*-NBE)s with the same cyclic olefin contents (Figure 15.1). The copolymerizations using metallocene catalysts generally exhibited low catalytic activities and or less efficient TCD incorporation [55, 56]. (^{t}BuC$_5$H$_4$)TiCl$_2$(N=C^{t}Bu$_2$) (**4**) exhibited high catalytic activities with efficient TCD incorporation in ethylene copolymerization in the presence of MAO cocatalyst, affording high molecular weight polymers with uniform compositions. In contrast, the Cp analogue, CpTiCl$_2$(N=C^{t}Bu$_2$) (**2**), which is effective catalyst for efficient α-olefin/TCD and ethylene/NBE copolymerizations [52, 53], showed low catalytic activities. Ethylene copolymers with *exo*-1,4,4a,9,9a,10-hexahydro-9,10(1′,2′)-benzeno-1,4-methanoanthracene (HBMN) prepared by CGC showed better elongation break (stress–strain behavior) than the ethylene/NBE copolymer sample possessing a similar T_g value [57]; the design of the monomer structure plays a role in the properties of the materials (e.g. transparency, thermal resistance, and mechanical performance in film).

As already described, there have been many reports for ethylene copolymerization with highly strained cyclic olefins, especially NBE for synthesis of COC. However, reports for the copolymerization with so called low strained cyclic olefins (cyclopentene [CPE], cyclohexene [CHE], cycloheptene [CHP], and *cis*-cyclooctene [COE]) have been limited. Although reports on the ethylene copolymerization with CPE were known [58–67], reports with CHE [68], CHP [62, 63], or COE [62,

63, 69, 70] have been limited [71]; one report was known for the synthesis of low molecular weight amorphous copolymers with CHP or COE by linked half-titanocene catalysts under specified conditions (ethylene 0.1–0.2 atm, M_n = 2300–8400) [62]. Ethylene copolymerization with CHE using Cp'TiCl$_2$(O-2,6-^{i}Pr$_2$C$_6$H$_3$) [Cp' ^{t}BuC$_5$H$_4$ (**1c**), 1,2,4-Me$_3$C$_5$H$_2$ (**1d**)] proceeded with 1,2-insertion, affording the copolymer with uniform composition [68] whereas CHE was not incorporated by the other catalysts (metallocenes, CGC, **1a**, **2**, **4**) under the similar conditions. The catalysts also incorporated 4-methyl-1-cyclohexene (with 1,2-insertion), and the copolymerization with 1-methylcyclopentene proceeded with 1,2- and 1,3-insertion [71].

It has been known that ethylene copolymerization with CPE by ordinary zirconocene (metallocene) catalysts proceeds via 1,3- (and 1,2-) insertion; the subsequent ethylene is inserted after isomerization of inserted CPE [58–60], whereas copolymerization by titanium catalysts proceeds via 1,2-CPE insertion [61–67]. The ^{t}BuC$_5$H$_4$-ketimide analogue (**4**) showed high catalytic activities (1.99–3.16 × 10^4 kg polymer/mol Ti·h) with efficient CPE incorporation in the copolymerization in the presence of MAO cocatalyst, affording high molecular weight copolymers with CPE content up to 43.6 mol% [64]. CPE was incorporated as isolated or alternating manner. Detailed microstructure analysis including tacticity and a linear relationship between the CPE content and the glass transition temperature (T_g) value was also reported [65].

The ethylene/COE copolymerization by the Cp*-phenoxide analogue (**1a**) yielded high molecular weight amorphous copolymers with efficient 1,2-COE incorporations (M_n = 1.08–12.6 × 10^5), whereas the Cp-ketimide analogue (**2**) showed higher catalytic activities affording ultrahigh molecular weight copolymers with unform compositions (but lower COE incorporations under the same conditions) [70]. In contrast, **CGC** and SBI-Zr gave (semi)crystalline copolymers with less COE incorporations (Scheme 15.5), and the resultant polymers by SBI-Zr possessed broad molecular weight distributions and COE was incorporated as 1,3-insertion [70]. The copolymerization with CHP by **1a** gave ultrahigh molecular weight amorphous copolymers (M_n = 1.32–3.08 × 10^6) with exclusive 1,2–CHP insertion. Moreover, **2** showed significant catalytic activities for the ethylene in the copolymerization with tricyclo[6.2.1.0(2,7)]undeca-4-ene (TCUE), affording high molecular weight copolymers (M_n = 6.4–22.0 × 10^5, TCUE content 9.4–40.7 mol%) with uniform compositions [70]. The activities in the copolymerizations with the phenoxide analogues (**1a,c,d**) were low compared to those by **2**, affording the copolymers with rather lower M_n values [70]. The resultant copolymers possessed high T_g values by introduction of an additional cyclic unit (Figure 15.1).

As demonstrated in Figure 15.1, linear relationships between their T_gs and the cyclic olefin contents were demonstrated in all cases, and their T_g values were affected by the ring size. Although it is not clear why the copolymers incorporating an eight-membered ring (COE) possessed higher T_g values than those incorporating five- and seven-membered rings (CPE and CHP, respectively) with the same cyclic olefin contents, the results clearly indicate that placing an additional norbornane ring into cyclohexene leads to a drastic increase in the T_g value in the resultant COCs [70].

15.3 Ethylene Copolymerization with Alken-1-ol for Introduction of Hydroxy Groups into Polylefins

Efficient synthesis of high molecular weight polyolefins containing a hydroxy group (introducting polar functionality into hydrophobic polyolefin) has been a fascinating goal, and development of molecular catalysts should play a key role. Polymerization with the early transition metal catalysts generally showed low catalytic activities to afford (rather) low molecular weight polymers, and both the activity and the molecular weight decreased upon increasing the hydroxy group content

[72–84]. These trends have been explained as due to a strong interaction of oxygen with the centered metal [72–84]; the trend does not change even though the OH group in alken-1-ol was protected with Al alkyls (Al^{i}Bu$_3$ etc.) [72–84] or trialkylsilyl group [74, 75] in advance. The direct ethylene copolymerization with polar monomers using late transition metal (Ni, Pd) catalysts generally afforded copolymers with certain branching (and the resultant polymers generally possess rather low molecular weights in most cases) [7–9, 85–90].

Table 15.1 summarizes selected results in the ethylene/1-decene (DC) copolymerizations by **1a,b** and CGC–MAO catalysts in the presence of 9-decen-1-ol (DC–OH) pretreated with Al^{i}Bu$_3$ *in situ* (Scheme 15.6). The phenoxide analogues (**1a,b**) gave rather high molecular weight copolymers containing OH group (runs 1,3,7) when these polymerizations were conducted at 25 °C [33]. The SiEt$_3$ analogue (**1b**) showed higher activities than **1a**, and the activity increased at 50–80 °C (runs 4,6)^3, whereas the activities of **1a** and CGC decreased at 50 °C (runs 2, 8). Note that the activity of **1b** at 50 °C increased with increases in the DC–OH concentration (run 5), whereas the polymer yield by CGC became negligible under the same conditions. A similar trend to **1b** was also observed in the ethylene/1-hexene copolymerization in the presence of 5-hexen-1-ol (HX–OH) at 50 °C. In all cases, apparent decreases in the molecular weights (M_n values) were not seen even with increase in the hydroxy group contents (up to 2.5 mol%) [33].

Moreover, **1b**–MAO catalyst also exhibited high catalytic activity in the ethylene/DC–OH copolymerization, yielding high molecular weight copolymer with a unimodal molecular weight

Table 15.1 Ethylene copolymerization with 1-decene (DC), 9-decen-1-ol (DC–OH) by Cp*TiCl$_2$(O-2,6-^{i}Pr$_2$-4-RC$_6$H$_2$) [R = H (**1a**), SiEt$_3$ (**1b**)], [Me$_2$Si(C$_5$Me$_4$)(N^{t}Bu)]TiCl$_2$ (CGC)–MAO catalysts[1] [33] / John Wiley & Sons.

Run	catalyst (μmol)	DC[2]/M	DC–OH[2]/M	temp/°C	yield/mg	activity[3]	M_n[4] × 10^{-4}	M_w/M_n[4]	DC–OH[5]/mol%	DC[5]/mol%
1	**1a** (0.025)	0.79	0.093	25	154	37000	13.8	1.83	0.8	15.2
2	**1a** (0.01)	0.79	0.093	50	53.0	31800	7.29	1.81	1.2	17.8
3	**1b** (0.025)	0.79	0.093	25	189	45400	11.7	1.81	1.3	15.2
4	**1b** (0.01)	0.79	0.093	50	79.9	47900	6.82	1.70	1.5	16.7
5	**1b** (0.01)	0.70	0.19	50	122	73200	6.70	1.69	2.5	14.1
6	**1b** (0.025)	0.79	0.093	80	216	51800	2.65	1.83	0.8	14.1
7	**CGC** (0.025)	0.79	0.093	25	61.0	14600	8.97	1.79	0.9	14.0
8	**CGC** (0.01)	0.79	0.093	50	11.1	6660	–	–		
9	**CGC** (0.01)	0.70	0.19	50	5.0	3060	–	–		
10	**1b** (0.01)	–	0.19	50	635	381000	6.55	1.83	2.6	–
11	**1b** (0.01)	–	0.28	50	212	127000	7.16	1.84	3.6	–
12	**1b** (0.01)	–	0.28^6	50	120	72000	10.1	2.01	2.4^7	–

1) Conditions: toluene and 1-decene (DC) and 9-decen-1-ol (DC–OH) total 30.0 mL, ethylene 6 atm, MAO 3.0 mmol, Al^{i}Bu$_3$ 3.0 (DC–OH 0.093 M) or 6.0 mmol (DC–OH 0.19 M), 10 min.
2) Initial monomer concentration.
3) Activity in kg-polymer/mol-Ti·h.
4) GPC data in *o*-dichlorobenzene *vs* polystyrene standards.
5) DC–OH, DC content (mol%) estimated by ^{1}H NMR spectra.
6) 5-Hexen-1-ol (HX–OH) were used.
7) HX–OH contents estimated by ^{1}H NMR spectra.

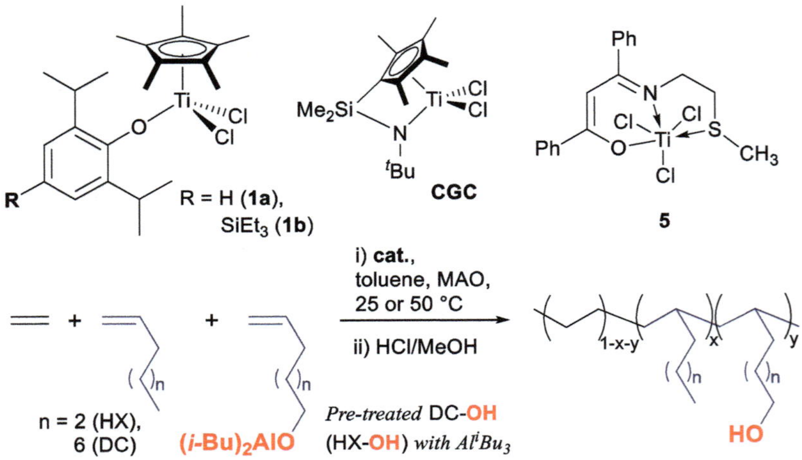

Scheme 15.6 Synthesis of ethylene copolymers with 1-decene (DC) and 9-decen-1-ol (DC–OH) [1-hexene (HX) and 5-hexen-1-ol (HX–OH)] [33] / John Wiley & Sons.

distribution (run 10) as well as with uniform composition confirmed by differential scanning calorimetry (DSC) thermogram. The high activity was maintained even with an increase in the DC–OH concentration and the trend was similar for the ethylene copolymerization with 5-hexen-1-ol. Significant decreases in M_n values were not observed under these copolymerization conditions [33]. Observed catalytic activities in the ethylene/DC–OH copolymerization (72,000–381,000 kg-polymer/mol-Ti·h) were apparently higher than those reported previously [72–84, 91] exemplified by **5** (1,200–10,100 kg-polymer/mol-T·h, ethylene 1 atm at 25 °C) [91], as well as those reported in the synthesis of ethylene copolymers containing polar functionalities [92, 93].

15.4 Synthesis of Biobased Ethylene Copolymers by the Incorporation of Linear and Cyclic Terpenes

As described in the introduction, synthesis of functional polymers from renewable feedstocks has been considered as the attractive and important subject [19–22]. There have thus been many reports of synthesis of the advanced polyesters [21, 22, 94–99] by ring-opening polymerization of cyclic monomers or condensation polymerization of bifunctional monomers derived from the oxygen-rich molecular biomass. Cyclic monoterpenes, consisting of two isoprene units (formula of $C_{10}H_{16}$; e.g. limonene, pinene, terpinene, and camphene) can also be considered as promising monomers obtained from abundant natural resources [100, 101]. However, the successful synthesis by metal catalyzed coordination insertion polymerization have been limited until recently [23, 24, 28], whereas there are reports in the synthesis by adopting ionic (cationic and radical) polymerization.

The ethylene/limonene copolymerization was conducted by $Me_2Si(\eta^3$-2,7-$^tBu_2$fluorenyl)(N-cyclo-$C_{12}H_{23}$)TiMe$_2$ (Flu-CGC–CDD, Scheme 15.7) in the presence of MMAO cocatalyst (ethylene <1 atm, 25 °C, overnight) gave the rather high molecular weight copolymer (800 kg-polymer/mol-Ti, $M_n =$ 57,000, $M_w/M_n = 1.5$, limonene c. 1 mol%) and the attempted copolymerization under low ethylene feed gave the polymer with broad molecular weight distribution ($M_w/M_n = 23.8$) [28]. In contrast, $(1,2,4$-$Me_3C_5H_2)TiCl_2(O$-$2,6$-$^iPr_2C_6H_3)$ (**1d**) gave the rather high molecular weight copolymers with

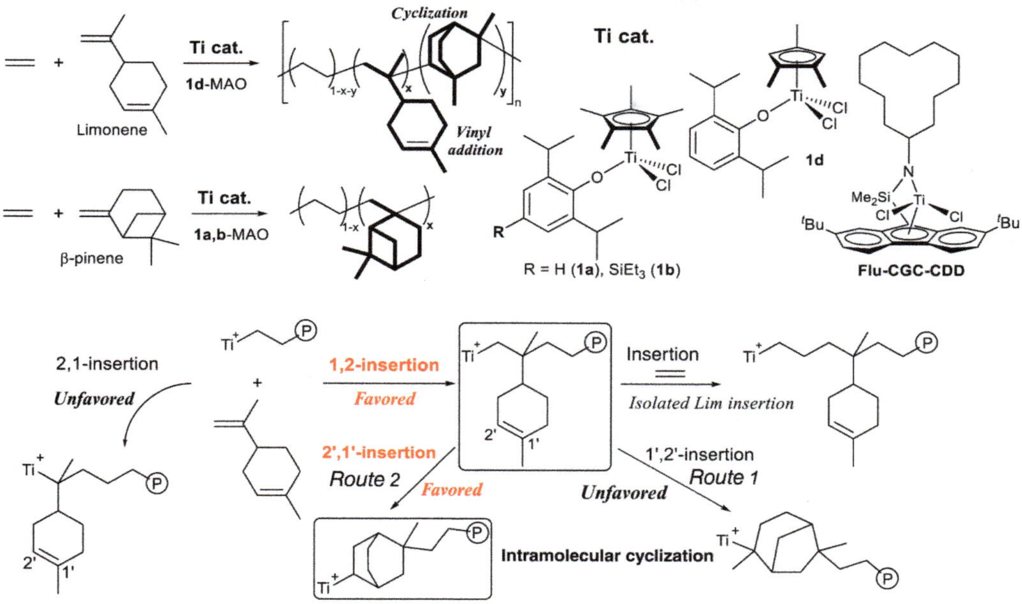

Scheme 15.7 Ethylene copolymerization with limonene, β-pinene. Synthesis of biobased polyolefin [23, 28].

unimodal molecular weight distributions (M_n = 4.09–12.6 × 10^4, M_w/M_n = 1.41–2.67) as well as with uniform compositions confirmed by DSC thermograms (observed a sole melting temperature) [22]. Both the catalytic activity and the M_n value in the copolymer decreased with increasing the limonene concentration charged (as well as with increase in the limonene content in the copolymer), and the activity increased at 50 °C. The Cp* analogues (**1a,b**) showed low limonene incorporation, and the attempted copolymerization by [Me$_2$Si(C$_5$Me$_4$)(N^{t}Bu)]TiCl$_2$ (CGC) gave polymers with negligible limonene incorporation. On the basis of the microstructural analysis by ^{13}C NMR spectra and the dept spectra in the copolymers (and the copolymers after olefin hydrogenation), resonances ascribed to 1,2-limonene insertion and the subsequent cyclization (favored 2′,1′-insertion, Scheme 15.7) were observed; degree of 1,2-insertion/cyclization (expressed as an S value) was affected by the Cp′ employed [22].

Synthesis of high molecular weight poly(ethylene-*co*-β-pinene)s with unimodal molecular weight distributions (M_n = 2.10–22.4 × 10^4, M_w/M_n = 1.53–2.32) could also be demonstrated by using Cp*TiCl$_2$(O-2,6-^{i}Pr$_2$-4-R-C$_6$H$_2$) [R = H (**1a**), SiEt$_3$ (**1b**)]–MAO catalyst systems [22]. The catalytic activity was affected by the comonomer concentration charged (and the ethylene pressure), Al/Ti molar ratio and polymerization temperature (at 50 °C), which did not affect both β-pinene incorporation and the M_n values significantly. In both cases, the comonomer incorporation was affected by the cyclopentadienyl fragment (Cp′) employed. The Me$_3$Cp analogue (**1d**) also incorporates β-pinene under the same conditions but showed rather lower catalytic activities than the Cp* analogues (**1a,b**); the M_n values in the resultant copolymers by **1d** possessed lower than those prepared by **1a,b** [22]. The attempted copolymerization by [Me$_2$Si(C$_5$Me$_4$)(N^{t}Bu)]TiCl$_2$ (CGC) gave polymers with negligible limonene incorporation.

β-Myrcene (myrcene, My), consisting of two isoprene units, has also been a promising monomer obtained from natural resources [100, 101], and there have been reports of polymer synthesis by radical, anionic, or metal-catalyzed coordination polymerization. However, there was only one report concerning the ethylene copolymerization (by scandium catalysts) [102], whereas there have been reports for the styrene/myrcene copolymerization have been known on relevance of styrene-butadiene (SBR) or styrene-isoprene rubbers [102–109]. The resultant polymers prepared by the scandium catalysts, however, possessed (multi)block (E-*bl*-My) microstructures (possessing melting temperature at 133 °C even with My 9 mol%) [102].

Synthesis of random ethylene/myrcene copolymers has been demonstrated using Cp*TiCl$_2$(O-2,6-^{i}Pr$_2$-4-RC$_6$H$_2$) [R = H (**1a**), SiEt$_3$ (**1b**)]–MAO catalysts [24]. The catalysts exhibited rather efficient myrcene (My) incorporation and enabled synthesis of the high molecular weight (semicrystalline or amorphous, depending on the My content) copolymers with unimodal molecular weight distributions as well as uniform compositions (confirmed by DSC thermograms). The microstructural analysis of the copolymers (and the copolymers after olefin hydrogenation) by NMR spectroscopy (^{13}C NMR spectra and the dept spectra) revealed that the copolymers possessed cyclopentane units containing My pendant arm (–CH$_2$CH=CMe$_2$); which are presumably formed by 2,1- or 1,4-My insertion and subsequent cyclization after ethylene insertion (Scheme 15.8). The resultant copolymer showed promising elastic properties, and the elongation at break increased upon increasing the My contents accompanied with decrease in the tensile strength and toughness.

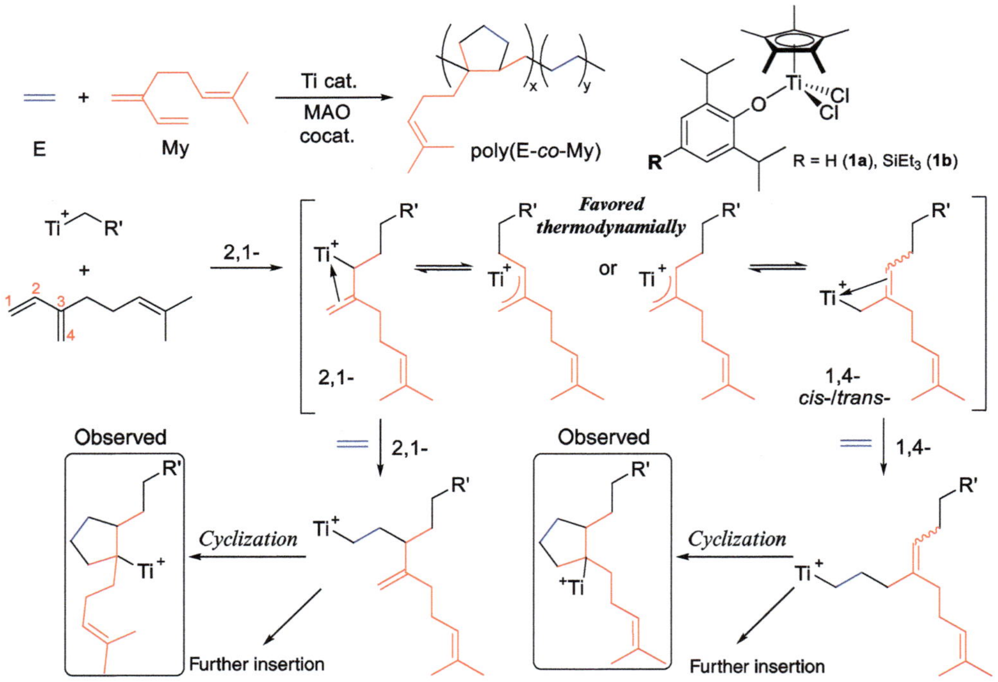

Scheme 15.8 Ethylene copolymerization with β-myrcene. Synthesis of biobased elastomer [24].

15.5 Concluding Remarks and Outlook

Transition metal catalyzed olefin polymerization by transition metal catalysis is the core technology for production of polyolefins, widely used synthetic polymers in the world. Development of new ethylene copolymers that have not been incorporated by conventional catalysts (Ziegler-Natta, metallocene etc.) has been a long-term subject, whereas isotactic PP has been produced by heterogeneous Ziegler-Natta catalysts. This is due to that unification of monomers should be considered in terms of reduction of undesirable by-product especially by using specified monomers (through many synthetic steps) and the recycling should be easier than the other specified polymers. Development of the molecular catalyst plays a key role for the success [1–18]. Although, as described in sections 15.2–4, the synthesis of functional polymers from renewable feedstocks (as alternative of fossil oil) has been considered an attractive and important subject worthy of many reports [19–22].

In this chapter, reported examples in synthesis of ethylene copolymers by incorporation of sterically encumbered olefins (previously called traditionally unreactive olefins in coordination polymerization) and cyclic olefins have been reviewed. Some of these efforts should contribute to significant reduction of additives (concern in recycling) often employed to improve material properties in polyolefins. However, reports in synthesis of ethylene copolymers with aromatic vinyl monomers (such as styrene) [2, 10, 15, 110, 111] were not introduced. Reports regarding the synthesis of polyolefin containing polar functionalities by adopting late transition metal catalysts have also been introduced; these are addressed in the reported reviews [9, 85–90]. Efficient synthesis of ethylene copolymers containing hydroxy groups should provide new possibilities, and the synthesis by incorporation of terpenes (limonene, pinene, and myrcene) could introduce the possibility of producing polyolefins from natural abundant resources (the plant kingdom). The importance in the development of molecular catalysts can be seen to allow these goals to be accomplished.

Moreover, the analysis of catalytically active species has been the central focus for understanding the catalysis mechanisms and information on the structural and electronic nature of species should be useful for better catalyst design as well as for basic understanding of catalysis in organometallic chemistry. In addition to conventional NMR technique and single crystal x-ray crystallography, solid methods in this field, we recently introduced solution synchrotron XAS (x-ray absorption spectroscopy) analysis as an efficient tool for this purpose. The method provides information concerning the oxidation state and the geometry through x-ray absorption near edge structure (XANES) and regarding their coordination atoms to the centered metal through extended x-ray absorption fine structure (EXAFS) spectra [112, 113] Since the method provides information concerning species that cannot be seen in NMR and/or ESR spectra, the method could be popular in the near future.

Through this chapter, we hope that we have introduced a clear picture for the synthesis of sustainable polymers enabled by new, efficient molecular catalysis.

Acknowledgements

The projects by KN were partly supported by Grant-in-Aid for Scientific Research from the Japan Society for the Promotion of Science (JSPS, No. 15H03812, 18H01982, 21H01942) and the Tokyo Metropolitan Government Advanced Research Grant Number R2-1.

References

1 For example, Kaminsky, W. (1996). *Macromol. Chem. Phys.* 197: 3907.
2 McKnight, A.L. and Waymouth, R.M. (1998). *Chem. Rev.* 98: 2587.
3 Nomura, K. and Zhang, W. (2014). *Organometallic Reactions and Polymerization* (ed. K. Osakada), Lecture Notes in Chemistry 85. 89–117. Berlin, Germany: Springer-Verlag.
4 Baier, M.C., Zuideveld, M.A., and Mecking, S. (2014). *Angew. Chem., Int. Ed.* 53: 9722.
5 Stürzel, M., Mihan, S., and Mülhaupt, R. (2016). *Chem. Rev.* 116: 1398.

6 Hoff, R. (ed.) (2018). *Handbook of Transition Metal Polymerization Catalysts*, 2e. Hoboken, NJ: Wiley.
7 For example, Ittel, S.D., Johnson, L.K., and Brookhart, M. (2000). *Chem. Rev.* 100: 1169.
8 Gibson, V.C. and Spitzmesser, S.K. (2003). *Chem. Rev.* 103: 283.
9 Chen, C. (2018). *Nat. Rev. Chem.* 2: 6.
10 Li, H. and Marks, T.J. (2006). *Proc. Nat. Acad. Sci. USA* 103: 15295.
11 Delferro, M. and Marks, T.J. (2011). *Chem. Rev.* 111: 2450.
12 McInnis, J.P., Delferro, M., and Marks, T.J. (2014). *Acc. Chem. Res.* 47: 2545.
13 Nomura, K., Liu, J., Padmanabhan, S., and Kitiyanan, B. (2007). *J. Mol. Catal. A: Chem.* 267: 1.
14 Nomura, K. (2009). *Dalton Trans.* 8811.
15 Nomura, K. and Liu, J. (2011). *Dalton Trans.* 40: 7666.
16 Suhm, J., Schneider, M.J., and Mülhaupt, R. (1998). *J. Mol. Catal. A: Chem.* 128: 215.
17 For example, Arriola, D.J., Bokota, M., Campbell, R.E., Jr. et al. (2007). *J. Am. Chem. Soc.* 129: 7065.
18 van Doremaele, G., van Duin, M., Valla, M., and Berthoud, A. (2017). *J. Poly. Sci. Part A: Polym. Chem.* 55: 2877.
19 Mülhaupt, R. (2013). *Macromol. Chem. Phys.* 214: 159.
20 Gandini, A. (2008). *Macromolecules* 41: 9491.
21 Coates, G.W. and Hillmyer, M.A. (2009). *Macromolecules* 42: 7987.
22 Gandini, A. and Lacerda, T.M. (2018). Monomers and polymers from chemically modified plant oils and their fatty acids. In: *Polymers from Plant Oils*, 2e. (eds. A. Gandini and T.M. Lacerda), 33–82. Beverly: Scrivener Publishing LLC.
23 Kawamura, K. and Nomura, K. (2021). *Macromolecules* 54: 4693.
24 Kitphaitun, S., Chaimongkolkunasin, S., Manit, J. et al. (2021). *Macromolecules* 54: 10049.
25 Kaminsky, W., Bark, A., Spiehl, R. et al. (1988). Isotactic polymerization of olefins with homogeneous zirconium catalysts. In: *Transition Metals and Organometallics as Catalysts for Olefin Polymerization*. (eds. W. Kaminsky and H. Sinn), 291–301. Berlin: Springer-Verlag.
26 Pino, P., Giannini, U., and Porri, L. (1987). Polymerization of vinyl acetate and related monomers. In: *Encyclopedia of Polymer Science and Engineering*, 2e. 8. (eds. H.F. Mark, N.M. Mikales, C.C. Overberger, and G. Menges), 155–179. New York: Wiley, Interscience.
27 Shaffer, T.D., Canich, J.A.M., and Squire, K.R. (1998). *Macromolecules* 31: 5145.
28 Nakayama, Y., Sogo, Y., Cai, Z., and Shiono, T. (2013). *J. Polym. Sci. Part A: Polym. Chem.* 51: 1223.
29 Li, H., Li, L., Marks, T.J. et al. (2003). *J. Am. Chem. Soc.* 125: 10788.
30 Li, H., Li, L., Schwartz, D.J. et al. (2005). *J. Am. Chem. Soc.* 127: 14756.
31 Nomura, K., Itagaki, K., and Fujiki, M. (2005). *Macromolecules* 38: 2053.
32 Itagaki, K., Fujiki, M., and Nomura, K. (2007). *Macromolecules* 40: 6489.
33 Kitphaitun, S., Yan, Q., and Nomura, K. (2020). *Angew. Chem. Int. Ed.* 59: 23072.
34 Kitphaitun, S., Yan, Q., and Nomura, K. (2021). *ChemistryOpen* 10: 867.
35 Nomura, K. and Itagaki, K. (2005). *Macromolecules* 38: 8121.
36 Khan, F.Z., Kakinuki, K., and Nomura, K. (2009). *Macromolecules* 42: 3767.
37 Nomura, K., Kakinuki, K., Fujiki, M., and Itagaki, K. (2008). *Macromolecules* 41: 8974.
38 Liu, J. and Nomura, K. (2008). *Macromolecules* 41: 1070.
39 Kakinuki, K., Fujiki, M., and Nomura, K. (2009). *Macromolecules* 42: 4585.
40 Kaminsky, W. (1994). *Angew. Makromol. Chem.* 223: 101.
41 Cherdron, H., Brekner, M.-J., and Osan, F. (1994). *Angew. Makromol. Chem.* 223: 121.
42 Tritto, I., Boggioni, L., and Ferro, R. (2006). *Coord. Chem. Rev.* 250: 212.
43 Li, X. and Hou, Z. (2008). *Coord. Chem. Rev.* 252: 1842.
44 Zhao, W. and Nomura, K. (2016). *Catalysts* 6: 175.

45 Boggioni, L. and Tritto, I. (2017). *MRS Bull.* 38: 245.
46 Hasan, T., Nishii, K., Shiono, T., and Ikeda, T. (2002). *Macromolecules* 35: 8933.
47 Hasan, T., Shiono, T., and Ikeda, T. (2004). *Macromolecules* 37: 7432.
48 Hasan, T., Ikeda, T., and Shiono, T. (2004). *Macromolecules* 37: 8503.
49 Cai, Z., Harada, R., Nakayama, Y., and Shiono, T. (2010). *Macromolecules* 43: 4527.
50 Cai, Z., Nakayama, Y., and Shiono, T. (2006). *Macromolecules* 39: 2031.
51 Yuan, H., Kida, T., Kim, H. et al. (2020). *Macromolecules* 53: 4323.
52 Nomura, K., Wang, W., Fujiki, M., and Liu, J. (2006). *Chem. Commun.* 25: 2659.
53 Zhao, W. and Nomura, K. (2016). *Macromolecules* 49: 59.
54 For example, JP2001-106730; JP2006-22266; JP2008-248171 (Mitsui Chemicals Co.).
55 For example, Kaminsky, W., Engehausen, R., and Kopf, J. (1995). *Angew. Chem. Int. Ed. Engl.* 34: 2273.
56 Apisuk, W., Ito, H., and Nomura, K. (2016). *J. Polym. Sci. Part A: Polym. Chem.* 54 (17): 2662.
57 Hong, M., Cui, L., Liu, S., and Li, Y. (2012). *Macromolecules* 45: 5397.
58 For examples Kaminsky, W. and Spiehl, R. (1989). *Makromol. Chem.* 190 (3): 515.
59 Jerschow, A., Ernst, E., Hermann, W., and Müller, N. (1995). *Macromolecules* 28 (21): 7095.
60 Naga, N. and Imanishi, Y. (2002). *Macromol. Chem. Phys.* 203: 159.
61 For examples, Lavoie, A.R., Ho, M.H., and Waymouth, R.M. (2003). *Chem. Commun.* 864.
62 Lavoie, A.R. and Waymouth, R.M. (2004). *Tetrahedron* 60: 7147.
63 Naga, N. (2005). *J. Polym. Sci. Part A: Polym. Chem.* 43 (6): 1285.
64 Liu, J. and Nomura, K. (2007). *Adv. Synth. Catal.* 349 (14–15): 2235.
65 For examples, Fujita, M. and Coates, G.W. (2002). *Macromolecules* 35: 9640.
66 Tang, L.-M., Duan, Y.-Q., Pan, L., and Li, Y.-S. (2005). *J. Polym. Sci. Part A: Polym. Chem.* 43: 1681.
67 Pennington, D.A., Coles, S.J., Hursthouse, M.B. et al. (2006). *Macromol. Rapid Commun.* 27: 599.
68 Wang, W., Fujiki, M., and Nomura, K. (2005). *J. Am. Chem. Soc.* 127: 4582.
69 Buchmeiser, M.R., Camadanli, S., Wang, D. et al. (2011). *Angew. Chem., Int. Ed.* 50: 3566.
70 Harakawa, H., Okabe, M., and Nomura, K. (2020). *Polym. Chem.* 11: 5590.
71 Harakawa, H., Patamma, S., Boccia, A.C. et al. (2018). *Macromolecules* 51 (3): 853.
72 Aaltonen, P. and Löfgren, B. (1995). *Macromolecules* 28 (15): 5353.
73 Aaltonen, P., Fink, G., Löfgren, B., and Seppälä, J. (1996). *Macromolecules* 29: 5255.
74 Goretzki, R. and Fink, G. (1998). *Macromol. Rapid Commun.* 19: 511.
75 a.) Goretzki, R. and Fink, G. (1999). *Macromol. Chem. Phys.* 200: 881. b) Marques, M. M., Correia, S. G., Asceso, J. R., Ribeiro, A. F. G., Gomes, P. T., Dias, A. R., Foster, P., Rausch, M. D., and Chien, J. C. W., *J. Polym. Sci. PartA: Polym. Chem.*, **1999**, *37*, 2457.
76 Hakala, K., Helaja, T., and Löfgren, B. (2000). *J. Polym. Sci. Part A: Polym. Chem.* 38: 1966.
77 Wang, W., Hou, L., Luo, S. et al. (2013). *Macromol. Chem. Phys.* 214: 2245.
78 Hou, L., Wang, W., Sheng, J., and Liu, C. (2015). *RSC Adv.* 5: 98929.
79 Imuta, J. and Kashiwa, N. (2002). *J. Am. Chem. Soc.* 124: 1176.
80 Kashiwa, N., Matsugi, T., Kojoh, S. et al. (2003). *J. Polym. Sci.: Part A: Polym. Chem.* 41: 3657.
81 Inoue, Y., Matsugi, T., Kashiwa, N., and Matyjaszewski, K. (2004). *Macromolecules* 37: 3651.
82 Hagihara, H., Murata, M., and Uozumi, T. (2001). *Macromol. Rapid Commun.* 22: 353.
83 Hagihara, H., Tsuchihara, K., Takeuchi, K. et al. (2004). *J. Polym. Sci.: Part A: Polym. Chem.* 42: 52.
84 Hagihara, H., Tsuchihara, K., Sugiyama, J. et al. (2004). *Macromolecules* 37: 5145.
85 For example, Boffa, L.S. and Novak, B.M. (2000). *Chem. Rev.* 100: 1479.
86 Nakamura, A., Ito, S., and Nozaki, K. (2009). *Chem. Rev.* 109: 5215.
87 Chen, E.Y.-X. (2009). *Chem. Rev.* 109: 5157.
88 Franssen, N.M.G., Reek, J.N.H., and Bruin, B.D. (2013). *Chem. Soc. Rev.* 42: 5809.
89 Nakamura, A., Anselment, T.M.J., Claverie, J. et al. (2013). *Acc. Chem. Res.* 46: 1438.

90   Tan, C. and Chen, C.L. (2019). *Angew. Chem. Int. Ed.* 58: 7192.
91   Yang, X.H., Liu, C.R., Wang, C. et al. (2009). *Angew. Chem. Int. Ed.* 48: 8099.
92   Terao, H., Ishii, S., Mitani, M. et al. (2008). *J. Am. Chem. Soc.* 130: 17636.
93   Chen, J., Motta, A., Wang, B. et al. (2019). *Angew. Chem. Int. Ed.* 58: 7030.
94   Hillmyer, M.A. and Tolman, W.B. (2014). *Acc. Chem. Res.* 47: 23902.
95   Fortman, D.J., Brutman, J.P., De Hoe, G.X. et al. (2018). *ACS Sustainable Chem. Eng.* 6: 111452.
96   Coulembier, O., Degée, P., Hedrick, J.L., and Dubois, P. (2006). *Prog. Polym. Sci.* 31: 723.
97   Meier, M.A.R., Metzger, J.O., and Schubert, U.S. (2007). *Chem. Soc. Rev.* 36: 1788.
98   Xia, Y. and Larock, R.C. (2010). *Green Chem.* 12: 1893.
99   Stempfle, F., Ortmann, P., and Mecking, S. (2016). *Chem. Rev.* 116: 4597.
100 Wilbon, P.A., Chu, F., and Tang, C. (2013). *Macromol. Rapid Comm.* 34: 8.
101 Thomsett, M.R., Storr, T.E., Monaphan, O.R. et al. (2016). *Green Materials* 4: 115.
102 Ren, X., Guo, F., Fu, H. et al. (2018). *Polym. Chem.* 9: 1223.
103 Georges, S., Touré, A.O., Visseaux, M., and Zinck, P. (2014). *Macromolecules* 47: 4538.
104 Liu, B., Li, L., Sun, G. et al. (2015). *Chem. Commun.* 51: 1039.
105 Naddeo, M., Buonerba, A., Luciano, E. et al. (2017). *Polymer* 131: 151.
106 Laur, E., Welle, A., Vantomme, A. et al. (2017). *Catalysts* 7: 361.
107 Li, W., Zhao, J., Zhang, X., and Gong, D. (2019). *Ind. Eng. Chem. Res.* 58: 2792.
108 González, Z.J.L., Enríquez, M.F.J., López, G.H.R. et al. (2020). *RSC Adv.* 10: 44096.
109 Lamparelli, D.H., Paradiso, V., Monica, F.D. et al. (2020). *Macromolecules* 53: 1665.
110 Nomura, K. (2010). Copolymerization of ethylene with styrene: design of efficient Transition Metal Complex Catalysts. In: *Syndiotactic Polystyrene - Synthesis, Characterization, Processing, and Applications*. (ed. J. Schellenberg), 60–91. Hoboken: John Wiley & Sons, Inc.
111 Aoki, H. and Nomura, K. (2021). *Macromolecules* 54: 83.
112 Nomura, K. (2019). *Catalysts* 9: 1016.
113 Yi, J., Nakatani, N., and Nomura, K. (2020). *Dalton Trans.* 49: 8008.

16

Catalytic Depolymerization of Plastic Waste

Noel Angel Espinosa-Jalapa[1] and Amit Kumar[2]

[1] *Institut für Anorganische Chemie, Universität Regensburg, Universitätsstraße 31, Regensburg, Germany*
[2] *School of Chemistry, University of St. Andrews, North Haugh, St. Andrews, UK*

16.1 Introduction

Plastics are integral parts of our society, and it is hard to imagine our lives without them. However, the robust and non-biodegradable nature of plastics have led to intense plastic pollution on both land and ocean raising a serious threat to our ecosphere [1, 2]. It is estimated that the world produces ~380 million tons of plastic waste annually and this number is supposed to be double by 2034 [3]. Most of the plastic waste goes straight to the landfill and only around 15% of the plastic waste is currently recycled [4, 5]. The current state-of-the-art technologies to recycle plastics are based on mechanical recycling where plastic waste is subjected to mechanical processing (e.g. separating, washing, drying, grinding, re-granulating, and compounding) to recover plastic products. However, each time a plastic waste is mechanically recycled, its quality degrades and on average a plastic can only be mechanically recycled 2–5 times, after which it ends up in a landfill [6–12]. Thus, mechanical recycling can not provide a sustainable solution to the current challenge of plastic pollution. Distinct from mechanical recycling, the approach of chemical recycling or closed-loop recycling is based on the chemical depolymerisation of plastics to form monomers or oligomers which can be used as a chemical feedstock or fuel or transformed into the virgin plastic thus closing the loop of plastic production [8, 10, 13–17]. Thus, chemical recycling is the only sustainable approach to recycling plastics. Another important effect of the chemical recycling process can be manifested in the substitution of the crude oil feedstock (for plastic production) with the products obtained from the depolymerisation process. This can potentially save ~3.5 billion barrels of oil which is equivalent to £130 billion [10]. Furthermore, this will also contribute to our targets of net-zero emissions because a significant amount of CO_2 is emitted into the atmosphere when crude oil is used to make plastic. It is estimated that ~1.87 tons of CO_2-equivalent greenhouse gas emissions can be reduced by the chemical recycling of 1 ton of plastic [18]. Thus, the chemical recycling of plastics is crucial to the circular economy. This chapter will discuss catalytic processes that have been demonstrated for the chemical recycling of various types of plastics under thermal conditions. The topic of the photocatalytic degradation of plastic waste is beyond the scope of this chapter and the main emphasis has been given on processes catalysed by homogeneous catalysts.

Catalysis for a Sustainable Environment: Reactions, Processes and Applied Technologies Volume 2, First Edition. Edited by Armando J. L. Pombeiro, Manas Sutradhar, and Elisabete C. B. A. Alegria.
© 2024 John Wiley & Sons Ltd. Published 2024 by John Wiley & Sons Ltd.

16.2 Pyrolysis

Pyrolysis is the thermal degradation of a material/chemical at a high temperature (300–900 °C) in the absence of O_2. A general chemical reaction occurring in a pyrolysis chamber can be described as:

$$C_nH_mO_p \rightarrow \sum_{liquid} C_xH_yO_z + \sum_{gas} C_aH_bO_c + H_2O + C\,(char).$$

Pyrolysis can be carried out in the absence of a catalyst, in which case it results in mostly gaseous products of large components. However, the use of a catalyst allows us to control the selectivity forming more liquid products avoiding char and ash, and enhance the degradation rate, especially in the range of 350–600 °C [19]. The amount of solid or gaseous products can be affected by various reaction conditions such as residence time, heating rate, and catalyst [20, 21].

Pyrolysis is a common technique used for the depolymerisation of polyolefins such as polyethylene (PE) [20, 22, 23], polystyrenes (PS) [20, 23–25], and poly(vinyl chloride) (PVC) [26–29]. Plastics such as PEs are like long alkane chains containing strong C–C and C–H bonds. The absence of any polarisable or unsaturated functional groups makes other processes such as aminolysis, and hydrogenation (*vide infra*) less suitable for the degradation of such plastics. Therefore, the approach of pyrolysis for the chemical recycling of such plastics has been investigated in much detail.

Several heterogeneous catalysts such as zeolite-based catalyst (ZSM-5) [30–35], CoMo oxide/Al_2O_3 [34–37], and ZnO [38] have been studied for this process. The catalytic pyrolysis usually results in a liquid rich in hydrocarbons (C_{11}–C_{20}) that has properties similar to that of diesel [22]. The catalysts used for polyolefins depolymerisation can be categorised into three parts: (i) fluid cracking catalysts such as zeolite, silica-alumina, and clay; (ii) reforming catalysts that involve transition-metals attached to silica-alumina; and (iii) activated carbon. Catalysts containing acidic functional groups have demonstrated high activity, presumably by protonating the defective site of polymers resulting in the formation of reactive carbonium ions [39, 40]. Although the approach of pyrolysis has been the most successful for the chemical recycling of polyolefins, this process has also been studied for the depolymerisation of other plastics such as polyesters [41], nylons [42], polycarbonates (PCs) [43, 44], and polyurethanes [45]. More detailed reports on the use of various catalysts for the depolymerisation of various plastics can be found in recently reported reviews [46–49].

There are a number of benefits of recycling plastics via the pyrolysis process. For example, (i) the residual output, char, can be used as a fuel or feedstock for other petrochemical processes [10, 16, 32, 37, 47], (ii) a fuel of high calorific values can be produced from plastic waste [22];,and (iii) the process can tolerate the presence of other plastics and municipal solid waste [19, 24, 49, 50]. The main disadvantage is that the selectivity toward the monomer (e.g., ethylene in the case of PE) is very low and usually a mixture of hydrocarbons is produced. Regardless, several commercial plants for the catalytic pyrolysis of plastic waste are currently in operation in various parts of the world. Recently, BASF has launched ventures to convert the oil resulting from the pyrolysis of plastic waste to feedstock useful for plastic and other materials supporting the vision of a circular economy [51].

16.3 Gasification

Gasification is a thermochemical process to convert a carbonaceous waste into a mixture of gases such as N_2, CH_4, CO, H_2, and CO_2 [52–54]. The process is interesting as it can result in the production of synthesis gas (Syngas) (CO+H_2) which is a useful chemical feedstock. Gasification is different from pyrolysis as: (i) gasification is carried out in the presence of O_2 whereas pyrolysis is carried out in the absence of O_2, and (ii) the products of gasification are only gases whereas pyrolysis results in a mixture of liquids and gases [55, 56]. The process can be carried out just in the presence of air, called air-gasification, or in the presence of both air and steam, called steam-gasification. The process can also be carried out in the presence of coal or biomass, called co-gasification [57]. The direct air-gasification process is the most studied one for the depolymerisation of plastics. The produced gas stream from the air-gasification is mainly used for energy applications. The main advantages of air-gasification are the simplicity of the process and the lack of requirement for external energy sources as in the case of steam-gasification. Furthermore, the tar-content (gaseous hydrocarbons) of the gas produced is lower in comparison to the steam-gasification process. Figure 16.1 depicts some typical examples of tar produced during the gasification process at varied temperature ranges. Polyolefins such as PEs [58, 59], polypropylenes (PP) [57, 59], PSs [59], and PVCs [60–62], as well as polyesters [63], have been studied using air-gasification process. The selectivity of produced syn gas and tar depends on various factors such as type of reactors, equivalence ratio (ratio of fuel/air), residence time, gas velocity, and heating rate. Several heterogeneous catalysts such as olivine, alumina, dolomite, zeolite, active carbon, and active carbon loaded with transition-metals have been studied for the air-gasification of polyolefins and polyesters [55, 58–62, 64, 65].

In comparison to air-gasification, steam gasification produces H_2-rich syngas with a high H_2/CO ratio that is more suitable as a feedstock for chemical synthesis. The main limitation of the steam-gasification process is the need for high heat input to carry the underlying steam-reforming process ($CH_4 + H_2O \rightarrow CO + 3H_2$) that is endothermic in nature ($\Delta H = 206$ kJ/mol). Detailed reports on the use of the catalytic gasification process to depolymerize plastics can be found in the recently reported review articles [55, 65].

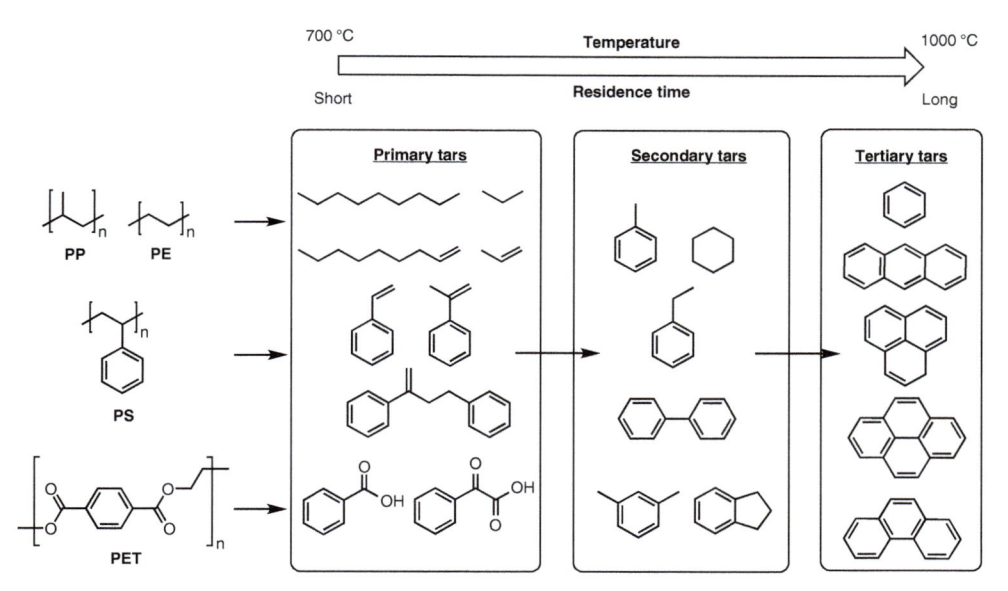

Figure 16.1 Tar formation and evolution pathways in the gasification of plastics of different nature (Reproduced with permission from Ref [55]).

16.4 Solvolysis

Solvolysis involves the degradation of materials such as polymeric plastics using solvents or reagents such as water, amines, or alcohols. Solvolysis is an effective process to depolymerize plastics containing carbonyl groups such as polyesters, nylons, PCs, and polyurethanes as nucleophilic solvents/reagents can attack the carbonyl groups and degrade the polymer chains. Depending on the reagents, the process can be divided into four types described below: hydrolysis, aminolysis, methanolysis, and glycolysis.

16.4.1 Hydrolysis

Plastics can be depolymerized via hydrolysis by reacting them with H_2O under thermal conditions. Hydrolysis can be carried out under basic/alkaline, or acidic conditions using a base (e.g. NaOH) or an acid (e.g. H_2SO_4). Alkaline hydrolysis of polyethylene terephthalate (PET) has been studied using an alkaline solution of NaOH, or KOH of a concentration of 4–20 wt% at 100–200 °C [66]. The reaction results in the formation of ethylene glycol and disodium(potassium) terephthalate that is reacted with H_2SO_4 to produce the terephthalic acid (Figure 16.2a). Phase-transfer catalysts have been used to carry out this transformation under milder conditions (temperature below 100 °C) [67–70]. Acidic hydrolysis of PET has been mostly studied using H_2SO_4 although nitric and phosphoric acids have also been employed for this process (Figure 16.2b) [71, 72]. Due to the corrosive nature of H_2SO_4 and the high cost associated with recycling H_2SO_4, hydrolysis of PET is more preferred under alkaline conditions. A Switzerland-based company, Gr3n, has commercialized the recycling of PET so that PET waste (e.g. bottles and textiles) is hydrolyzed in a microwave reactor using DEMETO technology to produce terephthalic acid and ethylene glycol in less than 10 minutes at 200 °C [73, 74]. A French company Carbios has developed a PET hydrolysis technology using enzymatic catalysts [75, 76].

Hydrolysis of PET under neutral conditions has also been reported although it requires relatively harsh conditions (e.g. 1–4 MPa and 200–300 °C) [77, 78]. Water-soluble salts as catalysts have been reported for this process [79, 80].

Hydrolysis of polyamides has also been studied with the aim to develop the chemical recycling of nylons. For example, Sifniades et al. [81], and Braun et al. [82] patented the hydrolysis of

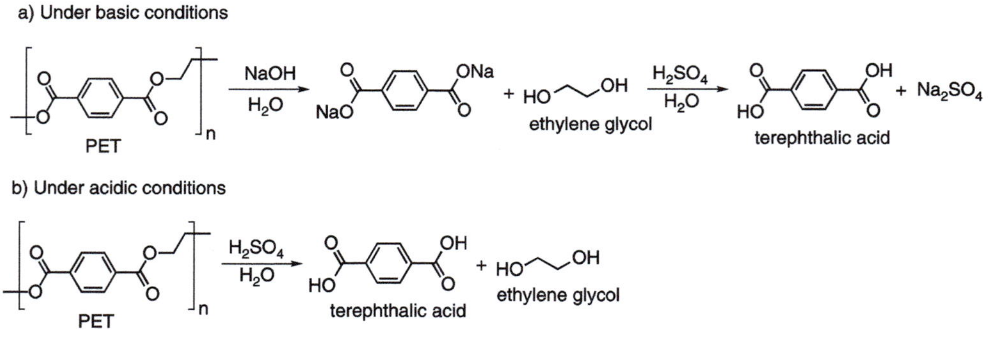

Figure 16.2 Polyethylene terephthalate (PET) hydrolysis under basic and acidic conditions and their respective final products.

nylon-6 from carpet waste to form caprolactam at temperature 290–340 °C, and pressure of 3–15 bar. At low-pressure relatively large amount of caprolactam-dimer was obtained. In another patent, Jenczewski et al. reported the depolymerisation of carpets to high-purity caprolactam (80%) at 250–350 °C and 32–186 bar pressure [83]. High pressure is required to keep the water in a liquid state. The reaction condition has a temperature limitation. At a high temperature, although the rate of hydrolysis of nylon-6 is high, the monomer caprolactam shows decomposition to give a low yield. Furthermore, at a high temperature corrosion becomes a serious problem making industrial application difficult. On the other hand, at a lower temperature, the decomposition of caprolactam is slow but at the same time, the rate of depolymerisation is slower making it less attractive for industrial applications [84]. Therefore, several catalysts such as HPA (heteropoly acid, $H_3PW_{12}O_{40}$) have been used to produce a higher yield of caprolactam [85, 86]. The advantages of the use of HPA as a catalyst include strong acidity, high hydrothermal stability, and low corrosion to most metals avoiding the formation of passivation layer [87]. A catalytic mechanism for this process is outlined in Figure 16.3. Protons from the HPA catalysts protonate the NH group of nylons making the carbonyl centre more nucleophilic that gets attacked by water degrading the amide linkage. Further degradation followed by dehydrative cyclization results in the formation of caprolactam [85]. Aquafil (Slovenia) has commercialized the hydrolysis technology for the chemical recycling of carpets (nylon-6) to make caprolactam that is converted to make fresh carpets [88, 89].

Due to the simple operative technique, the hydrolysis process has also been studied for the depolymerisation of other plastics such as PCs [90, 91], and polyurethanes [92, 93].

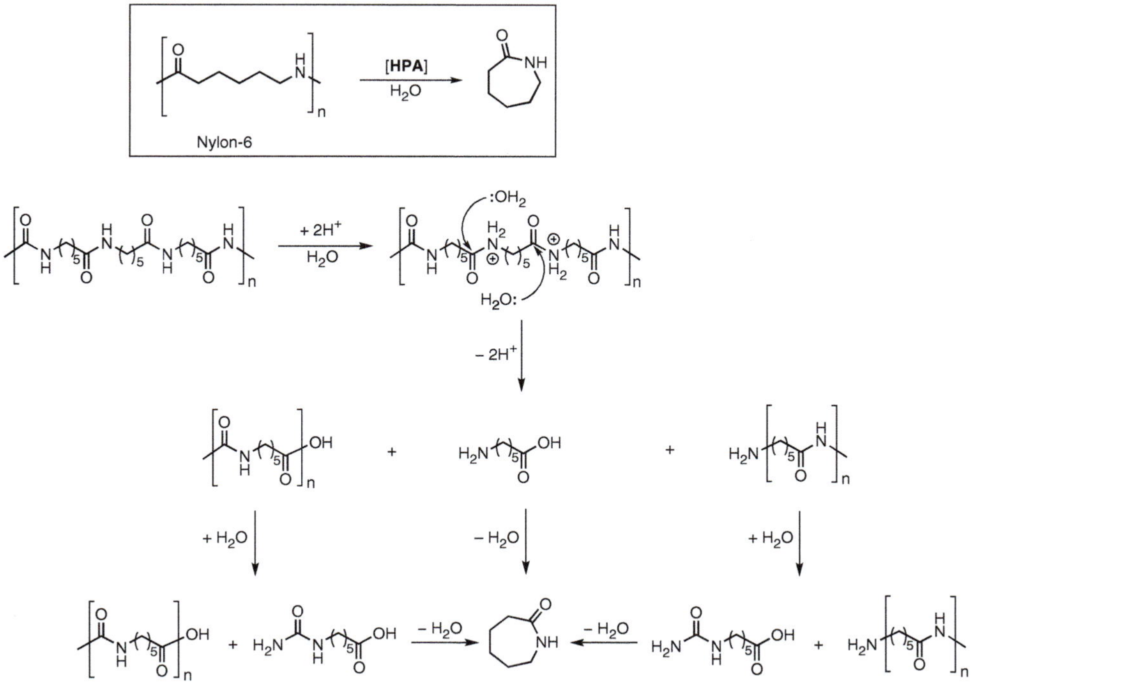

Figure 16.3 Pathway for the depolymerisation of nylon-6 via hydrolysis under acidic conditions catalyzed by heteropoly acid (HPA).

16.4.2 Glycolysis and Methanolysis

The glycolysis process involves the reaction of ethylene glycol with plastics such as PET to produce the monomer bis(2-hydroxyethyl)terephthalate (BHET) (Figure 16.4). The monomer can be then recycled back to produce fresh PET plastic [94]. The process is being (has been) commercialized by a number of companies including Garbo [95], IBM [96], Eastman [97], and Dupont-Teijin [98]. A Netherland-based company Ioniqa uses glycolysis technology to recycle PET at the scale of 10 kilotons/year [99, 100]. Another Netherland-based company CuRE [101] and an Indian company PerPETual [102] have commercialized PET recycling using glycolysis where PET is not depolymerized to the monomer (BHET) but to a mixture of oligomers which are then transformed into the fresh PET.

Similar to glycolysis, methanolysis has also been used for the chemical recycling of PET where PET waste is reacted with methanol to produce dimethyl terephthalate (DMT) and ethylene glycol that can be converted back to fresh PET (Figure 16.4) [103, 104]. In the case of methanolysis, a higher yield of the monomer (DMT, up to 90%) is obtained whereas glycolysis leads to much lower monomer recovery (~25% of BHET) [105]. However, harsh reaction conditions are used in the case of methanolysis, and purification of monomers is challenging due to the presence of complicated mixtures of glycols, phthalate derivatives, and alcohols [94, 106].

16.4.3 Aminolysis

The aminolysis process involves the reaction of amines with plastics under thermal conditions (Figure 16.4). Amines are more nucleophilic than alcohols and therefore aminolysis has a thermodynamic advantage over methanolysis or glycolysis [107]. The process is termed ammonolysis when ammonia is used as the amine. Ammonolysis of PET has been studied using liquid ammonia in the presence of catalysts such as $Zn(OAc)_2$ in the temperature range of 70–180 °C [108]. The process results in the formation of terephthalamide (TPA), and ethylene glycol. The process has not received much attention due to the limited demand for the TPA.

Primary amines such as methylamine, ethylamine, ethanolamine, and n-butylamine have been used for the aminolysis of PET. The process results in the formation of diamides containing the used

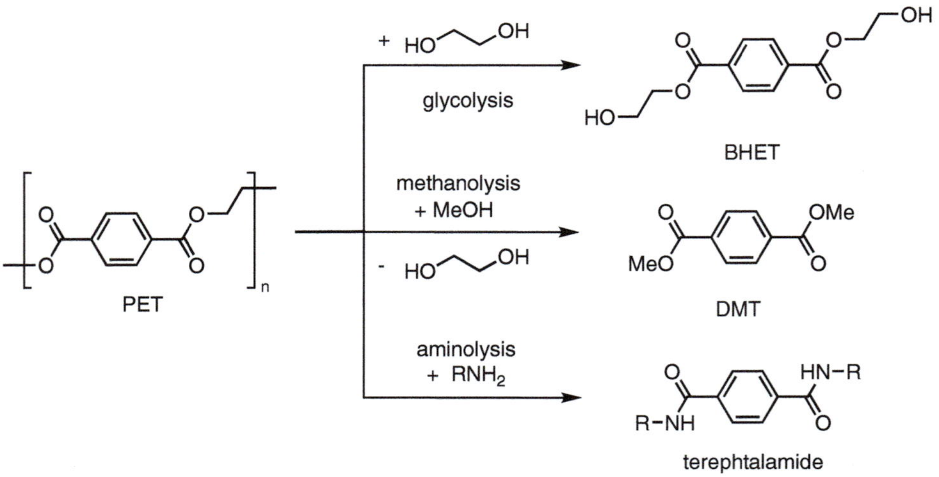

Figure 16.4 Glycolysis, methanolysis, and aminolysis of polyethylene terephthalate (PET).

amine and ethylene glycol. Several catalysts have been employed to enhance the selectivity and to carry out the reactions under mild conditions. For example, Fukushima et al. have reported the aminolysis of PET catalyzed by triazabicyclodecane (TBD) using several aliphatic, allylic, and aromatic amines [109]. The process results in 63–89% yield of monomer in 1–2 h at 110–120 °C. Other catalysts such as metal salts (e.g. sodium acetate [110, 111], sodium sulfate [111, 112]), β-zeolite acid, and montmorillonite KSF clay have been reported for the aminolysis of PET [113]. More details on these processes have been reported in recent review articles by Fieser [114] and George [115]. In comparison to other processes such as glycolysis, pyrolysis, and gasification, aminolysis of polyesters has been relatively less explored and currently, the process has not been commercialized.

16.5 Hydrogenation

Reactions based on catalytic hydrogenation are considered a sustainable approach for organic synthesis [116–119]. There are a number of advantages of this approach in comparison to the conventional reduction methods for reasons such as: (i) hydrogenation reactions are atom-economic, clean and do not generate any waste; and (ii) H_2 gas can be sourced renewably (e.g. from biomass or by the electrolysis of water). A wide variety of catalysts have been developed for the catalytic hydrogenation of various functional groups such as C=C, C=O, and C=N [120]. Hydrogenation of carbonyl groups is especially interesting as it could result in the degradation of molecules. For example, the hydrogenation of an ester molecule can lead to two alcohols via the cleavage of C–O bond present in the ester molecule, and the hydrogenation of an amide molecule can lead to the formation of an alcohol and an amine via the cleavage of the C–N bond. This presents opportunities to utilize the approach of catalytic hydrogenation for the depolymerisation of plastics especially those containing carbonyl groups such as polyesters, polyamides, PCs, polyurethanes, and polyureas. Homogeneous catalysts especially pincer catalysts are the current state-of-the-art catalysts for the hydrogenation of such compounds [121–123]. Inspired by this, a few molecular catalysts have been studied for the hydrogenative depolymerisation of such plastics as summarized in the following sections. An important challenge in the use of a homogeneous catalyst for the depolymerisation of plastic waste is to find an appropriate solvent that could dissolve the plastics under reaction conditions and also allow the catalytic reactions to occur. This is because plastics such as nylons or polyurethanes dissolve poorly in organic solvents (e.g. toluene, THF) that are conventionally used for hydrogenation reactions.

16.5.1 Hydrogenative Depolymerisation of Polyesters

A number of catalysts have been developed for the hydrogenation of esters to alcohols. Pincer catalysts exhibiting metal-ligand cooperation are the current state-of-the-art catalysts for the hydrogenation of esters [124, 125]. Molecular catalysts of various metals such as Ru [126], Ir [127], Fe [128], Mn [129], and Co [130] have been reported for the hydrogenation of esters to alcohols. The concept has also been expanded for the hydrogenative depolymerisation of polyesters using a few catalysts. The first example of hydrogenative depolymerization of polyesters was reported by Robertson where the Milstein's RuPNN complex **Ru-1** (Figure 16.5) was utilized for the hydrogenation of various polyesters such as polycaprolactone (PCL), poly(R-3-hydroxybutyric acid) [P(3HB)], poly(3-hydroxypropionic acid) [P(3HP)], PET and polylactic acid (PLA) [131, 132]. Using the Ru-PNN-bipy catalyst **Ru-1** (1 mol%, Figure 16.5), PET was hydrogenatively depolymerized to produce 1,4-phenylenedimethanol and ethylene glycol in almost quantitative

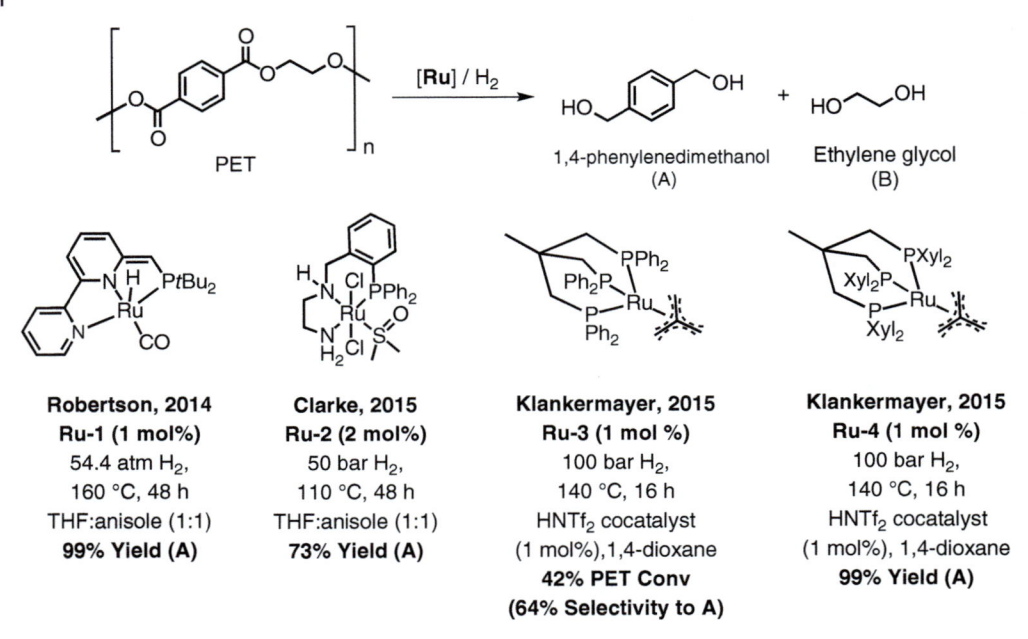

Figure 16.5 Hydrogenative depolymerisation of polyethylene terephthalate (PET).

yields at 54.4 atm H_2 and 160 °C. A combination of THF and anisole (1:1) solvent was used to ensure the solubility of PET and the catalyst (Figure 16.5). Under the same conditions, PLA was also hydrogenatively depolymerized to produce 1,2-propanediol in a quantitative yield. Clarke group, in 2015, reported a tridentate aminophosphine ruthenium complex (**Ru-2**, Figure 16.5) for the hydrogenative depolymerisation of PET flakes [133]. The catalytic conditions were optimised using various analogous ruthenium complexes and reaction conditions (temperature, time, loading of catalyst and base) on diesters and PET. Using the catalyst **Ru-2** (2 mol%, Figure 16.5) and KOtBu (40 mol%), PET was depolymerized to 1,4-phenylenedimethanol (73% yield) and ethylene glycol under 50 bar of H_2 pressure and at 110 °C in 48 h (Figure 16.5). A more active catalyst for the hydrogenative depolymerisation of a variety of polyesters such as PCL, PET, PLA, and polybutylene terephthalate (PBT) has been reported by Klankermayer using [Ru(triphos) tmm] (**Ru-3**, Figure 16.5) and [Ru(triphos-xyl)tmm] catalysts (**Ru-4**, Figure 16.5) [134]. [Ru(triphos)tmm] **Ru-3** catalyst (1 mol%, Figure 16.5) in the presence of the $HNTf_2$ cocatalyst was able to depolymerize PLA to 1,2-propanediol in a quantitative yield at 140 °C and 100 bar H_2 pressure using 1,4-dioxane as the solvent. However, a lower conversion (42%) was obtained in case of the hydrogenative depolymerisation of PET using the [Ru(triphos)tmm] **Ru-3** catalyst (Figure 16.5). Furthermore, a lower selectivity toward 1,4-phenylenedimethanol (64%) and ethylene glycol was also observed. It was suggested that the reduced selectivity is due to the formation of ethers favored by the presence of acid co-catalyst. Interestingly, using the modified ruthenium catalyst **Ru-4** [Ru(triphos-xyl)tmm] (1 mol%, Figure 16.5) and $HNTf_2$ as a cocatalyst, a quantitative conversion of PET was achieved to produce the 1,4-phenylenedimethanol and ethylene glycol at 140°C in 1,4-dioxane (Figure 16.5). Followed by this success, a variety of consumer products (such as PET flakes from a dyed soda bottle, synthetic pillow filling, and yogurt pots) were also depolymerized to produce diols with excellent selectivity using catalyst loading as low as 0.5 mol%.

Recently, Enthaler has reported the hydrogenative depolymerisation of the end-of-life PLA (sourced from transparent cups) to produce 1,2-propanediol (>99% yield as determined by nuclear magnetic resonance spectroscopy [NMR]) using the ruthenium-Macho-BH pincer catalyst (**Ru-5**, 0.5 mol%, Figure 16.6) in THF under the condition of H_2 (45 bar), 140°C, three hours [135]. Interestingly, the hydrogenative depolymerisation of PLA was also coupled with the degradation of poly(oxymethylene) (POM) to produce a cyclic acetal (Figure 16.6). 1,2-propanediol produced from the hydrogenative depolymerisation of PLA was used for the degradation of POM to produce cyclic acetal as previously reported by Klankermayer [136]. Enthaler has also recently reported the chemical recycling of PCL (polycaprolactone) using the approach of catalytic hydrogenation and dehydrogenation in the presence of the ruthenium catalyst **Ru-5** (Figure 16.6) [137]. The end-of-life PCL (M_n~80,000 g/mol) was hydrogenatively depolymerized to produce 1,6-hexanediol in almost quantitative yield in the THF solvent using 0.4 mol% catalyst, 45 bar H_2 in 24 hours. The produced monomer 1,6-hexanediol was dehydrogenatively polymerized to a mixture of polyesters [poly(ε-caprolactone) (M_w: 10,706) and 1,6-hexanediol-adipate copolymer (M_n: 3200)] exhibiting a high conversion of up to 92% and a turnover number (TON) of 230.

Recently, de Vries et al. reported the upcycling of polyesters to form polyethers using the ruthenium-triphos complex prepared in situ **Ru-3** (Figure 16.7) and a Lewis acid co-catalyst [138]. Authors utilized the ruthenium-triphos **Ru-3** complex (Figure 16.7) as they were earlier demonstrated for the direct reductive etherification of carboxylic acid esters to ethers [139, 140]. H_2O was

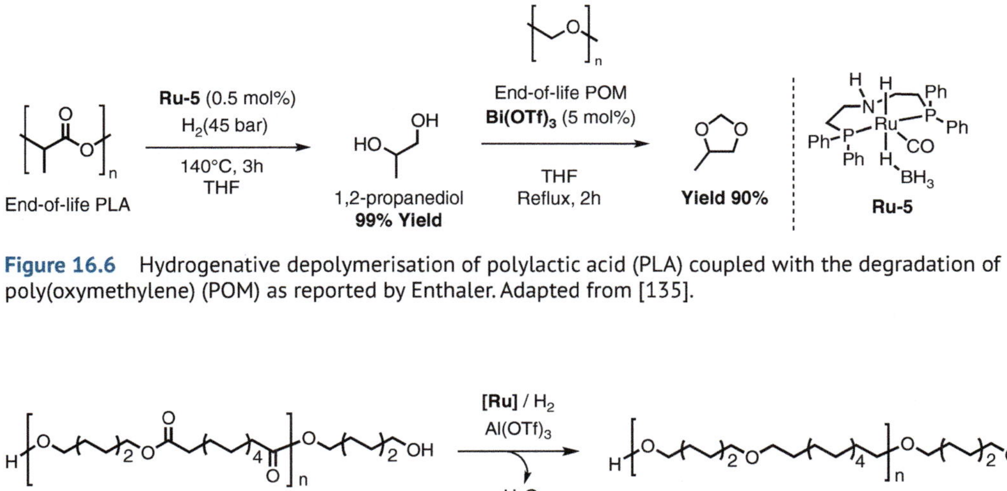

Figure 16.6 Hydrogenative depolymerisation of polylactic acid (PLA) coupled with the degradation of poly(oxymethylene) (POM) as reported by Enthaler. Adapted from [135].

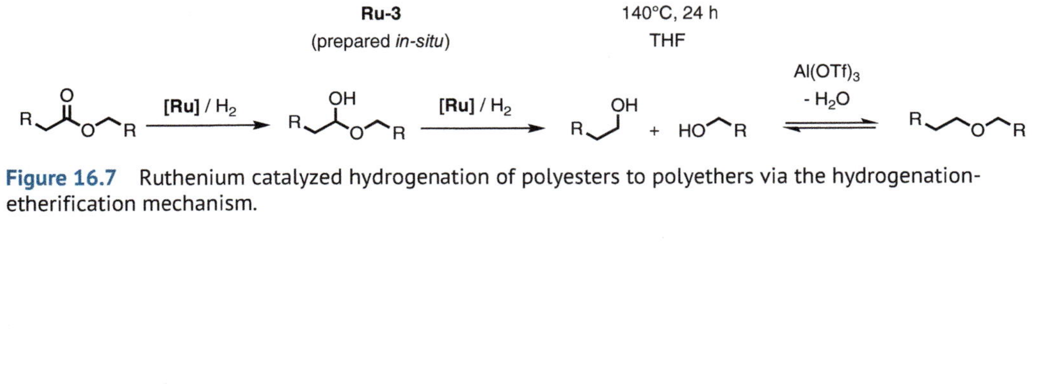

Figure 16.7 Ruthenium catalyzed hydrogenation of polyesters to polyethers via the hydrogenation-etherification mechanism.

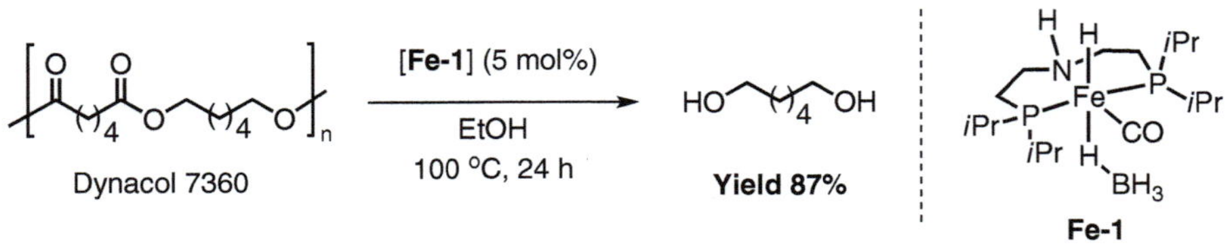

Figure 16.8 Iron catalyzed hydrogenation of polyesters to diols.

formed as the by-product in this reaction. Using the precatalyst Ru(acac)$_3$ (3 mol%), triphos ligand (4.5 mol%) and Al(OTf)$_3$ (7.5 mol%) as a co-catalyst, polyesters such as poly(hexene-1,12-dodecanate) (PHDD), poly(hexyl-1,6-adipate) (PHA), and poly[(2-(ethoxy)ethyl)phthalate] (PEGP) were transformed to polyethers in the yield range of 80–99%. The control experiments suggested that the Lewis acid co-catalyst is needed for the etherification step based on which a tandem hydrogenation-etherification mechanism as outlined in Figure 16.7 was suggested.

Significant progress has been made in the past few years on the use of a base-metal catalyst for the (de)hydrogenation reactions [141–143]. Along this line, de-Vries et al. have used an iron-macho-BH complex (**Fe-1**, Figure 16.8) for the depolymerisation of polyesters through transfer hydrogenation. Using EtOH as a source of hydrogen, Dynacol 7360 was depolymerized to form 1,6-hexanediol in 87% yield at 100 °C in 24 hours (Figure 16.8) [144].

16.5.2 Hydrogenative Depolymerisation of Polycarbonates

In 2011, catalytic hydrogenation of organic carbonates to methanol and alcohol was first reported by Milstein using the dearomatized ruthenium-PNN pincer catalyst **Ru-1** (Figure 16.9) [145]. Followed by this seminal report, a few ruthenium [146–148] and manganese [149–151] catalysts have been reported for this transformation in the past few years. The concept has also been expanded for the hydrogenative depolymerisation of PCs (Polycarbonates) to produce diols and methanol. This process was first demonstrated by Ding et al. in 2012 [148]. In the presence of the ruthenium-macho complex **Ru-6** (15.8 mg, 0.1 mol %, Figure 16.9) and KOtBu (0.1 mol%), poly(propylene carbonate) (PPC) (M$_w$ = 100,698) was depolymerized to form methanol and 1,2-propylene glycol in a quantitative yield under the conditions of 50 bar H$_2$, 140 °C, and 24 hours (Figure 16.9). This was followed by a report from Robertson et al. in 2014 in which the Milstein's RuPNN pincer catalyst **Ru-1** (1 mol% and 2 mol% KOtBu, Figure 16.9) was used for the hydrogenative depolymerisation of PPC and polyethylene carbonate (PEC) under 54.4 bar of H$_2$ at 160 °C to produce the corresponding diols and methanol in a quantitative yield (Figure 16.9) [131]. In 2018, Klankermayer reported a more active catalyst for the hydrogenative depolymerisation of various end-of-life PCs such as compact discs (CDs) and a regular beverage cup in the presence of the catalyst [Ru(triphos)tmm] (**Ru-3**, Figure 16.5) and the cocatalyst HNTf$_2$ in only 0.33 mol% catalytic loading [134]. A few catalysts based on earth-abundant metals have also been studied for the hydrogenative depolymerisation of PCs. In 2018, Milstein reported the first catalyst of earth-abundant metals for the hydrogenative depolymerisation of PCs using a manganese pincer complex (**Mn-1**, Figure 16.9) [149]. Using 2 mol% of **Mn-1** and 4 mol% of KH, PPC (Polypropylene Carbonate) was hydrogenated (110 °C, 50 bar H$_2$) to 1,2-propanediol and methanol in 68%, and 59% yields respectively. Around the similar time, Rueping reported another manganese catalyst for hydrogenative depolymerisation of PPC (**Mn-2**, Figure 16.9). Relatively higher yields of diol (91%) and methanol (84%) were obtained using a relatively lower catalytic loading (1 mol% Mn + 2.5 mol% KOtBu) albeit at a higher temperature of 140 °C (50 bar H$_2$) [151]. In 2019, Werner reported the transfer hydrogenation of PPC in the

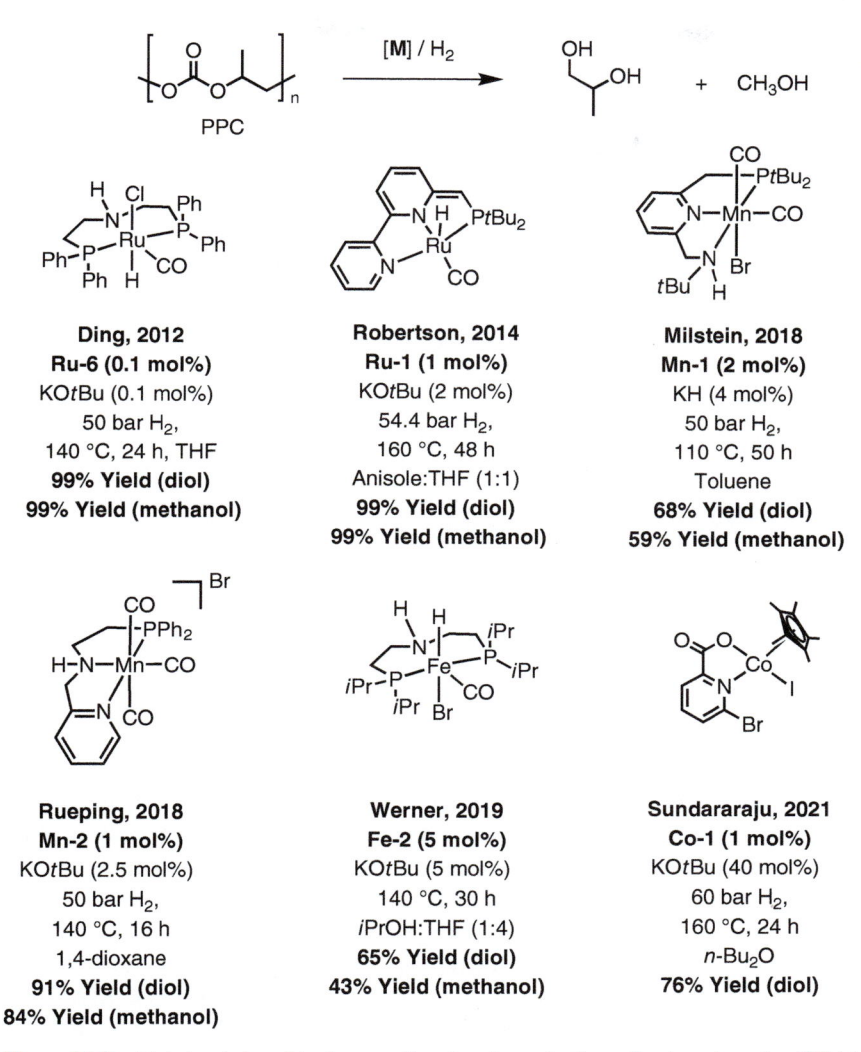

Figure 16.9 Metal-catalyzed hydrogenative depolymerisation of polycarbonates (PCs) to 1,2-propylene glycol and methanol.

Ding, 2012
Ru-6 (0.1 mol%)
KOtBu (0.1 mol%)
50 bar H_2,
140 °C, 24 h, THF
99% Yield (diol)
99% Yield (methanol)

Robertson, 2014
Ru-1 (1 mol%)
KOtBu (2 mol%)
54.4 bar H_2,
160 °C, 48 h
Anisole:THF (1:1)
99% Yield (diol)
99% Yield (methanol)

Milstein, 2018
Mn-1 (2 mol%)
KH (4 mol%)
50 bar H_2,
110 °C, 50 h
Toluene
68% Yield (diol)
59% Yield (methanol)

Rueping, 2018
Mn-2 (1 mol%)
KOtBu (2.5 mol%)
50 bar H_2,
140 °C, 16 h
1,4-dioxane
91% Yield (diol)
84% Yield (methanol)

Werner, 2019
Fe-2 (5 mol%)
KOtBu (5 mol%)
140 °C, 30 h
iPrOH:THF (1:4)
65% Yield (diol)
43% Yield (methanol)

Sundararaju, 2021
Co-1 (1 mol%)
KOtBu (40 mol%)
60 bar H_2,
160 °C, 24 h
n-Bu_2O
76% Yield (diol)

presence of an iron-macho pincer catalyst where iPrOH was used as a hydrogen source [152]. Using catalyst **Fe-2** in combination with KOtBu (5 mol% respectively) in iPrOH at 140 °C, PPC was depolymerized to propylene glycol and methanol in 65% and 43% yields respectively (Figure 16.9). Interestingly, the authors observed the depolymerisation reaction even in the absence of the iron-pincer catalyst and just in the presence of KOtBu where a mixture of cyclic carbonate and propylene glycol were obtained as the products. The authors suggest that the depolymerisation of PPC could also result from the transesterification by iPrOH catalyzed by KOtBu. Recently, Sundararaju et al. have reported the synthesis of Cp*Co(III) complexes bearing N,O-chelated ligand and utilised them for the hydrogenation of a variety of organic carbonates and PCs [153]. Using 1 mol% of catalyst **Co-1**, and 40 mol% KOtBu, commercially available PPC (M_n = 50,103 g/mol) was hydrogenatively depolymerized (60 bar H_2, 160 °C, dibutylether solvent) into their corresponding monomers methanol and propylene glycol (76% isolated yield, Figure 16.9).

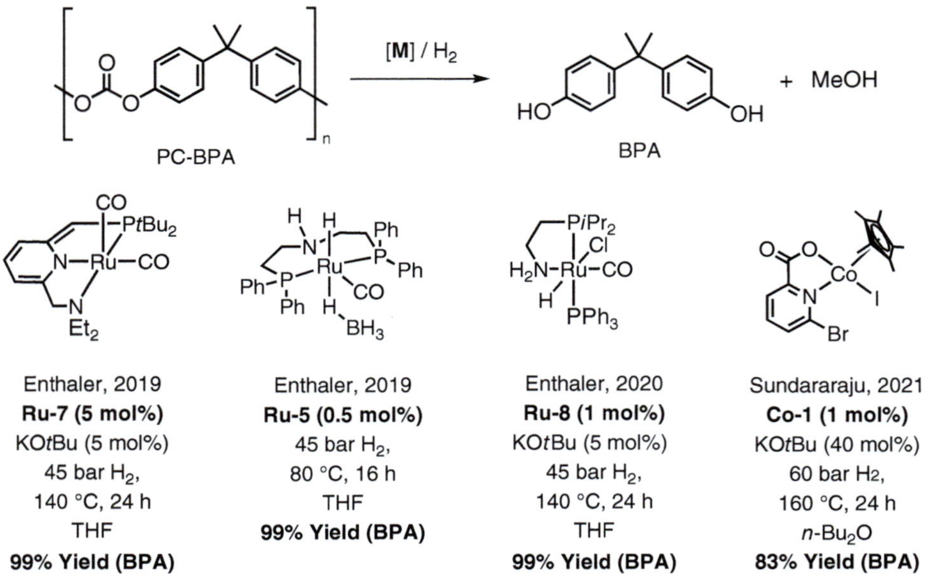

Figure 16.10 Metal-catalyzed hydrogenative depolymerisation of PC-BPA to bisphenol A (BPA) and methanol.

Enthaler, 2019	Enthaler, 2019	Enthaler, 2020	Sundararaju, 2021
Ru-7 (5 mol%)	**Ru-5 (0.5 mol%)**	**Ru-8 (1 mol%)**	**Co-1 (1 mol%)**
KOtBu (5 mol%)	45 bar H_2,	KOtBu (5 mol%)	KOtBu (40 mol%)
45 bar H_2,	80 °C, 16 h	45 bar H_2,	60 bar H_2,
140 °C, 24 h	THF	140 °C, 24 h	160 °C, 24 h
THF	**99% Yield (BPA)**	THF	n-Bu_2O
99% Yield (BPA)		**99% Yield (BPA)**	**83% Yield (BPA)**

Followed by this, in 2019, Enthaler reported the hydrogenative depolymerisation of poly(bisphenol A carbonate) (PC-BPA) using the dearomatized Milstein's RuPNN catalyst (**Ru-7**, Figure 16.10) to produce bisphenol A (BPA) and methanol. End-of-life BPA plastic sourced from CDs was also hydrogenatively depolymerized. However, a high catalytic loading (5 mol% of **Ru-7** and 5 mol% KOtBu) was needed for this reaction (condition: 140 °C, 45 bar H_2, 24 hours, Figure 16.10) [154]. Soon after, the same group reported that the same transformation can be achieved under milder conditions and low catalytic loading using a ruthenium-Macho-BH pincer catalyst **Ru-5** (0.5 mol% ruthenium catalyst, 80 °C, 45 bar H_2, 16 hours, Figure 16.10) [155]. Furthermore, the Enthaler group has recently reported the hydrogenative depolymerisation of PC-BPA using the in situ generated ruthenium catalyst (**Ru-8**). Using the combination of the precatalyst [RuClH(CO)(PPh$_3$)$_3$] (1 mol%), the PN ligand 2-(di-iso-propylphosphino)ethylamine (1 mol%), and KOtBu (5 mol%), commercial available PC-BPA was quantitatively hydrogenated (140 °C, 45 bar H_2) to BPA [156]. End-of-life PC-BPA sourced from DVDs and safety goggles were also hydrogenatively depolymerized to BPA in 29–85% yield using the same catalyst mixture described previously, in combination with the ligand diazabicycloundecene (DBU, 0–1 mol%) at 20 bar H_2 pressure, 120 °C and 24 hours. Recently, Sundararaju reported the depolymerisation of PC-BPA catalysed by the cobalt complex **Co-1** (Figure 16.10) [153]. Using 1 mol% of catalyst **Co-1**, and 40 mol% KOtBu, commercially available PC-BPA (Mn = 45,103 g/mol) was hydrogenatively depolymerized (60 bar H_2, 160 °C, dibutylether solvent) to BPA in 83% yield (Figure 16.10). Furthermore, end-of-life PC-BPA sourced from CDs was also depolymerized on a gram scale to produce BPA in 66% yield (2.98 g).

16.5.3 Hydrogenative Depolymerisation of Nylons and Polyurethanes

Catalytic hydrogenation of amides to form amines and alcohols has been extensively studied using transition-metal catalysts [157]. Similarly, hydrogenation of organic carbamates to form amines,

16.5 Hydrogenation

alcohols and methanol has been known since 2011 [145]. However, it is only recently that the hydrogenative depolymerisation of polyamides and polyurethanes have been demonstrated for the first time. In 2020, Milstein reported the first example of the hydrogenative depolymerisation of a variety of nylons using a Ru-PNNH pincer catalyst [158]. The key was the use of the dimethyl sulfoxide (DMSO) solvent that could dissolve nylons at the reaction condition (150 °C, 70 bar H_2) and allow the catalytic hydrogenation reaction to occur. A high conversion (60–99%) for the depolymerisation of the commercial beads of nylon-6 was obtained using 2 mol% ruthenium catalyst (**Ru-9** and **Ru-10**, Figure 16.11) and 8 mol% KOtBu under 70 bar of H_2 at 150 °C; however, the selectivity of the monomer (6-amino-1-hexanol) was relatively poor and the majority of products were detected as a mixture of oligomers (dimer-pentamer) (Figure 16.11). It was observed that if not for a matter of solubility, a higher catalytic activity was observed in 1,4-dioxane solvent in comparison to that of the DMSO. However, DMSO was needed to dissolve the nylon. Therefore, a dual hydrogenation approach was used where the hydrogenation reaction was carried out twice on the same sample. First, the reaction was carried out in DMSO to break down most of the nylons into oligomers. Then the DMSO was removed and the reaction was carried by adding the fresh catalyst and 1,4-dioxane solvent. This led to a higher selectivity of monomer in the range of 55–82% depending on the type of polyamides (Figure 16.11). A polyurethane was also hydrogenated using the ruthenium catalyst under the conditions of 150 °C, 70 bar H_2, 48 h to produce the corresponding diol and diamine in 70% and 66% respectively.

This work was followed by a report from Schaub et al. in which technical grades of nylon-66 (PA-66) and polyurethane were hydrogenatively depolymerized to their corresponding monomers (diamines and diols) using ruthenium-pincer catalysts [159]. Interestingly, the reaction was carried out in THF, which has advantages over DMSO as DMSO can degrade at a high temperature

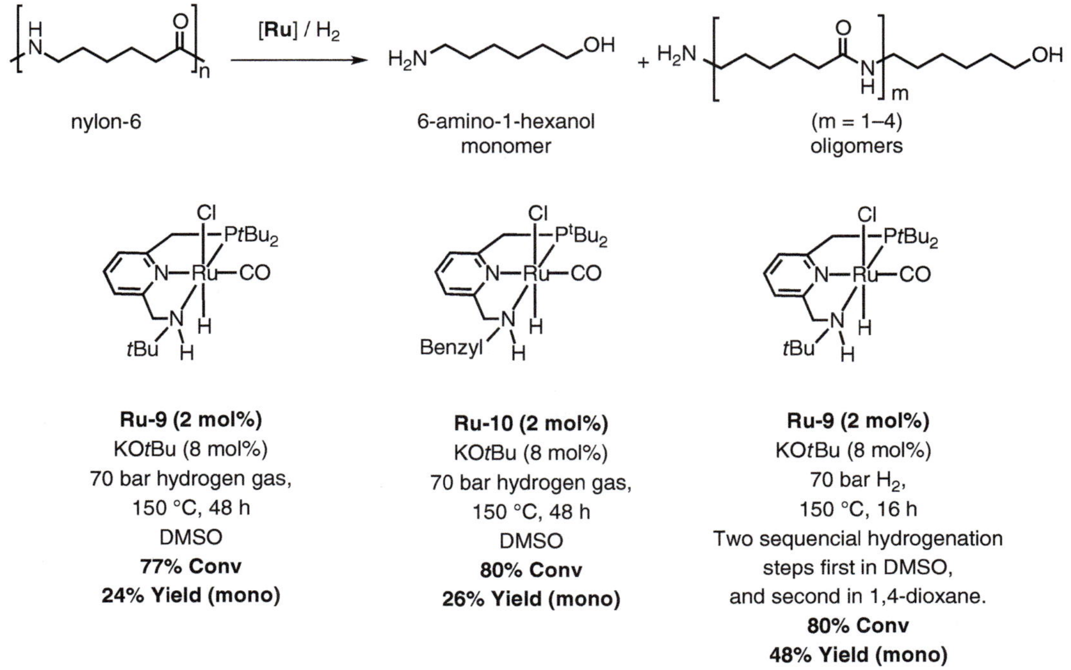

Figure 16.11 Ruthenium catalyzed hydrogenative depolymerisation of nylon-6 as reported by Milstein Adapted from [158].

raising safety concerns and DMSO can limit the catalytic activity as mentioned earlier. Using 1 mol% ruthenium catalyst **Ru-11** (Figure 16.12) and 4 mol% KOtBu, PA-66 ($M_w = 8750$) was hydrogenatively depolymerized to produce 1,6-diaminohexane and 1,6-hexanediol in 62% and 37% yields, respectively (condition: 50 bar H_2, 120 °C, 20 hours, Figure 16.12). Under these conditions, polyurethanes were also hydrogenatively depolymerized to give diamines and diols in 72–85% yields. Interestingly, technical grade nylon Ultramid® A 27, was also hydrogenatively depolymerized to give diamine and diol albeit with lower yields and under harsher reaction conditions. Remarkably, end-of-life toluenediamine (TDA)-based polyurethane sourced from a kitchen sponge was hydrogenatively depolymerized on a 10 g scale to produce a quantitative yield of TDA (2.37 g) exhibiting a TON of 970. Along this line, recently, Kristensen, Skrydstrup et al. have reported a study on the use of macho-pincer catalysts for the hydrogenative depolymerisation of end-of-life polyurethanes [160]. Initial studies were carried out on a model dicarbamate substrate (ethane-1,2-diyl bis[(4-benzylphenyl)carbamate]) using macho-pincer catalysts of ruthenium (**Ru-12**), iridium (**Ir-1**), manganese (**Mn-3**) and iron (**Fe-1**) (1 mol% metal-catalyst, 2 mol% base, 30 bar H_2, 150 °C, THF, Figure 16.12). Except for iron, all the three pincer catalysts exhibited almost quantitative hydrogenation of dicarbamate to form diol and amine. The iron-pincer catalyst showed ~37% yield of amine. Interestingly, ~37% of amine was formed even in the absence of the metal-catalyst and just in the presence of the base. Based on these observations, the authors speculate that an alternative pathway where carbamate can be deprotonated in the presence of a base leading to the formation of an isocyanate that could be hydrogenated to amine might be possible.

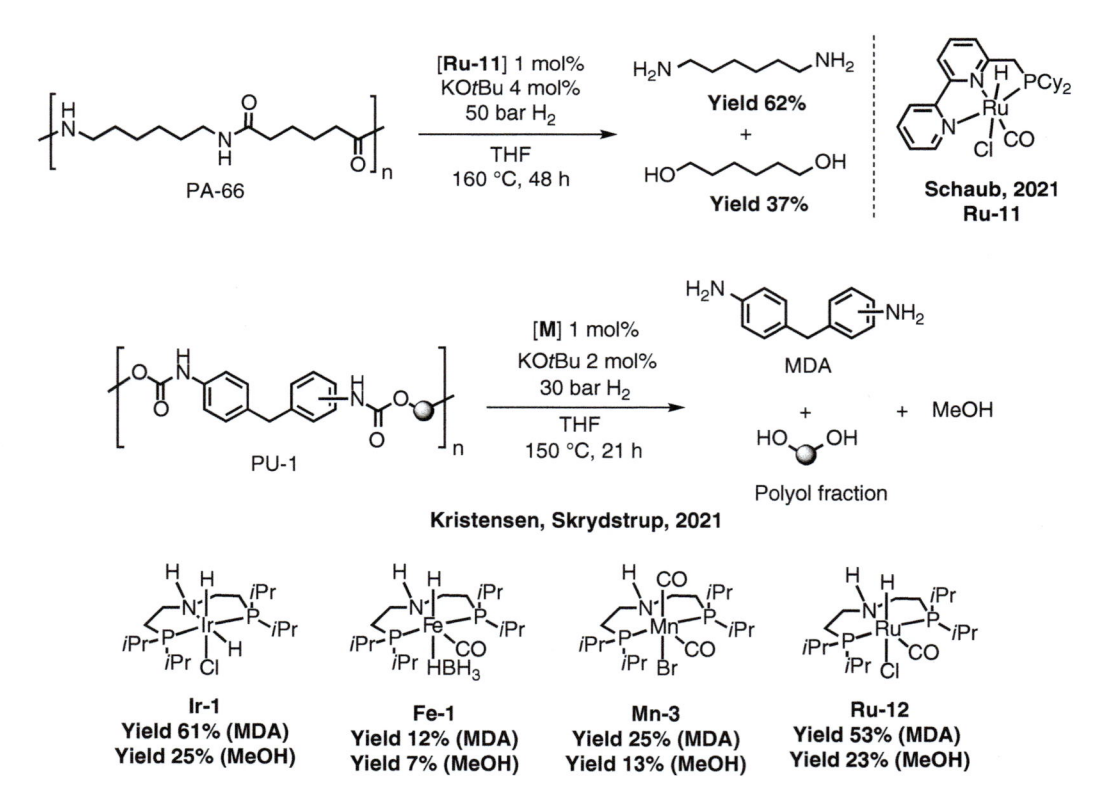

Figure 16.12 Hydrogenative depolymerisation of nylons and polyurethanes catalyzed by pincer complexes.

Interestingly, when catalysis was attempted for the hydrogenation of a polyurethane substrate (PU-1, prepared from the polymerization of an equimolar amount of dipropylene glycol and a mixture of 4,4-diphenylisocyanate and 2,4′-diphenylisocyanate), the best yield of diamine [methylenedianiline (MDA), 61%] was obtained using the catalyst **Ir-1** (Figure 16.12). This was followed by the catalyst **Ru-12** exhibiting 53% yield (Figure 16.12). The other two base-metal catalysts showed low yields of 12–25% (Figure 16.12). On the other hand, no reactivity was observed when no transition-metal was used even in the presence of KO*t*Bu unlike that of dicarbamate as discussed. Further optimization studies showed *i*PrOH to be a better solvent and K_3PO_4 to be a better base. Using the optimised conditions, a variety of end-of-life polyurethanes were hydrogenated to form diamines and diols. Optimization studies were also carried out to demonstrate the use of manganese pincer catalyst **Mn-3** (Figure 16.12) for the degradation of end-of-life polyurethane. At 180 °C, up to 45 mg of diamine was recovered from the hydrogenative depolymerisation of 260 mg of the end-of-life polyurethane foam.

A few other reports on the hydrogenative depolymerisation of polyurethanes using manganese-based catalysts have been published in the recent past. Werner has demonstrated the use of manganese-macho PNP (**Mn-4**, Figure 16.13) for the depolymerisation of commercial polyurethanes (M_n = 2200–3100) via a transfer hydrogenation route using *i*PrOH as a hydrogen source [161]. Using 2 mol% of the manganese complex **Mn-4** and 6 mol% KO*t*Bu, polyurethanes were degraded to diamines, diols, and methanol in the yield range of 35–65% in THF/ *i*PrOH at 150 °C (Figure 16.13). However, end-of-life polyurethane was not studied. Soon after this, Shaub reported a study on the use of several manganese pincer catalysts for the hydrogenative depolymerisation of polyurethanes and discovered that the MnPNN catalyst **Mn-5** (Figure 16.13) is the most active for this reaction [162]. Catalytic conditions were optimised using polymers made in the lab (Figure 16.13) and then further tested on a variety of end-of-life polyurethanes that were depolymerized to produce

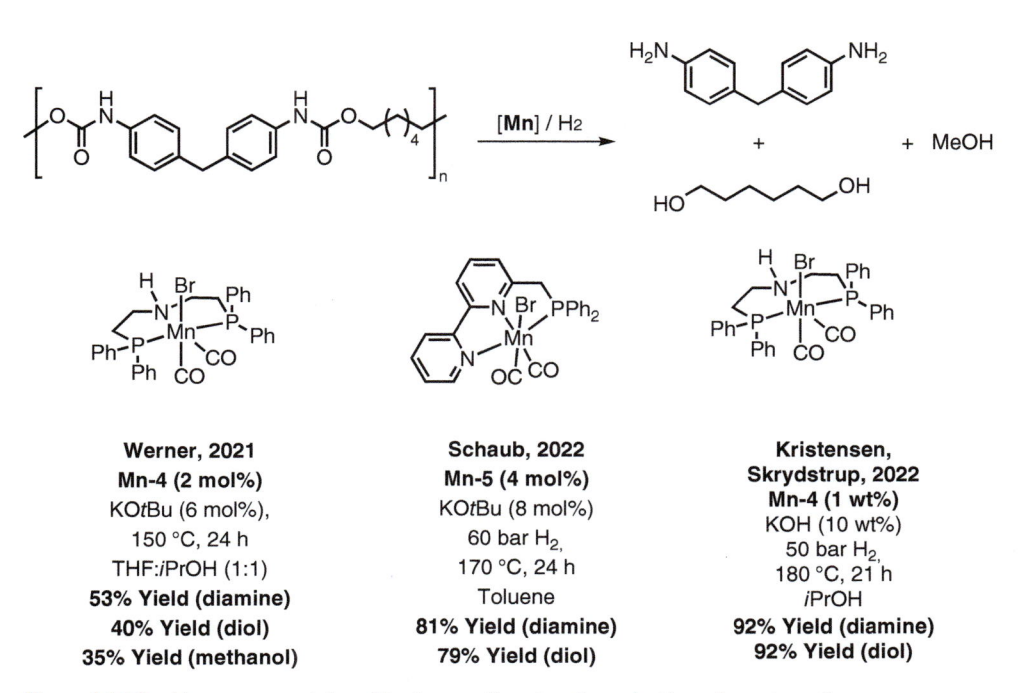

Figure 16.13 Manganese catalyzed hydrogenative depolymerisation of a polyurethane.

polyetherols, and TDA in yields of up to 89 % and 76 %, respectively (reaction condition: 200 °C, 60 bar H_2, THF, 72 hours). Around a similar time, Kristensen and Skrydstrup studied several manganese pincer catalysts for the same transformation and found that the **Mn-4** catalyst is the most effective one for the hydrogenative depolymerisation of polyurethanes (Figure 16.13) [163]. Under the optimised conditions (1 wt% **Mn-4**, 0.9 wt% KOH, 50 bar H_2, and 180 °C in iPrOH), end-of-life polyurethanes (260 mg) were depolymerized to produce diamines (44–52 mg) and diols (111–133 mg).

16.5.4 Hydrogenative Depolymerisation of Polyureas

In 2011, hydrogenation of urea derivatives to amines and methanol was first reported by Milstein using the dearomatized ruthenium pincer catalyst **Ru-1** (Figure 16.14) [164]. Milstein also reported

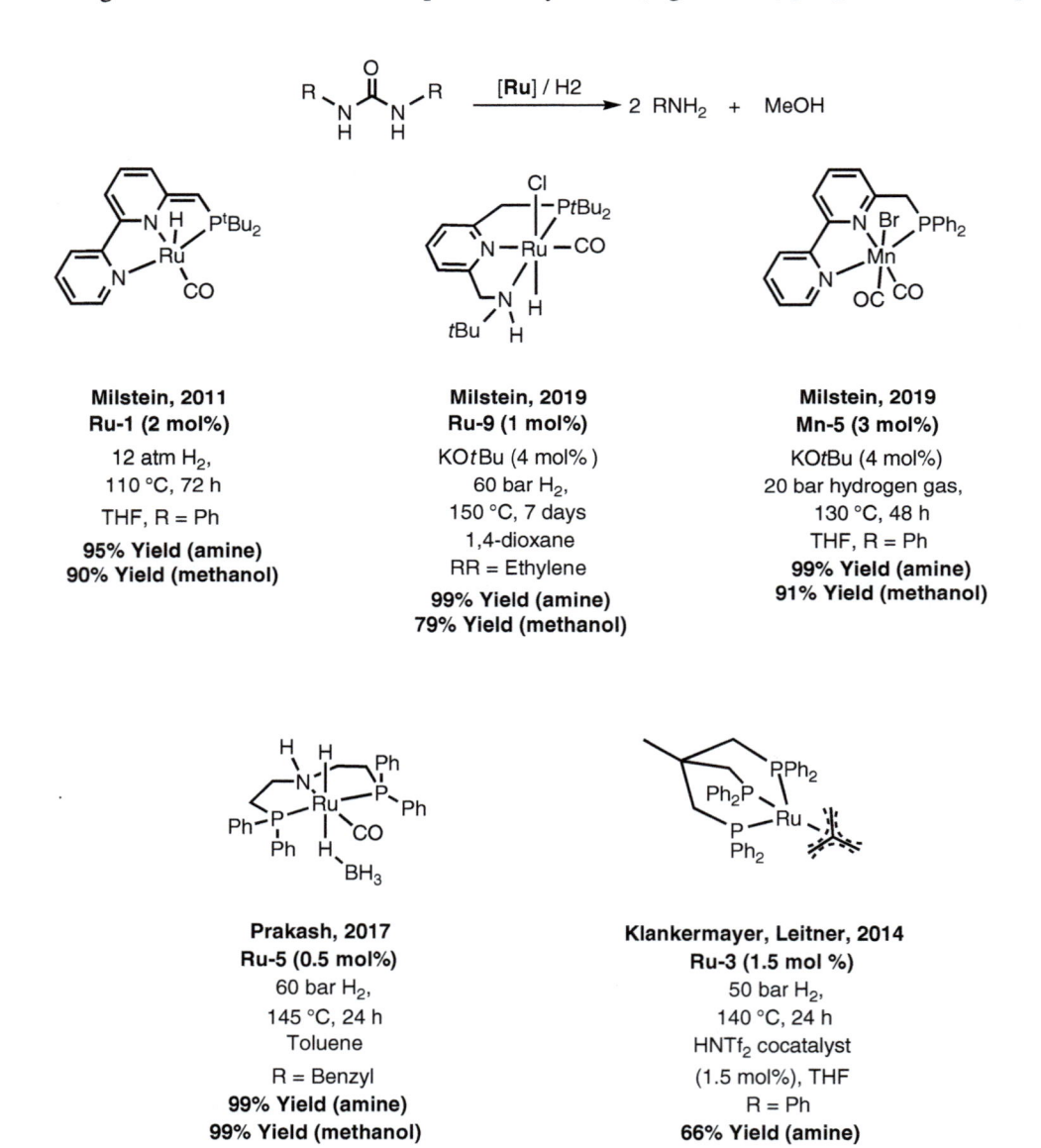

Milstein, 2011
Ru-1 (2 mol%)
12 atm H_2,
110 °C, 72 h
THF, R = Ph
95% Yield (amine)
90% Yield (methanol)

Milstein, 2019
Ru-9 (1 mol%)
KOtBu (4 mol%)
60 bar H_2,
150 °C, 7 days
1,4-dioxane
RR = Ethylene
99% Yield (amine)
79% Yield (methanol)

Milstein, 2019
Mn-5 (3 mol%)
KOtBu (4 mol%)
20 bar hydrogen gas,
130 °C, 48 h
THF, R = Ph
99% Yield (amine)
91% Yield (methanol)

Prakash, 2017
Ru-5 (0.5 mol%)
60 bar H_2,
145 °C, 24 h
Toluene
R = Benzyl
99% Yield (amine)
99% Yield (methanol)

Klankermayer, Leitner, 2014
Ru-3 (1.5 mol %)
50 bar H_2,
140 °C, 24 h
HNTf$_2$ cocatalyst
(1.5 mol%), THF
R = Ph
66% Yield (amine)

Figure 16.14 Metal catalyzed hydrogenation of urea derivatives to amine and methanol.

a new liquid organic hydrogen carrier (LOHC) based on the synthesis of a cyclic urea (ethylene urea) and its hydrogenation to ethylenediamine and methanol in the presence of the PNNH ruthenium pincer catalyst **Ru-9** (Figure 16.14) [165]. The manganese-based pincer catalyst **Mn-5** has also been reported by Milstein for the hydrogenation of urea derivatives to amines and methanol (Figure 16.14) [166]. Additionally, Prakash [167] and Klankermayer and Leitner [168] have also reported the hydrogenation of diphenyl urea to aniline and methanol in excellent yield using the ruthenium-macho pincer catalyst **Ru-5**, and the [Ru(Triphos)(TMM)] **Ru-3** respectively (Figure 16.14).

Recently, Kumar has reported the first example of the hydrogenative depolymerisation of polyureas using the ruthenium-macho **Ru-6** and the iridium-macho **Ir-1** pincer catalysts (Table 16.1) [169]. Two polyureas made from aliphatic diamines (M_n = ~4500) were depolymerized to produce

Table 16.1 Metal-catalyzed hydrogenative depolymerisation of polyureas to diamine and methanol.[a] Reproduced with permission from Ref [169].

General reaction: Polyurea + [M] (2 mol%), KOtBu (4 mol%), 140 °C, 24–72 h, 50 bar H_2, Solvent → $H_2N-R-NH_2$ + MeOH

Catalysts: **Ir-1** = iridium-macho pincer with $^iPr_2$P, Cl, H ligands; **Ru-6** = ruthenium-macho pincer with Ph_2P, Cl, CO, H ligands.

Substrate	Catalyst	Solvent	Conv	Diamine	Methanol
Polyurea with ether-linked aliphatic diamine	Ir-1	THF	45%	40%	27%
		Anisole	25%	23%	17%
		DMSO	50%	43%	34%
		THF	62%	60%[b]	41%[b]
Polyurea with ether-linked aliphatic diamine	Ru-6	THF	40%	35%	20%
Polyurea with xylylene diamine	Ir-1	THF	0%	0%	0%
Polyurea with xylylene diamine	Ru-6	THF	0%	0%	0%
Polyurea with bis(cyclohexyl)methane diamine	Ir-1	THF	38%	35%[c]	28%[c]
		THF	58%	51%[b]	43%[b]

a) Catalytic conditions: Polyurea (1 mmol), **Ru-6** or **Ir-1** (0.02 mmol), KOtBu (0.04 mmol), solvent (2 mL), H_2 (50 bar), 140 °C, 24 h.
b) Reaction time 72 h.
c) Reaction time 36 h.

diamines and methanol in yields up to 41% and 60% for methanol and diamine respectively as described in Table 16.1. DMSO and THF were found to be suitable solvents of choice for this transformation at 140 °C and 50 bar of H_2 pressure. No conversion was obtained in the case of the polyurea made from p-xylenediamine possibly due to its poor solubility.

16.6 Depolymerisation of Polyethylene Using Alkane Metathesis

Although pyrolysis is the most common method for the chemical recycling of polyolefins as discussed, an approach based on cross alkane metathesis between PE and a lower alkane (e.g. n-hexane or n-octane) has been reported by Guan and Huang for the depolymerisation of PE [170]. The concept of cross-alkane metathesis was originally discovered by Goldman and Brookhart and involves three steps and two catalysts [171]. The first catalyst is an iridium pincer catalyst (**Ir-2**, **Ir-3**, or **Ir-4**, Figure 16.15) that is used for the dehydrogenation of alkane to form an alkene as well as for the reverse reaction i.e. hydrogenation of alkenes to form alkanes. The second catalyst is an olefin metathesis catalyst (Grubbs or Schrock's catalyst, e.g. **Mo-1**, Figure 16.15). The reaction starts with the dehydrogenation of PE (i.e. a long alkane chain) and hexane to form internal alkenes. In the second step, the formed alkenes undergo olefin metathesis which results in the formation of smaller chain alkenes. Subsequent hydrogenation of alkenes results in the formation of alkanes in the final step (Figure 16.15). Further continuation of this process leads to the formation of a mixture of smaller alkanes. Guan and Huang used the iridium PCP catalysts such as **Ir-3**, **Ir-5**, and **Ir-6** for the (de)hydrogenation and Re_2O_7/γ-Al_2O_3 catalyst for the olefin metathesis (Figure 16.15) [170]. Using these catalysts HDPE (high-density polyethylene of M_w = 3,350) was depolymerized to alkanes of M_w = 680 (oil) using n-octane at 175 °C (four days). High molecular weight HDPE (M_w up to 105) were also depolymerized to oil products in 50–80% yields.

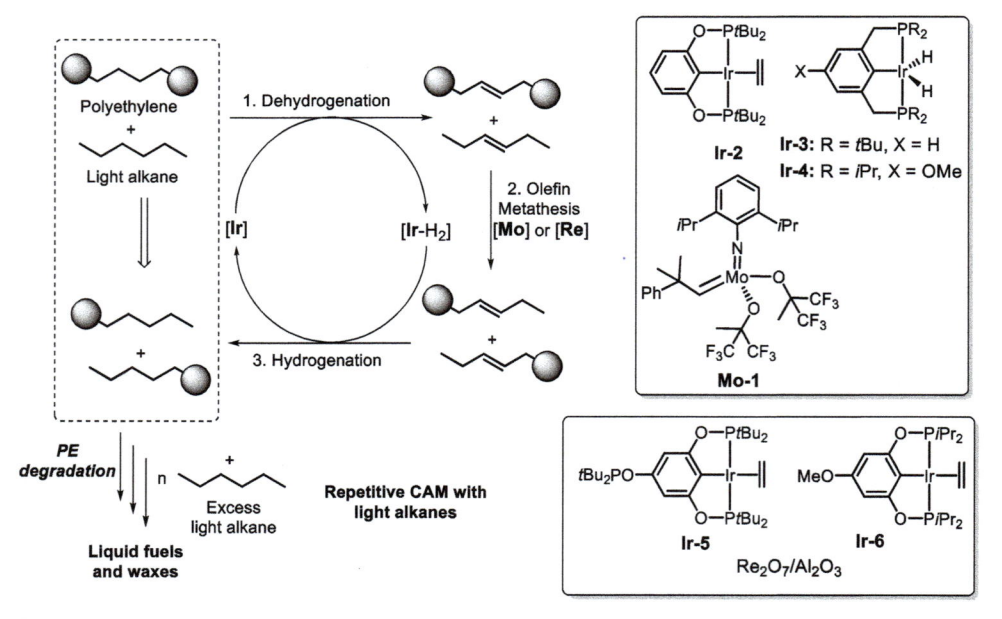

Figure 16.15 Cross-alkane metathesis between polyethylene (PE) and a light alkane (n-hexane) as reported by Guan and Huang. Adapted from [170].

End-of-life PE sourced from PE bottles, bags and films were also transformed into oil (C7–C38) and wax (minor product).

16.7 Hydrogenolysis

Hydrogenolysis refers to the cleavage of C–C or C–X (X = heteroatom e.g. O, N) bond via the addition of H_2. The process differs from hydrogenation (previously described) as hydrogenolysis refers to the degradation of saturated hydrocarbons whereas hydrogenation is carried on an unsaturated bond (e.g. C=C or C=O). Catalytic hydrogenolysis has been explored for the depolymerisation of polyolefins such as PE and PP. All the catalysts studied are heterogeneous in nature possibly due to the need for a high temperature to break the strong C–C or C–H bonds. Seminal studies on the hydrogenolysis of alkanes were carried out by Corker and Basset in 1996 where they reported that simple alkanes such as propane, butane, and pentane can be degraded to lower alkanes in the presence of H_2 and a zirconium hydride catalyst supported on silica [(XSiO)$_3$ZrH, **Zr-1a**, Figure 16.16] obtained by surface organometallic chemistry [172]. The highly electrophilic nature of the catalyst can cleave C–C bonds under mild conditions (H_2 pressure <1atm, and temperature 25–150 °C) [173]. This concept was further utilized by Dufaud and Basset for the hydrogenolysis of PE and PP to obtain diesel or lower alkanes [174]. The previous silica-supported zirconium hydride catalyst was modified by surface treatment to make the catalyst even more electrophilic. The catalyst was proposed to be a zirconium hydride supported on silica-alumina with the two possible structures **Zr-1a** and **Zr-1b** as shown in Figure 16.16. Heating PE (M_w up to 125,000) and PP (M_w up to 250,000) in the presence of H_2 (1 bar) and catalyst **Zr-1** at 150–190 °C resulted in the formation of saturated oligomers or lower alkanes. The statistical distribution of the produced alkanes was found to depend on the reaction time and the nature of polyolefin [174]. For example, a low molecular weight PE (C18–C50) was degraded to lower alkanes (C1–C4) in 84% yields with the

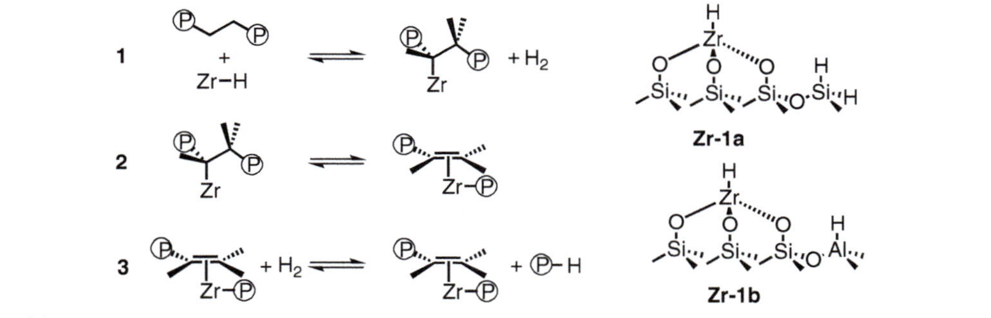

Figure 16.16 Zirconium catalyzed hydrogenolysis of polyethylene (PE). Proposed structures of the zirconium hydride catalyst supported on silica are shown on the right. Adapted from [174].

remaining product characterised as diesel (C10–C17) in five hours. A proposed pathway for the hydrogenolysis of polyolefins is outlined in Figure 16.16. The reaction starts with the C–H activation of the alkane chain via sigma-bond metathesis resulting in the formation of a Zr-alkyl complex and H_2 (Step 1). This is followed by a β-alkyl transfer to form a zirconium complex that is attached to the two fragments of the polymer via a π-bond and a sigma bond (Step 2). The final step involves the reaction of the formed complex with H_2 both at the terminal part (that eliminates the oligomeric fragment) and at the C=C centre (Step 3).

A few other heterogeneous catalysts have been reported since then for the hydrogenolysis of polyolefins. Nakagawa, Tomishige et al. have reported the hydrogenolysis of squalane (C30) in the presence of a Ru/CeO_2 catalyst [175]. The catalyst was found to be reusable and produced almost complete degradation of squalane (C30) at 6 MPa H_2 and 240 °C, 18 hours to produce tribranched C14–C16 and dibranched C9 and C10 hydrocarbons. A more selective degradation was later reported by the same group using Ru/SiO_2 in combination with other metals [176]. In 2019, Poeppelmeier, Sadow, Delferro et al. reported the hydrogenoysis of PE (Mn = 8000–158,000 Da) or a single-use plastic bag (Mn = 31,000 Da) into lower alkanes of high-values such as lubricants and waxes at 11.7 bar H_2 and 300 °C (up to 96 hours) under solvent-free conditions [177]. The catalyst is well-dispersed platinum nanoparticles supported on $SrTiO_3$. Interesting mechanistic studies carried out by spectroscopic techniques and density functional theory (DFT) computation suggest that the binding of PE to the catalyst surface (that is more favorable on the Pt sites than on the $SrTiO_3$ support) dictates the molecular weight and PDI of the degraded oligomers. A few other heterogeneous catalysts involving Pt/SiO_2 [178], Ru/C [179], and ruthenium supported on tungstated zirconia (Ru-WZr) [180] have been reported recently for the hydrogenolysis of polyolefins. Along this line, Kratish and Marks have recently reported the hydrogenolysis of polyesters such as PET using a combination of two catalysts: $Hf(OTf)_4$ and Pd/C, under 1 atm H_2 to produce terephthalic acid and ethane [181]. A mechanistic investigation was carried out using both experiments and DFT calculations that suggested that $Hf(Otf)_4$ catalyzes the cleavage of an alkoxy C–O bond to produce a carboxylic acid and ethene via the process called retro-hydroalkoxylation. This is coupled with hydrogenation of ethene in the presence of Pd/C to produce ethane as outlined in Figure 16.17.

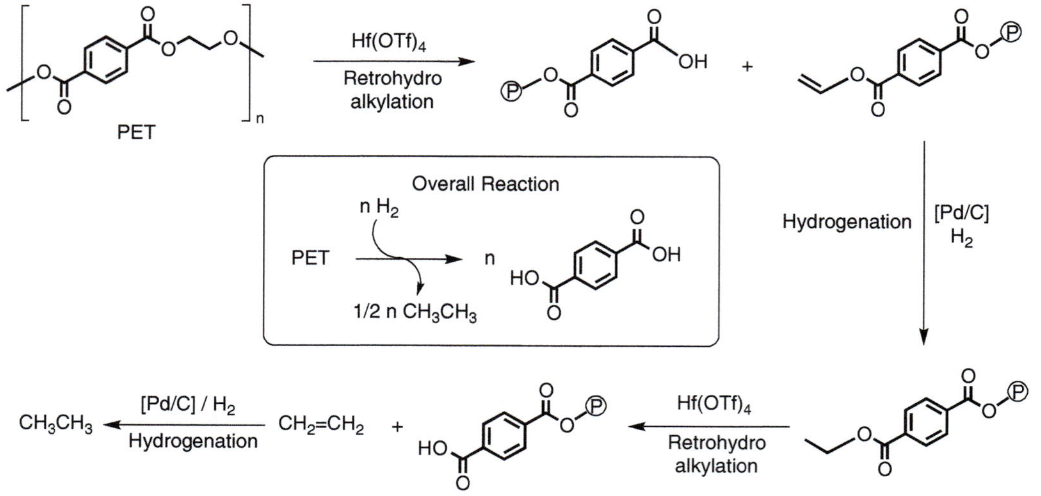

Figure 16.17 Proposed mechanism for the hydrogenolysis of polyethylene terephthalate (PET) to terephtalic acid and ethane catalyzed by a combination of Pd/C and $Hf(Otf)_4$ catalysts. Adapted from [181].

16.8 Hydrosilylation and Hydroboration

The approach of hydrosilylation and hydroboration have also been utilised for the reductive depolymerisation of plastic waste. The mild reduction potential of Si–H or B–H bonds in silanes or boranes together with the strong affinity of silicon or boron with an oxygen atom presents opportunities to use these reagents to depolymerize oxygen-containing plastics under mild conditions. The first example of the depolymerisation of plastic waste using such reducing agents was reported by Brookhart in 2008 and involved the use of an iridium pincer catalyst (**Ir-7**, Figure 16.18) to reductively depolymerize polyethyleneglycol (PEG) to $Et_3SiOCH_2CH_2OSiEt_3$ and ethane at 65 °C in the presence of triethylsilane after 4 hours [182]. At longer reaction times, the formed $Et_3SiOCH_2CH_2OSiEt_3$ was found to be fully converted to $Et_3SiOSiEt_3$ and ethane (Figure 16.18).

A mechanism for the cleavage of Et_2O was proposed based on kinetic and spectroscopic studies as outlined in Figure 16.19. Stoichiometric studies suggested that the complex **Ir-7** in the presence of triethylsilane forms a cationic complex where acetone is replaced by triethylsilane forming complex **Ir-8**. Complex **Ir-8** reacts with diethylether to form an iridium dihydride complex **Ir-9** and

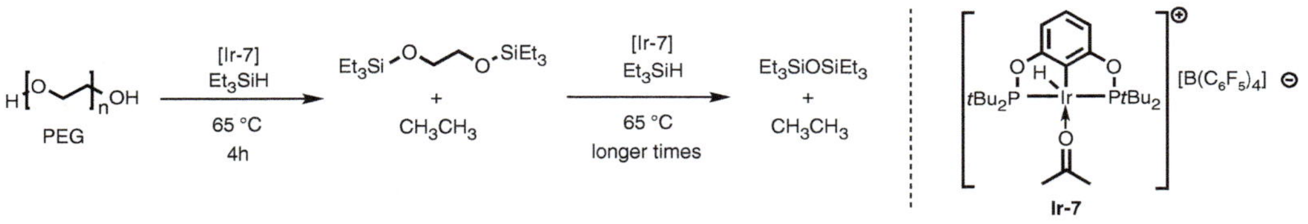

Figure 16.18 Iridium catalyzed hydrosilylative depolymerisation of polyethylene glycol (PEG). Adapted from [182].

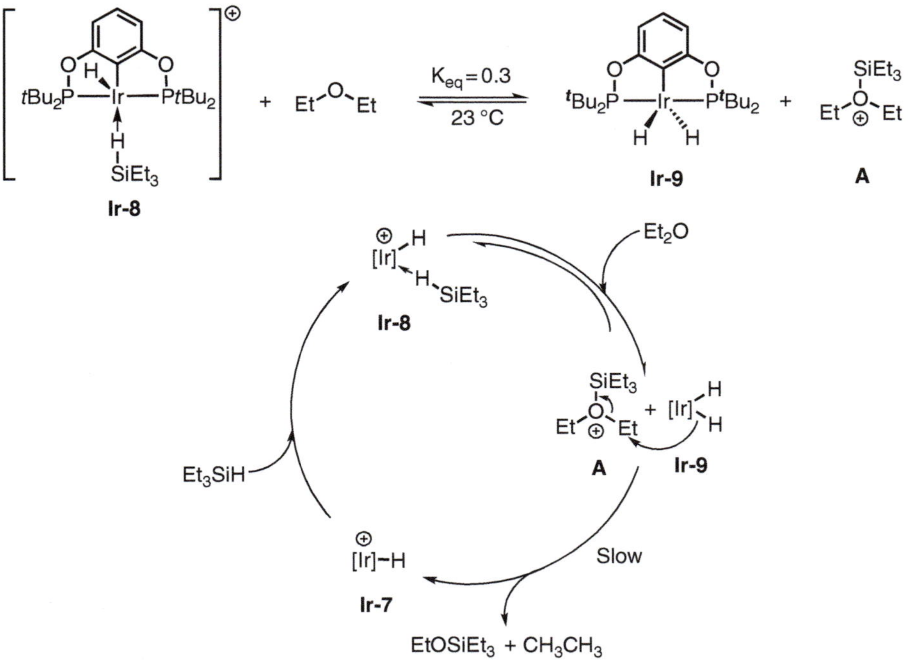

Figure 16.19 Proposed mechanism for the iridium catalyzed hydrosilylation of Et_2O to $EtOSiEt_3$ and ethane. Adapted from [182].

diethyl(triethylsilyl)oxonium ion **A** [182]. Complex **Ir-9** then reduces diethyl(triethylsilyl)oxonium ion **A** to produce EtOSiEt$_3$ and ethane with the concomitant regeneration of the complex **Ir-7**. A different cycle involving complex **Ir-8** and **Ir-9** was proposed for the reduction of EtOSiEt$_3$ to Et$_3$SiOSiEt$_3$ and ethane in the presence of Et$_3$SiH (Figure 16.19).

In 2015, Cantat reported the depolymerisation of polyethers, polyesters, and PCs using commercially available transition-metal free catalysts such as B(C$_6$F$_5$)$_3$ and (Ph$_3$C)[B(C$_6$F$_5$)$_4$] and reductants such as polymethylhydrosiloxane [Me$_3$Si(OSiMeH)$_n$OSiMe$_3$, PMHS] and tetramethyldisiloxane (Me$_2$SiHOSiHMe$_2$, TMDS) [183]. Using 2 mol% B(C$_6$F$_5$)$_3$ and 1.1 equivalents of TMDS, PEG (polyethylene glycol) was depolymerized to give ethane (and siloxane byproducts) in almost quantitative yields. This approach was also utilized for the reductive depolymerisation of other plastics such as polyesters, and PCs. PET sourced from plastic bottles were successfully depolymerized using Et$_3$SiH (4.3 equiv) and B(C$_6$F$_5$)$_3$ (2 mol%) at room temperature to produce two silyl ethers as shown in Figure 16.20a. Hydrolysis of silyl ethers using 2.1 equivalents of tetra-*n*-butylammonium fluoride (TBAF·3H$_2$O) resulted in the formation of ethylene glycol and 1,4-phenylenedimethanol that can be used as chemical feedstock for various organic compounds. Interestingly, the use of either 11.0 equivalents of PMHS or 6.0 equiv of TMDS as reductants in the presence of 5 mol% B(C$_6$F$_5$)$_3$ led to the formation of *p*-xylene (82% and 75% yield respectively) and ethane (Figure 16.20b). This approach was also utilized for the reductive depolymerization of PCs. In the other hand, the hydrosilylation of PCs such as PC-BPA was found to be more rapid in comparison to that of PET and polyether. In the presence of 2 mol% B(C$_6$F$_5$)$_3$ and 4.2 equivalents of Et$_3$SiH, PC-BPA was reductively depolymerized to produce disilylated BPA (82% yield) and methane. In the presence of TMDS (2.2 equivalents) that is a stronger reducing agent, PC-BPA was depolymerized to the disilylated BPA in 98% yield in one hour at room temperature (Figure 16.20c). Although interesting, this catalytic system has a number of limitations such as incompatibility with several solvents, high cost, and low selectivity. These drawbacks were later addressed in a report by Cantat et al. in 2018 that reported the reductive depolymerisation of various polyesters such as PCL, poly(dioxanone) (PDO), PLA, polyethylene succinate (PES), and PET as well as PCs such as PPC and PC-BPA using Brookharts catalyst **Ir-10** (Figure 16.20) [184]. Under a

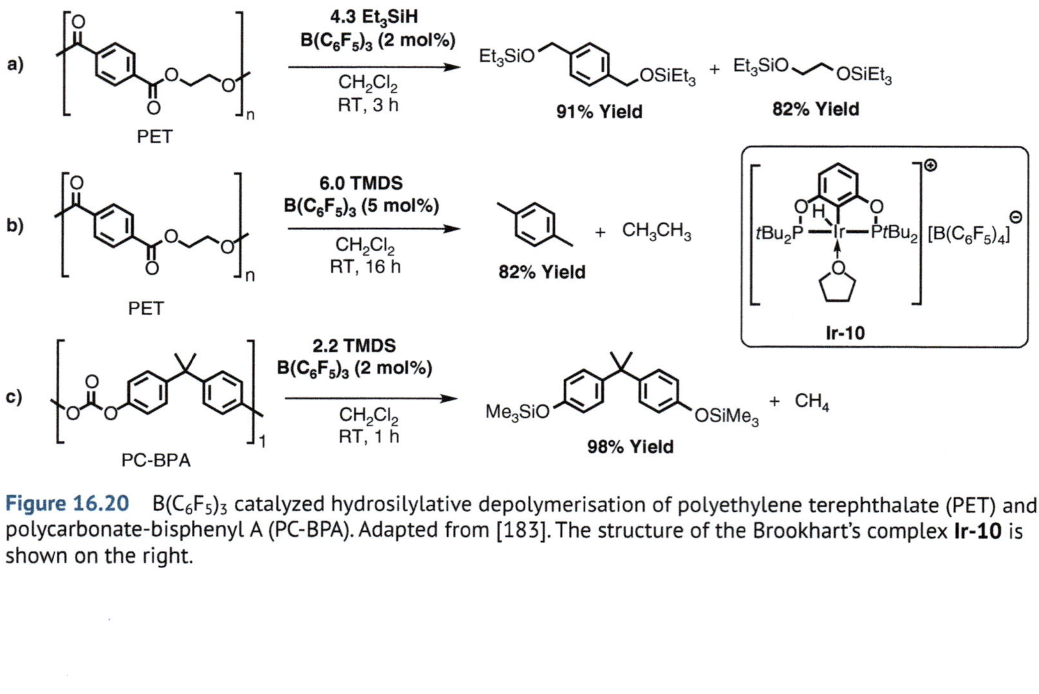

Figure 16.20 B(C$_6$F$_5$)$_3$ catalyzed hydrosilylative depolymerisation of polyethylene terephthalate (PET) and polycarbonate-bisphenyl A (PC-BPA). Adapted from [183]. The structure of the Brookhart's complex **Ir-10** is shown on the right.

relatively low catalytic loading of 0.3–1 mol% of complex **Ir-10** polyesters and PCs were depolymerized using triethylsilane in chlorobenzene to afford alkanes and silylated alcohols in moderate to excellent yields.

Along this line, Fernandes et al. reported the reductive depolymerisation of various polyesters using silanes and an air-stable dioxomolybdenum complex, $MoO_2Cl_2(H_2O)_2$ [185]. The catalyst was recycled 8 times and the production of propane from the depolymerisation of PLA was demonstrated on the gram scale. Fernandes has also reported reductive depolymerisation of various polyesters (e.g. PCL, PET, PBT, and PLA) using $Zn(OAc)_2 \cdot 2H_2O$ catalyst and silanes such as $(EtO)_2MeSiH$ (3 equiv.) or $PhSiH_3$ (2 equiv.) as reducing agents [186]. Aliphatic polyester (polycaprolactone) was successfully converted to 1,6-hexanediol in 98% yield in THF at 110 °C. The process was carried out on a gram scale with 79% yield and the reusability of the catalyst was also demonstrated (recycling up to 7 times). A mechanism was proposed based on the previous report by Beller et al. [187]. The first step is the activation of silane by $Zn(OAc)_2 \cdot 2H_2O$ that can form an intermediary species **B**. Species **B** subsequently reacts with the polyester to form the silyl acetal intermediate **C** that subsequently reacts with another equivalent of silane to form disilyl ether **D** in a similar pathway. Hydrolysis of disilyl ether leads to the formation of the corresponding diol (Figure 16.21).

Similar to silanes, boranes have also been used as reducing agents for the reductive depolymerisation of plastics. In 2019, Rueping reported the depolymerisation of polypropylene carbonate (PPC) using the commercially available di-n-butylmagnesium [Mg(n-Bu)$_2$] catalyst (**Mg-1**, Figure 16.22) in the presence of pinacolborane (HBPin) reductant [188]. Using 3 mol % of Mg(n-Bu)$_2$ (0.5 M in heptane), and HBpin (3.1 equiv), PPC (Mn = ~50,000) was reduced to form borylated diol in 91% yield at 65 °C in three hours. A mechanism involving three cycles was proposed based on the control experiments (Figure 16.22). In the first cycle, catalysis is initiated by the reaction of Mg(n-Bu)$_2$

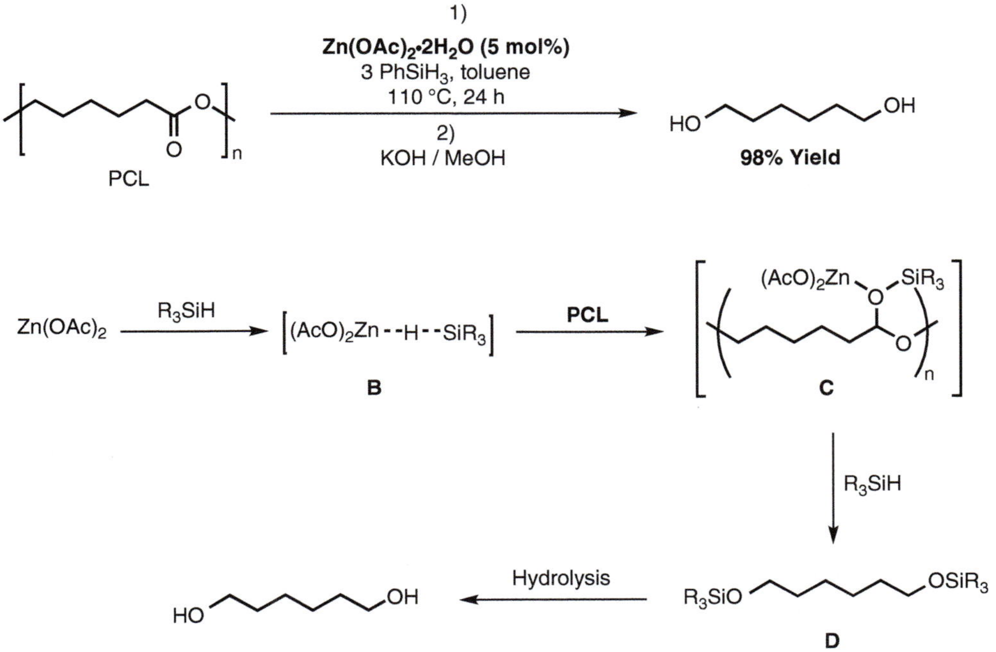

Figure 16.21 $Zn(OAc)_2$ catalyzed hydrosilylative depolymerisation of polycaprolactone (PCL) and a proposed mechanism. Adapted from [186].

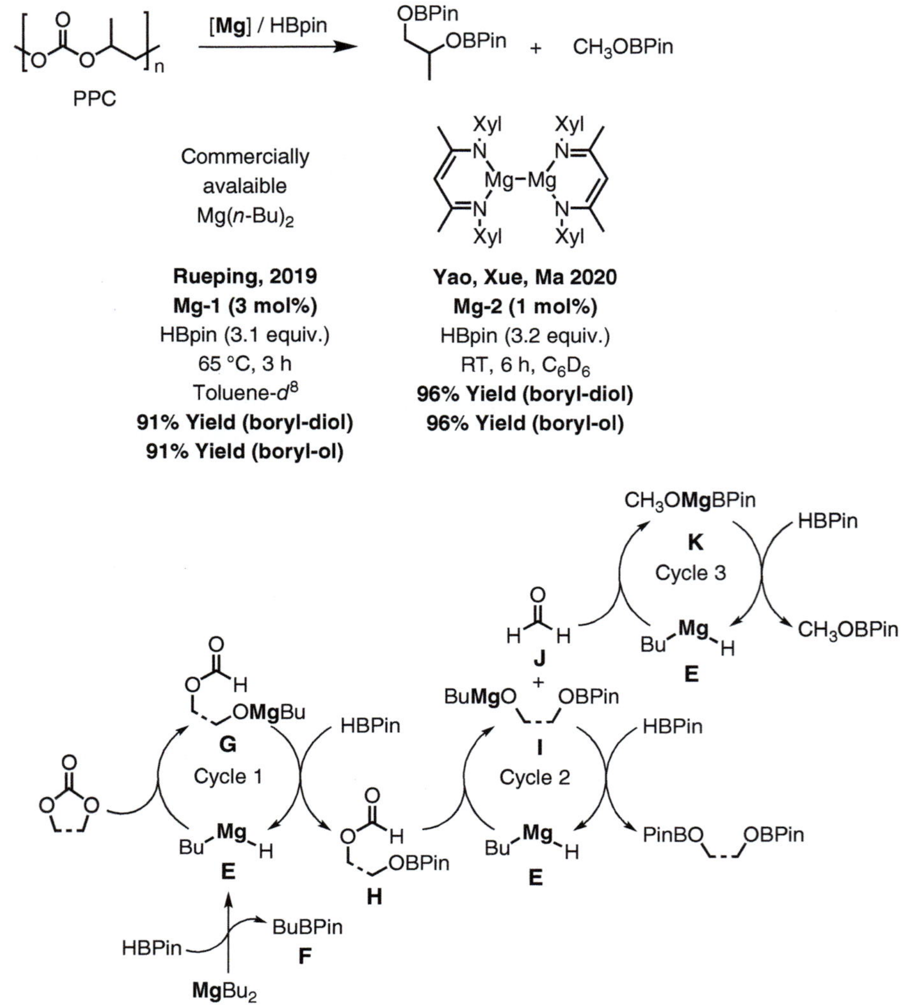

Figure 16.22 Magnesium catalyzed hydroborylation of polypropylene carbonate (PPC) and a proposed mechanism by Rueping et al. Adapted from [188].

with HBPin to form the catalytically active magnesium hydride species **E** that reacts with an organic carbonate to form magnesium alkoxide intermediate **G**. Intermediate **G** undergoes boron exchange with HBPin to form a formate intermediate **H** with the concomitant regeneration of the magnesium hydride species **E**. Formate **H** reacts with the magnesium hydride **E** in the second cycle to form formaldehyde (**J**) and intermediate **I** that goes through boron exchange with HBPin to form the borylated diol product. Formaldehyde gets reduced in the final cycle through a similar pathway to form CH_3OBPin. Followed by this, Yao, Xue, Ma et al. reported the reductive depolymerisation of PPC using a magnesium dimer catalyst **Mg-2** (1 mol%, Figure 16.22) and HBPin reductant (3.2 equiv) to produce borylated diol in 96% yield at room temperature [189].

Along this line, Berthet and Cantat have recently reported the reductive depolymerisation of polyesters using lanthanum(III) tris(amide) $La[N(SiMe_3)_2]_3$ catalyst and HBPin reducing agent [190]. Using 1 mol% of $La[N(SiMe_3)_2]_3$ catalyst and 2.2 equiv of HBPin various polyesters including end-of-life plastics such as PLA and PET were depolymerized to borylated diols in high yield (71–95%) (reaction condition: 100 °C, 24 hours). The advantage of hydrosilylation and hydroboration approach is the mild reaction condition (e.g. room temperature in some cases) but the use of an excess of reducing agents (e.g. 3–6 equivalents of silanes or boranes) produces significant waste and makes the process less attractive for commercialization.

16.9 Summary

In conclusion, we have reviewed here various methods for the catalytic depolymerization of plastics. Considering the urgent need for plastic recycling, this topic has been of high interest in both academia and industry in recent times. Plastic recycling technologies based on pyrolysis, gasification, and solvolysis (e.g. hydrolysis, glycolysis, and methanolysis) mainly for polyolefins and PET have been commercialized by several companies as previously mentioned and multiple investments for the launch of new plants have been made in the past five years. The studies show clearly that such chemical recycling processes can contribute to reducing climate change, terrestrial acidification, and fossil resource scarcity; however, these require higher capital investments and generate lower profit in comparison to those based on mechanical recycling [191]. The processes based on pyrolysis and gasification transform the plastic waste into smaller hydrocarbons (liquids) or gaseous products (e.g. $CO + H_2$) that can be used as fuel or chemical feedstock. Other methodologies such as solvolysis, and reductive depolymerisation can depolymerize plastics to the monomers that can be used to make fresh plastics of high quality. Several recent studies have been reported in the recent past for the reductive depolymerisation of various types of plastics using the approach of catalytic hydrogenation, hydrogenolysis, hydrosilylation, and hydroboration. The reductive depolymerisation processes based on hydroboration and hydrosilylation can be achieved under mild conditions (e.g. room temperature) but the processes generate significant waste making them less sustainable. The methodology based on catalytic hydrogenation and dehydrogenation presents an atom-economic approach for the closed-loop production/recycling of plastics such that the plastics can be depolymerized to make the monomers which can be then dehydrogenatively polymerized to make the fresh plastics. The proof of this concept has been demonstrated recently for nylons, polyesters, polycarbonates, polyurethanes and polyureas. However, low activity and selectivity as well as the recyclability issues of the reported homogeneous catalysts are the bottlenecks in the commercialization of such technologies. Thus, there is a significant scope for developing efficient catalysts for the (de)hydrogenative (de)polymerisation process to make the process cost-effective.

References

1 Roland, G., Jambeck, J.R., and Lavender, K.L. (2022). *Sci. Adv.* 3 (7): e1700782.
2 Ball, P. (2020). *Nat. Mater.* 19 (9): 938.
3 World Economic Forum. (2016). The new plastics economy: rethinking the future of plastics.
4 Plastics Europe. (2020). Plastics – the facts 2020. https://plasticseurope.org/knowledge-hub/plastics-the-facts-2020 (accessed 15 June 2023).
5 Kuczenski, B. and Geyer, R. (2010). *Resour. Conserv. Recycl.* 54 (12): 1161–1169.
6 Ellen MacArthur Foundation. (2017). The new plastics economy: rethinking the future of plastics & catalysing action. https://ellenmacarthurfoundation.org/the-new-plastics-economy-rethinking-the-future-of-plastics-and-catalysing (accessed 15 June 2023).

7 La Mantia, F.P. (1999). *Macromol. Symp.* 147 (1): 167–172.
8 Ragaert, K., Delva, L., and Van Geem, K. (2017). *Waste Manag.* 69: 24–58.
9 Singh, N., Hui, D., Singh, R. et al. (2017). *Compos. Part B Eng.* 115: 409–422.
10 Rahimi, A. and García, J.M. (2017) *Nat. Rev. Chem.* 1 (6): 46.
11 Lu, X.-B., Liu, Y., and Zhou, H. (2018). *Chem. – A Eur. J.* 24 (44): 11255–11266.
12 Coates, G.W. and Getzler, Y.D.Y.L. (2020). *Nat. Rev. Mater.* 5 (7): 501–516.
13 Worch, J.C. and Dove, A.P. (2020). *ACS Macro. Lett.* 9 (11): 1494–1506.
14 Garcia, J.M. and Robertson, M.L. (2017). *Science* 358 (6365): 870–872.
15 Vollmer, I., Jenks, M.J.F., Roelands, M.C.P. et al. (2020). *Angew. Chem. Int. Ed.* 59 (36): 15402–15423.
16 Thiounn, T. and Smith, R.C. (2020). *J. Polym. Sci.* 58 (10): 1347–1364.
17 Yin, S., Tuladhar, R., Shi, F. et al. (2015). *Polym. Eng. Sci.* 55 (12): 2899–2909.
18 U.S. Environmental Protection Agency (2023). Documentation chapters for greenhouse gas emission, energy and economic factors used in the Waste Reduction Model (WARM). https://www.epa.gov/warm/documentation-chapters-greenhouse-gas-emission-energy-and-economic-factors-used-waste (accessed 15 June 2023).
19 Miandad, R., Barakat, M.A., Aburiazaiza, A.S. et al. (2016). *Process Saf. Environ. Prot.* 102: 822–838.
20 Onwudili, J.A., Insura, N., and Williams, P.T. (2009). *J. Anal. Appl. Pyrolysis* 86 (2): 293–303.
21 Sasse, F. and Emig, G. (1998). *Chem. Eng. Technol. Ind. Chem. Equipment-Process Eng.* 21 (10): 777–789.
22 Sharma, B.K., Moser, B.R., Vermillion, K.E. et al. (2014). *Fuel Process. Technol.* 122: 79–90.
23 Lee, K.-H. (2012). Pyrolysis of waste polystyrene and high-density polyethylene. In: *Material Recycling* (ed. D. Achilias), 175–192. Rijeka: InTech Eur.
24 Nisar, J., Ali, G., Shah, A. et al. (2019). *Energy Fuels* 33 (12): 12666–12678.
25 Achilias, D.S., Kanellopoulou, I., Megalokonomos, P. et al. (2007). *Macromol. Mater. Eng.* 292 (8): 923–934.
26 Chen, Y., Zhang, S., Han, X. et al. (2018). *Energy Fuels* 32 (2): 2407–2413.
27 Ma, S., Lu, J., and Gao, J. (2002). *Energy Fuels* 16 (2): 338–342.
28 Ma, S., Lu, J., and Gao, J. (2004). *Energy Sources* 26 (4): 387–396.
29 Yu, J., Sun, L., Ma, C. et al. (2016). *Waste Manag.* 48: 300–314.
30 Del Remedio Hernández, M., Gómez, A., García, Á.N. et al. (2007). *Appl. Catal. A Gen.* 317 (2): 183–194.
31 Mastral, J.F., Berrueco, C., Gea, M., and Ceamanos, J. (2006). *Polym. Degrad. Stab.* 91 (12): 3330–3338.
32 Aguado, J., Serrano, D.P., San Miguel, G. et al. (2007). *J. Anal. Appl. Pyrolysis* 79 (1–2): 415–423.
33 Artetxe, M., Lopez, G., Amutio, M. et al. (2013). *Ind. Eng. Chem. Res.* 52 (31): 10637–10645.
34 Li, K., Lee, S.W., Yuan, G. et al. (2016). *Energies* 9 (6): 431.
35 Klaimy, S., Ciotonea, C., Dhainaut, J. et al. (2020). *ChemCatChem.* 12 (4): 1109–1116.
36 Chaianansutcharit, S., Katsutath, R., Chaisuwan, A. et al. (2007). *J. Anal. Appl. Pyrolysis* 80 (2): 360–368.
37 Neves, I.C., Botelho, G., Machado, A.V. et al. (2007). *Polym. Degrad. Stab.* 92 (8): 1513–1519.
38 Tekade, S.P., Gugale, P.P., Gohil, M.L. et al. (2020). *Energy Sources, Part A Recover. Util. Environ. Eff.* 1–15. doi: 10.1080/15567036.2020.1856976.
39 Rutenbeck, D., Papp, H., Freude, D., and Schwieger, W. (2001). *Appl. Catal. A Gen.* 206 (1): 57–66.
40 Li, L., Gao, J., Xu, C., and Meng, X. (2006). *Chem. Eng. J.* 116 (3): 155–161.
41 Cunliffe, A.M., Jones, N., and Williams, P.T. (2003). *Environ. Technol.* 24 (5): 653–663.

42 Czernik, S., Elam, C.C., Evans, R.J. et al. (1998). *J. Anal. Appl. Pyrolysis* 46 (1): 51–64.
43 Wang, J., Jiang, J., Meng, X. et al. (2020). *Environ. Sci. Technol.* 54 (13): 8390–8400.
44 Antonakou, E.V., Kalogiannis, K.G., Stefanidis, S.D. et al. (2014). *Polym. Degrad. Stab.* 110: 482–491.
45 Kumagai, S., Yabuki, R., Kameda, T. et al. (2019). *Chem. Eng. J.* 361: 408–415.
46 Sharuddin, S.D.A., Abnisa, F., Daud, W.M.A.W., and Aroua, M.K. (2016). *Energy Convers. Manag.* 115: 308–326.
47 Kumagai, S. and Yoshioka, T. (2016). *J. Japan Pet. Inst.* 59 (6): 243–253.
48 Kumar, P.A., Singh, R.K., and Mishra, D.K. (2010). *Renew. Sustain. Ener. Rev.* 14: 233–248.
49 Ryu, H.W., Kim, D.H., Jae, J. et al. (2020). *Bioresour. Technol.* 310: 123473.
50 Owusu, P.A., Banadda, N., Zziwa, A. et al. (2018). *J. Anal. Appl. Pyrolysis* 130: 285–293.
51 RECENSO (2023). Chemisches Recycling. https://www.recenso.eu/de/chemisches-recycling.html (accessed 15 June 2023).
52 Heidenreich, S. and Foscolo, P.U. (2015). *Prog. Energy Combust. Sci.* 46: 72–95.
53 Sikarwar, V.S., Zhao, M., Fennell, P.S. et al. (2017). *Prog. Energy Combust. Sci.* 61: 189–248.
54 Ranzi, E., Dente, M., Goldaniga, A. et al. (2001). *Prog. Energy Combust. Sci.* 27 (1): 99–139.
55 Lopez, G., Artetxe, M., Amutio, M. et al. (2018). *Renew. Sustain. Energy Rev.* 82: 576–596.
56 Arena, U. (2012). *Waste Manag.* 32 (4): 625–639.
57 Pohořelý, M., Vosecký, M., Hejdová, P. et al. (2006). *Fuel* 85 (17–18): 2458–2468.
58 Erkiaga, A., Lopez, G., Amutio, M. et al. (2013). *Fuel* 109: 461–469.
59 Janajreh, I., Adeyemi, I., and Elagroudy, S. (2020). *Sustain. Energy Technol. Assess.* 39: 100684.
60 Cho, M.-H., Choi, Y.-K., and Kim, J.-S. (2015). *Energy* 87: 586–593.
61 Slapak, M.J.P., Van Kasteren, J.M.N., and Drinkenburg, A.A.H. (2000). *Resour. Conserv. Recycl.* 30 (2): 81–93.
62 Borgianni, C., De Filippis, P., Pochetti, F., and Paolucci, M. (2002). *Fuel* 81 (14): 1827–1833.
63 Choi, M.-J., Jeong, Y.-S., and Kim, J.-S. (2021). *Energy* 223: 120122.
64 Mastellone, M.L. and Arena, U. (2008). *AIChE J.* 54 (6): 1656–1667.
65 Dogu, O., Pelucchi, M., Van de Vijver, R. et al. (2021). *Prog. Energy Combust. Sci.* 84: 100901.
66 Abdelaal, M.Y., Sobahi, T.R., and Makki, M.S.I. (2008). *Int. J. Polym. Mater.* 57 (1): 73–80.
67 Kosmidis, V.A., Achilias, D.S., and Karayannidis, G.P. (2001). *Macromol. Mater. Eng.* 286 (10): 640–647.
68 Polk, M.B., Leboeuf, L.L., Shah, M. et al. (1999). *Polym. Plast. Technol. Eng.* 38 (3): 459–470.
69 López-Fonseca, R., González-Marcos, M.P., González-Velasco, J.R., and Gutiérrez-Ortiz, J.I. (2009). *J. Chem. Technol. Biotechnol. Int. Res. Process. Environ. Clean Technol.* 84 (1): 92–99.
70 López-Fonseca, R., González-Velasco, J.R., and Gutiérrez-Ortiz, J.I. (2009). *Chem. Eng. J.* 146 (2): 287–294.
71 Yoshioka, T., Sato, T., and Okuwaki, A. (1994). *J. Appl. Polym. Sci.* 52 (9): 1353–1355.
72 Yoshioka, T., Okayama, N., and Okuwaki, A. (1998). *Ind. Eng. Chem. Res.* 37 (2): 336–340.
73 Parravicini, M., Crippa, M., and Bertele, M.V. (2013). Method and apparatus for the recycling of polymeric materials via depolymerization process. WO Pat. Appl. WO2013014650 A, 1.
74 Demeto (2020). www.demeto.eu (accessed 15 June 2023).
75 Desrousseaux, M.-L., Texier, H., Duquesne, S. et al. (2016). A process for degrading plastic products. France, Patent EP16305578.
76 Carbios. www.carbios.com/en (accessed 15 June 2023).
77 Liu, Y., Wang, M., and Pan, Z. (2012). *J. Supercrit. Fluids* 62: 226–231.
78 Campanelli, J.R., Cooper, D.G., and Kamal, M.R. (1994). *J. Appl. Polym. Sci.* 53 (8): 985–991.
79 Wang, Y., Zhang, Y., Song, H. et al. (2019). *J. Clean. Prod.* 208: 1469–1475.

80 Liu, F., Cui, X., Yu, S. et al. (2009). *J. Appl. Polym. Sci.* 114 (6): 3561–3565.
81 Sifniades, S., Levy, A.B., and Hendrix, J.A.J. (1997). Process for depolymerizing nylon-containing waste to form caprolactam. (AlliedSignal Inc.) US Patent, Patent 5,681,952.
82 Braun, M., Levy, A.B., and Sifniades, S. (1999). *Polym. Plast. Technol. Eng.* 38 (3): 471–484.
83 Jenczewski, T.J., Crescentini, L., and Mayer, R.E. (1995). Monomer recovery from multi-component materials. (AlliedSignal Inc.) US Patent, Patent 5,457,197.
84 Iwaya, T., Sasaki, M., and Goto, M. (2006). *Polym. Degrad. Stab.* 91 (9): 1989–1995.
85 Chen, J., Liu, G., Jin, L. et al. (2010). *J. Anal. Appl. Pyrolysis* 87 (1): 50–55.
86 Chen, J., Li, Z., Jin, L. et al. (2010). *J. Mater. Cycles Waste Manag.* 12 (4): 321–325.
87 Mizuno, N. and Misono, M. (1998). *Chem. Rev.* 98 (1): 199–218.
88 Robello, R., Calenti, F., Jahič, D., and Mikuž, M. (2013). Method of polyamide fiber recycling from elastomeric fabrics. (Aquafil S.P.A.) Patent WO 2013/03408 A1.
89 Aquafil. (2022). www.aquafil.com (accessed 15 June 2023).
90 Liu, F.-S., Li, Z., Yu, S.-T. et al. (2009). *J. Polym. Environ.* 17 (3): 208–211.
91 Ikeda, A., Katoh, K., and Tagaya, H. (2008). *J. Mater. Sci.* 43 (7): 2437–2441.
92 Dai, Z., Hatano, B., Kadokawa, J., and Tagaya, H. (2002). *Polym. Degrad. Stab.* 76 (2): 179–184.
93 Qinghua, S.Y.Z.X.Z. and Fengqiu, C. (2009). *Chem. React. Eng. Technol.* (1): 88–92.
94 Sinha, V., Patel, M.R., and Patel, J.V. (2010). *J. Polym. Environ.* 18 (1): 8–25.
95 Garbo. garbo.it/en/home (accessed 15 June 2023).
96 Allen, R.D., Bajjuri, K.M., Hedrick, J.L. et al. (2016). Methods and materials for depolymerizing polyesters. (International Business Machines Corporation) US Patent, Patent 9,255,194 B2.
97 Pecorini, T.J., Moskala, E.J., Lange, D., and Seymour, R.W. (2013). Process for the preparation of polyesters with high recycle content. (Eastman Chemical Company) US Patent, Patent US 2013/0041053 A1.
98 Petcore Europe. (2019). DuPont Teijin Films launches chemical recycling LuxCR™ depolymerisation process | petcore-europe.org. https://www.petcore-europe.org/news-events/227-dupont-teijin-films-luxcr.html (accessed 15 June 2023).
99 Van Berkum, S., Phillipi, V., Artigas, M.V. et al. (2016). Reusable capture complex. (Ioniqa Technologies B.V.) US Patent, Patent US 10,323,135 B2.
100 Ioniqa. ioniqa.com/company.
101 Cure Polyester Rejuvenation. (2023). https://curetechnology.com/about-cure-technology (accessed 15 June 2023).
102 Devraj, S. (2016). Flakes of ester mixtures and methods for their production. Patent WO 2013/175497.
103 Goto, M., Koyamoto, H., Kodama, A. et al. (2002). *J. Phys. Condens. Matter* 14 (44): 11427.
104 Genta, M., Iwaya, T., Sasaki, M. et al. (2005). *Ind. Eng. Chem. Res.* 44 (11): 3894–3900.
105 Al-Sabagh, A.M., Yehia, F.Z., Eshaq, G. et al. (2016). *Egypt. J. Pet.* 25 (1): 53–64.
106 Karayannidis, G.P. and Achilias, D.S. (2007). *Macromol. Mater. Eng.* 292 (2): 128–146.
107 Soni, R.K., Singh, S., and Dutt, K. (2010). *J. Appl. Polym. Sci.* 115 (5): 3074–3080.
108 Blackmon, K.P., Fox, D.W., and Shafer, S.J. (1990). Process for converting pet scrap to diamine monomers. (General Electric Company) US Patent, Patent 4,973,746.
109 Fukushima, K., Lecuyer, J.M., Wei, D.S. et al. (2013). *Polym. Chem.* 4 (5): 1610–1616.
110 Shukla, S.R. and Harad, A.M. (2006). *Polym. Degrad. Stab.* 91 (8): 1850–1854.
111 Pingale, N.D. and Shukla, S.R. (2009). *Eur. Polym. J.* 45 (9): 2695–2700.
112 Parab, Y.S., Pingale, N.D., and Shukla, S.R. (2012). *J. Appl. Polym. Sci.* 125 (2): 1103–1107.
113 Parab, Y., Shukla, S., and Shah, R. (2012). *Curr. Chem. Lett.* 1 (2): 81–90.
114 Kosloski-Oh, S.C., Wood, Z.A., Manjarrez, Y. et al. (2021). *Mater. Horizons* 8 (4): 1084–1129.

115 George, N. and Kurian, T. (2014). *Ind. Eng. Chem. Res.* 53 (37): 14185–14198.
116 Kumar, A. and Gao, C. (2021). *ChemCatChem.* 13 (4): 1105–1134.
117 Chen, H., Wan, K., Zhang, Y., and Wang, Y. (2021). *ChemSusChem.* 14 (19): 4123–4136.
118 Bai, S.-T., De Smet, G., Liao, Y. et al. (2021). *Chem. Soc. Rev.* 50 (7): 4259–4298.
119 Kumar, A., Daw, P., and Milstein, D. (2022). *Chem. Rev.* 122 (1): 385–441.
120 Blaser, H., Malan, C., Pugin, B. et al. (2003). *Adv. Synth. Catal.* 345 (1-2): 103–151.
121 Werkmeister, S., Neumann, J., Junge, K., and Beller, M. (2015). *Chem. Eur. J.* 21 (35): 12226–12250.
122 Lawrence, M.A.W., Green, K.-A., Nelson, P.N., and Lorraine, S.C. (2018). *Polyhedron* 143: 11–27.
123 Maser, L., Vondung, L., and Langer, R. (2018). *Polyhedron* 143: 28–42.
124 Khusnutdinova, J.R. and Milstein, D. (2015). *Angew. Chem. Int. Ed.* 54 (42): 12236–12273.
125 Gunanathan, C. and Milstein, D. (2014). *Chem. Rev.* 114 (24): 12024–12087.
126 Zhang, J., Leitus, G., Ben-David, Y., and Milstein, D. (2006). *Angew. Chem. Int. Ed.* 45 (7): 1113–1115.
127 Brewster, T.P., Rezayee, N.M., Culakova, Z. et al. (2016). *ACS Catal.* 6 (5): 3113–3117.
128 Zell, T., Ben-David, Y., and Milstein, D. (2014). *Angew. Chem. Int. Ed.* 53 (18): 4685–4689.
129 Espinosa-Jalapa, N.A., Nerush, A., Shimon, L.J.W. et al. (2017). *Chem. Eur. J.* 23 (25): 5934–5938.
130 Srimani, D., Mukherjee, A., Goldberg, A.F.G. et al. (2015). *Angew. Chem. Int. Ed.* 54 (42): 12357–12360.
131 Krall, E.M., Klein, T.W., Andersen, R.J. et al. (2014). *Chem. Commun.* 50 (38): 4884–4887.
132 Milstein, D., Balaraman, E., Gunanathan, C. et al. (2012). Novel ruthenium complexes and their uses in processes for formation and/or hydrogenation of esters, amides and derivatives thereof. US Patent, Patent US 2013/0281664 A1.
133 Fuentes, J.A., Smith, S.M., Scharbert, M.T. et al. (2015). *Chem. Eur. J.* 21 (30): 10851–10860.
134 Westhues, S., Idel, J., and Klankermayer, J. (2018). *Sci. Adv.* 4 (8): eaat9669.
135 Kindler, T., Alberti, C., Fedorenko, E. et al. (2020). *ChemistryOpen* 9 (4): 401–404.
136 Beydoun, K. and Klankermayer, J. (2020). *ChemSusChem.* 13 (3): 488–492.
137 Alberti, C. and Enthaler, S. (2021). . *ChemistrySelect* 6 (41): 11244–11248.
138 Stadler, B.M., Hinze, S., Tin, S., and de Vries, J.G. (2019). *ChemSusChem.* 12 (17): 4082–4087.
139 Li, Y., Topf, C., Cui, X. et al. (2015). *Angew. Chem. Int. Ed.* 54 (17): 5196–5200.
140 Erb, B., Risto, E., Wendling, T., and Gooßen, L.J. (2016). *ChemSusChem.* 9 (12): 1442–1448.
141 Filonenko, G.A., Van Putten, R., Hensen, E.J.M., and Pidko, E.A. (2018). *Chem. Soc. Rev.* 47 (4): 1459–1483.
142 Mukherjee, A. and Milstein, D. (2018). *ACS Catal.* 8 (12): 11435–11469.
143 Alig, L., Fritz, M., and Schneider, S. (2018). *Chem. Rev.* 119 (4): 2681–2751.
144 Farrar-Tobar, R.A., Wozniak, B., Savini, A. et al. (2019). *Angew. Chem. Int. Ed.* 58 (4): 1129–1133.
145 Balaraman, E., Gunanathan, C., Zhang, J. et al. (2011). *Nat. Chem.* 3 (8): 609–614.
146 Wu, X., Ji, L., Ji, Y. et al. (2016). *Catal. Commun.* 85: 57–60.
147 Kim, S.H. and Hong, S.H. (2014). *ACS Catal.* 4 (10): 3630–3636.
148 Han, Z., Rong, L., Wu, J. et al. (2012). *Angew. Chem. Int. Ed.* 51 (52): 13041–13045.
149 Kumar, A., Janes, T., Espinosa-Jalapa, N.A., and Milstein, D. (2018). *Angew. Chem. Int. Ed.* 57 (37): 12076–12080.
150 Kaithal, A., Hölscher, M., and Leitner, W. (2018). *Angew. Chem. Int. Ed.* 57 (41): 13449–13453.
151 Zubar, V., Lebedev, Y., Azofra, L.M. et al. (2018). *Angew. Chem. Int. Ed.* 57 (41): 13439–13443.
152 Liu, X., de Vries, J.G., and Werner, T. (2019). *Green Chem.* 21 (19): 5248–5255.
153 Dahiya, P., Gangwar, M.K., and Sundararaju, B. (2021). *ChemCatChem.* 13 (3): 934–939.
154 Alberti, C., Eckelt, S., and Enthaler, S. (2019). *ChemistrySelect* 4 (42): 12268–12271.

155 Kindler, T., Alberti, C., Sundermeier, J., and Enthaler, S. (2019). *ChemistryOpen* 8 (12): 1410–1412.
156 Alberti, C., Kessler, J., Eckelt, S. et al. (2020). *ChemistrySelect* 5 (14): 4231–4234.
157 Cabrero-Antonino, J.R., Adam, R., Papa, V., and Beller, M. (2020). *Nat. Commun.* 11 (1): 1–18.
158 Kumar, A., von Wolff, N., Rauch, M. et al. (2020). *J. Am. Chem. Soc.* 142 (33): 14267–14275.
159 Zhou, W., Neumann, P., Al Batal, M. et al. (2021). *ChemSusChem.* 14 (19): 4176–4180.
160 Gausas, L., Kristensen, S.K., Sun, H. et al. (2021). *JACS Au* 1 (4): 517–524.
161 Liu, X. and Werner, T. (2021). *Chem. Sci.* 12 (31): 10590–10597.
162 Zubar, V., Haedler, A.T., Schütte, M. et al. (2022). *ChemSusChem.* 15 (1): e202101606.
163 Gausas, L., Donslund, B.S., Kristensen, S.K., and Skrydstrup, T. (2022). *ChemSusChem.* 15 (1): e202101705.
164 Balaraman, E., Ben-David, Y., and Milstein, D. (2011). *Angew. Chem. Int. Ed.* 50 (49): 11702–11705.
165 Xie, Y., Hu, P., Ben-David, Y., and Milstein, D. (2019). *Angew. Chem. Int. Ed.* 58 (15): 5105–5109.
166 Das, U.K., Kumar, A., Ben-David, Y. et al. (2019). *J. Am. Chem. Soc.* 141 (33): 12962–12966.
167 Kothandaraman, J., Kar, S., Sen, R. et al. (2017). *J. Am. Chem. Soc.* 139 (7): 2549–2552.
168 vom Stein, T., Meuresch, M., Limper, D. et al. (2014). *J. Am. Chem. Soc.* 136 (38): 13217–13225.
169 Kumar, A. and Luk, J. (2021). *European J. Org. Chem.* 2021 (32): 4546–4550.
170 Jia, X., Qin, C., Friedberger, T. et al. (2016). *Sci. Adv.* 2: e1501591.
171 Goldman, A.S., Roy, A.H., Huang, Z. et al. (2006). *Science* 312 (5771): 257–261.
172 Corker, J., Lefebvre, F., Lécuyer, C. et al. (1996). *Science* 271 (5251): 966–969.
173 Lecuyer, C., Quignard, F., Choplin, A. et al. (1991). *Angew. Chem. Int. Ed.* 30 (12): 1660–1661.
174 Dufaud, V. and Basset, J. (1998). *Angew. Chem. Int. Ed.* 37 (6): 806–810.
175 Oya, S., Kanno, D., Watanabe, H. et al. (2015). *ChemSusChem.* 8 (15): 2472–2475.
176 Nakaji, Y., Nakagawa, Y., Tamura, M., and Tomishige, K. (2018). *Fuel Process. Technol.* 176: 249–257.
177 Celik, G., Kennedy, R.M., Hackler, R.A. et al. (2019). *ACS Cent. Sci.* 5 (11): 1795–1803.
178 Tennakoon, A., Wu, X., Paterson, A.L. et al. (2020). *Nat. Catal.* 3 (11): 893–901.
179 Jia, C., Xie, S., Zhang, W. et al. (2021). *Chem. Catal.* 1 (2): 437–455.
180 Wang, C., Xie, T., Kots, P.A. et al. (2021). *JACS Au* 1 (9): 1422–1434.
181 Kratish, Y. and Marks, T.J. (2022). *Angew. Chem. Int. Ed.* 61 (9): e202112576.
182 Yang, J., White, P.S., and Brookhart, M. (2008). *J. Am. Chem. Soc.* 130 (51): 17509–17518.
183 Feghali, E. and Cantat, T. (2015). *ChemSusChem.* 8 (6): 980–984.
184 Monsigny, L., Berthet, J.-C., and Cantat, T. (2018). *ACS Sustain. Chem. Eng.* 6 (8): 10481–10488.
185 Nunes, B.F.S., Oliveira, M.C., and Fernandes, A.C. (2020). *Green Chem.* 22 (8): 2419–2425.
186 Fernandes, A.C. (2021). *ChemSusChem.* 14 (19): 4228–4233.
187 Das, S., Addis, D., Zhou, S. et al. (2010). *J. Am. Chem. Soc.* 132 (6): 1770–1771.
188 Szewczyk, M., Magre, M., Zubar, V., and Rueping, M. (2019). *ACS Catal.* 9 (12): 11634–11639.
189 Cao, X., Wang, W., Lu, K. et al. (2020). *Dalt. Trans.* 49 (9): 2776–2780.
190 Kobylarski, M., Berthet, J.-C., and Cantat, T. (2022). *Chem. Commun.* 58 (17): 2830–2833.
191 Voss, R., Lee, R.P., and Fröhling, M. (2022). *Circ.Econ.Sust.* 2: 1369–1398.

17

Bioinspired Selective Catalytic C-H Oxygenation, Halogenation, and Azidation of Steroids

Konstantin P. Bryliakov

Zelinsky Institute of Organic Chemistry RAS, Leninsky Pr. 47, Moscow, Russian Federation

17.1 Introduction

Complex natural products have been coined a quintessential interdisciplinary nexus, with the the studies of bioactivity of natural products having resulted in a plethora of novel pharmaceuticals. Complex molecules of natural origin have been the cornerstone of studies for substrate diversification as structurally complex scaffolds, which is considered crucial for the rapid generation of chemical libraries of new biological targets and which provides tools to probe the pharmacological profile of these targets [1]. Furthermore, these efforts have provided a significant test-bed for the identification of site-selective catalysts, and have brought about the concept of LSF [2, 3], focused on achieving the facile and direct functionalization of structurally complex molecules through fewer synthetic steps. The latter suggests revolutionary strategies for the generation of chemical libraries and modulation of their key pharmacological properties (such as efficacy, bioavailability, and metabolism), without the need for developing or redesigning multistep synthetic methods [3–5].

Among chemical approaches to substrate diversification, selective and deliberate functionalization of non-activated $C(sp^3)$-H groups, has been a long sought after goal in organic chemistry [6, 7]. Drug-like molecules such as steroids possess a number of different 3º and 2º C-H bonds, as well as other functional groups, and therefore provide particularly challenging and fruitful platforms and testing areas for developing practical LSF methods and approaches [8, 9].

In addition to the extensively studied enzymatic oxidative functionalizations of complex molecules at C-H bonds [10–15], complementary synthetic approaches had emerged in the last two decades that are focused on exploiting earth-abundant transition metal based catalysts mimicking the reactivity of natural metalloenzymes [16–24]. Such *biomimetic* approaches do not require directing and/or protecting groups, allowing them to effectively modulate chemo-, regio-, and stereoselectivity of C-H functionalization by catalyst architecture and, in some cases, by applying specific or supramolecular catalyst-substrate recognition [25, 26], in line with the principle of biomimetic control of chemical selectivity [27]. These approaches potentially offer an effective solution for avoiding the major drawbacks of microbial processes, such as poor solubility of steroids in aqueous media and hence low volume yield of the process, low steroid conversions, and substrate/products toxicity to microbial cells. On the other hand, the state-of-the-art of bioinspired catalytic C-H oxygenations assumes the use of environmentally benign oxidant H_2O_2 as the terminal oxidant of choice, which is another step forward

Catalysis for a Sustainable Environment: Reactions, Processes and Applied Technologies Volume 2, First Edition. Edited by Armando J. L. Pombeiro, Manas Sutradhar, and Elisabete C. B. A. Alegria.
© 2024 John Wiley & Sons Ltd. Published 2024 by John Wiley & Sons Ltd.

compared with the stoichiometric C-H oxidations by dioxiranes or toxic metal oxides, or oxidative transformations using high-molecular-weight inorganic and organic oxidants [28–30].

Besides catalyzed oxygenations at C-H bonds, a variety of selective aliphatic C-H heterofunctionalizations (such as halogenations, azidations, and aminations) could be accomplished through the use of synthetic transition metal based catalysts [31–34]. However, the discussion of such processes is limited to bioinspired ones, sharing the common C-centered radical intermediates in cytochrome P450-like oxygen-rebound [35] or oxygen non-rebound [36] C-H oxidation mechanisms. Synthetic approaches relying on non-catalytic reactions, or involving hypervalent iodine reagents, or requiring directing groups, and others are beyond this scope.

The next section is dedicated to the selective C-H oxyfunctionalization reactions. First, fundamentals of the state-of-the-art mechanistic landscape of aliphatic C-H oxygenations are provided, followed by discussing particular catalyst systems, generally adhering to chronological order.

17.2 The Mechanistic Frame of Bioinspired Oxidative C-H Functionalizations

Typically, the mechanisms of C-H oxidation in the presence of transition metal based catalysts resemble the two-step oxidations mediated by enzymes of the CYP superfamily [35]. C-H activation at the first step occurs via hydrogen atom transfer (HAT) to the metal-oxido active species, generating a hydroxometal species and a carbon-centered radical in the solvent cage (Scheme 17.1). The fate of the radical may be different, depending on its stability under the conditions. If the radical is short-lived, it is rapidly trapped by the M-OH intermediate (oxygen rebound pathway [35]) to form an alcohol and metal species in lower oxidation state (Scheme 17.1a). In principle, there are no fundamental restrictions for the short-lived C-radical to be trapped by another ligand X in the 1st coordination sphere of the metal based catalyst without cage escape, which is considered an alternative rebound pathway (Scheme 17.1b) [37]. The sufficiently short lifetime of the radical is crucial for achieving stereoselective/stereospecific C-H oxygenation, such that epimerization does not take place between the H abstraction and rebound steps.

On the other hand, the lifetime of the carbon radical may be long enough to escape from the solvent cage into the bulk solution, which is considered "non-rebound" pathway(s) (Scheme 17.1c, d) [36]. In fact, this escape gives rise to further opportunities, including radical trapping by dissolved dioxygen (with formation of organic peroxides, Scheme 17.1d) or by externally added source of heteroatom(s) (Scheme 17.1c) to form corresponding products with novel C-X bonds. The latter opportunity is exploited in selective C-H halogenations and azidations; in this case, the reaction mixture is typically kept degassed to suppress undesired dioxygen capture.

When discussing the mechanistic scenarios shown in Scheme 17.1, several considerations should be kept in mind. First, the overall regioselectivity is determined by that of the first HAT step and hence governed by the regioselectivity of H abstraction as performed by the electrophilic metal-oxo species. Clearly, the latter depends on the nucleophilicity of the particular C-H groups and, considering electronic effects only, gives preference to more electron-rich $3°$ carbons over their $2°$ counterparts [38]. To surmount this natural substrate bias, one should either apply steric/chirality constraints on the catalyst's architecture, or introduce substrate recognition elements in the ligand scaffold [26, 39].

Moreover, in case of C-H oxidations at $2°$ carbons, the product of hydroxylation, *sec*-alcohol, has weaker and more electron rich C-H bond than the substrate itself. This entails its higher reactivity towards the breaking of the C-H bond by electrophilic oxidants, eventually leading to ketone formation (Scheme 17.1a). However, the selectivity for the desired *sec*-alcohol could be effectively improved by

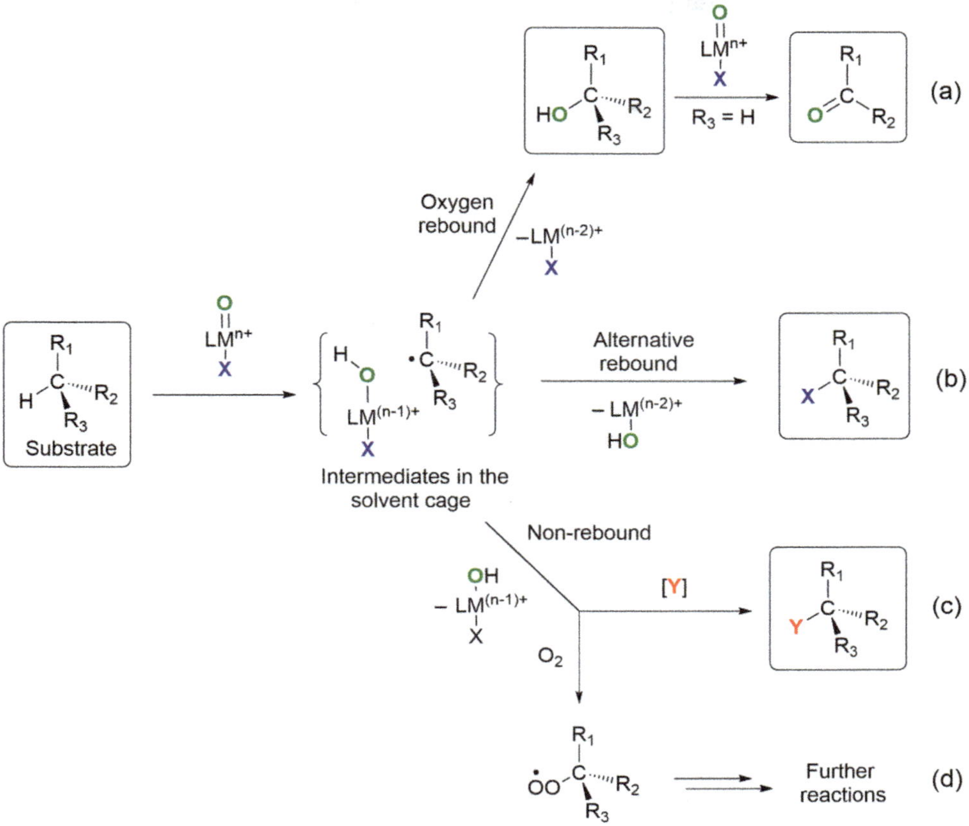

Scheme 17.1 Mechanistic alternatives for C-H oxidative functionalizations by metal-oxo species. L is a chelating polydentate ligand, X is a labile ligand, and [Y] stands for the source of heteroatom (or group).

suppressing overoxidation to ketone by virtue of the so-called polarity reversal approach that consists of conducting the reaction in the media of strong hydrogen-bond donor solvents [40, 41].

Another challenge is C-H functionalization stereoselectivity. Steroidal substrates of natural origin are chiral molecules, which suggests that the reactivity of the metal-oxo species toward stereochemically non-equivalent H atoms could be tuned by changing the catalyst's absolute chirality; actually, the catalyst's chirality can affect or guide the oxidation regioselectivity as well. Further, in the case of non-rebound pathways, carbon radical epimerization may take place, leading to erosion of the original stereochemistry. Significant loss of the original stereochemistry at the 3° carbons may be considered as an indication that trapping of the radical occurs outside the solvent cage [42]. In the case of complex substrates like steroids, however, the epimerization may be hampered in many cases by the overall conformational non-flexibility. It should be noticed that within the non-rebound reaction pathways, the radical lifetime should be long enough for escape from the solvent cage. In the case of metalloporphyrin based catalysts, the oxygen rebound step could be effectively facilitated or suppressed by varying the labile catalyst's ligand X (Scheme 17.1) [31].

The next section surveys historical examples (mostly reported before the mid-2010s) of transition metal mediated selective oxygenation of complex molecules of natural origin. Because mechanistic studies, in most cases, are performed on relatively simple model substrates, the following discussion that focuses mostly on steroids and terpenoids does not go deeply into mechanistic details.

17.3 Transition Metal Catalyzed C-H Oxyfunctionalizations: Early Examples

For the purposes of this contribution, catalyst systems based on, mostly, porphyrinic or related macrocyclic metal complexes, exploiting 2-electron oxidants such as iodosylbenzene (PhIO), m-chloroperoxybenzoic acid (mCPBA), 2,6-dichloropyridine-N-oxide (2,6-Cl$_2$PyNO), will be considered as early examples, even though this grading does not fully match the chronological order. The oxidants listed have long been known to convert porphyrinic (L)M^{n+} complexes into (L)Mn$^{(n+2)+}$=O ctive oxygen-transferring species [16, 17].

The first example of bioinspired selective remote C-H oxidation of steroids was reported by Grieco and Stuk in 1990 [43]. Upon adding iodosylbenzene to androstanyloxy manganese(III) tetraphenylporhyrins of type **1**, the authors managed to direct the oxidation at the C12, C14, or C17 positions, depending on the length of the tether linkage (Figure 17.1). An analogous approach allowed redirection of the oxidation to, preferentially, the C9 and C12 positions [44]. Subsequently, the same group studied manganese(III)salen-attached androstan-3α-ol complexes of the type **2** (Figure 17.1) that, depending on the bulkiness of the substituents, preferentially afforded either androstan-3α,14α-diol or androstan-3α,12α,14α-triol [45]. Strictly speaking, the oxidations reported by Grieco may be considered as quasi-catalytic at best, because the metal site performed only one (or perhaps two, in case of C17 ketonization) turnovers.

Truly catalytic intermolecular hydroxylation of steroidal substrates was first achieved by Breslow et al., who designed β-cyclodextrin attached Mn-porphyrin complex **3**, (Figure 17.1) that catalyzed the hydroxylation of protected androstane-3β,17β-diol **I** into, after deprotection, the corresponding 6,3β,17β-triol, with iodosylbenzene (5 equiv.) [46]. The system required 10 mol.% catalyst loadings; to ensure sufficient substrate solubility in water, a hydrophilic SO$_3$H-containing protecting group was used. The observed regioselectivity was achieved, presumably, owing to binding of properly engineered protecting groups into two *trans* cyclodextrin rings of **3**. Later, a broader set of substrates and porphyrin catalysts was examined, with, for example, the hexadecafluorinated analog of **3** allowing direction of the hydroxylation to some extent at the 15α position, or production of the tetrahydroxylated androstane derivative [47–49].

On the other hand, it was shown that, even in the absence of substrate recognition motifs (e.g. cyclodextrins), *tetrakis*(pentafluorophenyl)porphyrin ruthenium(III) catalyst **4** [(PFPP)RuI] (0.1 mol. %) can catalyze the oxidation of equilenin acetate **II** with PhIO to a mixture of 14α-alcohol and C11-ketone (Figure 17.1), with the catalyst performing total 570 turnovers [50]. Both C14 and C11 are "activated" benzylic positions, so the observed selectivity fully agrees with the innate substrate reactivity.

Iida et al. reported remote aliphatic C-H oxidation of a series of steroids (such as estrone acetate, cholestane and cholane derivatives, and 5β-bile acids) or terpenoids (penta- and tetracyclic triterpenoids), catalyzed by ruthenium [51–53] and osmium [54–56] 5,10,15,20-tetramesitylporphyrin (TMP) complexes, using (2,6-Cl$_2$PyNO), and tBuOOH, respectively, as terminal oxidants. Typically, a mixture of oxygenated regioisomers resulted from this reaction with the oxidation tending to occur at less sterically encumbered and more electron-rich 3° or benzylic positions. The catalyst loadings for Ru and Os porphyrins were 0.4 mol.% and 0.5 mol.%, respectively, and the excess of oxidant was 3-fold in the case of 2,6-Cl$_2$pyNO and 11–20-fold in the case of tBuOOH. Curiously, the authors discounted metal-oxo active species in the case of Os-catalyzed oxidations, but lacked solid experimental evidence. It is notable to mention the first example of γ-lactonization in the course of oxidation of methyl ester-peracetylated derivatives of (5β)-3-oxobile acids of type **III** (Figure 17.2), which was rationalized in terms of, presumably, HBr-facilitated direct acyl substitution upon interaction with the initially inserted OH groups at C20 [52].

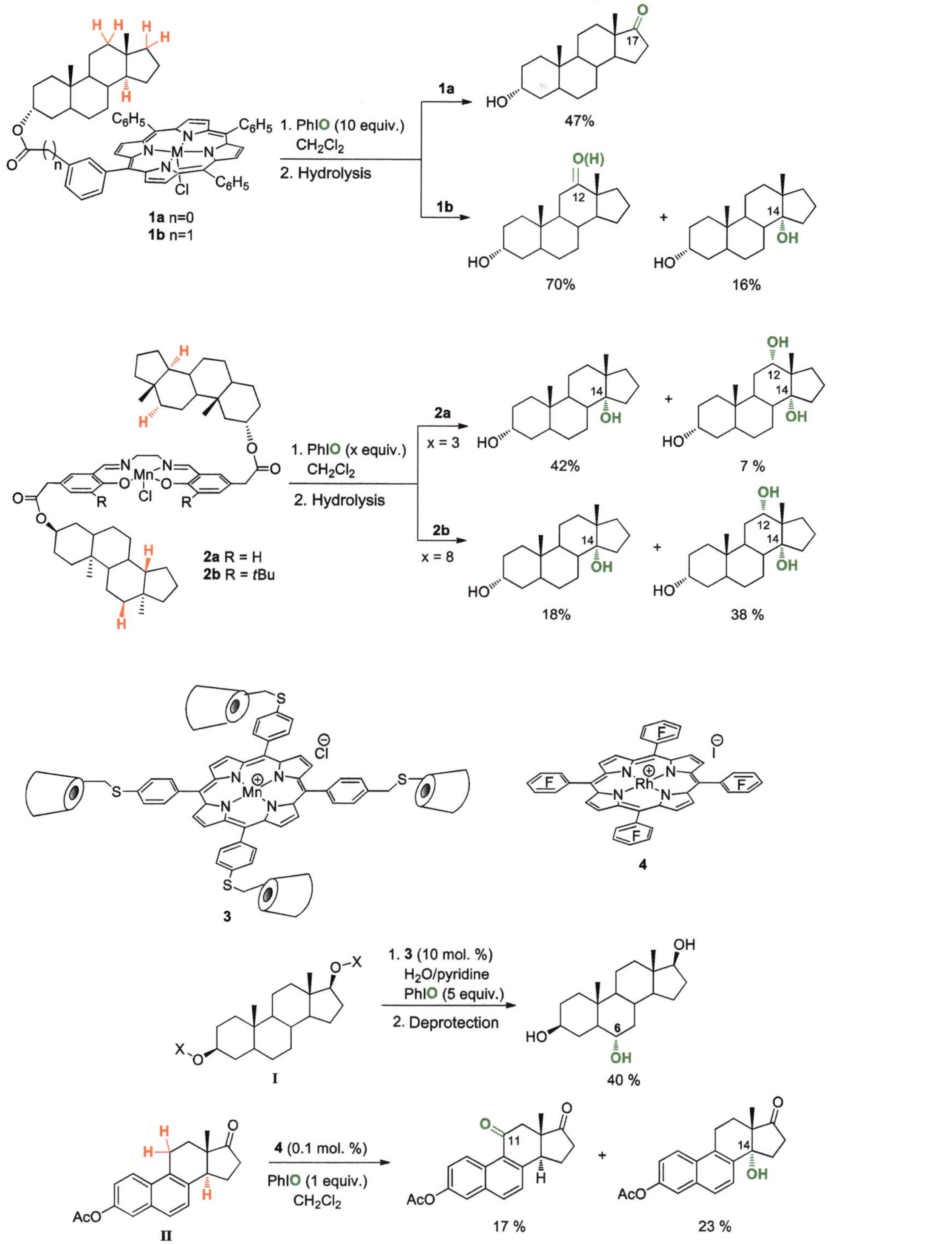

Figure 17.1 Metalloporphyrin-catalyzed oxidation of steroids with iodosylbenzene (PhIO; isolated yields given).

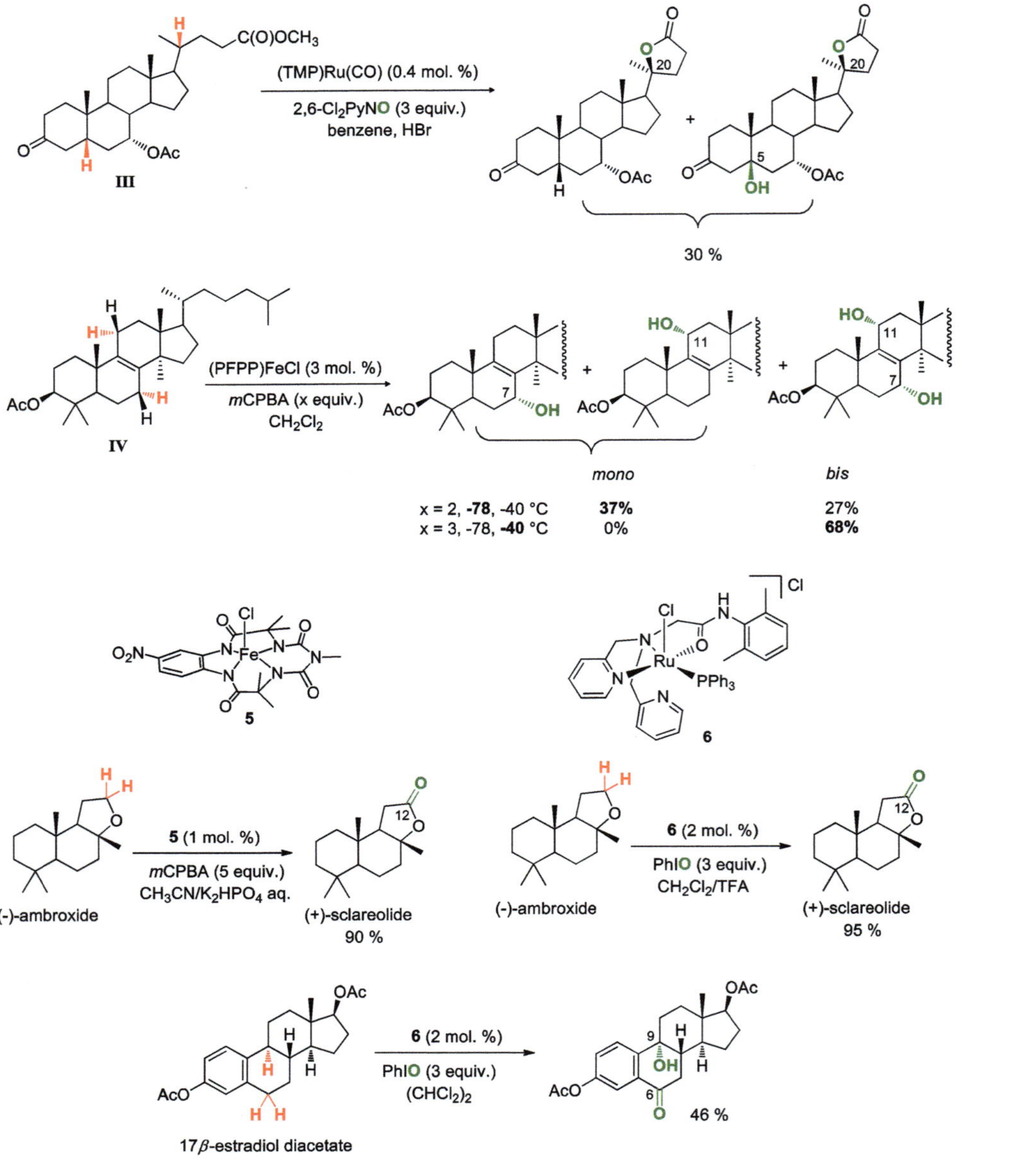

Figure 17.2 Metalloporphyrin-catalyzed oxidation of steroids with various oxidants. TFA = trifluoroacetic acid.

Konoike et al. developed an allylic hydroxylation protocol relying on iron(III) *tetrakis* (pentafluorophenyl)porphyrin chloride [(PFPP)FeCl] as a catalyst (3 mol.%), and *m*CPBA as an oxidant [57]. Depending on the conditions (excess of oxidant, temperature regime), dihydrolanosterol (**IV**) could be converted to *mono*- or *bis*-hydroxylated derivatives (Figure 17.2). The reaction was proposed to proceed via an oxygen rebound mechanism, driven by Fe-oxo porphyrin active species [57]. In another early example, Vijayarahavan and Chauhan reported the oxidation (hydroxylation and ketonization) of androstenedione at C19 with cumyl hydroperoxide in the presence of *tetrakis*(2,6-dichlorophenyl)porphinato iron(III) chloride [58]; however, the observed selectivity for methyl group, unusual for electrophylic metal-oxo species, may reflect a free radical driven rather than biomimetic metal based oxidation mechanism.

More recently, Gupta et al. reported a logical continuation of the previously-mentioned studies: using Fe(III) complex **5** with N_4-donor ligand ^{NO2}bTAML as the catalyst (1 mol.%) and *m*CPBA as the terminal oxidant, a diterpenoid of natural origin, (−)-ambroxide, was oxidized to (+)-sclareolide in high isolated yield (Figure 17.2) [59]. The reaction occurred exclusively at the activated C12 position, without any further oxidation even after adding 10 equiv. *m*CPBA. Subsequently, an electrocatalytic version of this synthetic protocol was reported, which demonstrated inferior yields [60]. Even higher (+)-sclareolide yield was reported for the oxidation with PhIO (3 equiv.) in the presence of 2 mol.% of [Ru(bpga)(Cl)(PPh$_3$)]Cl catalyst **6** (Figure 17.2) [61]: this value appears to be the highest ever reported for ambroxide oxidations (cf. [59, 60, 62–65] and Section 17.4). The same system allowed oxidizing 17β-estradiol diacetate selectively at the benzylic positions to the corresponding 9-hydroxo-6-keto derivative in moderate isolated yield (Figure 17.2).

These early examples of bioinspired oxidations primarily relied on synthetically elaborate transition metal porphyrin complexes, used practically unsuitable oxidants, and showed moderate predictability and tunability of the oxidation selectivity. However, these studies laid a fundamental and practical basis that was embodied in the creation of new generation of bioinspired catalysts, mostly focusing around nonheme (non-porhyrinic) complexes of earth-abundant metals such as Mn and Fe and described in the next section.

17.4 Transition Metal Catalyzed C-H Oxyfunctionalizations with H_2O_2

The pioneering contribution by White and Chen in 2007 demonstrated the possibility of performing the selective oxidations of aliphatic C-H groups with the "green" oxidant hydrogen peroxide in a predictably selective manner, governed primarily by electronic factors [66]. Subsequently, the authors reported the application of their nonheme iron based catalyst [(*R,R*)-PDPFe(CH$_3$CN)$_2$](SbF$_6$)$_2$ ((*R,R*)-**7**) for the selective oxidation of complex substrates such as (−)-ambroxide (Figure 17.3) [67]. The system required the presence of acetic acid as a co-catalytic additive, which facilitated the heterolytic O-O bond cleavage in the initially formed LFe(III)-OOH intermediate to generate the active perferryl species (Scheme 17.2); this is actually the so-called carboxylic acid assisted pathway of heterolytic H_2O_2 activation, archetypical of nonheme iron and manganese catalytically active complexes [26, 68–70]. The oxidation of (−)-ambroxide first occurred selectively at the C12 methylenic group, "activated" by the presence of ethereal oxygen; however, the resulting (+)-sclareolide could be involved in further oxidation by the same system at the most sterically accessible C2 position, to yield (+)-2-oxo sclareolide (Figure 17.3).

The general reactivity rule as described, with the electron-rich C-H groups oxidizing first, followed by the less electron-rich but sterically more accessible one, could not be escaped until the authors introduced bulky 2,6-bis(trifluoromethyl) groups at the PDP ligand that restricted the substrate approach to the metal center in the resultant complex **8** (Figure 17.3) [71]. This modification

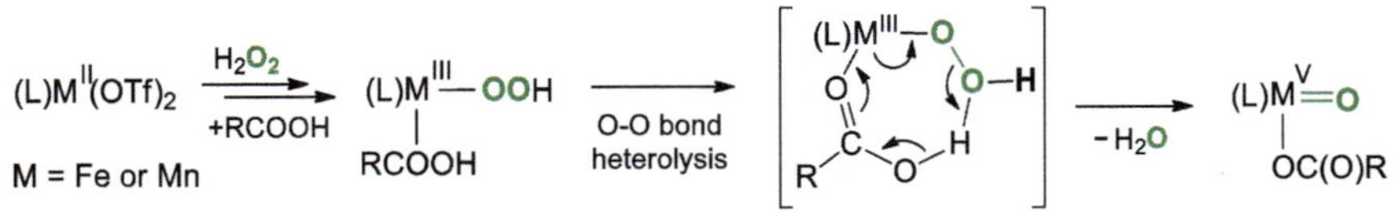

Scheme 17.2 Carboxylic-acid assisted formation of reactive metal-oxo species.

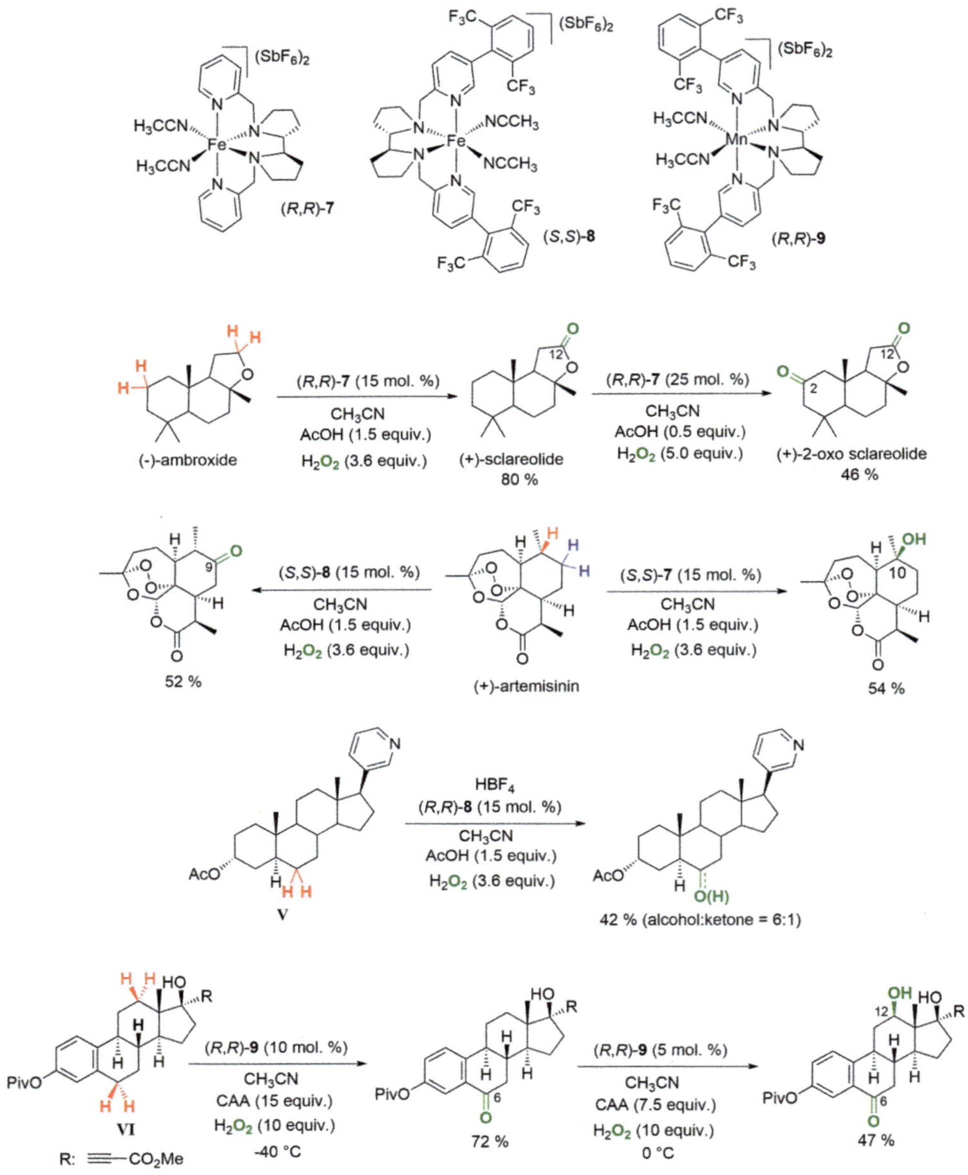

Figure 17.3 Selective oxidation of complex substrates with H_2O_2 in the presence of nonheme iron complexes. (CAA = chloroacetic acid).

allowed manipulation of oxidation selectivity, overriding the natural substrate bias. For example, while in the presence of the parent catalyst (S,S)-**7**, sesquiterpene lactone (+)-artemisinin was preferentially hydroxylated at the C10 methine carbon, replacing the catalyst with (S,S)-**8** diverted the selectivity towards ketonization at C9 (Figure 17.3). Oxidation of abiraterone acetate analogue **V** in the presence of (R,R)-**8** was accomplished preferentially at the C6 methylenic group (Figure 17.3); in the latter case, HBF_4 was added to protect the pyridinic nitrogen by virtue of protonation [72]. In a similar manner, (R,R)-**9**, the Mn counterpart of the iron prototype (R,R)-**8**, enabled the oxidation of remote 2° groups in ethynylestradiol derivative **VI** (Figure 17.3), leaving more nucleophilic but sterically less accessible 3° sites intact [73]. In a sequential oxidation protocol, selective ketonization at C6 and further C12β-hydroxylation was achieved [73]. Crucially, the Mn based catalyst operated at lower loadings (5–10 mol.%) compared to the Fe analogue in the presence of a chloroacetic acid additive.

Klein Gebbink et al. reported the oxidation of (−)-ambroxide in acetonitrile in the presence of a mixture of stereoisomeric iron complexes called mix-**10** (Figure 17.4), which yielded a mixture of (+)-sclareolide and the hydroxy acid [74]. In turn, (+)-sclareolide could be involved in further reaction, affording a mixture of products of ketonization at C1, C2, and C3 of ring A (Figure 17.4).

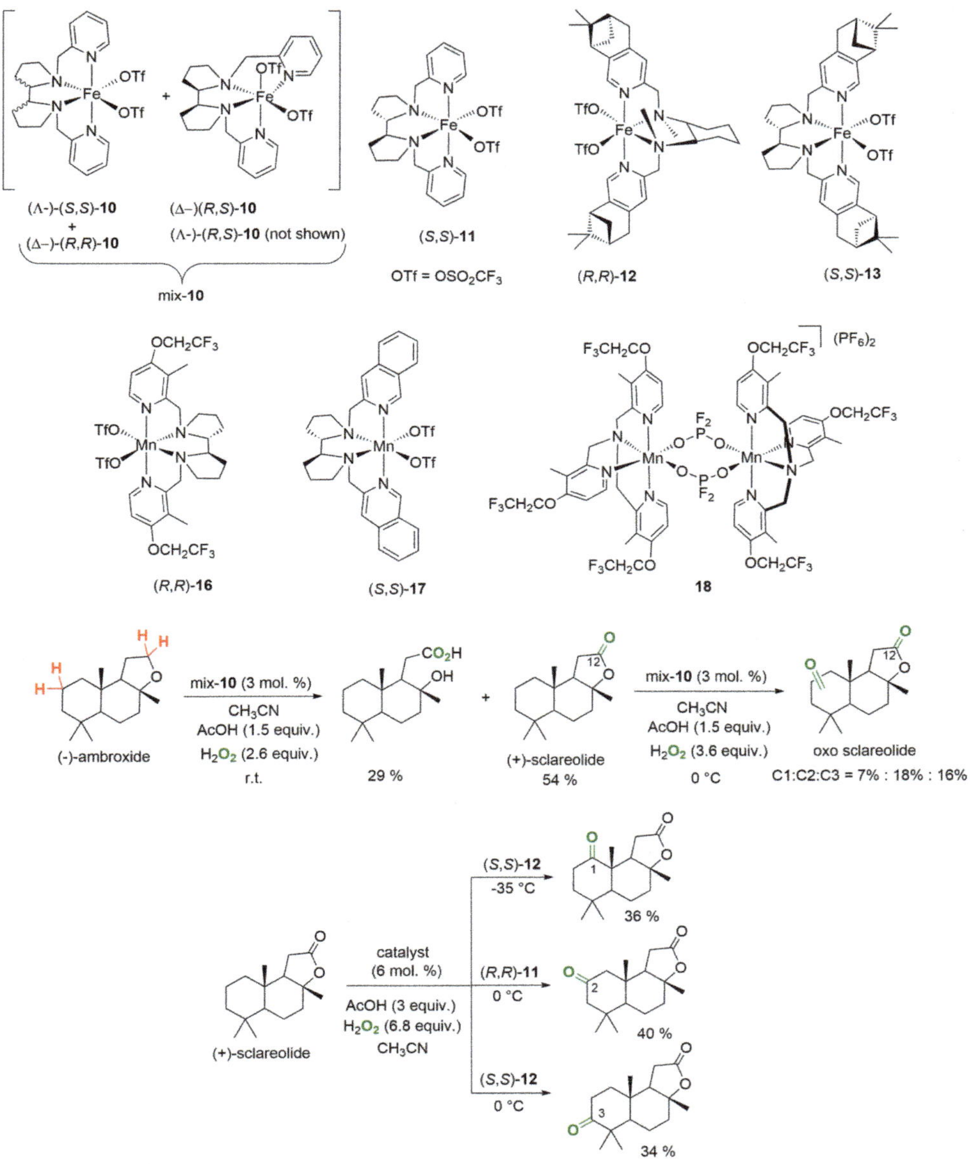

Figure 17.4 Chemo-, regio- and stereoselective oxidation of (-)-ambroxide and its derivatives with H_2O_2, and structures of nonheme catalysts used for this purpose.

Costas et al. scurinized the combined effect of ligand architecture (including chirality) on the selectivity of oxidation of complex substrates in the presence of Fe(PDP) catalysts of the type **11**, **12**, and **13** (Figure 17.4) [75–77]. In agreement with the results of White (mentioned previously), increased steric hindrance at the iron site (complexes **12** and **13** vs. **11**) was found to divert the oxidation selectivity toward more sterically accessible methylenic groups, hampering the oxidation of more nucleophilic 3° groups [75–77]. Interestingly, by varying the reaction temperature and ligand structure, (+)-sclareolide could be preferentially ketonized at that C1, C2, or C3 sites of ring A. Complexes of types **12** and **13** showed higher catalytic efficiency compared with complexes **7** and **8** (Figure 17.3), which was also explained in terms of steric hindrance of the central atom.

A few years later, the authors developed a series of catalysts based on iron complexes of types **14** and **15**, with *bis*-amino-*bis*-pyridylmethyl ligands bearing bulky *tris*(isopropylsilyl) substituents, that exhibited enhanced selectivity toward methylenic groups of *trans*-androsterone acetate **VII** [78]. Moreover, the sign of the catalyst chirality (Δ or Λ) was the crucial factor for switching between preferential oxidation at the C12 or C6 atoms, respectively (Figure 17.5). An analogous selectivity trend was documented for the oxidation of *cis*-androsterone, yet in somewhat lower yields [78].

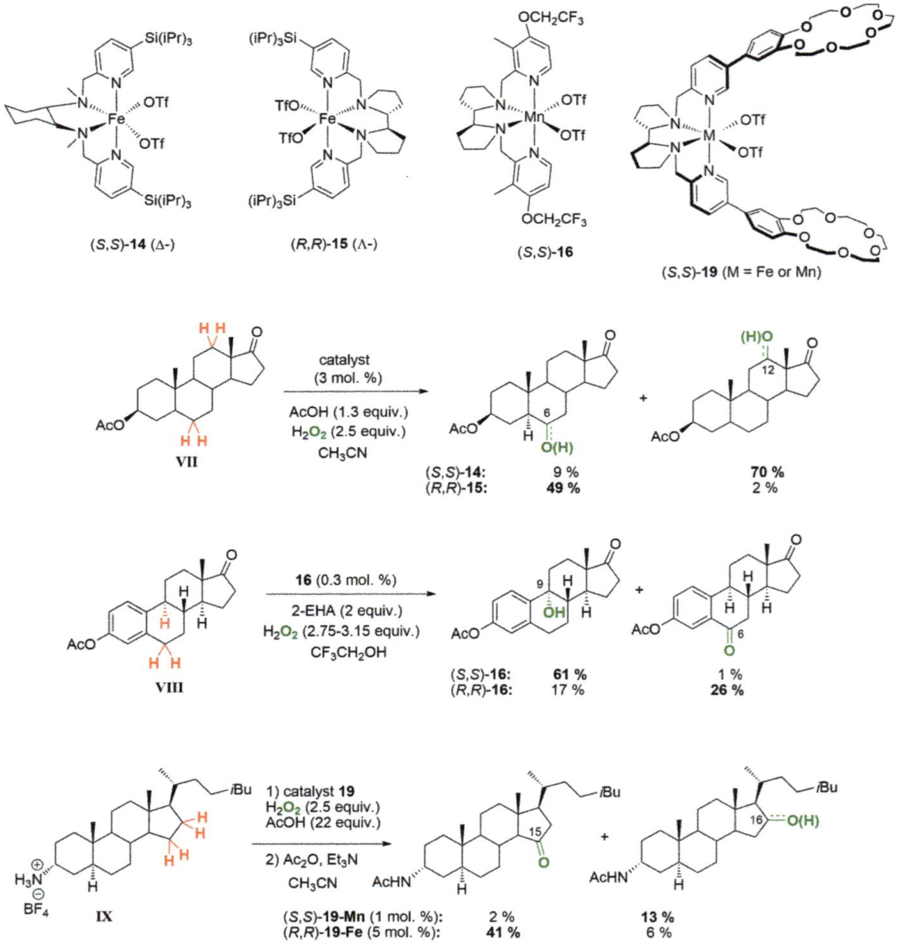

Figure 17.5 Effect of ligand architecture, chirality, or supramolecular recognition on the regioselectivity of C-H oxygenations with H_2O_2. 2-EHA = 2-ethylhexanoic acid.

Bryliakov et al. reported another example of a catalyst chirality effect on the regioselectivity of estrone acetate (**VIII**) oxidation. In the presence of Mn complexes of type **16**, the C6 or C9 positions could be oxidized preferentially in the presence of 2-ethylhexanoic acid as an additive (Figure 17.5) [41]. Mn based complexes showed excellent efficiency, such that 0.3 mol. % catalyst loadings were sufficient, and using 2,2,2-trifluoroethanol as reaction solvent instead of CH_3CN was found to cause beneficial effect on the keto-C6/hydroxo-C9 ratio [41].

Costas et al. achieved the elusive oxygenation at the D ring of cholestane derivative **IX** owing to supramolecular binding (via the protonated amino group) with the active species derived from catalysts of type **19**, bearing benzocrown moieties (Figure 17.5) [79]. Interestingly, by simultaneous change of the ligand chirality and the metal (Fe or Mn), the oxygenation could be directed toward the C15 or C16 methylene (Figure 17.5). A similar switch of C15/C16 selectivity was demonstrated for a couple of other androstane derivatives [79].

The combined effect of catalyst architecture and the reaction solvent allowed the authors to switch the selectivity of (−)-ambroxide oxidation between C12-ketonization (biosynthetic-like oxidation mode) and C2α-hydroxylation (bioorthogonal oxidation mode), with complexes **17** and **18** (Figure 17.4), in combination with chloroacetic acid additive, showing the best performance (in terms of yield and selectivity) [80]. In the medium of CH_3CN/CH_2Cl_2 (with minor additives of β-fluorinated alcohols), the oxidation was governed mostly by electronic factors with (−)-ambroxide preferentially converting to (+)-sclareolide (Figure 17.6). This effect was most pronounced in

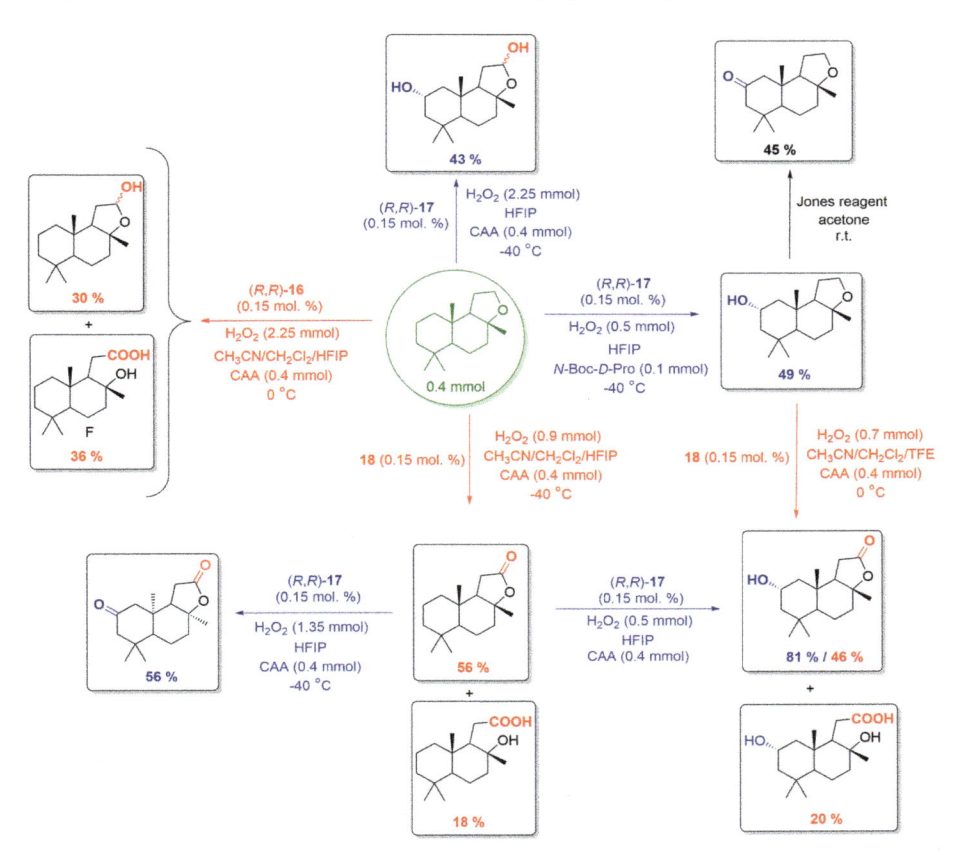

Figure 17.6 Synthetic approaches to oxidized metabolites of (−)-ambroxide. HFIP = hexafluoroisopropanol, TFE = trifluoroethanol, CAA = chloroacetic acid, N-Boc-D-Pro = N-Boc protected D-proline. Reprinted with permission from Ref [80]. Reproduced with permission from Ref [81] Ottenbacher et.al., 2021 / Elsevier

the case of the least sterically encumbered catalytically active sites stemming from catalyst **18**. In contrast, using strongly hydrogen bond donating hexafluoroisopropanol (HFIP) as the reaction solvent effectively suppressed the oxidation at the ethereal C12 methylenic group, as well as ketonization of the initially formed C2α-hydroxo ambroxide (cf. Ref. [81]). Applying the biosynthetic-like and bioorthogonal oxidation protocols, a number of doubly oxygenated ambroxide derivatives were obtained (Figure 17.6).

Hydroxylation at C2 occurred stereoselectively with generation of 2α- (equatorial) C-OH groups, which was rationalized in terms of a release of 1,3-diaxial strain in the transition state for C-H bond cleavage, determined by the presence of two axial methyl groups. Talsi et al. studied the oxidation of (+)-sclareolide with H_2O_2 in the presence of nonheme iron complexes of the types **11**, **20**, **21** (Figure 17.7), and revealed that reducing the electrophilicity of the catalyst enhanced the stereodiscrimination between 2α- and 2β-hydroxylation: for example, the ΔE_a for the k^{ax}/k^{eq} ratio in the case of **21** was about 1.5 kJ/mol higher than for **20** [82, 83].

Overall, bioinspired catalyst systems for the selective oxygenation of steroids and terpenoids have demonstrated remarkable progress over the last decade, providing various novel practical recipes to tuning the oxidation efficiency and selectivity. Some of these approaches involve changing the central (metal) atom, tuning the ligand sterics and switching its chirality, replacing the carboxylic acid additive and the reaction solvent, or exploiting substrate-catalyst binding/recognition elements. Examining the generality of such practical approaches to modifications of various bioactive molecules of natural origin may be a potentially highly rewarding future research direction.

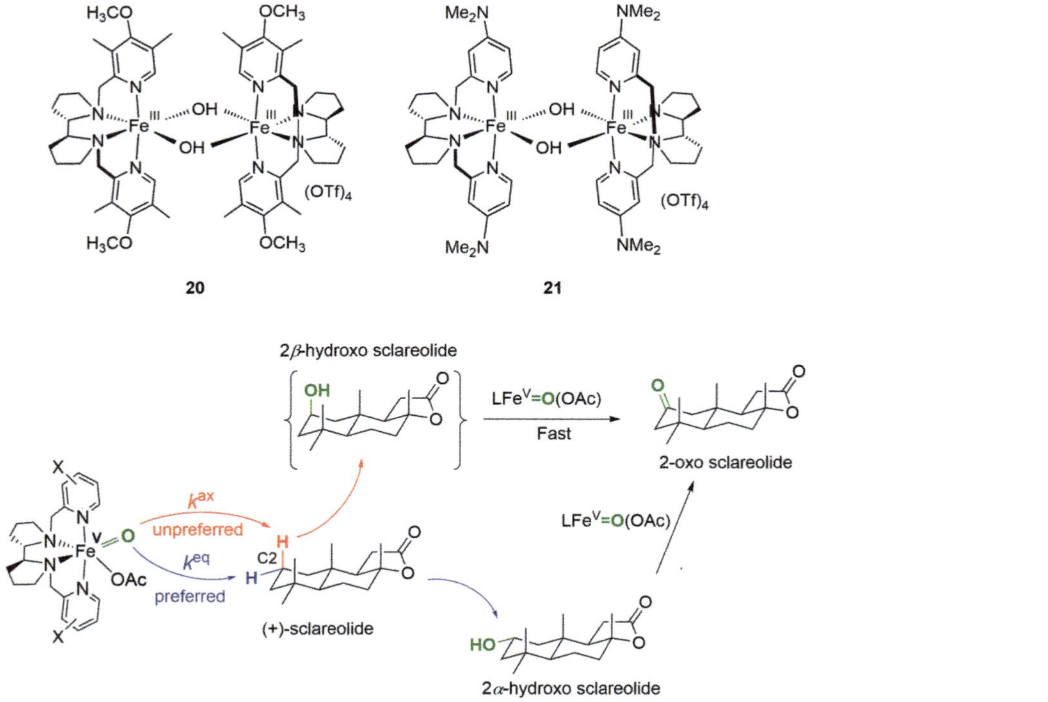

Figure 17.7 Illustration of the stereochemistry of (+)-sclareolide hydroxylation at C2 in the presence of iron based catalysts.

17.5 Transition Metal Catalyzed Halogenations and Azidations

In 2010, Liu and Groves reported that the oxidation of C-H groups in alkanes with sodium hypochlorite in the presence of manganese tetraphenylporphyrin complex (TPP)MnCl **22** resulted in formation of alkyl chlorides with high selectivity (Figure 17.8) [84]. This was the first example of regioselective catalytic chlorination of non-activated methylenic groups without the need for internal directing groups. Reaction of 5α-cholestane **X** yielded only products of chlorination at C2 and C3 methylenes of the A ring, the least sterically hindered positions, with preference for α-chlorides, with (+)-sclareolide preferentially converting to the corresponding 2α-chloride (Figure 17.8). The reaction was conducted under nitrogen atmosphere in biphasic medium, with tetrabutylammonium chloride as a phase transfer catalyst (PTC); the role of NaOCl as the source of Cl in the product was firmly established, because replacing NaOCl with NaOBr resulted in selective bromination instead of chlorination [84].

The authors revealed that the first step was hydrogen atom abstraction by the initially formed (TPP)Mn(V)=O(L) complex (Figure 17.8). The key issue was that at the next step, instead of oxygen rebound, the reaction followed non-rebound pathway (Scheme 17.1), which ended up with trapping the C-centered radical by (HO)(TMP)Mn^{IV}-OCl complex, yielding the chloride [31]. This hypothesis was supported by the observed radical rearrangement of radical-clock substrate norcarane, accompanying its chlorination [84, 85]. Decreasing the rate of the oxygen rebound step (such that it became slower than cage escape) was achieved due to strong donor axial ligand, such as HO$^-$ 31] or F$^-$ [86].

Subsequently, selective C-H fluorination was reported, using (5,10,15,20-tetramesitylporphyrin) manganese (TMP)MnCl complex **23** as a catalyst, an excess of iodosylbenzene as an oxidant, and AgF as a fluoride source (Figure 17.8) [86]. Fluorination of (+)-sclareolide and 5α-androstane-17-one (**XI**), predictably (because it is determined, mostly, by the first, H abstraction, step), showed regio- and stereoselectivity similar to those reported for chlorination of these and similar substrates [84, 86]. The authors proposed the catalytic cycle for the fluorination mechanism (Figure 17.8), taking into account that, according to DFT modeling, replacing axial HO with F should further lower the barrier for F rebound by ca. 3 kcal/mol [86]. A similar selective C-H fluorination protocol was subsequently developed for the selective introduction of ^{18}F label for positron emission tomography imaging, based on (Salen)MnCl-type catalyst [87]; in the latter case, however, the reaction occurred selectively at benzylic C-H groups. On the other hand, Mn-porphyrinic catalyst **23** was recently found to mediate selective 16β-fluorination in progesterone **XII** (Figure 17.8) and 5α-dihydrotestosterone by cyclotrone-produced [^{18}F]fluoride [88].

Remarkably, Groves'd=s methods of selective C-H halogenation as described allowed the development of a complementary manganese-catalyzed C-H azidation reaction, relying, in the case of non-activated methylenic groups, on the (TMP)MnCl catalyst **23** [89]. Replacing AgF with NaN$_3$ resulted in trapping the azide (instead of F), thus giving access to, for example, 2α-N$_3$-sclareolide and 9-N$_3$-estrone acetate (Figure 17.9). A plausible reaction mechanism, conceptually similar to that for the fluorination reaction, was proposed [89].

The advances reported for manganese-porphyrin catalysts, with such environmentally undesirable oxidants as sodium hypochlorite and iodosylbenzene, encourage the search for more efficient (e.g. nonheme, metal based) catalyst systems that so far have not been achieved (see [90] and references therein), as well as for greener oxidants. To date, we know very little about the possibilities to tune the regioselectivity of such functionalizations, but logic suggests that essentially planar porphyrin ligands may provide few opportunities for manipulating the regioselectivity through the (ligand-controlled) H abstraction step. Another challenge remaining is securing high stereoselectivity while conducting the desired C-H functionalization, which, because of the possibility of

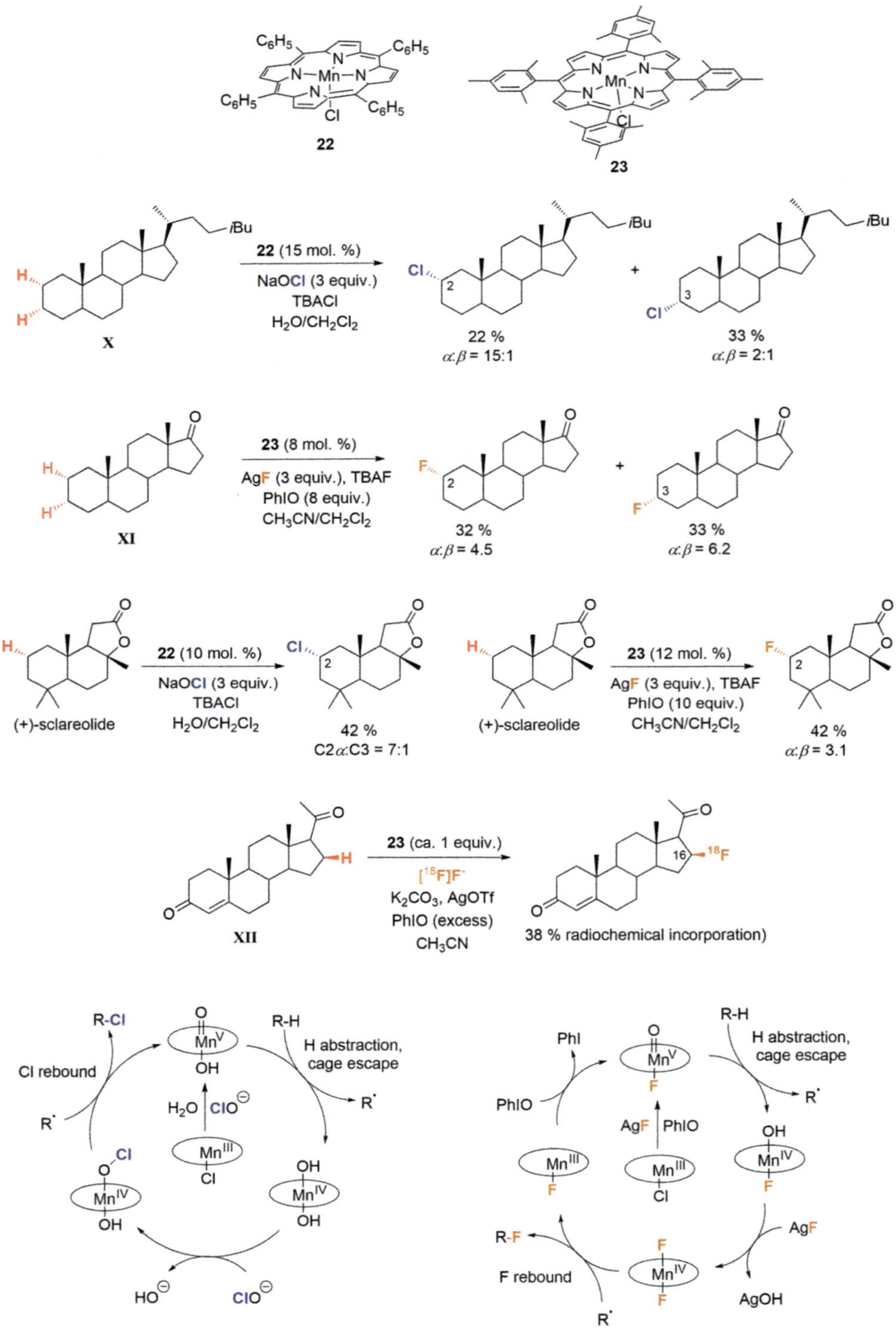

Figure 17.8 Metalloporphyrin-catalyzed halogenation, structures of catalysts, and proposed mechanisms. TBACl, TBABr, TBAF = tetrabutylammonium chloride, bromide and fluoride, respectively.

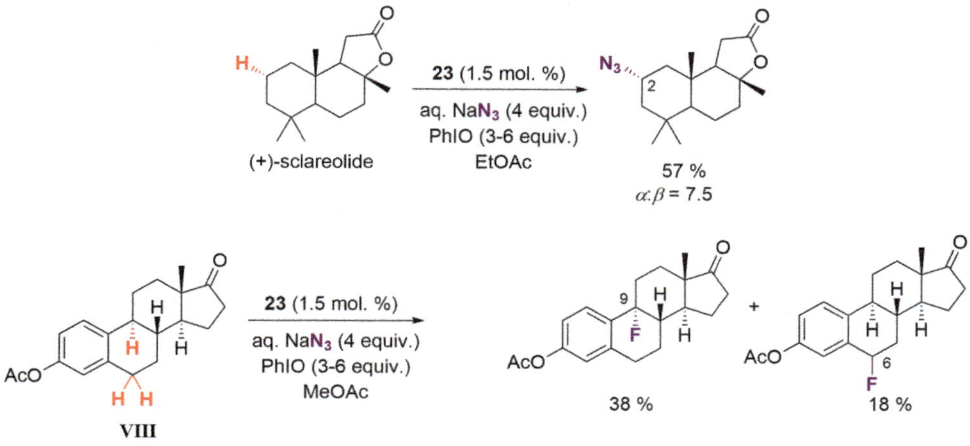

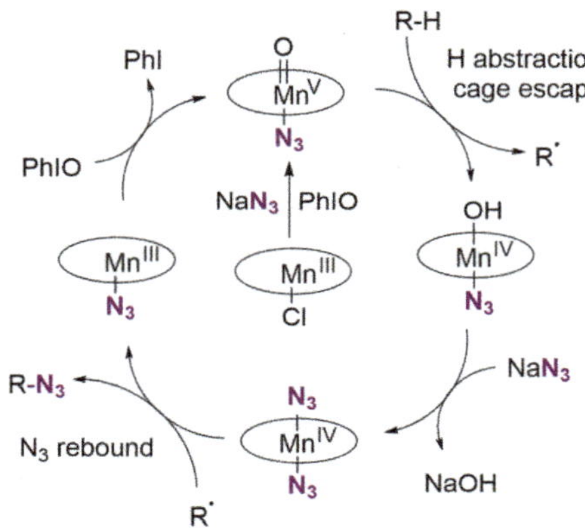

Figure 17.9 Proposed mechanism for metalloporphyrin-catalyzed azidations.

C-radical epimerization after cage escape (Scheme 17.1), would possibly require designing sophisticated ligands with so far unimaginable architectures. And, of course, it would be highly desirable to broaden the scope of heteroatoms that could be used for bioinspired C-H heterofunctionalization of compex substrates.

17.6 Conclusions and Outlook

Developing bioinspired approaches to selective oxidative functionalization of non-activated C(sp^3)-H groups has demonstrated great progress in the last decade [19–26, 31], with drug-like molecules, such as steroids and terpenoids, possessing a number of different 3° and 2° C-H bonds that provide fruitful platforms and testing areas for developing practical LSF methods and approaches [8, 9].

Conceptually, the catalytic processes discussed in this chapter share the same mechanism of the first, C-H activation, step (via direct H abstraction by electrophilic metal-oxo species), just as natural metalloenzymes do [12, 14, 35]. At the same time, further scenarios for the oxygenations and halogenations/azidations are different. In the former case, H abstraction is followed by oxygen rebound step (Scheme 17.1), whereas, in the latter case, there should be cage escape with subsequent halogen or azide rebound, which leads to the desired heterofunctionalization products.

To date, various approaches to chemo-, regio-, stereoselective oxygenation of steroids and terpenoids have been proposed, relying on environmentally benign oxidant H_2O_2 and highly efficient nonheme Fe and Mn complexes capable of performing up to several hundreds of turnovers. By changing the central atom, the ligand sterics and chirality, additive(s), and reaction solvent, or by exploiting substrate-catalyst binding/recognition elements, the reaction selectivity can be diverted toward the desired C-H group in some cases, effectively overriding the natural substrate bias. In fact, at the moment this area could be regarded as expecting a switch from exploratory research to practical implementation, which could potentially make valuable contributions to developing novel synthetic methods of late-stage direct oxidative functionalization of structurally complex molecules of natural origin.

Selective C-H halogenations and azidations have so far been less elaborate, exploiting sophisticated metalloporphyrins as catalysts and environmentally undesirable high-molecular-weight oxidants. At the same time, even at the state-of-the-art level, some degree of variability in the functionalization selectivity has been demonstrated. Though relatively underdeveloped, this area has demonstrated practical promise for applications such as aliphatic [^{18}F]radiofluorination of complex organic molecules for positron emission tomography imaging [88]. Highly challenging seems to be designing catalyst systems, capable of using H_2O_2 as oxidant for the selective C-H halogenations and azidations, with the C-X bond forming step occurring in the first coordination sphere of the metal center (without cage escape): this would prevent the C-radical epimerization and the corresponding loss of stereochemistry.

Further directions of progress in this area may be envisioned as examining the generality of such approaches to modifications of various bioactive molecules of natural origin/broadening the substrate scope; designing more selective, efficient and environmentally benign catalyst systems; searching for novel approaches to tuning the oxidation selectivity, and others. Developing selective aerobic C-H oxidations have so far faced fundamental obstacles (owing to absence of heterolytic dioxygen activation mechanism on synthetic bioinspired catalysts); however, it would be extremely tempting to elaborate effective electro- [60] or photocatalytic protocols for such oxidations.

Overall, this rapidly developing area holds great practical promise for engineering novel synthetic approaches for the direct alteration of natural products through C-H heterofunctionalization, which could find use in today's drug discovery, streamlining late-stage diversification of pharmaceutically relevant molecules, facilitating rapid generation and identification of metabolites and selectively labeled radiotracers, as well as other natural products of biomedical relevance.

Acknowldegment

This work was part of the project supported by the Russian Science Foundation, #20–13–00032.

References

1 Chambers, R.K., Zaho, J., Delaney, C.P., and White, M.C. (2020). *Adv. Synth. Catal.* 362: 417–423. doi: 10.1002/adsc.201901472.
2 Cernak, T., Dykstra, K.D., Tyagarajan, S. et al. (2016). *Chem. Soc. Rev.* 45: 546–576. doi: 10.1039/C5CS00628G.
3 Moir, M., Danon, J.J., Reekie, T.A., and Kassiou, M. (2019). *Expert Opin. Drug Discov.* 14: 1137–1149. doi: 10.1080/17460441.2019.1653850.
4 Shugrue, C.R. and Miller, S.J. (2017). *Chem. Rev.* 117: 11894–11951. doi: 10.1021/acs.chemrev.7b00022.

5 Hong, B., Luo, T., and Lei, X. (2020). *ACS Cent. Sci.* 6: 622–635. doi: 10.1021/acscentsci.9b00916.
6 Arndtsen, B.A., Bergman, R.G., Mobley, T.A., and Peterson, T.H. (1995). *Acc. Chem. Res.* 28: 154–162. doi: 10.1021/ar00051a009.
7 Gutekunst, W.R. and Baran, P.S. (2011). *Chem. Soc. Rev.* 40: 1976–1991. doi: 10.1039/C0CS00182A.
8 Hartwig, J.F. and Larsen, M.A. (2016). *ACS Cent. Sci.* 2: 281–292. doi: 10.1021/acscentsci.6b00032.
9 Karimov, R. and Hartwig, J.F. (2018). *Angew. Chem. Int. Ed.* 57: 4234–4241. doi: 10.1002/anie.201710330.
10 Pellisier, H. and Santelli, M. (2001). *Org. Prep. Proced. Int.* 33: 1–58. doi: 10.1080/00304940109356574.
11 Lewis, J.C., Coelho, P.S., and Arnold, F.H. (2010). *Chem. Soc. Rev.* 40: 2003–2021. doi: 10.1039/c0cs00067a.
12 Kille, S., Zilly, F.E., Acevedo, J.P., and Reetz, M.T. (2011). *Nat. Chem.* 3: 738–743. doi: 10.1038/NCHEM.1113.
13 Donova, M.V. and Egorova, O.V. (2012). *Appl. Microbiol. Biotechnol.* 94: 1423–1447. doi: 10.1007/s00253-012-4078-0.
14 Fessner, N.D. (2019). *ChemCatChem*. 11: 2226–2242. doi: 10.1002/cctc.201801829.
15 Chakrabarty, S., Wang, Y., Perkins, J.C., and Narayan, A.R.H. (2020). *Chem. Soc. Rev.* 49: 8137–8155. doi: 10.1039/d0cs00440e.
16 Che, C.-M. and Huang, J.-S. (2009). *Chem. Comm.* 3996–4015. doi: 10.1039/b901221d.
17 Che, C.-M., Lo, V.K.-Y., Zhou, C.-Y., and Huang, J.-S. (2011). *Chem. Soc. Rev.* 40: 1950–1975. doi: 10.1039/C0CS00142B.
18 McMurray, L., O'Hara, F., and Gaunt, M.J. (2011). *Chem. Soc. Rev.* 40: 1885–1898. doi: 10.1039/c1cs15013h.
19 Talsi, E.P. and Bryliakov, K.P. (2012). *Coord. Chem. Rev.* 256: 1418–1434. doi: 10.1016/j.ccr.2012.04.005.
20 Cusack, K.P., Koolman, H.F., Lange, U.E.W. et al. (2013). *Bioorg. Med. Chem. Lett.* 23: 5471–5483. doi: 10.1016/j.bmcl.2013.08.003.
21 Hartwig, J.F. (2016). *J. Am. Chem. Soc.* 138: 2–24. doi: 10.1021/jacs.5b08707.
22 Qiu, Y. and Gao, S. (2016). *Nat. Prod. Rep.* 33: 562–581. doi: 10.1039/c5np00122f.
23 Liu, Y., You, T., Wang, H.-X. et al. (2020). *Chem. Soc. Rev.* 49: 5310–5358. doi: 10.1039/d0cs00340a.
24 Doiuchi, D. and Uchida, T. (2021). *Synthesis* 53: 3235–3246. doi: 10.1055/a-1525-4335.
25 Vicens, L., Olivo, G., and Costas, M. (2020). *ACS Catal.* 15: 8611–8631. doi: 10.1021/acscatal.0c02073.
26 Costas, M. (2021). *Chem. Rec.* 21: 4000–4014. doi: 10.1002/tcr.202100227.
27 Breslow, R. (1980). *Acc. Chem. Res.* 13: 170–177. doi: 10.1021/ar50150a002.
28 Salvador, J.A.R., Silvestre, S.M., and Moreira, V.M. (2006). *Curr. Org. Chem.* 10: 2227–2257. doi: 10.2174/138527206778742641.
29 Salvador, J.A.R., Silvestre, S.M., and Moreira, V.M. (2012). *Curr. Org. Chem.* 16: 1243–1276. doi: 10.2174/138527212800564204.
30 Junrong, H., Min, Y., Chuan, D. et al. (2021). *Front. Chem.* 9: 737530. doi: 10.3389/fchem.2021.737530.
31 Liu, W. and Groves, J.T. (2015). *Acc. Chem. Res.* 48: 1727–1735. doi: 10.1021/acs.accounts.5b00062.
32 Karimov, R.R., Sharma, A., and Hartwig, J.F. (2016). *ACS Cent. Sci.* 2: 715–724. doi: 10.1021/acscentsci.6b00214.
33 Park, Y., Kim, Y., and Chang, S. (2017). *Chem. Rev.* 117: 9247–9301. doi: 10.1021/acs.chemrev.6b00644.

34 Sala, R., Loro, C., Foschi, F., and Broggini, G. (2020). *Catalysts* 10: 1173. doi: 10.3390/catal10101173.
35 Huang, X. and Groves, J.T. (2017). *J. Biol. Inorg. Chem.* 22: 185–207. doi: 10.1007/s00775-016-1414-3.
36 Cho, K.B., Hirao, H., Shaik, S., and Nam, W. (2016). *Chem. Soc. Rev.* 45: 1197–1210. doi: 10.1039/C5CS00566C.
37 Ottenbacher, R.V., Bryliakova, A.A., Shashkov, M.V. et al. (2021). *ACS Catal.* 11: 5517–5524. doi: 10.1021/acscatal.1c00811.
38 Bryliakov, K.P. (2021). Manganese-catalyzed C-H oxygenation reactions. In: *Manganese Catalysis in Organic Synthesis* (ed. J.B. Sortais), 183–202. Wiley-VCH GmbH. doi: 10.1002/9783527826131.ch6.
39 Vidal, D., Olivo, G., and Costas, M. (2018). *Chem. Eur. J.* 24: 5042–5054. doi: 10.1021/10.1002/chem.201704852.
40 Dantignana, V., Milan, M., Cussó, O. et al. (2017). *ACS Cent. Sci.* 3: 1350–1358. doi: 10.1021/acscentsci.7b00532.
41 Ottenbacher, R.V., Talsi, E.P., Rybalova, T.V., and Bryliakov, K.P. (2018). *ChemCatChem.* 10: 5323–5330. doi: 10.1002/cctc.201801476.
42 Chen, K. and Que, L., Jr. (2001). *J. Am. Chem. Soc.* 123: 6327–6337. doi: 10.1021/ja010310x.
43 Grieco, P.A. and Stuk, T.L. (1990). *J. Am. Chem. Soc.* 112: 7799–7801. doi: 10.1021/ja00177a052.
44 Stuk, T.L., Grieco, P.A., and Marsh, M.M. (1991). *J. Org. Chem.* 56: 2957–2959. doi: 10.1021/jo00009a006.
45 Kaufman, M.D., Grieco, P.A., and Bougie, D.W. (1993). *J. Am. Chem. Soc.* 115: 11648–11649. doi: 10.1021/ja00077a094.
46 Breslow, R., Zhang, X., and Huang, Y. (1997). *J. Am. Chem. Soc.* 119: 4535–4536. doi: 10.1021/ja9704951.
47 Yang, J. and Breslow, R. (2000). *Angew. Chem. Int. Ed.* 39: 2692–2694. doi: 10.1002/1521-3773(20000804)39:15<2692::AID-ANIE2692>3.0.CO;2-3.
48 Yang, J., Gabriele, B., Belvedere, S. et al. (2002). *J. Org. Chem.* 67: 5057–5067. doi: 10.1021/jo020174u.
49 Fang, Z. and Breslow, R. (2006). *Org. Lett.* 8: 251–254. doi: 10.1021/ol052589i.
50 Yang, J. and Breslow, R. (2000). *Tetrahedron Lett.* 41: 8063–8067. doi: 10.1016/S0040-4039(00)01402-7.
51 Iida, T., Ogawa, S., Shiraishi, K. et al. (2003). *ARKIVOC* viii: 171–179. doi: 10.3998/ark.5550190.0004.817.
52 Ogawa, S., Iida, T., Goto, T. et al. (2004). *Org. Biomol. Chem.* 2: 1013–1018. doi: 10.1039/B314965J.
53 Iida, T., Ogawa, S., Miyata, S. et al. (2004). *Lipids* 39: 873–880. doi: 10.1007/s11745-004-1309-0.
54 Ogawa, S., Hosoi, K., Iida, T. et al. (2007). *Eur. J. Org. Chem.* 3555–3563. doi: 10.1002/ejoc.200700158.
55 Iida, T., Ogawa, S., Hosoi, K. et al. (2007). *J. Org. Chem.* 72: 823–830. doi: 10.1021/jo061800g.
56 Ogawa, S., Wakatsuki, Y., Makino, M. et al. (2010). *Chem. Phys. Lipids* 163: 165–171. doi: 10.1016/j.chemphyslip.2009.10.012.
57 Konoike, T., Araki, Y., and Kanda, Y. (1999). *Tetrahedron Lett.* 40: 6971–6974. doi: 10.1016/S0040-4039(99)01437-9.
58 Vijayarahavan, B. and Chauhan, S.M.S. (1999). *Tetrahedron Lett.* 31: 6223–6226. doi: 10.1016/S0040-4039(00)97030-8.
59 Jana, S., Ghosh, M., Ambule, M., and Sen Gupta, S. (2017). *Org. Lett.* 19: 746–749. doi: 10.1021/acs.orglett.6b03359.
60 Chandra, B., Hellan, K.M., Pattanayak, S., and Sen Gupta, S. (2020). *Chem. Sci.* 11: 11877–11885. doi: 10.1039/d0sc03616a.
61 Doiuchi, D., Nakamura, T., Hayashi, H., and Uchida, T. (2020). *Chem. Asian J.* 15: 762–765. doi: 10.1002/asia.202000134.

62 Zhou, M., Hintermair, U., Hashiguchi, B.G. et al. (2013). *Organometallics* 32: 957–965. doi: 10.1021/om301252w.
63 Hao, B., Gunaratna, M.J., Zhang, M. et al. (2016). *J. Am. Chem. Soc.* 138: 16839–16848. doi: 10.1021/jacs.6b12113.
64 Laudadio, G., Govaerts, S., Wang, Y. et al. (2018). *Angew. Chem. Int. Ed.* 57: 4078–4082. doi: 10.1002/anie.201800818.
65 Thapa, P., Hazoor, S., Chouhan, B. et al. (2020). *J. Org. Chem.* 85: 9096–9105. doi: 10.1021/acs.joc.0c01013.
66 Chen, M.S. and White, M.C. (2007). *Science* 318: 783–787. doi: 10.1126/science.1148597.
67 Chen, M.S. and White, M.C. (2010). *Science* 327: 566–571. doi: 10.1126/science.1183602.
68 Mas-Ballesté, R. and Que, Jr, L. (2007). *J. Am. Chem. Soc.* 129: 15964–15972. doi: 10.1021/ja075115i.
69 Ottenbacher, R.V., Talsi, E.P., and Bryliakov, K.P. (2015). *ACS Catal.* 5: 60–69. doi: 10.1021/cs5013206.
70 Zima, A.M., Lyakin, O.Y., Ottenbacher, R.V. et al. (2017). *ACS Catal.* 7: 60–69. doi: 10.1021/acscatal.6b02851.
71 Gormisky, P.E. and White, M.C. (2013). *J. Am. Chem. Soc.* 135: 14052–14055. doi: 10.1021/ja407388y.
72 Howell, J.M., Feng, K., Clark, J.R. et al. (2015). *J. Am. Chem. Soc.* 137: 14590–14593. doi: 10.1021/jacs.5b10299.
73 Zhao, J., Nanjo, T., de Lucca, E.C., Jr., and White, M.C. (2019). *Nat. Chem.* 11: 213–221. doi: 10.1038/s41557-018-0175-8.
74 Yazerski, V.A., Spannring, P., Gatineau, D. et al. (2014). *Org. Biomol. Chem.* 12: 2062–2070. doi: 10.1039/c3ob42249f.
75 Gymez, L., Canta, M., Font, D. et al. (2013). *J. Org. Chem.* 78: 1421–1433. doi: 10.1021/jo302196q.
76 Prat, I., Gymez, L., Canta, M. et al. (2013). *Chem. Eur. J.* 19: 1908–1913. doi: 10.1002/chem.201203281.
77 Canta, M., Font, D., Gymez, L. et al. (2014). *Adv. Synth. Catal.* 356: 818–830. doi: 10.1002/adsc.201300923.
78 Font, D., Canta, M., Milan, M. et al. (2016) *Angew. Chem. Int. Ed.* 55: 5776–5779. doi: 10.1002/anie.201600785.
79 Olivo, G., Capocasa, G., Ticconi, B. et al. (2020). *Angew. Chem. Int. Ed.* 59: 12703–12708. doi: 10.1002/anie.202003078.
80 Ottenbacher, R.V., Samsonenko, D.G., Nefedov, A.A. et al. (2021). *J. Catal.* 399: 224–229. doi: 10.1016/j.jcat.2021.05.014.
81 Borrell, M., Gil-Gaballero, S., Bietti, M., and Costas, M. (2020). *ACS Catal.* 10: 4702–4709. doi: 10.1021/acscatal.9b05423.
82 Zima, A.M., Babushkin, D.E., Lyakin, O.Y. et al. (2022). *ChemCatChem* e202101430. doi: 10.1002/cctc.202101430.
83 Zima, A.M., Lyakin, O.Y., Bryliakova, A.A. et al. (2022). *Chem. Rec.* e202100334. doi: 10.1002/tcr.202100334.
84 Liu, W. and Groves, J.T. (2010). *J. Am. Chem. Soc.* 132: 12847–12849. doi: 10.1021/ja105548x.
85 Liu, W., Nielsen, R.J., Goddard, W.A., III, and Groves, J.T. (2017). *ACS Catal.* 6: 4182–4188. doi: 10.1021/acscatal.7b00655.
86 Liu, W., Huang, X., Cheng, M.-J. et al. (2012). *Science* 337: 1322–1325. doi: 10.1126/science.1222327.
87 Huang, X., Liu, W., Ren, H. et al. (2014). *J. Am. Chem. Soc.* 136: 6842–6845. doi: 10.1021/ja5039819.
88 Cesarec, S., Robson, J.A., Carroll, L.S. et al. (2021). *Curr. Radiopharm.* 14: 101–106. doi: 10.2174/1874471013666200907115026.
89 Huang, X., Bergsten, T.M., and Groves, J.T. (2015). *J. Am. Chem. Soc.* 137: 5300–5303. doi: 10.1021/jacs.5b01983.
90 Rana, S., Biswas, J.P., Sen, A. et al. (2018). *Chem. Sci.* 9: 7843–7858. doi: 10.1039/c8sc02053a.

18

Catalysis by Pincer Compounds and Their Contribution to Environmental and Sustainable Processes

Hugo Valdés and David Morales-Morales

Universidad Nacional Autónoma de México, Circuito Exterior S/N. Ciudad Universitaria, Mexico

18.1 Introduction

The development of catalytic processes has grown together with the chemistry of pincer metal complexes. The first metal pincer complex was described by Shaw in the late 1970s [1, 2], but was not until 1989 that the term pincer was used to describe a tridentate ligand [3]. At that time, a pincer ligand had to contain at least one M-C bond between the pincer ligand and the metal fragment. Currently, this definition has broadened and, practically, any three-dentated ligand could be called a pincer if it occupies adjacent binding sites in a metal complex adopting a meridional geometry, although tripodal ligands such as triphos should not be considered as pincer ligands. The structure of pincer ligands is unique; the pincer ligand forms two fused metallocycles that generally contain five- or six-membered rings around the metal fragment. Although pincer complexes with two different metallocycles are also known, they are far less common. Their great thermal and chemical stability is in part due to the formation of these two metallocycles. Here, the chelate effect produced by the pincer ligand plays an important role in the formation of their related pincer complexes.

Besides their stability, pincer complexes can be easily functionalized, allowing their electro- and stereo-properties to be fine-tuned [4]. These features have allowed the preparation of a plethora of pincer complexes with different architectures [5, 6], and more importantly, have allowed the development of highly active catalysts based on these complexes. In fact, pincer complexes have been used to catalyze several transformations, including cross-coupling reactions, cross-coupling dehydrogenative reactions, and polymerizations, among others [7–27]. All of these transformations have contributed enormously to the establishment of more sustainable methodologies to obtain high-valued organic compounds.

In this context, the activation and catalytic transformation of small molecules, such as H_2, CO_2, and biomass, is very attractive from different points of view. The activation of hydrogen opens the door to develop novel hydrogenation processes [28], which may result in the establishment of hydrogen-based fuel technologies. It is well known that hydrogen-based fuel technologies are more convenient than fossil-based fuel technologies because their only initial residue is H_2O.

Catalysis for a Sustainable Environment: Reactions, Processes and Applied Technologies Volume 2, First Edition. Edited by Armando J. L. Pombeiro, Manas Sutradhar, and Elisabete C. B. A. Alegria.
© 2024 John Wiley & Sons Ltd. Published 2024 by John Wiley & Sons Ltd.

The use of CO_2 as feedstock is highly convenient. Each year, human industrial activities produce several tons of CO_2 that are released to the atmosphere and are the main cause of climate change. Thus, the idea of transforming CO_2 into valuable products might contribute to decrease the amount of CO_2 released, and, consequently, palliate the effects of climate change. For instance, CO_2 can be catalytically transformed to valuable chemicals, such as methanol, formic acid, formaldehyde, or polymers [29]. Of these, the conversion of CO_2 to methanol has potential applications in hydrogen storage, which results in a highly convenient strategy to capture the CO_2 and convert it to hydrogen-based fuel.

The transformation of biomass into biofuel is a technology that has been employed to obtain fuel from different natural and renewable sources, such as vegetable oil, animal fats, or recycled restaurant grease [30, 31]. One of the drawbacks of this approach is that some processes require high temperatures; for example, the *Guebert reaction* requires temperatures around 200 °C to obtained butanol isomers[32]. However, with the correct selection of catalysts, this reaction can be carried out at lower temperature [33–35], which is more suitable from the environmental point of view.

Therefore, this chapter deals with the use of pincer complexes to catalyze some reactions that are important for the development of hydrogen-based fuel technologies and the use of biomass to produce *advanced fuels*. Specifically, we describe advances in the hydrogenation of CO_2 to produce methanol, the dehydrogenation of methanol to produce hydrogen gas, and the use of biomass to produce biofuels. All of these processes are catalyzed by pincer metal complexes.

18.2 Hydrogenation of CO_2 to Methanol and Related Processes

Hydrogenation reactions are ubiquitous in organic chemistry, allowing large-scale production of several pharmaceuticals, flavors, and fragrances. The early stages of this transformation involved the use of stoichiometric amounts of dangerous metal hydrides. The main drawbacks of these transformations are strongly associated with safety issues and laborious workup procedures that employ large amounts of toxic solvents. Consequently, such procedures generate high amounts of waste, and, consequently, have low atom economy. All of these factors have a negative impact on the environment and increase the cost of the final product. The most versatile strategy to avoid all these disadvantages is hydrogenation catalyzed by transition metals [36, 37]. In this sense, pincer complexes have shown an extraordinary performance, being active toward the hydrogenation of esters, carbonates, amides, imines, nitriles, alkenes, and alkynes [14, 38–57].

In 2006, Milstein et al. described the use of a NNP-Ru(II) pincer complex (Milstein's catalyst) for the hydrogenation of esters to primary alcohols, under neutral conditions and in the presence of molecular hydrogen [58] (Scheme 18.1). The catalyst performed well toward different esters bearing alkyl and benzyl functional groups, but the most important thing was to understand the role of the

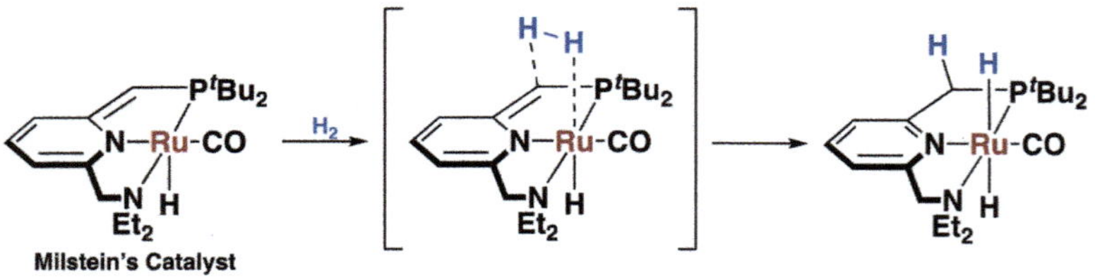

Scheme 18.1 Hydrogenation activation by Milstein's catalyst.

pincer ligand in the reaction mechanism. Traditionally, ligands were considered as simply spectators in some catalytic processes, but the NNP pincer ligand participates directly in the reaction mechanism in this case by helping with H-H bond activation together with the metal center [59]. The activation of molecular dihydrogen was studied by density functional theory (DFT) calculations [60], which showed that it proceeds through a transition state in which one atom of hydrogen is coordinated to Ru, and the other interacts with the C=C bond of one arm. After activation of the H-H bond a dihydride-aromatized pincer complex is formed. The energy gained by the aromatization of the pyridine stabilizes the transition state and, consequently, lowers the activation barrier. In fact, the overall reaction is formally a [2+2+2] addition of hydrogen to the C_{sp}^2 of the pincer arm and the Ru center. Thus, the electrons of the C=C bond, the electron lone pair of the nitrogen from the dearomatized pyridine, and two electrons of the dihydrogen are involved. This behavior not only works to activate H-H bond, but also to activate other σ-bonds, such C-H, N-H, and O-H [61–66]. At the end of the H_2 activation, the metal-hydride species is able to hydrogenate some C_{sp}^2 and C_{sp} species [67].

Milstein's catalyst has been used to catalyzed the hydrogenation of dimethyl carbonate to methanol and carbamates to alcohols and amines [68]. The synthesis of dimethyl carbamate from CO_2 is a well-stablished procedure and, in the context of "methanol economy", is convenient to convert CO_2 into other valuable chemical feedstocks such as methanol. The hydrogenation of dimethyl carbonate to methanol was achieved using an NNP-Ru pincer complex. The reaction was tested in the presence of tetrahydrofuran (THF) and under neat conditions (without solvent). Using THF at 10 atm of hydrogen, a yield of 96% in 48h was achieved with a TON of 960. Under neat conditions, a similar TON (890) was achieved after only two hours. A quantitative conversion was observed after eight hours under neat conditions. This reaction also works with other organic carbonates, such as diethyl carbonate, which was hydrogenated to produce ethanol and methanol under neat conditions. Remarkably, the generation of alkyl formates was not observed in these transformations. Furthermore, under similar reaction conditions, the hydrogenation of methyl formate to methanol and carbamate to alcohols and amines was achieved.

The reaction mechanism of the hydrogenation of dimethyl carbonate was studied using experimental and theoretical studies [60, 68, 69]. The stoichiometric reaction of the dihydride species with dimethyl carbonate generates methyl formate, methanol, and a non-aromatic-dicarbonyl species, while the reaction with formate generates methanol and a metoxy-Ru pincer-complex. The formation of the non-aromatic-dicarbonyl species takes place in the absence of hydrogen, probably due to the decomposition of dimethyl carbonate to methanol, CO and H_2. It is worth mentioning that the authors did not observe the formation of CO during the catalytic reaction. The second reaction corroborates that the dihydride species is capable of hydrogenating not only dimethyl carbonate, but also methyl formate to produce methanol. According to the DFT computational studies, the overall process is a three-stage sequential reaction (Scheme 18.2). In the first stage, the dimethyl carbonate is hydrogenated to methyl formate and methanol, then methyl formate affords formaldehyde and methanol, and, finally, formaldehyde is hydrogenated to produce methanol. The first two stages proceed through: i) hydrogen activation by the catalyst; ii) stepwise double hydrogen transfer; and iii) decomposition of hemiacetal intermediates. The last stage is a typical hydrogenation process. It is noteworthy that the oxidation state of the metal remains unchanged through all the reaction. Milstein's catalyst has been the inspiration for other related catalysts that are active in the hydrogenation of esters, amides, carbonates, carbamates, and formates [68, 70–73].

The pincer complex named MACHO is probably one of the most versatile catalysts. MACHO-type ligands are composed of an aliphatic framework bearing a phosphine atom in each arm of the

Scheme 18.2 Hydrogenation of CO_2 catalyzed by Milstein's catalyst.

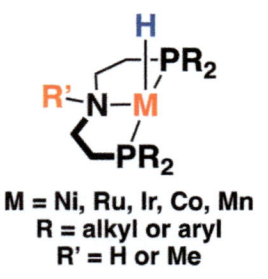

Scheme 18.3 MACHO-type pincer complexes.

M = Ni, Ru, Ir, Co, Mn
R = alkyl or aryl
R' = H or Me

pincer and a nitrogen atom in the center (Scheme 18.3). This ligand is capable of forming stable complexes with different transition metals, such as Ni, Ru, Ir, Co, and Mn. The main application of MACHO-type metal complexes is to catalyze hydrogenation reactions of ester, amides, nitriles, and alkenes. Usually, these reactions are carried out just in the presence of the catalyst and a pressure of dihydrogen, without the addition of base or any other additive, representing a green strategy to synthesize alcohols, amines, or alkanes [42, 74–82].

Recently, Avasare performed a comprehensive density study of six MACHO-type Mn(I) complexes and their possible application to CO_2 reduction to methanol [83]. The MACHO-type ligands were based on NNP-pincer ligands, where the terminal nitrogen atom contains different heterocycles, such as adenine, 2,3-diaminopyridine, 2,5-diaminopyridine, 2-aminopyridine, 2-aminopyrimidine, and 2-aminopurine. According to the DFT calculations, the complex with the 2,3-diaminopyridine moiety should give a TOF of 2 372 400 h^{-1} in the formation of methanol at 50 °C. The higher electrophilicity at the metal center produce a higher interaction energy, and consequently higher TOFs.

Further, Sandford et al. described and elegant method for the reduction of CO_2 to methanol through a tandem CO_2 capture/hydrogenation sequence [84]. First, the reaction of CO_2 with a dimethylamine affords dimethylammonium dimethylcarbamate (DMC), which can be hydrogenated with a MACHO-Ru pincer complex to generate methanol and dimethylformamide (DMF). The DMF is formed by the reaction of the in situ formed formic acid and dimethylamine. Overall, the conversion of CO_2 to the product mixture was achieved in 96%, which is considered a high carbon efficiency process. Following this strategy, Prakash developed an efficient method for the conversion of CO_2 to methanol, including CO_2 from air [85]. The catalytic system consisted in a biphasic reaction mixture of water and 2-methyltetrahydrofurane (2-MTHF). In the aqueous phase, the capture of the CO_2 occurs by the presence of an amine. They tested some amines and polyamines, finding that pentaethylenehexamine (PEHA) was the best. In the organic phase, the hydrogenation of the amide-derivate was achieved by the presence of a MACHO-based catalyst. Among the catalysts they evaluated, the commercially available Ru-MACHO-BH was the most active. Interestingly, using simulated air ([CO_2] = 408 ppm) the catalysts reached 96 turnover number (TON). This approach also works using an organic solvent (1,4-dioxane or triglyme), reaching up to 79% yield [86]. In a subsequent report, Prakash reported the immobilization of polyethylenimine onto a solid-silica support [87]. The amine source and the catalysts can be easily recovered and recycled multiple times.

Later, Prakash et al. reported mechanistic and structural-activity studies [88] (Scheme 18.4). They found that the phosphine substituents (PR_2) have an impact in the formation of methanol and the trend

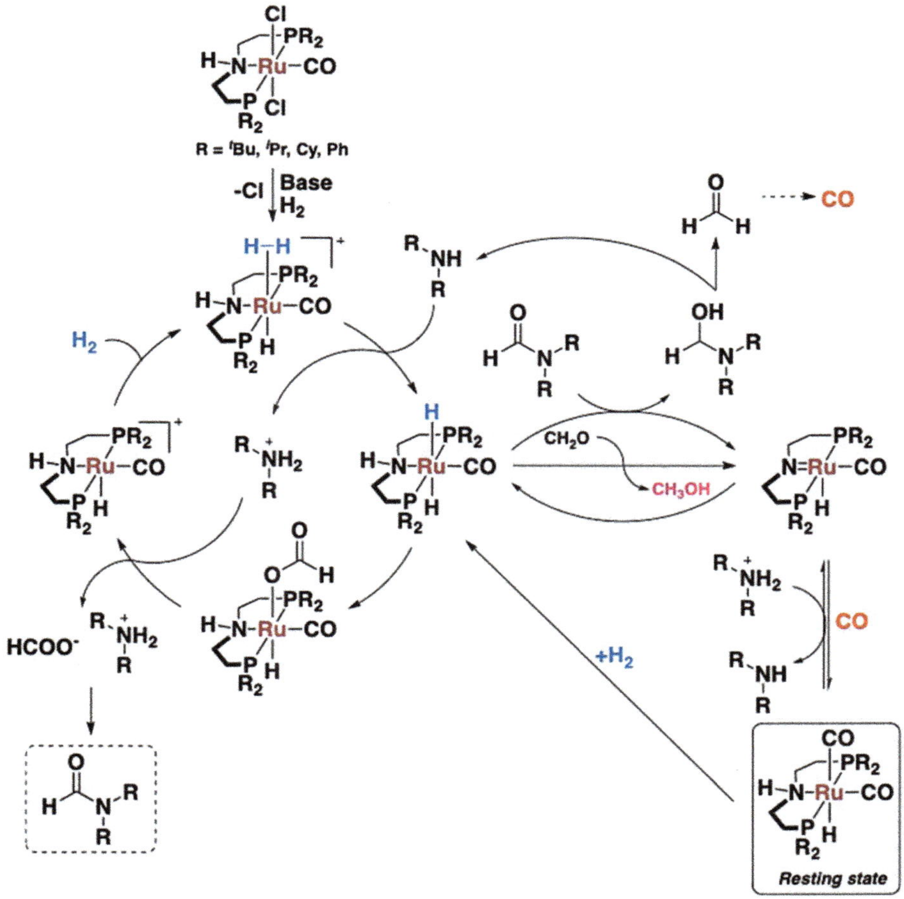

Scheme 18.4 Reaction mechanism for the hydrogenation of CO_2 catalyzed by MACHO-Ru a pincer complex.

from higher to lower methanol yield was $PPh_2 > P^iPr_2 > PCy_2 > P^tBu_2$. The reaction mechanism starts with the formation of a dihydride species promoted by the amine. Then CO_2 is hydrogenated and released as formic acid. The latter can then react with the ammonium salt to generate an amide and the catalyst is re-hydrogenated. The amide is hydrogenated by the dihydride Ru-MACHO complex, forming an hemiaminal and a Ru amido species. The hemiaminal decomposes to formaldehyde and amine. Finally, formaldehyde is hydrogenated to form methanol in the same cycle.

Some related strategies have been developed by different research groups [89]. For instance, Ding et al. employed a Ru-MACHO catalyst and morpholine to perform the reduction of CO_2 to methanol [90]. The reaction was carried out in two stages. First, the formylation of the amine takes place, reaching a TON of 1,940,000. The subsequent hydrogenation of the amide to produce methanol was achieved in 94% yield. The reaction also was tested in a one-pot experiment, reaching 36% yield of methanol. Following a similar approach, Prakash et al. described that an air stable Mn-MACHO catalyst may catalyze the reduction of CO_2 to methanol in the presence of morpholine or benzylamine in 71% and 84% yield, respectively [91]. Similarly, Bernskoetter et al. described the use of a Fe-MACHO catalyst for the two-step reduction of CO_2 in the presence of morpholine [92]. The formation of formylmorpholine was achieved in a TON of 1160, whereas a TON of 590 was obtained in the following hydrogenation. In the first stage of the reaction, a pressure of 250/1150 psi of CO_2/H_2 was employed, whereas the second stage was performed in the absence of CO_2 and a hydrogen pressure of 1150 psi.

On the other hand, Milstein et al. reported the capture of CO_2 using aminoethanols and its subsequent hydrogenation to produce methanol [93]. The CO_2 is captured with valinol in the presence of a catalytic amount of Cs_2CO_3 to produce the corresponding oxazolidone in high yields (65–90%). Then the NNP-Ru catalyst is used to hydrogenate the oxazolidone to give methanol in 95–100% yield. Interestingly, the capture of CO_2 was achieved at low pressure (1–3 bar), and the reduction of the oxazolidone was performed using a hydrogen pressure of 60 bar.

Other strategies to reduce CO_2 to methanol involves the use of alcohols instead of amines. The first example of such transformation was described by Sanford et al. [94]. The reaction was performed in a cascade catalysis involving the use of three different catalysts. In the first step, complex $[(PMe_3)_4Ru(Cl)(OAc)]$ catalyzes the reduction of CO_2 to formic acid, then the formic acid reacts with an alcohol to produce an ester. The latter reaction was promoted by the Lewis acid $[Sc(OTf)_3]$. Finally, the ester compound was hydrogenated by an NNP-Ru catalyst to form methanol. The reaction was further optimized by selecting other catalysts. The selected mix of catalysts was $[Ru(H)_2(P(CH_2CH_2Ph_2)_3]/[Sc(OTf)_3]$/PCP-Ir pincer complex. Using this selection, an overall TON of 428 was achieved in 40 hours at 155 °C [95].

One of the most active systems was described recently by Byers and Tsung. They reported a cascade hydrogenation of CO_2 to methanol using an encapsulated PNP-Ru pincer complex in MOF UiO-66 and a NNP-Ru catalyst [96]. The PNP-Ru/UiO-66 catalyst is a very active system for the reduction of CO_2 to formate. Actually, this is a robust catalyst due to the presence of the MOF that prevents the decomposition of the catalytic active site and its poisoning, and it can be recycled [97]. Additionally, the authors studied the effect of different alcohols in the TON, finding that increasing the acidity of the alcohol produced a higher TON. Using a sub-stoichiometric amount of 2,2,2-trifluoroethanol (TFE), the TON was 6,600, whereas a TON of 4,710 was obtained when using ethanol instead of TFE. Interestingly, the catalytic system was recycled five times, reaching an overall TON of 21,000 at the end of the cycles.

In 2021, Leitner et al. described the hydrogenation of CO_2 to methanol in the presence of alcohol and Lewis acids, using a MACHO-Mn(I) complex as catalyst [98]. The role of the Lewis acid was

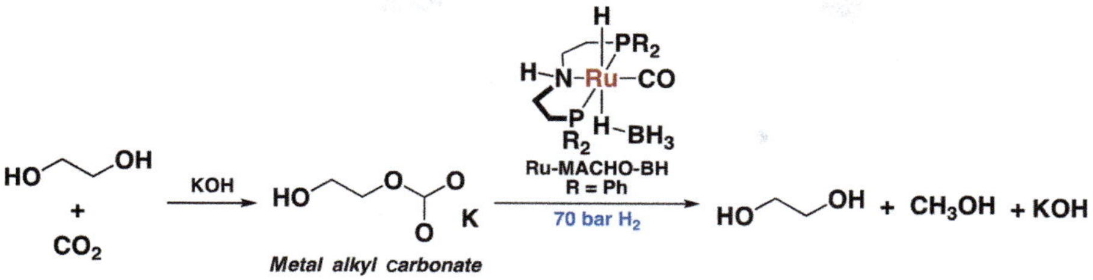

Scheme 18.5 Reduction of CO_2 in the presence of ethylene glycol catalyzed by Ru-MACHO-BH.

very interesting because it helped unlock the limiting resting state of the catalytic cycle. The limiting resting state consists of a MACHO-Mn-formate complex, which was isolated and characterized, so the Lewis acid helps to transfer the formate ligand to an ester intermediate, which can be easily hydrogenated. Under the optimized reaction conditions (methanol, [Ti(O^{i}Pr)$_4$], 5 bar CO, 160 bar H_2, 150 °C), the catalytic system reached a TON of 160.

The Ru-MACHO-BH catalyst has also been used for the hydrogenation of CO_2 using alcohols. In this line, Prakash et al. described the CO_2 capture from ambient air and its transformation to methanol (Scheme 18.5) [99]. Their approach consisted of the reaction of CO_2 with ethylene glycol under basic conditions to produce a metal alkyl carbonate, which could be further hydrogenated by Ru-MACHO-BH. They captured CO_2 by bubbling indoor air in a solution of 5 mmol of KOH in 10 mL of ethylene glycol for 48 hours. The CO_2 captured was around 3.3 mmol in the form of carbonate and alkyl carbonate salts. These salts were further hydrogenated using a dihydrogen pressure of 70 bar. After 20 hours, a 25 % yield was observed, and a full conversion was achieved after 72 hours. However, one of the drawbacks of this strategy is the deactivation of the base, which occurs during hydrogenation conditions, leading to the dehydrogenation of ethylene glycol in the presence of the base and catalyst to the glycolate salt. To avoid this problem, the use of tertiary-amines instead of KOH was investigated [100]. Tetramethylethylenediamine (TMEDA) and tetramethylbutanediamine (TMBDA) were the best options to produce methanol from CO_2. It was remarkable that CO_2 from simulated flue gas (10% CO_2/N_2) was captured in the ethylene glycol/amine system and hydrogenated to methanol. The simulated flue gas was bubbled overnight in a solution of 10 mmol of the amine in 10 mL of ethylene glycol. Using TMEDA, 3.2 mmol of CO_2 was captured whereas 7 mmol of CO_2 was obtained employing TMBDA. The subsequent hydrogenation of the resulting formate was achieved in 94 % yield for TMBDA and 72 % yield for TMEDA.

Another interesting strategy to reduce CO_2 to methanol involves the use of boranes and silanes instead of dihydrogen and a transition metal catalyst. In principle, the reaction of CO_2 with Si- or B-based compounds is thermodynamically more favorable due to formation of the strong Si-O or B-O bonds. In this sense, some pincer-based complexes have successfully catalyzed this transformation [101–105]. For instance, a PNCNP-Ni(II) pincer complex has been employed as catalyst for the hydrosilylation of CO_2 to methanol (Scheme 18.6) [106]. The reaction has a strong dependence of the solvent nature; hence, the hydrosilane is fully consumed in 12 hours when using very polar solvents such as DMF, whereas the reaction showed slower reaction rates when other, less polar solvents such as THF and CH_3CN were used. In non-polar solvents (toluene and CH_2Cl_2), methanol was not formed. Interestingly, in independent work, Li and Lei described the catalyst-free formylation of amines using CO_2 and hydrosilanes in polar solvents, so the solvent

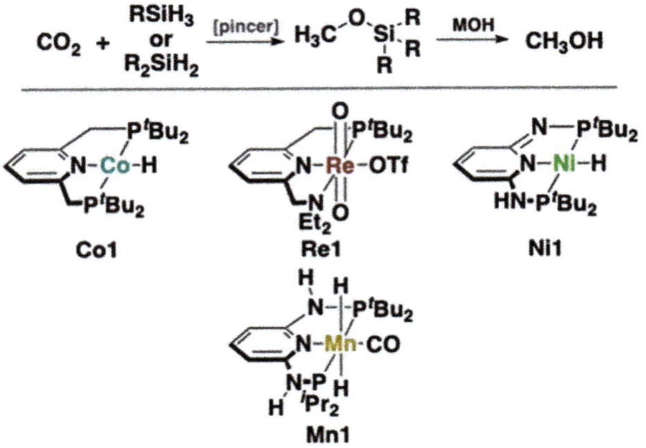

Scheme 18.6 Reduction of CO_2 with silanes catalyzed by pincer complexes.

may help in the reduction of CO_2 to methanol [107]. Here, under the optimized conditions, the reaction reached a TON of 4,900.

Kirchner and Gonsalvi studied the reduction of CO_2 to methanol catalyzed by a PNNNP-Mn(I) pincer complex [108]. It is noteworthy that this catalyst is active in other related processes, such as reduction of CO_2 to formate and coupling of alcohols and amines [19, 109, 110]. The catalytic CO_2 hydrosilylation was performed under mild conditions (80 °C, 1 bar CO_2). In the first step, the reaction of CO_2 with $RSiH_3$ affords silylformates very quickly (~5 min), which are then reduced to methoxysilyl derivatives. The hydrolysis of the latter affords methanol. Other silanes such as monohydrosilanes (Ph_3SiH or $PhMe_2SiH$, Et_3SiH) showed poor activity, whereas $PhSiH_3$ or Ph_2SiH_2 afford the methoxysilyl species in >99% yield after six hours at 80 °C. The proposed reaction mechanism starts with the insertion of CO_2 into the Mn-H bond, then $RSiH_3$ attacks the coordinated O atom, followed by a hydrogen transfer, which results in the regeneration of the initial pincer hydride complex. Next, the hydride attacks the carbonyl carbon to form a metal-coordinated silyl hemiacetal, which reacts with a second molecule of silane, forming a silyl ether and a methoxide-Mn complex. Finally, the methoxide is released in the form of methoxy-silane, and the catalyst is regenerated by reacting with another silane molecule.

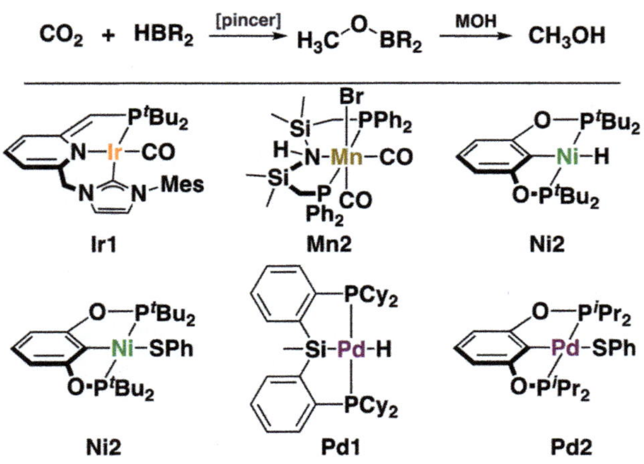

Scheme 18.7 CO_2 reduction with boranes catalyzed by pincer complexes.

Analogously, the reduction of CO_2 to methanol can be also performed using boranes instead of silanes (Scheme 18.7) [111]. POCOP-Ni [112–117] and -Pd [21, 118] pincer complexes have been used to catalyze such transformations. When the POCOP-Pd pincer complex was employed, the reduction of CO_2 to methanol was achieved using 1 atm of CO_2, 500 equivalents of catecholborane (HBcat), and benzene at room temperature. Under these conditions, a TOF of 1780 h^{-1} was observed [118]. Under similar conditions, the POCOP-Ni complex achieved a TOF of 495 h^{-1}. The different substituents at the phosphorous atom influence the catalytic activity of the POCOP-Ni complex. The bulkier the substituent is, the more activity the catalyst shows. For example, the ^{t}Bu substituent gave a TON of 100 in 45 minutes, whereas the ^{i}Pr derivative needs two hours to reach a similar TON, and the less active cyclopentyl group gave a TON of 30 in 12 hours [114, 115]. Additionally, the coordination of a thiolate ligand to the Ni-POCOP complex produces a more active species, reaching a TOF of 2,400 h^{-1} [116].

The first example of a Mn-pincer complex (Mn$_2$) for the reduction of CO_2 to methoxy boronate was described by Leitner et al. in 2018 [119]. The reaction was carried out in the absence of solvent and using a catalytic amounts of base (NaO^{t}Bu). In this case, the borane source was pinacolborane (HBpin). After 14 hours at 100 °C, the reaction achieved a TON of 883. Very recently, two research groups have independently studied the reaction mechanism using theoretical DFT calculations. The reaction consists in three essential stages: (i) hydroboration of CO_2 to HCOOBpin, (ii) hydroboration of HCOOBpin to HCHO, and (iii) hydroboration of HCHO to CH_3OBpin. Lei and Cao proposed that, after the deprotonation of the pincer ligand, HBpin reacts with the complex, forming an Mn-H and N-B species, then the boron atom interacts with one oxygen atom of the CO_2, favoring the approach of the carbonyl carbon atom to the hydride, and consequently the reduction of the carbonyl. The energy span of this proposal is 27.0 kcal/mol [120]. In contrast, Schaefer III et al. found that the boron is abstracted by the base (NaO^{t}Bu), forming an N-Na interaction and ^{t}BuO-Bpin species. The energy span of this proposal is 22.5 kcal/mol [121]. Thus, there is difference of 4.5 kcal/mol between the two proposals and it is possible that both mechanisms may be operating at the same time. By changing the pincer by a PNNNP ligand and adding a co-catalyst (B(OPh)$_3$) as a Lewis acid, the formation of the boryl-protected methanol was achieved in high yields [122].

In addition, the use of CO as feedstock is also attractive. In principle, the hydrogenation of CO to methanol is more thermodynamically favored than the hydrogenation of CO_2 to methanol (ΔH_{298K} = −90.7 kJ/mol vs −49.5 kJ/mol), so the former should be carried out under softer reaction conditions. However, CO shows a high affinity to several transitions metals, forming very stable compounds that sometimes result in inactive species on some catalytic cycles. This is particularly problematic in the case of homogeneous catalysts. Nevertheless, there are reports that describe the transformation of CO to methanol (Scheme 18.8) using metal pincer complexes [123–125] and other transition metal-based catalysts [126]. MACHO-Ru-BH type complexes are able to fully convert CO into methanol in the presence of a base and dialkylamine in a sequential stepwise reaction [123]. Analogously to CO_2, CO can be captured in the presence of an amine and a base, forming an amide that is subsequently hydrogenated by the catalyst. When the reaction is carried out using 10 bar CO, 70 bar H_2 at 145 °C, for 168 hours, methanol is obtained in 59% yield, giving a TON of 539. Using a MACHO-Mn catalyst and indoles as nitrogen promoters, the formation of methanol was achieved in TON's up to 3170 [124].

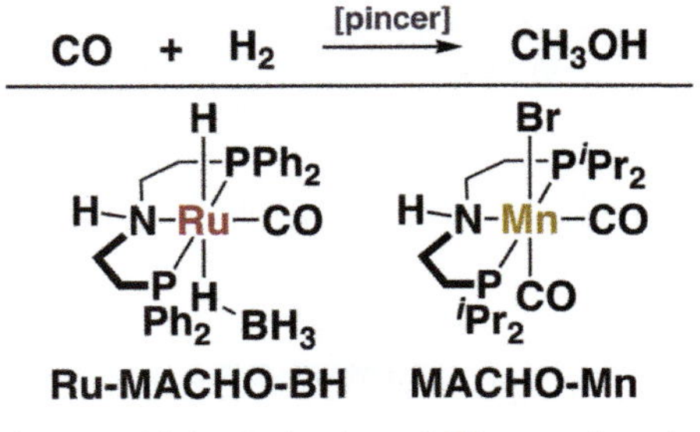

Scheme 18.8 Reduction of CO to methanol catalyzed by pincer metal complexes.

It is worth mentioning that the reaction was performed using lower hydrogen pressures (40 bar). Some alcohols have also promoted the conversion of CO to methanol [125]. In this case, the MACHO-Mn catalyst gives a TON of 4,023, using 5 bar CO, 50 bar H_2, and ethanol, at 150 °C for 12 hours. The reaction mechanism was also studied, and follows a typical ligand-assisted mechanism in which hydrogen is first activated by the presence of the N in the pincer and the metal center, forming a hydride species. Then the ester or the amide is hydrogenated, generating formaldehyde, which is hydrogenated by a hydride-Mn species. All of these achievements represent a great breakthrough if we consider that methanol synthesis from syngas (CO/H_2 mixtures) requires high temperatures (200–300 °C) and pressures (50–100 atm). Thus, MACHO pincer complexes have helped to develop procedures that require less energy consumption and that are consequently considered greener.

Following a similar approach, the MACHO-Mn catalyst was also active in the β-methylation of alcohols using 5 bar CO, 15 bar H_2, and a two-fold excess of base (NaO^tBu) [127]. This interesting reactivity can be explained by the fact that the formed formaldehyde from the hydrogenation of CO can react with the aldehyde formed by the dehydrogenation of the alcohol under these reaction conditions, in a typical aldol condensation under basic conditions. The scope of the reaction was wide, including functionalities such as aromatic rings, heterocycles, and alkenes.

The synthesis of ethylene glycol from CO/H_2 catalyzed by pincer complexes has also been explored in recent years (Scheme 18.9) [128–130]. The reaction proceeds in two stages. (i) The CO is captured in the form of a dione. Dione formation is catalyzed by a Pd catalyst ([Pd(acac)$_2$] or Pd/C) and consists of the reaction of CO with piperidine (forming an oxamide), ethanol (forming an oxalate), or piperidine/ethanol mixture (forming an oxamate). This first stage was inspired by previous work regarding oxidative couplings of CO and amines or alcohols [131, 132]. (ii) The second stage consists of the hydrogenation reaction of the dione using a pincer complex. Hence, the Ru-MACHO-BH complex was able to convert an oxamide (1,2-di(piperidin-1-yl)ethane-1,2-dione) to ethylene glycol in 96%, using a 60 bar H_2, KO^tBu (catalytic) in toluene at 160 °C for 12 hours [128]. Milstein's catalyst (NNP-Ru pincer complex) showed good activity toward the hydrogenation of oxamate (ethyl 2-oxo-2-(piperidin-1-yl) acetate), reaching 92% yield under optimized conditions (toluene, KO^tBu, 60 bar H_2, 160 °C, 10 hours) [129]. Additionally, Milstein's catalyst was also active toward the hydrogenation of oxalate (diethyl oxalate). In this case, the reaction required a lower temperature (100 °C) and hydrogen pressure (40 bar) to afford >90% yield [130].

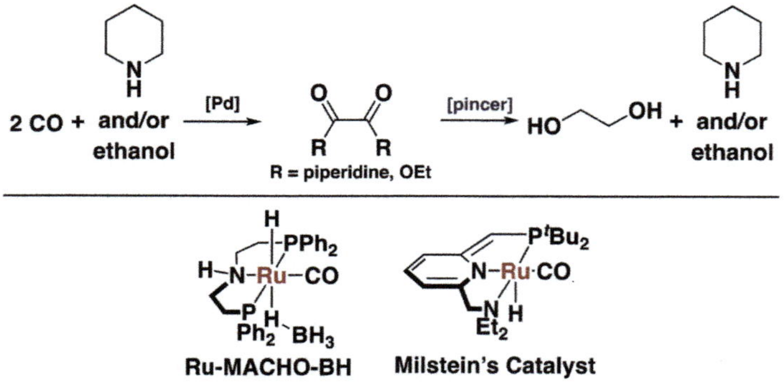

Scheme 18.9 Synthesis of ethylene glycol using CO as feedstock.

18.3 Hydrogen Production from Methanol and Water

Methanol reforming consists of the production of molecular H_2 and CO_2 from methanol. This reaction is achieved in three successive dehydrogenation reactions, starting with the dehydrogenation of methanol to formaldehyde, which reacts with water, to further undergo another dehydrogenation, forming formate. The latter is then dehydrogenated to produce CO_2. In each dehydrogenation step, a molecule of hydrogen is produced, hence each molecule of methanol has the potential to form three molecules of hydrogen (Scheme 18.10). At present, this process is of particular interest due the high content of hydrogen in methanol, allowing the transport of hydrogen in a safer way than using the traditional available methods that involves cryogenic temperatures, high pressure, and inefficient tanks to contain hydrogen. In fact, the content of hydrogen in methanol is 12.6%, which is adequate for some mobile applications. Furthermore, methanol is a liquid that can be easily transported by pipelines, facilitating its distribution. Therefore, the simple idea of transporting chemical hydrogen in the form of methanol is highly attractive to extend the use of hydrogen-based technologies.

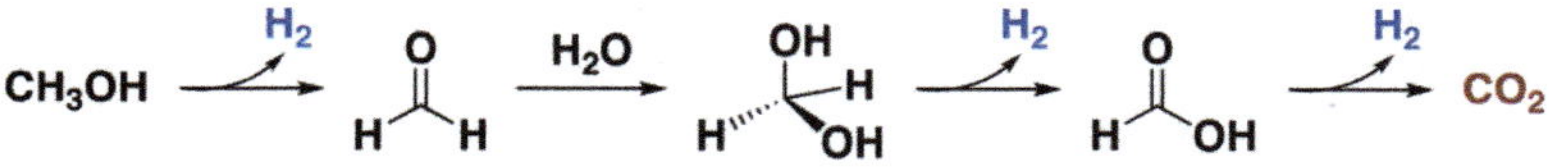

Scheme 18.10 Dehydrogenation reaction of methanol to CO_2 and H_2.

The methanol dehydrogenation reaction must be very selective to hydrogen and CO_2 to feed a hydrogen fuel cell because the presence of small contaminants such as CO or NH_3, poisons the catalyst, reducing the production of energy that the fuel cell is able to achieve. This reaction has been heterogeneously catalyzed and high temperatures are required (>200 °C) [133, 134]. Pincer catalysts have shown a remarkable activity toward methanol reforming processes at low temperatures. In 2013, Beller's group described major breakthroughs of aqueous-methanol reforming catalyzed by pincer complexes [135–137]. Ru-MACHO pincer complexes catalyzed the production of hydrogen from methanol at 72 °C, reaching a TOF of 124 h^{-1} in one hour. The phosphorous substituents play an important role in the activity of the catalyst, being more active with aromatic substituents; TOF = 124 h^{-1} for PPh_2 vs 45 h^{-1} for P^iPr_2. Remarkably, the production of CO was very low ranging on the ppm scale (<10 ppm). The reaction was further improved by using the Ru-MACHO-BH catalyst and the addition of triglyme to the catalytic reaction. It was found that triglyme increases the solubility of the catalyst, and consequently a higher activity was recorded in the presence of KOH (TOF > 900 h^{-1}, TON = 2,743). The reaction also proceeds in the absence of base, reaching a TON of 59, and gas evolution of 61 mLh^{-1}. Interestingly, a positive synergistic effect was observed by the addition of an extra Ru-based catalyst in the absence of base. When the reaction was carried out in the presence of $[Ru(H)_2(dppm)_2]$, the production of hydrogen gas reached was 239 h^{-1} in the first hour, but then a deacceleration of the reaction was observed, decreasing the TOF to 138 h^{-1} after three hours, and 83 h^{-1} after seven hours. A more stable hydrogen evolution was achieved by using a Ru-based bi-catalyst with a bigger bite angle. Using $[Ru(H)_2(dppm)_2]$, a TOF of ~90 h^{-1} was observed (TOF = 87 h^{-1} [one hour], 94 h^{-1} [three hours], 93 h^{-1} [seven hours]).

Because of the high activity of Ru-MACHO pincer complexes, they have been immobilized using the supported liquid phase (SLP) concept [138, 139]. This approach consists in dispersing a thin film of liquid containing the catalyst onto a large inner surface area, allowing a continuous vapor-phase process that ideally can be directly used in an hydrogen fuel cell. Hence, Ru-MACHO catalysts together with KOH were immobilized in two ionic liquids and dispersed onto alumina. Charging 2 wt% of Ru-MACHO in 10 wt% of KOH on porous alumina the activity was 14 $mol_{H_2}mol_{Ru}^{-1}h^{-1}$ at 150 °C and 1 bar and this catalytic system maintained its activity over 70 h time on stream with negligible deactivation. Further improvements of this catalytic system were later described. The use of CsOH instead of KOH and the presence of a second Ru-MACHO-Me catalyst enhanced the catalytic activity, reaching a TOF of 91 h^{-1}.

Furthermore, the reaction mechanism employing Ru-MACHO and Ru-MACHO-Me catalyst was studied in detail [140–142] (Scheme 18.11). The Ru-MACHO starts reacting with the base, forming a Ru-amido species that is the "real" catalyst. Next, an HRO^- compound (R = CH_2, CHOH, CO) is coordinated to the Ru-amido species. This ligand is first coordinated by the oxygen atom, and subsequently is shifted to the hydrogen atom. According to DFT calculations focusing on the frontier orbital interaction between the Ru-amido species and the methoxide, gem-diolate, or formate, coordination through O or H atom is possible and exists in a dynamic equilibrium. The coordination of the H atom favors the formation of the Ru-hydride species and the formation of the RO^- compound, which can further react with ^-OH to capture formaldehyde, formic acid, or CO_2 by-products. The high concentration of base promotes this step and drives the reaction forward. Because of the basic character of the N atom in the MACHO ligand, it can deprotonate a RHOH species, forming a neutral pincer ligand. Next, methanol promotes the formation of dihydrogen and the regeneration of the catalyst.

Other Ru-based pincer catalysts have been also used for the methanol reforming reaction [143, 144]. For instance, Milstein's catalyst is active in the hydrogen production from methanol and water [143]. In this case, the reaction was carried out in the biphasic system of methanol-toluene-water, using 0.025 mol% of catalyst and a base (NaOH) at 100–105 °C. Under these conditions, a 77 % yield was obtained after 9 days. Interestingly, the organic phase was reused two more times, reaching similar yields (82% the second time and 80% the third time). Very recently, Milstein

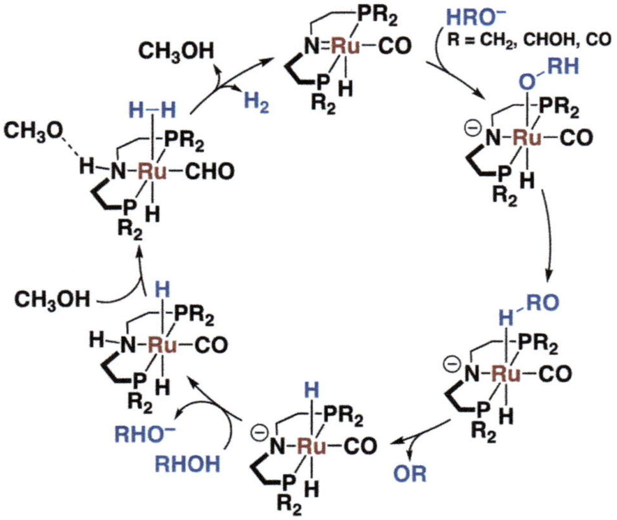

Scheme 18.11 Reaction mechanism of the methanol dehydrogenation catalyzed by pincer complexes.

described a base-free procedure for reforming methanol [144]. The catalyst was a Ru pincer complex with a dearomatized acridine-based ligand. This type of pincer ligands offers a unique platform to obtain highly active catalysts [145]. After heating at 150°C for 24 hours, a mixture of methanol and water in a 4:1 volumetric ratio, and the Ru-acridine pincer complex, a TOF of 4 h^{-1} was obtained. However, the addition of 1 equivalent, corresponding to the catalyst, of hexanethiol produce a tremendous increase in activity, reaching a TOF of 337 h^{-1}. Furthermore, by increasing the amount of methanol (9:1 methanol:water), a TOF of 643 was obtained after six hours, but the TOF decreased to 480 h^{-1} after 12 hours. The presence of a higher amount of methanol increases the solubility of the catalyst, which has a positive effect in the catalytic performance.

Fe-MACHO-BH catalyst was also active in the hydrogenation of methanol under mild conditions, but the stability of the catalyst and its activity decreases over time [137]. The TON$_{max}$ was 9,834 in 46 hours, whereas the TON$_{max}$ was 9184 in 111 hours using five extra-equivalents of catalyst, so the catalytic system is capable to produce hydrogen gas up to five days. In a further report, Bernskoetter, Hazari and Holthausen described that the addition of a Lewis acid produced significant enhancement in the catalytic activity of Fe-MACHO-OOCH complexes [146]. The addition of LiBF$_4$ increases the yield and the TON of the reaction, going from 58% and 350 (20 min) to >99% and >599 (25 minutes), respectively. Under optimized conditions (4:1 methanol/ H$_2$O (160 µL/ 18µL), 10 mol % LiBF$_4$, 10 mL ethyl acetate, 0.006 mol % [Fe], reflux, 94 hours), the reaction produced a TON of 51,000 and yield of 50%. Using a higher catalyst loading (0.01 mol%), the full conversion of methanol was achieved in 52 hours.

The use of manganese pincer complexes has produced attractive results with respect to dehydrogenation reactions. Hence, Beller et al. described the catalytic activity of MACHO-Mn toward the dehydrogenation of methanol under basic conditions [147]. The catalyst reached a TON > 20,000 using a mixture of methanol/water (20 mL, 9:1), triglyme (20 mL), 8 M KOH, and 10 equivalents of MACHO ligand at 92 °C for > 800 h, which means that the catalyst is stable for over a month.

The catalytic activity of precious metals has been explored as well. The Ir-MACHO based catalyst was active in the dehydrogenation of methanol under basic conditions. Long-term experiments revealed that the catalyst is affected by the concentration of the base. Using 0.5 M of KOH, the conversion to hydrogen stops after 16 hours, giving a TON of 1,400, whereas the catalyst was stable for >60 hours using a KOH concentration of 8.0 M, achieving a TON of 1,900. Interestingly, the addition of a potential catalyst poison such as potassium carbonate did not affect the activity of the catalyst.

Another attractive source of hydrogen must be water through a process called water splitting, which produces H$_2$ and O$_2$, with the advantage that the production of CO is theoretically not possible. As can be noted, this is the reverse of the reaction that occurs in a hydrogen fuel cell. The water splitting reaction has been promoted by several pincer complexes under different reaction conditions, mostly in the presence of sacrificial oxidants, reductants, or different additives [24]. In 1997, Bauer et al. described the catalytic activity of the catalyst [Co(tpp)$_2$]Cl$_2$ to produce hydrogen from water [148]. The reaction was induced by light and produced initial quantum yields up to 20%. Also the electrocatalytic reduction of H$^+$ in the presence of [Co(terpy)$_2$]-type complexes was studied [149, 150]. Other pincer metal complexes with Cu [151], Ni [152–160], Ru [161, 162] have been active in the electrocatalytic and photocatalytic reduction of water. Also, some precious transition metal-based pincer catalysts have been used to promote water splitting and related processes [163–169].

A major breakthrough in the water splitting field, was the production of hydrogen and oxygen from water catalyzed by Milstein's catalyst [63, 170]. The reaction proceeds in two stages. Initially, Milstein's catalyst react with water at 25 °C, yielding a Ru(II) hydride-hydroxo complex. The production of hydrogen comes after the latter species was heated at 100 °C in water, together with the formation of a *cis*-dihydroxo complex, which by irradiation in the 320 to 420 nm range releases molecular oxygen and regenerates the Ru(II) hydride-hydroxo complex. Presumably, hydrogen peroxide (H_2O_2) is formed during this step through reductive elimination of two hydroxo ligands, then H_2O_2 disproportionates very quickly to H_2O and oxygen. Remarkably, t the addition of sacrificial oxidants or reductants was not necessary in this reaction.

The reaction mechanism has been studied by different research groups [171–175]. The activation of water is promoted by the dearomatized arm of the pincer, leading to an aromatized pincer ligand and a Ru(II) hydride-hydroxo species that was characterized by X-ray diffraction analysis; other Ru(II) hydride-hydroxo with different pincer ligands have also been characterized [140, 176]. Ru(II) hydride-hydroxo species may follow two routes as follows: (i) it can undergo a heterolytic coupling of the proton of the side arm with the hydride ligand, forming a dearomatized Ru pincer complex, which is actually the rate limiting step; (ii) a second molecule of water can react with the Ru(II) hydride-hydroxo species, forming hydrogen gas and a *cis*-dihydroxo complex. The light irradiation of the latter complex generates O_2 and water. Fang and Chen proposed that two *cis*-dihydroxo complexes form a dimer, and successive hydrogen atom transfer (HAT) processes occur by photoexcitation, forming a triplet O_2 molecule, water, and a carbonyl Ru(II)-complex [175]. According to their calculations, the formation of H_2O_2 was highly endothermic, and consequently, it is not formed during the reaction. Furthermore, all experimental findings described agree with this mechanism.

18.4 Biomass Transformation

Biomass can be transformed to valuable organic compounds, but one of the most important transformations is in the generation of biofuels [45, 177–181]. The first generation of biofuels was obtained from vegetable oil, animal fats, or recycled restaurant grease [30, 31], as well as from fermentation of sugar-containing materials by yeasts. In the first case, it is possible to obtain so-called biodiesel, whereas bioethanol is produced in the second case. The second generation of biofuels was based on the decomposition of lignocellulosic biomass to produce a mixture of alkanes, alkenes, and alcohols by the Fischer-Tropsch method. At present, the use of butanol isomers as fuel has attracted much attention because these have a higher energy density compared to ethanol and are insoluble in water. In contrast to butanol isomers, ethanol forms an azeotrope with water that is difficult to purify and for direct application as a fuel. Also, butanol isomers have similar properties to gasoline; for example the caloric value of gasoline is 32.5 MJ/L and the value for butanol is 29.2 MJ/L. The research octane number of gasoline ranges from 91 to 99, compared with 96 for butanol. Thus, the use of butanol isomers is attractive. In fact, butanol is considered an "advanced biofuel" [182, 183].

Based on the above, pincer complexes appeared to be a good option for some key steps in the synthesis of biofuels. Williams et al. described a bicatalytic strategy to the conversion of vegetable-derived triglycerides (corn or soybean oil) to fatty acid methyl esters (FAME), which is the basis of the generation of biodiesel fuels [184]. The catalytic system was composed by two catalysts: a N-heterocyclic carbene Ir(I) complex and a Fe-MACHO compound. The iridium complex enables the catalytic hydrogenation of polyunsaturation in corn and soybean oils and the glyceride dehydrogenation in the presence of methanol as a hydrogen source and a base. The presence of the Fe-MACHO complex allows completion of the hydrogenation of the fatty acids, reaching a 90% yield.

The synthesis of butanol isomers from ethanol has been the target to prepare advanced biofuels [185–190]. Ru-pincer complexes have shown good activity. For instance, an amide-derived

N,N,N-Ru(II) pincer and an acridine-based Ru(II)-pincer have been evaluated in the transformation of ethanol to butanol (Scheme 18.12). The N,N,N-Ru(II) pincer complex reached a 31% yield (TON = 420) using 10 mol% of NaOEt, 0.1 mol % of catalyst, and heating the reaction at 150 °C for two hours [185]. An improvement was observed using an acridine-based Ru(II)-pincer complex, which produced a conversion of 73.4%, giving a TON of 3671 [186]. The reaction conditions were similar: 0.03 mol% of catalyst, 20 mol % of NaOEt, at 150 °C for 40 hours.

Mn-MACHO type complexes have been used for the synthesis of butanol isomers. The MACHO-Mn(I) complex reached an impressive TON of 114,120 with 92% of butanol selectivity at 160 °C and in the presence of NaOEt (12 mol%) [187]. Interestingly, the selectivity of the product formation can be tuned by the temperature and the addition of others alcohols; for example, the addition of methanol generates the formation of iso-butanol at 200 °C [190]. The addition of an extra functionality in the N atom of the MACHO pincer ligand (Me-MACHO-Mn(I)) produced a negative effect in the catalytic activity.

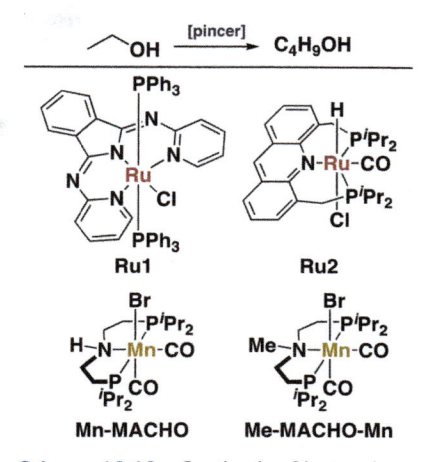

Scheme 18.12 Synthesis of butanol isomers using pincer metal complexes.

18.5 Conclusions

Pincer metal complexes have contributed enormously to the development of greener alternatives to produce energy sources and commodity products. They have been used in several reactions that involve hydrogenation and dehydrogenation reactions, allowing the usage and storage of H_2. Also, they have been used to transform CO_2 into valuable chemicals such as methanol that can be used as liquid organic hydrogen carrier.

In addition, MACHO-type pincer complexes have shown unique reactivities toward the activation of different molecules. The presence of the central nitrogen atom in the ligand plays an active role in the activation of small molecules, such as dihydrogen, allowing the development of highly active catalysts for hydrogenation reactions. When the MACHO-type ligands are coordinated to Mn or Ru, they produce incredibly active catalysts capable of achieving challenging transformations. Among these, the hydrogenation of CO_2 is currently probably one of the most important transformations and might help to palliate the effects of the climate change produced by CO_2. The idea of transforming CO_2 into valuable chemicals is very attractive, but it is even more relevant if it can be used to transport chemical hydrogen in the form of methanol. In a similar manner, Milstein's catalyst has exhibited an incredibly catalytic performance in such a transformation.

Finally, methanol and water are two compounds that have a high content of hydrogen, so their use as liquid hydrogen carriers is very convenient. MACHO-type and Milstein's catalysts have shown an outstanding performance toward the production of molecular hydrogen using methanol and water as starting materials. Both types of complexes have exhibited a good selectivity to produce hydrogen and CO_2, and a negligible amount of CO, during the overall process. Other alcohols such as ethanol can be transformed to advanced biofuels, which might result in novel technologies and applications toward more sustainable processes.

It is expected that the advances in the chemistry of pincer compounds and their uses in catalytic transformations may continue follow this lane and impact the further development of the processes discussed in this chapter, taking advantage of their well-known robustness, enhanced reactivities, and facile synthesis and structural modifications of these pincer species. This without a doubt represents a wide-open window of opportunities for future research in this field.

Acknowledgments

H.V. thanks CONACYT (CVU: 410706) and Generalitat de Catalunya (Beatriu de Pinós MSCA-Cofund 2019-BP-0080). D.M.-M. thanks the generous financial support of PAPIIT-DGAPA-UNAM (PAPIIT IN223323) and CONACYT A1-S-033933.

References

1 Moulton, C.J. and Shaw, B.L. (1976). *J. Chem. Soc., Dalton Trans.* 1976: 1020–1024.
2 van Koten, G., Timmer, K., Noltes, J.G., and Spek, A.L. (1978). *J. Chem. Soc., Chem. Commun.* 1978: 250–252.
3 van Koten, G. (1989). *Pure Appl. Chem.* 61: 1681–1694.
4 Roddick, D.M. (2013). *Top. Organomet. Chem.* 40: 49–88.
5 Valdés, H., González-Sebastián, L., and Morales-Morales, D. (2017). *J. Organomet. Chem.* 845: 229–257.
6 Valdés, H., Germán-Acacio, J.M., and Morales-Morales, D. (2018). In: *Organic Materials as Smart Nanocarriers for Drug Delivery* (ed. A.M. Grumezescu), 245–291. Elsevier - William Andrew. doi: 10.1016/B978-0-12-813663-8.00007-5.
7 Valdés, H., García-Eleno, M.A., Canseco-Gonzalez, D., and Morales-Morales, D. (2018). *ChemCatChem.* 10: 3136–3172.
8 Morales-Morales, D. and Jensen, C. (2007). *The Chemistry of Pincer Compounds.* Amsterdam: Elsevier.
9 Morales-Morales, D. (2008). *Mini-Rev. Org. Chem.* 5: 141–152.
10 Morales-Morales, D. (2004). *Rev. Soc. Quim. Mex.* 48: 338–346.
11 Serrano-Becerra, J.M. and Morales-Morales, D. (2009). *Curr. Org. Synth.* 6: 169–192.
12 Valdés, H., Rufino-Felipe, E., van Koten, G., and Morales-Morales, D. (2020). *Eur. J. Inorg. Chem.* 2020: 4418–4424.
13 Benito-Garagorri, D. and Kirchner, K. (2008). *Acc. Chem. Res.* 41: 201–213.
14 Zell, T. and Milstein, D. (2015). *Acc. Chem. Res.* 48: 1979–1994.
15 Choi, J., MacArthur, A.H.R., Brookhart, M., and Goldman, A.S. (2011). *Chem. Rev.* 111: 1761–1779.
16 Gunanathan, C. and Milstein, D. (2014). *Chem. Rev.* 114: 12024–12087.
17 Selander, N. and Szabó, K.J. (2011). *Chem. Rev.* 111: 2048–2076.
18 Van Der Boom, M.E. and Milstein, D. (2003). *Chem. Rev.* 103: 1759–1792.
19 Bertini, F., Glatz, M., Gorgas, N. et al. (2017). *Chem. Sci.* 8: 5024–5029.
20 Kang, P., Meyer, T.J., and Brookhart, M. (2013). *Chem. Sci.* 4: 3497–3502.
21 Suh, H.-W., Guard, L.M., and Hazari, N. (2014). *Chem. Sci.* 5: 3859–3872.
22 Peris, E. and Crabtree, R.H. (2018). *Chem. Soc. Rev.* 47: 1959–1968.
23 Szabó, K.J. (2013). *Top. Organomet. Chem.* 40: 203–242.
24 Zhang, H.-T. and Zhang, M.-T. (2021). *Top. Organomet. Chem.* 68: 379–449.
25 Valdés, H., Rufino-Felipe, E., and Morales-Morales, D. (2019). *J. Organomet. Chem.* 898: 120864.

26 Valdes, H., German-Acacio, J.M., van Koten, G., and Morales-Morales, D. (2022). *Dalton Trans.* 51: 1724–1744.
27 Kumar, A., Bhatti, T.M., and Goldman, A.S. (2017). *Chem. Rev.* 117: 12357–12384.
28 Alig, L., Fritz, M., and Schneider, S. (2019). *Chem. Rev.* 119: 2681–2751.
29 Wang, W.H., Himeda, Y., Muckerman, J.T. et al. (2015). *Chem. Rev.* 115: 12936–12973.
30 Zhou, Y., Remón, J., Jiang, Z. et al. (2022). *Green Energy Environ*. doi: 10.1016/j.gee.2022.03.001.
31 Kolet, M., Zerbib, D., Nakonechny, F., and Nisnevitch, M. (2020). *Catalysts* 10: 1189.
32 Falbe, J., Bahrmann, H., Lipps, W. et al. (2013). *Ullmann's Encyclopedia of Industrial Chemistry*. Wiley—VCH. doi: 10.1002/14356007.a01_279.pub2.
33 Gregorio, G., Pregaglia, G.F., and Ugo, R. (1972). *J. Organomet. Chem.* 37: 385–387.
34 Matsu-ura, T., Sakaguchi, S., Obora, Y., and Ishii, Y. (2006). *J. Org. Chem.* 71: 8306–8308.
35 Gabriëls, D., Hernández, W.Y., Sels, B. et al. (2015). *Cat. Sci. Tech.* 5: 3876–3902.
36 Xue, W. and Tang, C. (2022). *Energies* 15: 2011.
37 Piccirilli, L., Lobo Justo Pinheiro, D., and Nielsen, M. (2020). *Catalysts* 10: 773.
38 Kar, S., Rauch, M., Kumar, A. et al. (2020). *ACS Catal.* 10: 5511–5515.
39 Wang, Y., Wang, M., Li, Y., and Liu, Q. (2021). *Chem* 7: 1180–1223.
40 Cabrero-Antonino, J.R., Adam, R., Papa, V., and Beller, M. (2020). *Nat. Comm.* 11: 3893.
41 Ekanayake, D.A. and Guan, H. (2020). *Top. Organomet. Chem.* 63: 263–320.
42 Elangovan, S., Topf, C., Fischer, S. et al. (2016). *J. Am. Chem. Soc.* 138: 8809–8814.
43 Csendes, Z., Brunig, J., Yigit, N. et al. (2019). *Eur. J. Inorg. Chem.* 2019: 3503–3510.
44 Weber, S., Brunig, J., Zeindlhofer, V. et al. (2018). *ChemCatChem*. 10: 4386–4394.
45 Padilla, R., Koranchalil, S., and Nielsen, M. (2020). *Green Chem.* 22: 6767–6772.
46 Garbe, M., Junge, K., and Beller, M. (2017). *Eur. J. Org. Chem.* 2017: 4344–4362.
47 Maji, B. and Barman, M. (2017). *Synthesis* 49: 3377–3393.
48 Mukherjee, A. and Milstein, D. (2018). *ACS Catal.* 8: 11435–11469.
49 Filonenko, G.A., van Putten, R., Hensen, E.J.M., and Pidko, E.A. (2018). *Chem. Soc. Rev.* 47: 1459–1483.
50 Kallmeier, F. and Kempe, R. (2018). *Angew. Chem. Int. Ed.* 57: 46–60.
51 Kaithal, A., Holscher, M., and Leitner, W. (2018). *Angew. Chem. Int. Ed.* 57: 13449–13453.
52 Zubar, V., Lebedev, Y., Azofra, L.M. et al. (2018). *Angew. Chem. Int. Ed.* 57: 13439–13443.
53 Kumar, A., Janes, T., Espinosa-Jalapa, N.A., and Milstein, D. (2018). *Angew. Chem. Int. Ed.* 57: 12076–12080.
54 Elangovan, S., Garbe, M., Jiao, H. et al. (2016). *Angew. Chem. Int. Ed.* 55: 15364–15368.
55 Espinosa-Jalapa, N.A., Nerush, A., Shimon, L.J.W. et al. (2017). *Chem. Eur. J.* 23: 5934–5938.
56 Widegren, M.B. and Clarke, M.L. (2018). *Org. Lett.* 20: 2654–2658.
57 Das, K., Waiba, S., Jana, A., and Maji, B. (2022). *Chem. Soc. Rev.* doi: 10.1039/d2cs00093h.
58 Zhang, J., Leitus, G., Ben-David, Y., and Milstein, D. (2006). *Angew. Chem. Int. Ed.* 45: 1113–1115.
59 Polukeev, A.V. and Wendt, O.F. (2018). *J. Organomet. Chem.* 867: 33–50.
60 Li, H., Wen, M., and Wang, Z.X. (2012). *Inorg. Chem.* 51: 5716–5727.
61 Gunanathan, C. and Milstein, D. (2011). *Acc. Chem. Res.* 44: 588–602.
62 Ben-Ari, E., Leitus, G., Shimon, L.J., and Milstein, D. (2006). *J. Am. Chem. Soc.* 128: 15390–15391.
63 Kohl, S.W., Weiner, L., Schwartsburd, L. et al. (2009). *Science* 324: 74–77.
64 Gunanathan, C., Gnanaprakasam, B., Iron, M.A. et al. (2010). *J. Am. Chem. Soc.* 132: 14763–14765.
65 Khaskin, E., Iron, M.A., Shimon, L.J. et al. (2010). *J. Am. Chem. Soc.* 132: 8542–8543.
66 Zeng, H. and Guan, Z. (2011). *J. Am. Chem. Soc.* 133: 1159–1161.
67 Waldie, K.M., Ostericher, A.L., Reineke, M.H. et al. (2018). *ACS Catal.* 8: 1313–1324.
68 Balaraman, E., Gunanathan, C., Zhang, J. et al. (2011). *Nat. Chem.* 3: 609–614.

69 Hasanayn, F., Baroudi, A., Bengali, A.A., and Goldman, A.S. (2013). *Organometallics* 32: 6969–6985.
70 del Pozo, C., Iglesias, M., and Sánchez, F. (2011). *Organometallics* 30: 2180–2188.
71 Sun, Y., Koehler, C., Tan, R. et al. (2011). *Chem. Commun.* 47: 8349–8351.
72 Kim, D., Le, L., Drance, M.J. et al. (2016). *Organometallics* 35: 982–989.
73 Le, L., Liu, J., He, T. et al. (2018). *Organometallics* 37: 3286–3297.
74 Kuriyama, W., Matsumoto, T., Ogata, O. et al. (2012). *Org. Process Res. Dev.* 16: 166–171.
75 Otsuka, T., Ishii, A., Dub, P.A., and Ikariya, T. (2013). *J. Am. Chem. Soc.* 135: 9600–9603.
76 Junge, K., Wendt, B., Jiao, H., and Beller, M. (2014). *ChemCatChem.* 6: 2810–2814.
77 Yuwen, J., Chakraborty, S., Brennessel, W.W., and Jones, W.D. (2017). *ACS Catal.* 7: 3735–3740.
78 Choi, J.H. and Prechtl, M.H.G. (2015). *ChemCatChem.* 7: 1023–1028.
79 Chakraborty, S. and Berke, H. (2014). *ACS Catal.* 4: 2191–2194.
80 Chakraborty, S. and Milstein, D. (2017). *ACS Catal.* 7: 3968–3972.
81 Chakraborty, S., Leitus, G., and Milstein, D. (2017). *Angew. Chem. Int. Ed.* 56: 2074–2078.
82 Vasudevan, K.V., Scott, B.L., and Hanson, S.K. (2012). *Eur. J. Inorg. Chem.* 2012: 4898–4906.
83 Avasare, V.D. (2022). *Inorg. Chem.* 61: 1851–1868.
84 Rezayee, N.M., Huff, C.A., and Sanford, M.S. (2015). *J. Am. Chem. Soc.* 137: 1028–1031.
85 Kar, S., Sen, R., Goeppert, A., and Prakash, G.K.S. (2018). *J. Am. Chem. Soc.* 140: 1580–1583.
86 Kothandaraman, J., Goeppert, A., Czaun, M. et al. (2016). *J. Am. Chem. Soc.* 138: 778–781.
87 Kar, S., Goeppert, A., and Prakash, G.K.S. (2019). *ChemSusChem.* 12: 3172–3177.
88 Kar, S., Sen, R., Kothandaraman, J. et al. (2019). *J. Am. Chem. Soc.* 141: 3160–3170.
89 Bai, S.-T., Zhou, C., Wu, X. et al. (2021). *ACS Catal.* 11: 12682–12691.
90 Zhang, L., Han, Z., Zhao, X. et al. (2015). *Angew. Chem. Int. Ed.* 54: 6186–6189.
91 Kar, S., Goeppert, A., Kothandaraman, J., and Prakash, G.K.S. (2017). *ACS Catal.* 7: 6347–6351.
92 Lane, E.M., Zhang, Y., Hazari, N., and Bernskoetter, W.H. (2019). *Organometallics* 38: 3084–3091.
93 Khusnutdinova, J.R., Garg, J.A., and Milstein, D. (2015). *ACS Catal.* 5: 2416–2422.
94 Huff, C.A. and Sanford, M.S. (2011). *J. Am. Chem. Soc.* 133: 18122–18125.
95 Chu, W.-Y., Culakova, Z., Wang, B.T., and Goldberg, K.I. (2019). *ACS Catal.* 9: 9317–9326.
96 Rayder, T.M., Adillon, E.H., Byers, J.A., and Tsung, C.-K. (2020). *Chem* 6: 1742–1754.
97 Li, Z., Rayder, T.M., Luo, L. et al. (2018). *J. Am. Chem. Soc.* 140: 8082–8085.
98 Kuß, D.A., Hölscher, M., and Leitner, W. (2021). *ChemCatChem.* 13: 3319–3323.
99 Sen, R., Goeppert, A., Kar, S., and Prakash, G.K.S. (2020). *J. Am. Chem. Soc.* 142: 4544–4549.
100 Sen, R., Koch, C.J., Goeppert, A., and Prakash, G.K.S. (2020). *ChemSusChem.* 13: 6318–6322.
101 Scheuermann, M.L., Semproni, S.P., Pappas, I., and Chirik, P.J. (2014). *Inorg. Chem.* 53: 9463–9465.
102 Mazzotta, M.G., Xiong, M., and Abu-Omar, M.M. (2017). *Organometallics* 36: 1688–1691.
103 Sanchez, P., Hernandez-Juarez, M., Rendon, N. et al. (2018). *Dalton Trans.* 47: 16766–16776.
104 Ng, C.K., Wu, J., Hor, T.S., and Luo, H.K. (2016). *Chem. Commun.* 52: 11842–11845.
105 Espinosa, M.R., Charboneau, D.J., Garcia de Oliveira, A., and Hazari, N. (2018). *ACS Catal.* 9: 301–314.
106 Li, H., Goncalves, T.P., Zhao, Q. et al. (2018). *Chem. Commun.* 54: 11395–11398.
107 Lv, H., Xing, Q., Yue, C. et al. (2016). *Chem. Commun.* 52: 6545–6548.
108 Bertini, F., Glatz, M., Stöger, B. et al. (2018). *ACS Catal.* 9: 632–639.
109 Mastalir, M., Glatz, M., Gorgas, N. et al. (2016). *Chem. Eur. J.* 22: 12316–12320.
110 Mastalir, M., Glatz, M., Pittenauer, E. et al. (2016). *J. Am. Chem. Soc.* 138: 15543–15546.
111 Kostera, S., Peruzzini, M., and Gonsalvi, L. (2021). *Catalysts* 11: 58.
112 Ma, N., Tu, C., Xu, Q. et al. (2021). *Dalton Trans.* 50: 2903–2914.

113 Chakraborty, S., Zhang, J., Krause, J.A., and Guan, H. (2010). *J. Am. Chem. Soc.* 132: 8872–8873.
114 Chakraborty, S., Patel, Y.J., Krause, J.A., and Guan, H. (2012). *Polyhedron* 32: 30–34.
115 Chakraborty, S., Zhang, J., Patel, Y.J. et al. (2013). *Inorg. Chem.* 52: 37–47.
116 Liu, T., Meng, W., Ma, Q.Q. et al. (2017). *Dalton Trans.* 46: 4504–4509.
117 Zhang, J., Chang, J., Liu, T. et al. (2018). *Catalysts* 8: 508.
118 Ma, Q.Q., Liu, T., Li, S. et al. (2016). *Chem. Commun.* 52: 14262–14265.
119 Erken, C., Kaithal, A., Sen, S. et al. (2018). *Nat. Comm.* 9: 4521.
120 Zhang, L., Zhao, Y., Liu, C. et al. (2022). *Inorg. Chem.* 61: 5616–5625.
121 Jia, Z., Li, L., Zhang, X. et al. (2022). *Inorg. Chem.* 61: 3970–3980.
122 Kostera, S., Peruzzini, M., Kirchner, K., and Gonsalvi, L. (2020). *ChemCatChem.* 12: 4625–4631.
123 Kar, S., Goeppert, A., and Prakash, G.K.S. (2019). *J. Am. Chem. Soc.* 141: 12518–12521.
124 Ryabchuk, P., Stier, K., Junge, K. et al. (2019). *J. Am. Chem. Soc.* 141: 16923–16929.
125 Kaithal, A., Werle, C., and Leitner, W. (2021). *JACS Au* 1: 130–136.
126 Mahajan, D. (2005). *Top. Catal.* 32: 209–214.
127 Kaithal, A., Holscher, M., and Leitner, W. (2020). *Chem. Sci.* 12: 976–982.
128 Dong, K., Elangovan, S., Sang, R. et al. (2016). *Nat. Comm.* 7: 12075.
129 Satapathy, A., Gadge, S.T., and Bhanage, B.M. (2017). *ChemSusChem.* 10: 1356–1359.
130 Satapathy, A., Gadge, S.T., and Bhanage, B.M. (2018). *ACS Omega* 3: 11097–11103.
131 Pri-Bar, I. and Alper, H. (1990). *Can. J. Chem.* 68: 1544–1547.
132 Gadge, S.T. and Bhanage, B.M. (2013). *J. Org. Chem.* 78: 6793–6797.
133 Palo, D.R., Dagle, R.A., and Holladay, J.D. (2007). *Chem. Rev.* 107: 3992–4021.
134 Salman, M.S., Rambhujun, N., Pratthana, C. et al. (2022). *Ind. Eng. Chem. Res.* 61: 6067–6105.
135 Nielsen, M., Alberico, E., Baumann, W. et al. (2013). *Nature* 495: 85–89.
136 Monney, A., Barsch, E., Sponholz, P. et al. (2014). *Chem. Commun.* 50: 707–709.
137 Alberico, E., Sponholz, P., Cordes, C. et al. (2013). *Angew. Chem. Int. Ed.* 52: 14162–14166.
138 Schwarz, C.H., Agapova, A., Junge, H., and Haumann, M. (2020). *Catal. Today* 342: 178–186.
139 Schwarz, C.H., Kraus, D., Alberico, E. et al. (2021). *Eur. J. Inorg. Chem.* 2021: 1745–1751.
140 Alberico, E., Lennox, A.J.J., Vogt, L.K. et al. (2016). *J. Am. Chem. Soc.* 138: 14890–14904.
141 Yang, X. (2014). *ACS Catal.* 4: 1129–1133.
142 Lei, M., Pan, Y., and Ma, X. (2015). *Eur. J. Inorg. Chem.* 2015: 794–803.
143 Hu, P., Diskin-Posner, Y., Ben-David, Y., and Milstein, D. (2014). *ACS Catal.* 4: 2649–2652.
144 Luo, J., Kar, S., Rauch, M. et al. (2021). *J. Am. Chem. Soc.* 143: 17284–17291.
145 Kar, S. and Milstein, D. (2022). *Chem. Commun.* 58: 3731–3746.
146 Bielinski, E.A., Förster, M., Zhang, Y. et al. (2015). *ACS Catal.* 5: 2404–2415.
147 Anderez-Fernandez, M., Vogt, L.K., Fischer, S. et al. (2017). *Angew. Chem. Int. Ed.* 56: 559–562.
148 Konigstein, C. (1997). *Int. J. Hydrog. Energy* 22: 471–474.
149 Abe, T. and Kaneko, M. (2001). *J. Mol. Catal. A: Chem.* 169: 177–183.
150 Guadalupe, A.R., Usifer, D.A., Potts, K.T. et al. (2002). *J. Am. Chem. Soc.* 110: 3462–3466.
151 Wang, J., Li, C., Zhou, Q. et al. (2016). *Dalton Trans.* 45: 5439–5443.
152 Luo, S., Siegler, M.A., and Bouwman, E. (2018). *Organometallics* 37: 740–747.
153 Yap, C.P., Chong, Y.Y., Chwee, T.S., and Fan, W.Y. (2018). *Dalton Trans.* 47: 8483–8488.
154 Ramakrishnan, S., Chakraborty, S., Brennessel, W.W. et al. (2016). *Chem. Sci.* 7: 117–127.
155 Mondragon, A., Flores-Alamo, M., Martinez-Alanis, P.R. et al. (2015). *Inorg. Chem.* 54: 619–627.
156 Jing, X., Wu, P., Liu, X. et al. (2015). *New J. Chem.* 39: 1051–1059.
157 Luca, O.R., Konezny, S.J., Hunsinger, G.B. et al. (2014). *Polyhedron* 82: 2–6.
158 Luca, O.R., Konezny, S.J., Blakemore, J.D. et al. (2012). *New J. Chem.* 36: 1149–1152.
159 Luca, O.R., Blakemore, J.D., Konezny, S.J. et al. (2012). *Inorg. Chem.* 51: 8704–8709.

160 Levina, V.A., Rossin, A., Belkova, N.V. et al. (2011). *Angew. Chem.* 123: 1403–1406.
161 Boyer, J.L., Polyansky, D.E., Szalda, D.J. et al. (2011). *Angew. Chem. Int. Ed.* 50: 12600–12604.
162 Chen, Z., Glasson, C.R., Holland, P.L., and Meyer, T.J. (2013). *Phys. Chem. Chem. Phys.* 15: 9503–9507.
163 Du, P., Schneider, J., Li, F. et al. (2008). *J. Am. Chem. Soc.* 130: 5056–5058.
164 Elvington, M., Brown, J., Arachchige, S.M., and Brewer, K.J. (2007). *J. Am. Chem. Soc.* 129: 10644–10645.
165 Yamauchi, K., Masaoka, S., and Sakai, K. (2009). *J. Am. Chem. Soc.* 131: 8404–8406.
166 Rabbani, R., Saeedi, S., Nazimuddin, M. et al. (2021). *Chem. Sci.* 12: 15347–15352.
167 Bakir, M., Lawrence, M.A.W., Ferhat, M., and Conry, R.R. (2017). *J. Coord. Chem.* 70: 3048–3064.
168 Lawrence, M.A.W. and Mulder, W.H. (2018). *ChemistrySelect* 3: 8387–8394.
169 Viertl, W., Pann, J., Pehn, R. et al. (2019). *Faraday Discuss.* 215: 141–161.
170 Sandhya, K.S., Remya, G.S., and Suresh, C.H. (2015). *Inorg. Chem.* 54: 11150–11156.
171 Li, J., Shiota, Y., and Yoshizawa, K. (2009). *J. Am. Chem. Soc.* 131: 13584–13585.
172 Yang, X. and Hall, M.B. (2010). *J. Am. Chem. Soc.* 132: 120–130.
173 Sandhya, K.S. and Suresh, C.H. (2011). *Organometallics* 30: 3888–3891.
174 Ma, C., Piccinin, S., and Fabris, S. (2012). *ACS Catal.* 2: 1500–1506.
175 Chen, Y. and Fang, W.H. (2010). *J. Phys. Chem.* 114: 10334–10338.
176 Gibson, D.H., Pariya, C., and Mashuta, M.S. (2004). *Organometallics* 23: 2510–2513.
177 Lao, D.B., Owens, A.C.E., Heinekey, D.M., and Goldberg, K.I. (2013). *ACS Catal.* 3: 2391–2396.
178 Ahmed Foskey, T.J., Heinekey, D.M., and Goldberg, K.I. (2012). *ACS Catal.* 2: 1285–1289.
179 Luque, R., Lin, C.S.K., Wilson, K. (2016). *Handbook of Biofuels Production*. Woodhead Publishing.
180 Li, W., Xie, J.-H., Lin, H., and Zhou, Q.-L. (2012). *Green Chem.* 14: 2388–2390.
181 Kirchhecker, S., Spiegelberg, B., and de Vries, J.G. (2021). *Top. Organomet. Chem.* 69: 341–395.
182 Durre, P. (2007). *Biotechnol. J.* 2: 1525–1534.
183 Harvey, B.G. and Meylemans, H.A. (2011). *J. Chem. Technol. Biotechnol.* 86: 2–9.
184 Lu, Z., Cherepakhin, V., Kapenstein, T., and Williams, T.J. (2018). *ACS Sustain. Chem. Eng.* 6: 5749–5753.
185 Tseng, K.N., Lin, S., Kampf, J.W., and Szymczak, N.K. (2016). *Chem. Commun.* 52: 2901–2904.
186 Xie, Y., Ben-David, Y., Shimon, L.J., and Milstein, D. (2016). *J. Am. Chem. Soc.* 138: 9077–9080.
187 Fu, S., Shao, Z., Wang, Y., and Liu, Q. (2017). *J. Am. Chem. Soc.* 139: 11941–11948.
188 Kulkarni, N.V., Brennessel, W.W., and Jones, W.D. (2018). *ACS Catal.* 8: 997–1002.
189 Rawat, K.S., Mandal, S.C., Bhauriyal, P. et al. (2019). *Cat. Sci. Tech.* 9: 2794–2805.
190 Liu, Y., Shao, Z., Wang, Y. et al. (2019). *ChemSusChem* 12: 3069–3072.

19

Heterometallic Complexes

Novel Catalysts for Sophisticated Chemical Synthesis

*Franco Scalambra, Ismael Francisco Díaz-Ortega, and Antonio Romerosa**

Área de Química Inorgánica-CIESOL, Universidad de Almería, Almería, Spain
* Corresponding author

19.1 Introduction

With the aim of obtaining new more economic and eco-benign synthetic methodologies, researchers have gradually shifted toward new procedures to produce organic compounds with high yield, high stereoselectivity, and a large atom- and step-economy [1, 2]. Additionally, to meet the criteria suggested by green chemistry, the new synthetic processes must minimize waste and by-products to achieve high overall efficiency. Natural enzymes, and the synthetic reactions in which they are involved, are examples of efficient catalytic systems, and have inspired a wide variety of artificial metal catalysts and cooperative organocatalysts over the last two decades [3–10]. In this respect, natural catalysts can transform substrates through the cumulative influence of multiple non-covalent bonds and within tailored cavities, and these features are tentatively emulated by synthetic sophisticated macromolecular catalysts [11–18]. However, this goal requires laborious covalent modification of functional groups [19–21], which is one of the grand challenges of enzyme-inspired synthesis.

Natural enzymes inspired also the design of heterometallic coordination complexes, which attempt to mimic the behaviour of natural catalysts. Most of the structures of these catalysts are the consequence of considering that the reaction rate will rise with increasing substrate concentration in a reduced space. In addition, it is thought that the selectivity of the reaction can be improved by placing the substrate in a suitable position close to the catalytic center. Multiple metal centers can provide the necessary microenvironment to adapt to the substrate and also act as independent or cooperative catalytic units to selectively transform a substrate by more selective procedure. Importantly, a suitable environment around the metal centers is desirable to achieve efficient dual activation of nucleophiles and electrophiles through metal cooperation. Therefore, the concept of bimetallic cooperative catalysis is an appropriate strategy, as it is also important to provide a suitable environment to the metals, achieving efficient dual activation of the target substrates.

Thus, one of the key challenges in this field is to assemble the desired structures around the metal centers in a controllable way. In recent years, various synthetic procedures have been developed to construct heterometallic molecular architectures that mimic the functions of natural catalysts, and the two most important strategies are the so-called metal-ligand and self-sorting

Catalysis for a Sustainable Environment: Reactions, Processes and Applied Technologies Volume 2, First Edition. Edited by Armando J. L. Pombeiro, Manas Sutradhar, and Elisabete C. B. A. Alegria.
© 2024 John Wiley & Sons Ltd. Published 2024 by John Wiley & Sons Ltd.

approaches. A metal-ligand can be described as a coordination fragment that features attached functional groups with the ability to coordinate other metal centers. Potential advantages of this approach include the precise control of the different metals as well as the structural rigidity, functionality, and location of the appended functional groups in the final heterometallic complex. Examples of the successful use of this strategy include the self-assembly of discrete heterometallic units and the synthesis of MOFs [22–24]. The self-sorting approach is based on the ability of different molecules to recognize their mutual homologues, working as a discrimination process to prepare a variety of binding motifs with high efficiency [25–27]. The term self-sorting is described as a network of competing recognition events defined by the binding constants between all possible pairs. Therefore, the followed strategy involves the formation of specific pairs rather than the creation of a library of all the possible non-covalent complexes that might form in the reaction mixture. This clearly distinguishes self-sorting from self-assembly strategies, in which generally identical building blocks are used repeatedly. Self-sorting systems are often based on the same type of binding motifs, which allows cross-interactions to occur between non-matching building blocks. Finally, there are other important factors to consider that are not directly related to the binding motifs themselves and can help to control the outcome of self-sorting. Of these factors, stoichiometry, spacer size and shape, or spacer-spacer interactions can be considered among the most important.

In this chapter, we will describe the most interesting procedures in which heterometallic complexes have been used recently as catalysts and the advantages of these particular compounds over previously studied monometallic complexes, justifying the suitability of these catalysts despite their synthetic complexity.

19.2 C-X Formation (X = C, N, O, Metal)

The synthesis of organic molecules by functionalization of the C-H bond is one of the keystones of catalysis, especially if the processes exhibit high efficiency and atom economy. A step forward in such reactions is the use of catalytic processes mediated by diheterometallic complexes, usually detected as reaction intermediates, which show new catalytic pathways [28] that differ from conventional catalytic processes due to cooperative activation of reactants, giving rise to the enhancement of reactivity and selectivity of reactions. Nevertheless, achieving precise control and understanding of the relationship between two given metals are current challenges [29, 30].

Due to their selectivity, low energy demand, chemical economy, and green approach, some of the most attractive reactions involve the photochemical generation of C-C bonds. In 2021, the modular piece-by-piece synthesis of new decorated mono- and heterobimetallic Au(I)/Ru(II) complexes containing π-conjugated [2.2]paracyclophane (PCP) (**1–4**) was presented [31]. This ligand is a useful rigid spacer that maintains the metals in proper spatial orientations and also the metal-to-metal distances in such a way that, depending on the ligand architecture, can impose different arrangements of the Ru and Au moieties (Figure 19.1).

It is important to note that, beyond its interesting structural properties, this ligand is able to transfer electrons between the two metals due to the innate π-transannular communication through the cyclophane scaffold. Additionally, this ligand, both freely and coordinated to metal, is photosensitive to visible light. An interesting compound was the pseudo-*para*-AuRu complex (**1**), which shows non-additive as well as dynamic M-M interactions, as indicated by ultrafast transient absorption spectroscopy. Also, studies on isolated **1**$^+$ with ultrafast gas-phase photodissociation dynamics confirmed the intrinsic nature of the ultrafast formation of a long-lived electronically excited and reactive state.

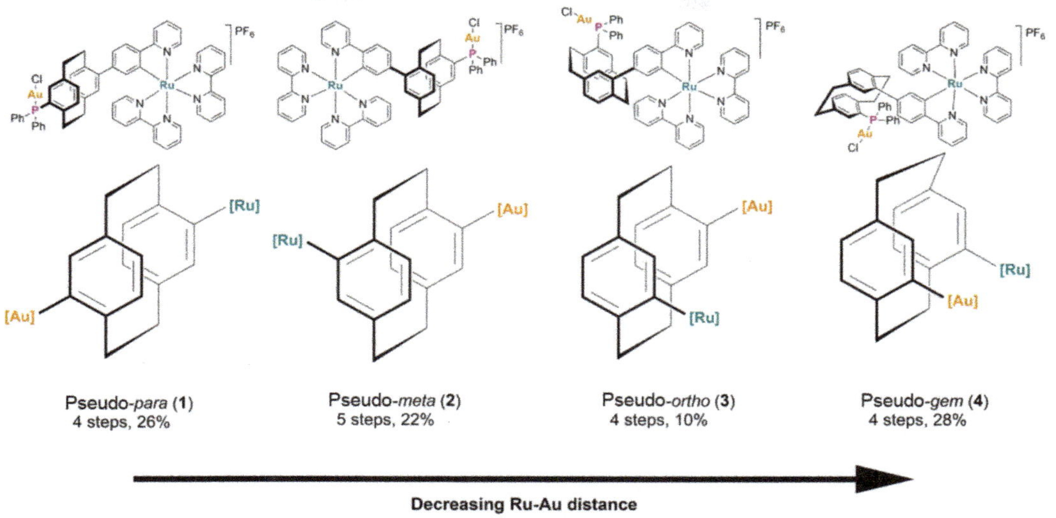

Figure 19.1 [2.2] Paracyclophane substitution pattern for bimetallic Au(I)/Ru(II) complexes **1–4**.

Another interesting feature of the heterodimetallic complexes presented in this study is their selective behaviour depending on the excitation wavelength: under irradiation at 312 nm, the pseudo-*para*-Au-Ru complex **1** experiences a clear influence on the dynamics of the Au moiety, but irradiation at lower energies (405 or 495 nm) only gives rise to dynamics analogous to Ru-bpy complexes, such as [Ru-(bpy)$_2$(ppy)]. Based on the results obtained, complexes **1–4** were evaluated as visible photocatalysts for the arylative Meyer-Schuster rearrangement (Scheme 19.1). Only some of the obtained dimetallic complexes were found to be active catalysts for this reaction. The complexes display a series of steric and electronic qualities, but their most important property is probably the different Au(I)-Ru(II) distances, which define whether the desired electronic interaction between metals can take place.

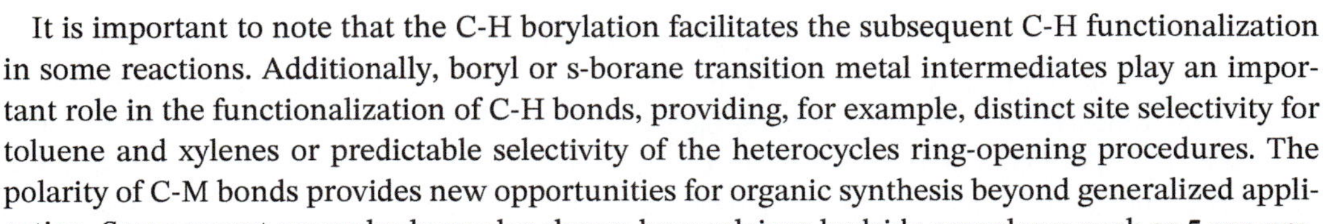

Scheme 19.1 The Meyer-Schuster rearrangement reaction catalyzes the Au-Ru complexes **1–4**.

It is important to note that the C-H borylation facilitates the subsequent C-H functionalization in some reactions. Additionally, boryl or s-borane transition metal intermediates play an important role in the functionalization of C-H bonds, providing, for example, distinct site selectivity for toluene and xylenes or predictable selectivity of the heterocycles ring-opening procedures. The polarity of C-M bonds provides new opportunities for organic synthesis beyond generalized application. Some recent examples have also shown how calcium hydride complexes such as **5** can promote, upon reaction with a diketominate Al species (**6**), the catalytic oxidative alumination of the sp^2 C-H bonds of inactivated benzene, toluene, and xylene to give **7** at room temperature (Scheme 19.2) [32]. It was proposed that the cleavage of the C-H bond is accomplished by activation of

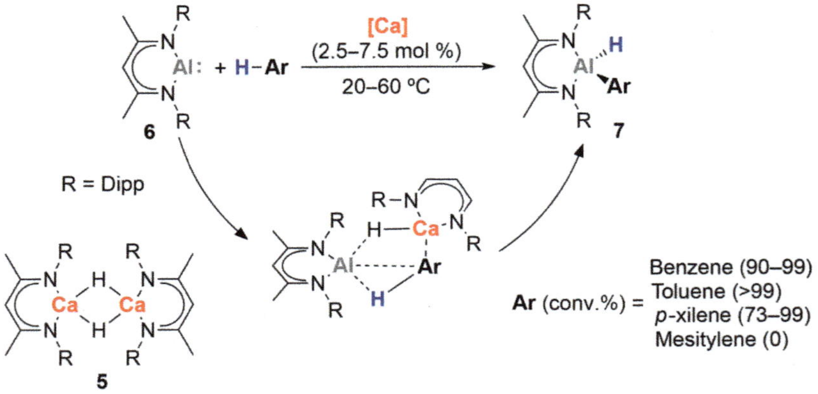

Scheme 19.2 Ca-catalyzed alumination of arenes.

the arene by π-coordination to a Ca, acting as a Lewis acid, which is favored by simultaneous coordination to an Al center. Calculations of the relevant Al-H-Ca species showed that there is an unusual low positive charge on the Al center, similar to that found for the anionic alumanyl complexes that are able to cleave the strong C-H bond of benzene [33, 34]. Nevertheless, being very interesting from a synthetic point of view, the Ca-catalyzed alumination of toluene occurs almost selectively at the *meta*-position of the substrate. The proposed mechanism involves a nucleophilic attack of the aluminum reagent on an arene bearing an electron-donating substituent [35, 36].

In addition, recent studies have revealed the important role of the C-H activation step in reactions involving the cleavage of strong C-O and C-F bonds. When the palladium catalyst [Pd(PCy$_3$)$_2$] was introduced in the reaction, the bimetallic intermediate formed with **6** enables the formation of C-O and C-F via a C-H activation process. This intermediate increases the selectivity of the C-H bond functionalization by properly arranging the substrate and finally allowing the interaction with the metal centers (Scheme 19.3) [37, 38]. According to density functional theory (DFT) calculations the mechanism starts with a (4+1)-cycloaddition of the furan with **6**, followed by a rearrangement of the intermediate. The selectivity of the reaction is correlated to the electronic influence of the substituent R′, which weakens the adjacent C-O sp^2 bond. Deeper in the mechanism, the presence of the [Pd(PCy$_3$)$_2$] complex provides further selectivity to the reaction. At room temperature, the aluminum insertion at the -C-H in position 2 of the substrate gives rise to the kinetic product **8** (Scheme 19.3). Next, under more strained conditions and by reaction with [Pd(PCy$_3$)$_2$], the scission of the furan C-O bond occurs, which implies the attack of an aluminum-based metalloligand on the 2-palladate heterocycle, giving the intermediate **9**. Therefore, the reactivation of **8** with palladium follows an alternative high-energy pathway that finally leads to the thermodynamic products **10a** and **10b**, with **10a** being the major product. Nevertheless, when [Pd(PCy$_3$)$_2$] is not introduced in the reaction, the complex **10b** is the major isomer. In the presence of [Pd(PCy$_3$)$_2$], also 3,4-dihydropyran and 2,3-dihydrofuran, which do not react cleanly with **6**, can undergo this reaction. Finally, at room temperature and in the absence of [Pd(PCy$_3$)$_2$], the C-F alumination reaction of fluoroarenes with **6** (Scheme 19.3) does not occur [39], but C-F alumination of mono-, di-, or trifluorobenzenes in the presence of the palladium complex takes place with high selectivity. The observed regioselectivity has been further investigated through mechanistic studies in combination with DFT calculations. These suggest a stepwise C-H to C-F functionalization process bringing the catalyst to a C-F bond adjacent to a reactive C-H site, giving the intermediate **12** and finally the isomers **13a** and **13b**.

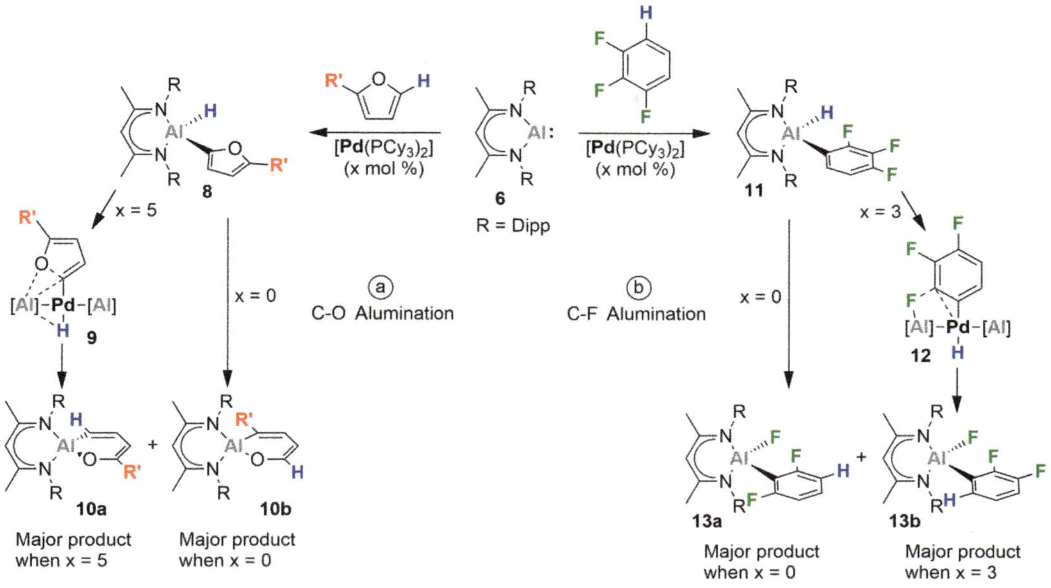

Scheme 19.3 Pd-catalyzed C-O alumination of furans and Pd-catalyzed C-F alumination of fluoroarenes.

Finally, it is interesting to point out that the mentioned C-H functionalization was experimentally supported by determining a 100% atomic efficient palladium-catalyzed isomerization of the kinetic C-H alumination product **14** to the thermodynamic C-F alumination product **15** (Scheme 19.4).

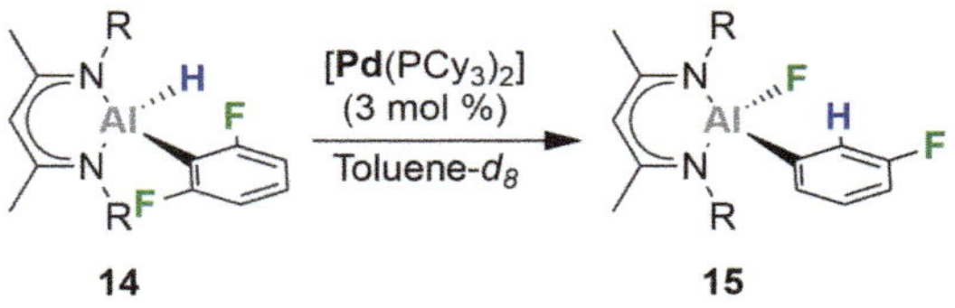

Scheme 19.4 Pd-catalyzed isomerization of **14** to **15**.

The heterometallic Os-Cu complex **16** [40], which contains a pincer ligand, was presented in 2022 as a catalyst for the selective difunctionalization of unactivated aliphatic alkenes with nucleophiles, being the first example of a synthetic strategy based on cooperative heterometallic-metalloaromatic catalysis (Scheme 19.5). More than 80 substrates (including monosubstituted, 1,1-disubstituted, 1,2-disubstituted, 1,1,2-trisubstituted unactivated alkenes, dienes and trienes, and various O- or N-nucleophiles) were reacted. Results showed how good the system is for selective amino- or oxyselenation, providing a wide variety of functionalized products with up to 50:1 rr and 20:1 dr.

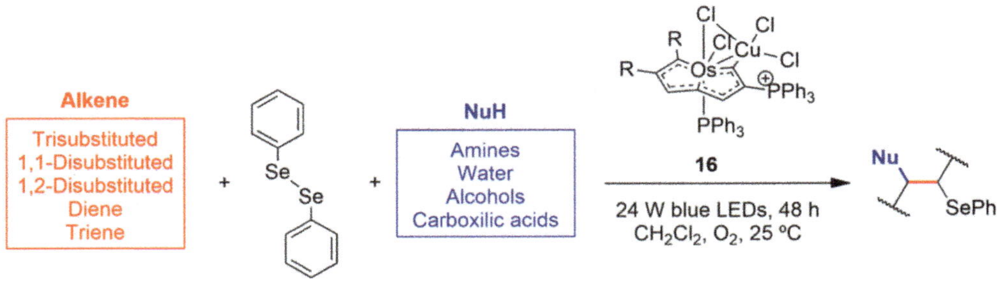

Scheme 19.5 Scope of the difunctionalization of unactivated aliphatic alkenes using **16**.

The experimental results supported that the system gives dual-site activation on the reaction substrates and enforces effective selectivity control, providing a robust catalytic framework and the ability to balance charges during reactions. Also, it was shown that the normally inactive osmium center enables the activation of N or O nucleophiles. (Scheme 19.6) Additionally, the metalloaromatic moiety provides further stabilization of the reaction intermediate and the Cu-Os bond and its cooperative effects determine the selectivity of the reaction by bringing the substrates close together.

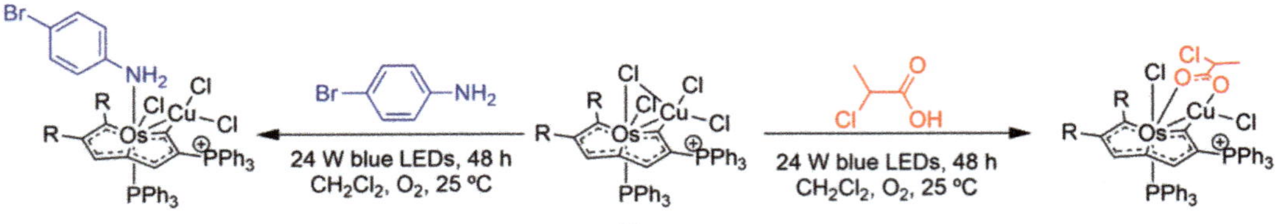

Scheme 19.6 Reactivity of **16** with O- and N-nucleophiles.

In conclusion, cooperative metal-metalloaromatic catalysis significantly improves the reactivity and selectivity of the reactions in which the difunctionalization of unactivated alkenes occurs. The published results prove the important synergistic effects between two metals and show how a proper design of the bimetallic catalyst can promote new, more selective, and efficient, synthetic methods.

19.3 Oxidation Processes

Interesting heterometallic complexes for oxidation reactions containing transition metal/lanthanides complexes were published recently. These complexes are receiving increasing attention due to their substantial activity as catalysts in a variety of synthetic processes, particularly in oxidation synthesis. Three isostructural tetranuclear heterometallic Co(II)–Ln(III) were described by Das et al. [41] The complexes $[Co_2Gd_2L_2(\mu^4\text{-}CO_3)_2(NO_3)_2]$ (**17**), $[Co_2Tb_2L_2(\mu^4\text{-}CO_3)_2(NO_3)_2]$ (**18**), and $[Co_2Dy_2L_2(\mu^4\text{-}CO_3)^2(NO_3)_2]$ (**19**), containing carbonato-bridges, were obtained by reaction of the Mannich ligand (H_2L = N,N′-dimethyl l-N,N′-bis(2-hydroxy-3-methoxy-5-methylbenzyl)-ethylenediamine) with $Co(OAc)_2 \cdot 4H_2O$ and Ln(III) nitrate salts (Ln = Gd, Tb and Dy) under atmospheric CO_2 and basic conditions. The crystal structures of the obtained compounds are characterized as containing two dinuclear $[(Co^{II}L)Ln^{III}(NO_3)]$ moieties linked through two μ^4-carbonato groups forming the tetranuclear $\{Co^{II}{}_2Ln^{III}{}_2\}$ core (Scheme 19.7).

The combination of Co(II) with the Ln(III) center led to heterometallic systems with moderate catalytic catecholase-like activity (Scheme 19.8) and quite high phenoxazinone-synthase-like activity (Scheme 19.9). In contrast, the analogue homometallic tetranuclear Co(II) complexes were inactive in the studied synthetic processes. The turnover frequency (TOF) values obtained for the catecholase-like reaction catalyzed by complexes **17–19** were respectively 254.5 h^{-1}, 272.4 h^{-1} and 291.3 h^{-1}, whereas for the phenoxazinone-synthase-like reaction they were 2930.6 h^{-1}, 2965.2 h^{-1}, and 2998.5 h^{-1}.

Results obtained from the study of the reactions with mass spectrometry suggested that the dinuclear $[(Co^{II}L)Ln^{III}CO_3]^{2+}$ ions, which form upon dissociation of the tetranuclear complexes,

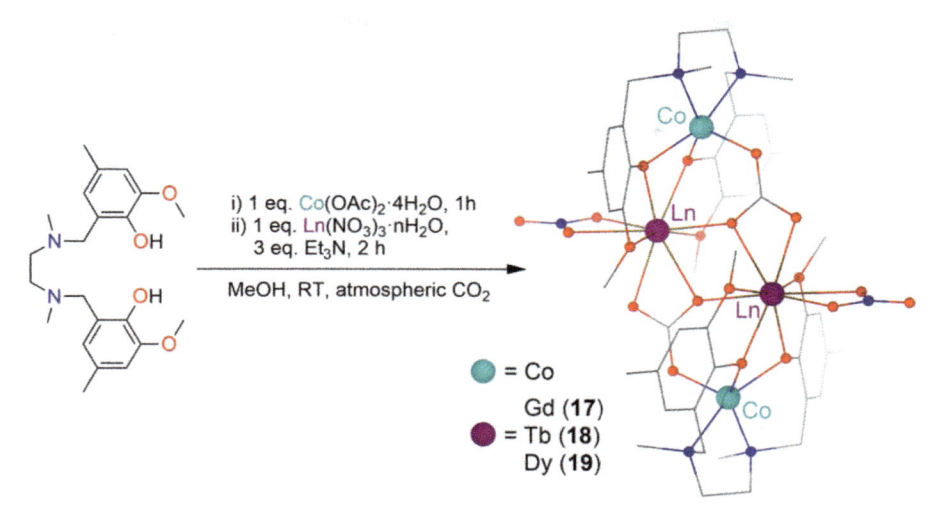

Scheme 19.7 Synthesis of complexes **17–19**. The general structure of the products is depicted from the molecular structure of **19** as found by single-crystal X-ray diffraction. Hydrogen atoms were removed for clarity. Adapted from [41].

Scheme 19.8 Oxidation of 3,5-DTBC to 3,5-DTBQ catalyzed by **17–19**.

TOF = 254.5 h^{-1} (**17**)
272.4 h^{-1} (**18**)
291.3 h^{-1} (**19**)

TOF = 2930.6 h^{-1} (**17**)
2965.2 h^{-1} (**18**)
2998.5 h^{-1} (**19**)

Scheme 19.9 Catalytic oxidation of o-aminophenol to phenoxazinone mediated by **17–19**.

are the catalytically active species. The authors proposed that the interaction between the substrate and catalyst is favored by the higher oxophilicity of Ln(III) ions, whereas the cobalt ion oxidizes the substrate by exchanging its oxidation state between +2 and +3, assisted by molecular oxygen in the electronic give-and-take. Therefore, the catalytic oxidase activities of the studied heterometallic complexes are favored by the presence of two metal centers.

19.4 CO$_2$ Activation

The activation of CO$_2$ is a process of paramount importance in chemistry to obtain a large variety of important organic molecules with industrial and biological relevance, and currently of particular interest for the sequestration of atmospheric CO$_2$. With respect to the activation of CO$_2$ controlled by heterometallic catalysts, a pinacolborane oxidation parallel to an interesting CO$_2$ reduction was described by Bagherzadeh and Mankad in 2015 [42]. The authors employed heterometallic Cu-M complexes (**20–25**) [43], consisting of a {NHC-Cu} unit bonded to the carbonyl half sandwiches [MoCp(CO)$_3$], [WCp(CO)$_3$], and [FeCp(CO)$_2$]$^-$ (Figure 19.2), as catalysts to react CO$_2$ with (pin)B-B(pin).

The reaction gave CO and HCO$_2$B(pin) (compound **23** in Figure 19.2), as well as the boryl ether (pin)B-O-B(pin) (**24**, Figure 19.2) (pin = pinacole), through deoxygenation of CO$_2$. On the other hand, if the product (pin)B-OH (**25**) is afforded, which is also generated by fulfilling complete CO$_2$ deoxygenation, it would react with H-B(pin) to generate **27** and H$_2$, and therefore should be difficult to be isolated and determined as the reaction product. The lability of the [M]$^-$(Bpin) seems related to the rate of decarbonylation, and follows the [M] leaving group ability ([MoCp(CO)$_3$]> [WCp(CO)$_3$]>[FeCp(CO)$_2$]), which also correlates with their relative pK$_a$ values [44]. Despite the studies performed, the authors have not been able to determine if the decarbonylation reactivity could also come from the polarity inversion during the catalyst activation step, which gives rise to the pair (NHC)Cu(Bpin) + [M]H, where the {(NHC)Cu(Bpin)} intermediate is the catalyst for the deoxygenation of CO$_2$ (NHC = N-heterocyclic carbene) [45]. The proposed mechanism for this process is shown in (Scheme 19.10): initially, the NHC-Cu-[M] complex undergoes self-activation and the Cu-component reacts with CO$_2$. The resulting Cu-formate transfers CO$_2$ to the HB(pin) upon insertion on the B-H bond, forming **26**. Then, through rate-determining, electrophilic activation by [M]-E (E = (Bpin) at the first turnover and then H throughout the process) the transfer of E$^+$ to **26** initiates the decarbonylation pathway, which proceeds through the intermediate **29**. In this reaction step, the reversible activation should be conducted through the sequestration of [M]$^-$ by CO$_2$ [46–48].

Based on the known catalytic properties of the monometallic {(NHC)Cu} moiety for CO$_2$ reduction, the heterometallic complexes (NHC)Cu-[M] **20–25** (NHC = N-heterocyclic carbenes) were synthesized and their catalytic activity studied for this reaction [49]. In contrast to monometallic Cu complexes, which catalyze exclusively the hydroboration of CO$_2$ with pinacolborane to

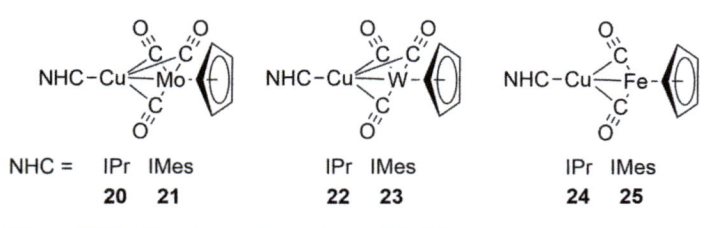

NHC =	IPr	IMes		IPr	IMes		IPr	IMes
	20	**21**		**22**	**23**		**24**	**25**

Figure 19.2 Structures of complexes **20–25**.

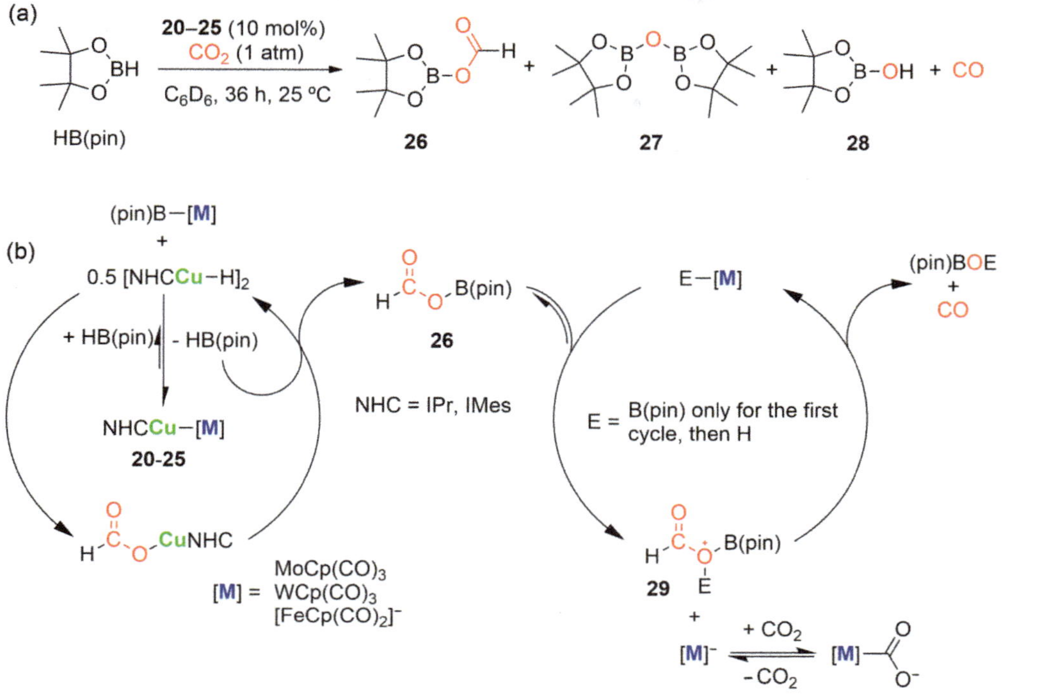

Scheme 19.10 (a) Catalytic pinacolborane oxidation via CO_2 reduction using **20–25**. (b) Proposed catalyst activation and hypothetical catalytic mechanism.

produce formate, analogous diheterometallic complexes containing Cu–Fe, Cu–W and Cu–Mo give rise to a mixture of pinacolborane-formate and CO. The reaction selectivity on CO vs. formate was controlled by tuning the electronic nature of the Cu/M pair, being the Cu-Mo complex which generates catalytically the largest amount of CO.

Heterometallic complexes as catalytic activators of CO_2 have been extended to the polymerization reaction. Polyolefins, cheap and easy to produce, are extensively used in industry for a wide variety of applications, being frequently found in packaging, automotive and electrical components, lubricants, and medical devices, among other products. One of the most interesting features of polyolefins is how their properties can be deeply changed by the catalyst used in their synthesis. Modifications of ligands, metal and geometry of the complex, and the employed co-catalyst can deeply modify the molecular weight and branching of the final product. The inclusion of a second catalytic center usually results in important alterations that affect the catalytic activity, polymer mass and structure, and comonomer enchainment selectivity, offering an additional point to tune the polymerization. These effects are as strong as short the M···M distance is, which can be modulated by ligation, counterion and solvent effects [50].

Ring-opening copolymerization (ROCOP) of epoxides and heterocumulenes is an attractive strategy to recycle CO_2, transforming it directly into useful materials such as polycarbonates/ester copolymers and also polycarbonate polyols, which can be further processed to synthesize polyurethanes [51–66]. So far, homodinuclear complexes have been used for epoxide and carbon dioxide/anhydride ring-opening copolymerization (ROCOP) but a few examples of heterodinuclear catalysts are known and were reported only recently. The first example of a catalyst for ROCOP was the Zn(II)/Mg(II) complex (**30**) (Figure 19.3), which is based on a tetraiminodiphenol macrocyclic ligand presented in 2015 [67].

Figure 19.3 Structures of complexes **30–32** and cyclohexene oxide/CO_2 catalyzed copolymerization.

M^1 = Mg, M^2 = Zn, X = Br (**30**)
M^1 = M^2 = Mg, X = Br (**31**)
M^1 = M^2 = Zn, X = Br (**32**)

TOF = 34 h^{-1} (**30**)
     15 h^{-1} (**31**)
     0 (**32**)

Conditions: 30–32 0.1 mol%, THF, 1 bar, 80 °C, 10 h

Complex **30** performs the polymerization reaction significantly better than the equivalent homodimetallic **31** and **32**, which contain Mg and Zn, respectively. For example, in the copolymerization of CO_2 and cyclohexene oxide at 1 bar, the TOF and turnover number (TON) obtained with **30** were more than double the values obtained with **31**, while **32** has not shown catalytic activity for the reaction. Also, **30** showed excellent efficiency for CO_2 uptake and high reaction control, giving polymers with narrow molecular weight distribution and an almost unitary ratio of carbonate linkages. The high degree of control of the polymerization mediated by **30** was also supported by a higher molecular weight of the obtained polymer and a marked increase of TOF (624 h^{-1}) at lower catalyst loadings, which is significantly competitive with results obtained with mononuclear catalysts containing Cr and Co [68–70].

In general, complex **30** showed a higher polymerization selectivity, control, and activity in comparison with the homodinuclear analogues and also combinations of them. The advantages offered by the heteronuclearity seemed related to the augmented epoxide coordination offered by the Zn, together with a faster carbonate attack provided by the Mg (Scheme 19.11). Additionally, this synergy between the different metal centers also activates the halide co-ligand to start the polymerization, which was found innocent when the homometallic di-zinc complex **32** was used [71–74].

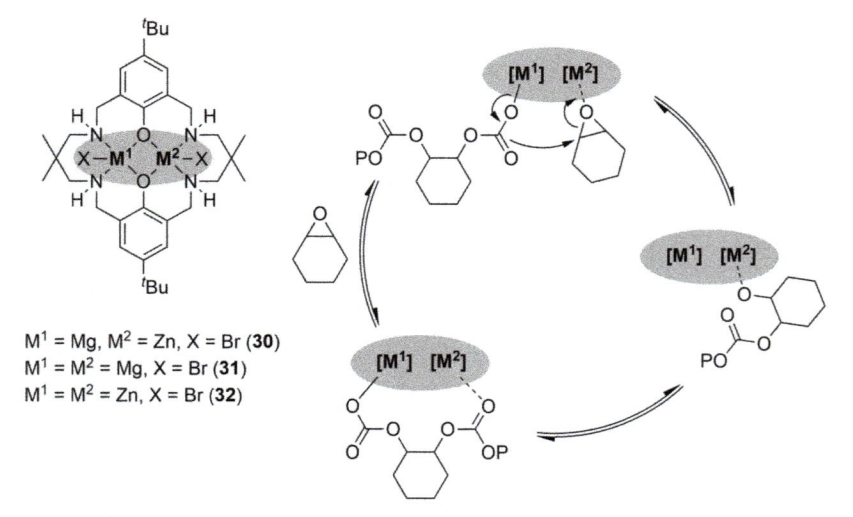

Scheme 19.11 Proposed mechanism for CO_2/epoxide copolymerization catalyzed by the dinuclear complexes **30–32**.

Another interesting approach was the use of heterobimetallic catalysts containing the {(salen)Al} moiety for the lactide ring-opening polymerization (ROP) [75–77], which underlined the real importance of choosing the adequate metal centers. The monometallic complexes containing Al (**34**), Mg (**35a, 35b**) and Zn (**36**) based on the salen-type ligand (**33**) were prepared and compared with their heterometallic derivatives containing Al/Mg (**37**), Al/Zn (**38**), Al/Li (**39**) and Al/Ca (**40**) (Scheme 19.12). Whereas monometallic Mg and Zn complexes were found to be inactive in rac-lactide ROP (rac-LA ROP), Al/Mg and Al/Zn combinations speeded up the catalytic reaction to 11 times compared with the mono-Al (**34**). On the other hand, Al/Mg and Al/Zn combinations inhibited the polymerization rates. Based on ab initio molecular dynamics (AIMD) calculations, structural studies and reaction kinetics, the authors propose that the improvement of the catalytic activity obtained with the Al/Mg and Al/Zn heterometallic complexes can arise from various factors: one of the reasons is the capacity of the second metal to provide additional coordination

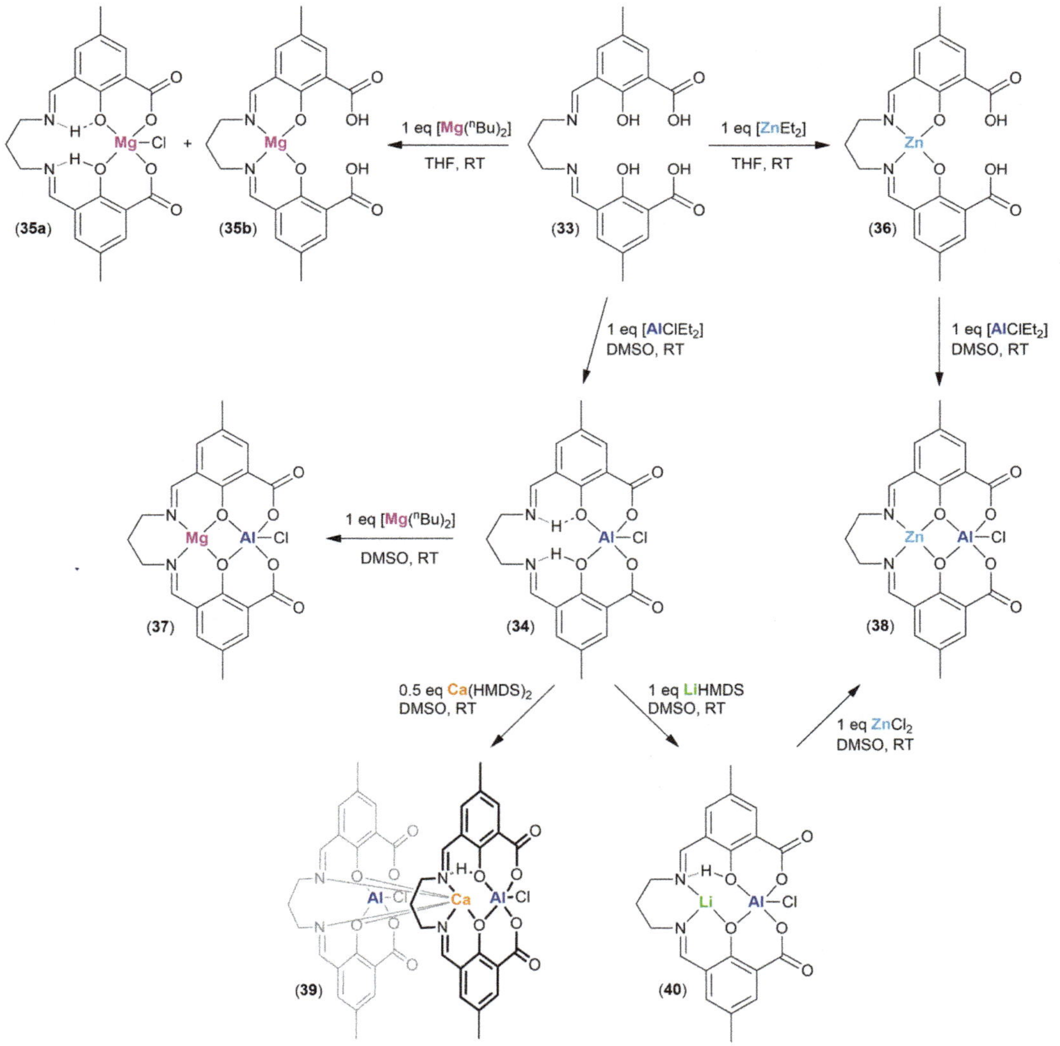

Scheme 19.12 Synthesis of complexes **34–40**.

sites for monomer binding, exhibiting additional Lewis acid sites and also weakening the Al–Cl bond, speeding up the reaction initiation. Also, the chloride bridge causes the Al center to adopt a square pyramidal geometry, that augments the vacant sites at the aluminium, thus shortening the induction period. Finally, the proximity of the metal center seems to compensate for the structural rigidity of the Al/Mg and Al/Zn complexes. It is important to point out that the use of AIMD calculations (applied for the first time to ROP by the authors) revealed fundamental details about the reactivity, offering a deeper and detailed understanding of the features that drive the activity of the studied catalysts.

19.5 O_2 and H_2 Generation from Water

The world's increasing fuel demand is directing many scientific efforts toward the development of practical processes that allow energy production from the greener and cheapest source at our disposal: water. In this sense, water can be split to cleanly generate oxygen and hydrogen, which can be subsequently used as fuel. So far, many electrode systems and heterogeneous catalysts have been proposed for electrochemical and photoelectrochemical water splitting. With respect to homogeneous catalysis, many transition metal complexes were found to be practical for the generation of oxygen and hydrogen from water, but only in the last decade the scientific efforts, taking example from the natural photosystem II (PS-II), started to consider multimetallic systems as an attractive concept to solve some of the problems inherent to water splitting [78–81]. An ideal artificial system inspired by PS-II should be composed of an anode consisting of a photosensitizer that allows the oxidation of two water molecules by a water oxidation catalyst (WOC) (Eq. 19.1), while, at the cathode side, four electrons are employed by a hydrogen evolving catalyst (HEC) to reduce four protons and generate two molecules of hydrogen (Eq. 19.2). Due to its complexity, the water oxidation step (Eq. 19.1) is the most problematic process.

$$2H_2O \rightarrow O_2 + 4H^+ + 4e^- \tag{19.1}$$

$$4H^+ + 4e^- \rightarrow 2H_2 \tag{19.2}$$

Among the proposed methods to obtain an efficient WOC, the radical coupling between the oxygen of a hydroxocerium(IV) ion and a metal oxo complex is an attractive path. The ruthenium(III) complex [Ru(L)(pic)$_3$] (**41**) (H$_3$L = 2,2'-iminodibenzoic acid, pic = 4-methylpyridine) [82], is an interesting example of a catalyst designed by taking into account and putting together various beneficial factors targeted to oxidation of liquid water: (i) the presence of a negatively charged ligand to stabilize high valent metal intermediate species; (ii) the choice of a redox non-innocent ligand that can cooperate with the metal in the electron transfers; (iii) water coordination to the metal to support proton-coupled electron transfer. In the presence of sacrificial ceric ammonium nitrate at pH = 1, complex **41** can generate O$_2$ with TON = 200 and very high TOF (0.168 s^{-1}).

The mechanism of the water oxidation mediated by **41** was investigated by kinetic studies and DFT calculations. What emerged is that the oxidation of water occurs at the ruthenium center, where the metal-ligand cooperation allows for the storage of many oxidative equivalents at a single site, which is mandatory for the required four electrons transfer to generate one O$_2$ molecule. The formation of the O-O bond proceeds via formal high-valent Ru(VII) species involving a multimetallic Ru-Ce intermediate (Scheme 19.13) with the participation of the anionic tridentate ligand in the electron-transfer process.

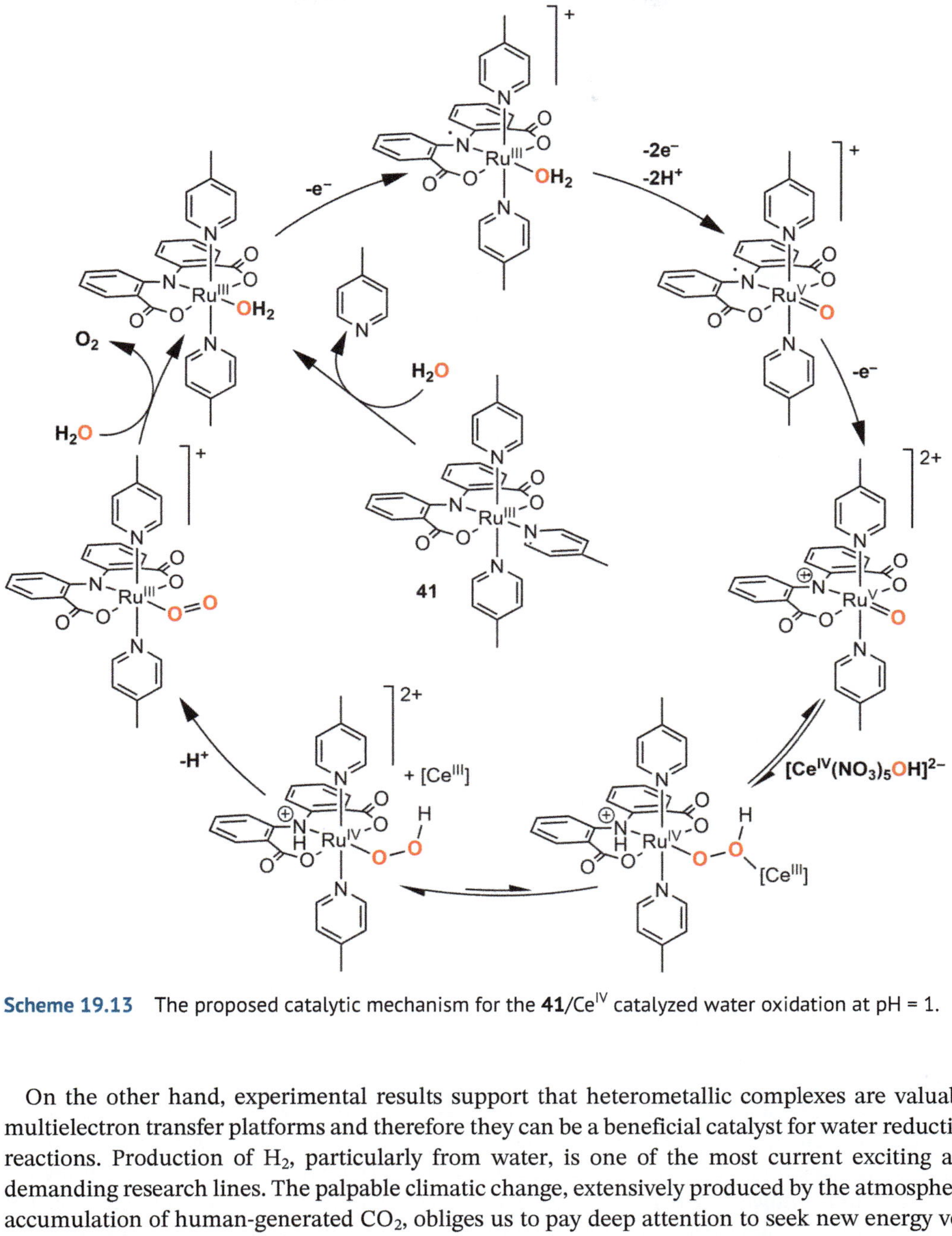

Scheme 19.13 The proposed catalytic mechanism for the **41**/Ce^{IV} catalyzed water oxidation at pH = 1.

On the other hand, experimental results support that heterometallic complexes are valuable multielectron transfer platforms and therefore they can be a beneficial catalyst for water reduction reactions. Production of H_2, particularly from water, is one of the most current exciting and demanding research lines. The palpable climatic change, extensively produced by the atmospheric accumulation of human-generated CO_2, obliges us to pay deep attention to seek new energy vectors different to fossil fuels, being hydrogen one of the most promising. Due to the well-known reaction of H_2 with O_2 gives rise to H_2O and a large amount of energy. Nevertheless, one of the big limits to the general use of H_2 as a substitute for non-renewable energy sources is developing ways to synthesize it by a practical and economical procedure. Water should be the natural favorite source for generating H_2 but the process requires a large amount of energy, which could be obtained from a renewable source such as the sun. Therefore, very intense research efforts are

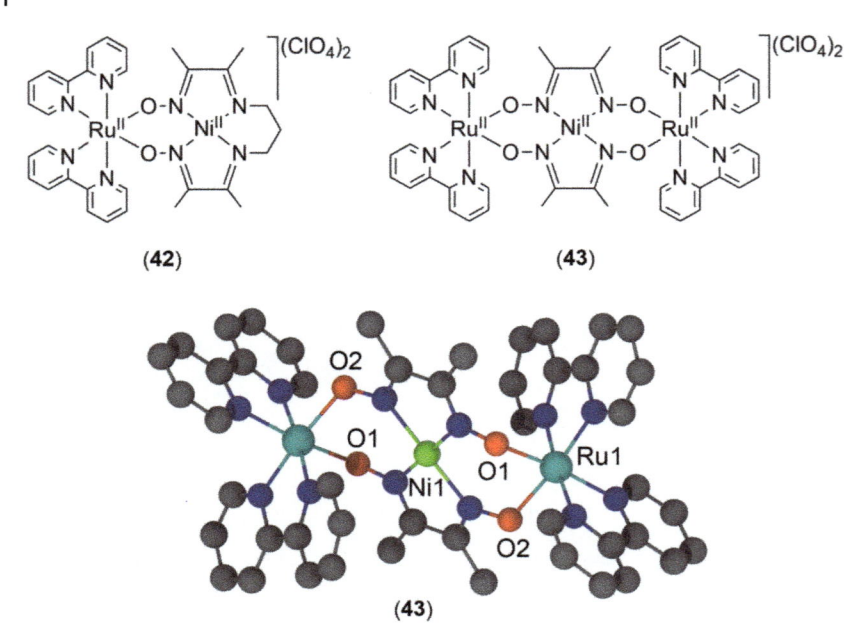

Figure 19.4 Top panel: structures of complexes **42** and **43**. Bottom panel: representation of the molecular structure of **43** as found by single-crystal X-ray diffraction; hydrogen atoms and anion were omitted for clarity.

ongoing to obtain catalysts whose reactivity is driven by light. Although several mononuclear species have been reported up to date, the exploration of multimetallic molecular photocatalysts started only recently. In 2021, Verani et al. presented a very interesting approach to water photoreduction [83]. The idea, supported by DFT, was to covalently bind two Ru(II) photosensitizer to a [Ni^{II}(oxime)] complex. The Ru(II) domains, serve as photoactivators, providing to the Ni center the electrons needed for proton reduction. To prove and support this hypothesis, the dimetallic [(bpy)$_2$Ru^{II}Ni^{II}(L$_1$)](ClO$_4$)$_2$ (**42**) and a trimetallic [(bpy)$_2$Ru^{II}Ni^{II}(L$_2$)$_2$Ru^{II}(bpy)$_2$](ClO$_4$)$_2$ (**43**) (Figure 19.4) were synthesized. Both complexes are able to generate the low-valent precursor involved in hydride formation before generating dihydrogen.

Nevertheless, complex **42** was found ineffective as a catalyst for the process due to the high energy barrier required for the protonation of the Ni(I) center after photoactivation. Further calculations and experimental clues obtained using **42** in presence of [Ru(bpy)$_3$]$^{2+}$ revealed that a second electron was necessary to trigger the reaction. Thus, complex **43** was synthesized and under blue light, as expected, the predicted protonation of the Ni(I) with the formation of a [Ru^{II}(Ni^{II}-H$^-$)Ru^{II}] species was achieved. Finally, this species was able to react with one proton to generate hydrogen. The entire catalytic process required 24 h, being the reaction TON of 49 in H$_2$O/CH$_3$CN at pH = 11, using triethylamine as a sacrificial electron donor (Scheme 19.14). The experimental and DFT studies suggested that in the catalytic process monometallic Ni(0) species were not formed and the second electron coming from the photoexcitation presumably remains on a bpy ring of one of the Ru(II) domains.

One of the most recent works published concerning catalytic H$_2$ addresses photo-generation using a bimetallic complex (**44–46**), showing that complexes containing noble metals are still components of reliable strategies to develop useful catalysts for this process, even if exploiting abundant metals should be the target for designing new catalysts [84]. Among the published

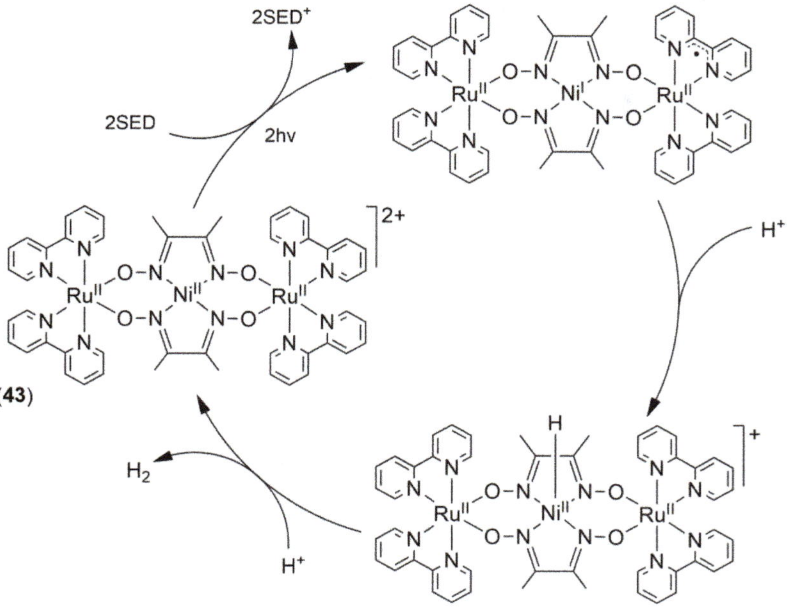

Scheme 19.14 Proposed mechanism for the photocatalyzed proton reduction, being catalyzed by **43**.

complexes, the most active for photocatalytic H_2 production was the Ru(II)-Rh(III) complex **46**, which was found to be able also to photocatalyze the generation of H_2 more efficiently than its parts and other similar complexes (Figure 19.5).

Only a small quantity, almost a trace amount, of H_2 was photo-generated by irradiation at $\lambda =$ 480 nm from an Ar-saturated mixture of dimethylacetamide–triethanolamine (4:1, v/v) together with the reductant 1,3-dimethyl-2-phenyl-2,3-dihydro-1H-benzo-[d]imidazole. Nevertheless, when a proton source was introduced in the reaction (3,5-difluorophenol (3,5-F_2–PhOH) or 2-(1,3-dimethyl-2,3-dihydro-1H-benzo[d]imidazol-2-yl)phenol) the generation of H_2 increased significantly. Interestingly, when the reaction was carried out under CO_2 without acid, the main product generated was H_2 (30.9 μmol; TON_{H2} =157.5), producing small amounts of HCOOH and CO. It is important to stress that the high quantum yield of the reaction (Φ_{H2} = 16.4%) was supported by the fact that the reaction of triethanolamine with CO_2 generates protons [85] and gives rise to a zwitterionic alkyl carbonate that acts as an acid, being the H^+ source. Under an Ar atmosphere and without acid, there are not enough protons for the photoproduction of H_2.

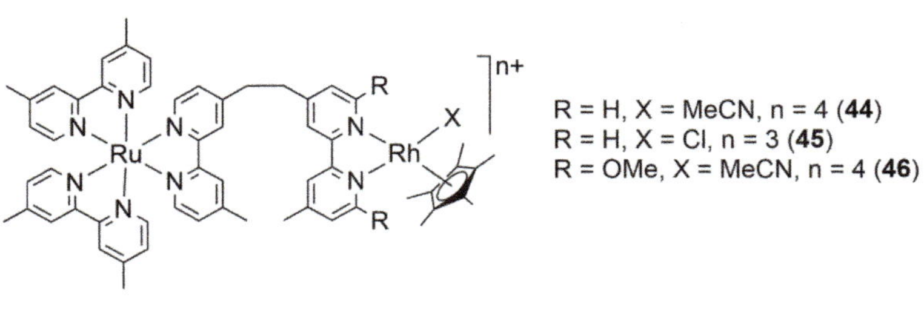

R = H, X = MeCN, n = 4 (**44**)
R = H, X = Cl, n = 3 (**45**)
R = OMe, X = MeCN, n = 4 (**46**)

Figure 19.5 Structure of complexes **44–46**.

The high photochemical H_2 generation mediated by **46** was justified as a consequence of the faster electron transfer from the Ru photosensitizer unit to the Rh catalyst center [86]. The photocatalytic ability of the Ru(II)–Rh(III) heterometallic complexes **44–46** is dependent on the diimine ligand coordinated to the Rh, which determines the final product under the same reaction conditions. A strong electron-donating methoxy substituent at the 6,6'-positions of the bipyridine bridging ligand induces a high and selective formation of H_2, while an electro-attracting substituent mostly leads to the formation of HCOOH, which was previously observed for Ru-mononuclear complexes containing bpy. The complex **44**, in which the bridging ligand bipyridine does not support substitutions, is not active for producing photocatalytically H_2. This was explained by the authors on the basis that Ru^{II}–Rh^{III}(H) and/or Ru^{II}–Rh(Cp*H) species do not react with a proton, even under the CO_2 atmosphere, producing an excited Ru unit that reacts with CO_2, giving rise finally to HCOOH. In contrast, an electron-donating substituted at the 6,6'-positions of the diimine ligand, which is also close to the Rh metal center, could help a mutual interaction between the proton source and "hydride reduced" species in the intermediates Ru^{II}–Rh^{III}(H)–OMe and/or Ru^{II}–Rh(Cp*H)–OMe and the faster generation of H_2 [87].

19.6 Conclusions

Recent results in processes catalyzed by heterometallic complexes are well-defined evidences of the huge advantages that the synergic collaboration between metals can provide to homogeneous catalysis when they are forming part of a single complex. Nevertheless, the introduction of additional metallic centers in the structure of a catalyst complicates enormously its synthesis and the interpretation of the mechanisms, increasing the effort needed to improve its performance. Some indications of the structure-activity-relationship of multimetallic catalysts can be extracted from the obtained results, which give some general indications about their design and possible properties. Firstly, the chosen metals need to display an affinity with the substrate in the starting complex and/or reaction intermediates, modulating the balance between activation and stabilization. Then, the distance between the metals is extremely important and should be weighed based on the substrate size. Also, for reactions involving electron transfers, groups connecting metals electronically are necessary. Finally, for photocatalytic reactions, an accurate evaluation of the combined excited state properties is essential to drive the reactivity far from unproductive pathways. In addition to all of these factors, the designed catalyst must be stable and robust to air and light and preferably built with non-noble metals.

Note

The authors declare no conflict of interest.

Acknowledgments

The authors thank Junta de Andalucía for funding the group PAI FQM-317 and the project PY20_00791, and the University of Almería for the project UAL2020-RNM-B2084 (both projects co-funded by the European Commission FEDER program).

References

1. Wender, P.A. and Miller, B.L. (2009). *Nature* 460 (7252): 197–201.
2. Trost, B. (1991). *Science* 254 (5037): 1471–1477.
3. Sawamura, M. and Ito, Y. (1992). *Chem. Rev.* 92 (5): 857–871.
4. Ma, J.-A. and Cahard, D. (2004). *Angew. Chem. Int. Ed.* 43 (35): 4566–4583.
5. Yamamoto, H. and Futatsugi, K. (2005). *Angew. Chem. Int. Ed.* 44 (13): 1924–1942.
6. Kanai, M., Kato, N., Ichikawa, E., and Shibasaki, M. (2005). *Synlett* 2005 (10): 1491–1508.
7. Taylor, M.S., Jacobsen, E.N., Jacobsen, E.N., and Taylor, M.S. (2006). *Angew. Chem. Int. Ed.* 45 (10): 1520–1543.
8. Mukherjee, S., Yang, J.W., Hoffmann, S., and List, B. (2007). *Chem. Rev.* 107 (12): 5471–5569.
9. Shibasaki, M., Kanai, M., Matsunaca, S., and Kumagai, N. (2009). *Acc. Chem. Res.* 42 (8): 1117–1127.
10. Park, J. and Hong, S. (2012). *Chem. Soc. Rev.* 41 (21): 6931–6943.
11. Ward, M.D., Hunter, C.A., and Williams, N.H. (2018). *Acc. Chem. Res.* 51 (9): 2073–2082.
12. Jongkind, L.J., Caumes, X., Hartendorp, A.P.T., and Reek, J.N.H. (2018). *Acc. Chem. Res.* 51 (9): 2115–2128.
13. Brown, C.J., Toste, F.D., Bergman, R.G., and Raymond, K.N. (2015). *Chem. Rev.* 115 (9): 3012–3035.
14. Gianneschi, N.C., Masar, M.S., and Mirkin, C.A. (2005). *Acc. Chem. Res.* 38 (11): 825–837.
15. Yoshizawa, M., Klosterman, J.K., and Fujita, M. (2009). *Angew. Chem. Int. Ed.* 48 (19): 3418–3438.
16. Otte, M. (2016). *ACS Catal.* 6 (10): 6491–6510.
17. Pluth, M.D., Bergman, R.G., and Raymond, K.N. (2009). *Acc. Chem. Res.* 42 (10): 1650–1659.
18. Raynal, M., Ballester, P., Vidal-Ferran, A., and van Leeuwen, P.W.N.M. (2014). *Chem. Soc. Rev.* 43 (5): 1660–1733.
19. Hasell, T. and Cooper, A.I. (2016). *Nat. Rev. Mater.* 1 (9): 1–14.
20. Gramage-Doria, R., Armspach, D., and Matt, D. (2013). *Coord. Chem. Rev.* 257 (3–4): 776–816.
21. Gale, P., Gunnlaugsson, T., and Ballester, P. (2010). *Chem. Soc. Rev.* 39 (10): 3810–3830.
22. Jansze, S.M. and Severin, K. (2018). *Acc. Chem. Res.* 51 (9): 2139–2147.
23. Wise, M.D., Ruggi, A., Pascu, M. et al. (2013). *Chem. Sci.* 4 (4): 1658–1662.
24. Kumar, G. and Gupta, R. (2013). *Chem. Soc. Rev.* 42 (24): 9403–9453.
25. Safont-Sempere, M.M., Fernández, G., and Würthner, F. (2011). *Chem. Rev.* 111 (9): 5784–5814.
26. Lal Saha, P.M., Schmittel, M., Saha, M.L., and Schmittel, M. (2012). *Org. Biomol. Chem.* 10 (24): 4651–4684.
27. Ghosh, S. and Isaac, L. (2009). Complex self-sorting systems. In: *Dynamic Combinatorial Chemistry: In Drug Discovery, Bioorganic Chemistry, and Materials Science* (ed. B.L. Miller), 118–154. John Wiley & Sons, Ltd.
28. Batuecas, M., Gorgas, N., and Crimmin, M.R. (2021). *Chem. Sci.* 12 (6): 1993–2000.
29. Campos, J. (2020). *Nat. Rev. Chem.* 4 (12): 696–702.
30. Deacy, A.C., Kilpatrick, A.F.R., Regoutz, A., and Williams, C.K. (2020). *Nat. Chem.* 12 (4): 372–380.
31. Zippel, C., Israil, R., Schüssler, L. et al. (2021). *Chem. Eur. J.* 27 (61): 15188–15201.
32. Brand, S., Elsen, H., Langer, J. et al. (2019). *Angew. Chem. Int. Ed.* 58 (43): 15496–15503.
33. Kurumada, S., Takamori, S., and Yamashita, M. (2019). *Nat. Chem.* 12 (1): 36–39.
34. Grams, S., Eyselein, J., Langer, J. et al. (2020). *Angew. Chem. Int. Ed.* 59 (37): 15982–15986.
35. Hicks, J., Vasko, P., Heilmann, A. et al. (2020). *Angew. Chem. Int. Ed.* 59 (46): 20376–20380.
36. Kurumada, S., Sugita, K., Nakano, R. et al. (2020). *Angew. Chem. Int. Ed.* 59 (46): 20381–20384.
37. Rekhroukh, F., Chen, W., Brown, R.K. et al. (2020). *Chem. Sci.* 11 (30): 7842–7849.
38. Hooper, T.N., Brown, R.K., Rekhroukh, F. et al. (2020). *Chem. Sci.* 11 (30): 7850–7857.
39. Kysliak, O., Görls, H., and Kretschmer, R. (2020). *Chem. Commun.* 56 (57): 7865–7868.

40 Cui, F.H., Hua, Y., Lin, Y.M. et al. (2022). *J. Am. Chem. Soc.* 144 (5): 2301–2310.
41 Das, A., Goswami, S., Sen, R., and Ghosh, A. (2019). *Inorg. Chem.* 58 (9): 5787–5798.
42 Bagherzadeh, S. and Mankad, N.P. (2015). *J. Am. Chem. Soc.* 137 (34): 10898–10901.
43 Banerjee, S., Karunananda, M.K., Bagherzadeh, S. et al. (2014). *Inorg. Chem.* 53 (20): 11307–11315.
44 King, R.B. (1970). *Acc. Chem. Res.* 3 (12): 417–427.
45 Laitar, D.S., Müller, P., and Sadighi, J.P. (2005). *J. Am. Chem. Soc.* 127 (49): 17196–17197.
46 Lee, G.R. and Cooper, N.J. (1985). *Organometallics* 4 (4): 794–796.
47 Zieliński, M., Zielińska, A., and Papiernik-Zielińska, H. (1994). *J. Radioanal. Nucl. Chem.* 183 (2): 301–311.
48 Halpern, J. and Kemp, A.L.W. (1966). *J. Am. Chem. Soc.* 88 (22): 5147–5150.
49 Zhang, L. and Hou, Z. (2013). *Chem. Sci.* 4 (9): 3395–3403.
50 McInnis, J.P., Delferro, M., and Marks, T.J. (2014). *Acc. Chem. Res.* 47 (8): 2545–2557.
51 Takeda, N. and Inoue, S. (1978). *Makromol. Chem.* 179 (5): 1377–1381.
52 Moore, D.R., Cheng, M., Lobkovsky, E.B., and Coates, G.W. (2003). *J. Am. Chem. Soc.* 125 (39): 11911–11924.
53 Luinstra, G.A. (2008). *Polym. Rev.* 48 (1): 192–219.
54 Kember, M.R., Knight, P.D., Reung, P.T.R., and Williams, C.K. (2009). *Angew. Chem. Int. Ed.* 48 (5): 931–933.
55 Nakano, K., Hashimoto, S., and Nozaki, K. (2010). *Chem. Sci.* 1 (3): 369.
56 Vagin, S.I., Reichardt, R., Klaus, S., and Rieger, B. (2010). *J. Am. Chem. Soc.* 132 (41): 14367–14369.
57 Klaus, S., Lehenmeier, M.W., Herdtweck, E. et al. (2011). *J. Am. Chem. Soc.* 133 (33): 13151–13161.
58 Jutz, F., Buchard, A., Kember, M.R. et al. (2011). *J. Am. Chem. Soc.* 133 (43): 17395–17405.
59 Kember, M.R. and Williams, C.K. (2012). *J. Am. Chem. Soc.* 134 (38): 15676–15679.
60 Wu, G.-P., Darensbourg, D.J., and Lu, X.-B. (2012). *J. Am. Chem. Soc.* 134 (42): 17739–17745.
61 Lehenmeier, M.W., Kissling, S., Altenbuchner, P.T. et al. (2013). *Angew. Chem. Int. Ed.* 52 (37): 9821–9826.
62 Liu, Y., Ren, W.-M., Liu, J., and Lu, X.-B. (2013). *Angew. Chem. Int. Ed.* 52 (44): 11594–11598.
63 Childers, M.I., Longo, J.M., van Zee, N.J. et al. (2014). *Chem. Rev.* 114 (16): 8129–8152.
64 Ellis, W.C., Jung, Y., Mulzer, M. et al. (2014). *Chem. Sci.* 5 (10): 4004.
65 Liu, Y., Ren, W.-M., He, -K.-K., and Lu, X.-B. (2014). *Nat. Commun.* 5 (1): 5687.
66 Kissling, S., Altenbuchner, P.T., Lehenmeier, M.W. et al. (2015). *Chem. Eur. J.* 21 (22): 8148–8157.
67 Garden, J.A., Saini, P.K., and Williams, C.K. (2015). *J. Am. Chem. Soc.* 137 (48): 15078–15081.
68 Darensbourg, D.J. and Fitch, S.B. (2007). *Inorg. Chem.* 46 (14): 5474–5476.
69 Sugimoto, H. and Kuroda, K. (2008). *Macromolecules* 41 (2): 312–317.
70 Nakano, K., Nakamura, M., and Nozaki, K. (2009). *Macromolecules* 42 (18): 6972–6980.
71 Darensbourg, D.J., Lewis, S.J., Rodgers, J.L., and Yarbrough, J.C. (2003). *Inorg. Chem.* 42 (2): 581–589.
72 Darensbourg, D.J. and Billodeaux, D.B. (2005). *Inorg. Chem.* 44 (5): 1433–1442.
73 Darensbourg, D.J., Ulusoy, M., Karroonnirum, O. et al. (2009). *Macromolecules* 42 (18): 6992–6998.
74 Buchard, A., Kember, M.R., Sandeman, K.G., and Williams, C.K. (2011). *Chem. Commun.* 47 (1): 212–214.
75 Isnard, F., Lamberti, M., Lettieri, L. et al. (2016). *Dalton Trans.* 45 (40): 16001–16010.
76 Chen, L., Li, W., Yuan, D. et al. (2015). *Inorg. Chem.* 54 (10): 4699–4708.
77 Gaston, A.J., Greindl, Z., Morrison, C.A., and Garden, J.A. (2021). *Inorg. Chem.* 60 (4): 2294–2303.
78 Yoshida, M., Masaoka, S., Abe, J., and Sakai, K. (2010). *Chem. Asian J.* 5 (11): 2369–2378.
79 Kimoto, A., Yamauchi, K., Yoshida, M. et al. (2012). *Chem. Commun.* 48 (2): 239–241.
80 Yoshida, M., Kondo, M., Torii, S. et al. (2015). *Angew. Chem.* 127 (27): 8092–8095.
81 Yoshida, M., Kondo, M., Torii, S. et al. (2015). *Angew. Chem. Int. Ed.* 54 (27): 7981–7984.

82 Kundu, A., Dey, S.K., Dey, S. et al. (2020). *Inorg. Chem.* 59 (2): 1461–1470.
83 El Harakeh, N., de Morais, A.C.P., Rani, N. et al. (2021). *Angew. Chem. Int. Ed.* 60 (11): 5723–5728.
84 Ghosh, D., Fabry, D.C., Saito, D., and Ishitani, O. (2021). *Energy Fuels* 35 (23): 19069–19080.
85 Sampaio, R.N., Grills, D.C., Polyansky, D.E. et al. (2020). *J. Am. Chem. Soc.* 142 (5): 2413–2428.
86 Yamazaki, Y., Ohkubo, K., Saito, D. et al. (2019). *Inorg. Chem.* 58 (17): 11480–11492.
87 Wang, W.H., Hull, J.F., Muckerman, J.T. et al. (2012). *Energy Environ. Sci.* 5 (7): 7923–7926.

20

Metal-Organic Frameworks in Tandem Catalysis

Anirban Karmakar* and Armando J.L. Pombeiro

Centro de Química Estrutural, Institute of Molecular Sciences, Instituto Superior Técnico, Universidade de Lisboa, Av. Rovisco Pais, Lisboa, Portugal
* Corresponding author

20.1 Introduction

Catalysis plays a fundamental role in chemical industry. In traditional catalysis, the substrate is activated by a specific catalyst by minimising the reaction energetic barrier [1]. Over the last few decades the concept of multicatalysis, in which multiple reactions can be performed in a single pot using multifunctional catalysts, has evolved rapidly [2, 3]. In this context, tandem or cascade reactions that comprise two or more catalytic reactions performed in a single pot without any purification or separation of the intermediates are relevant [4]. The tandem reactions are often catalysed by two different catalysts or a bifunctional catalyst having multiple active sites. The reactant(s) is (are) activated by an active site of the catalyst to yield an intermediate, which is further activated by the same or another active site present in the catalyst to produce the final product (Figure 20.1) [5, 6]. Thus, waste, time, energy, solvents, and reagents are saved. In the context of sustainable chemistry, the tandem reactions can provide effective and ecologically friendly chemical synthesis methods [7].

The design of multifunctional MOFs to catalyse tandem reactions remains challenging. In fact, to obtain an effective catalytic process, a pathway should be attained in which the catalyst can promote the different reactions in a one-pot process. Moreover, in multifunctional catalysts the active sites, e.g. acidic and basic sites, tend to neutralize each other. Several heterogenous catalysts, such as $Pd^0/CaCO_3$, $Pd^0/BaCO_3$, and Au NPs supported on ZrO_2, Pd-Co NPs on silica, Pd-Os/MgO catalysts, zeolites, covalent organic frameworks (COFs), MOFs, and others have been reported [5, 8–11]. Among them, MOFs are attracting significant consideration, consistent with their favourable structural properties. The possibility of introducing various acid and basic sites in a single MOF affords an excellent prospect to explore MOFs as heterogenous catalysts in the area of tandem reactions [12]. Thus, recently several functionalized MOFs have been reported that can effectively catalyse various tandem reactions [13].

Catalysis for a Sustainable Environment: Reactions, Processes and Applied Technologies Volume 2, First Edition. Edited by Armando J. L. Pombeiro, Manas Sutradhar, and Elisabete C. B. A. Alegria.
© 2024 John Wiley & Sons Ltd. Published 2024 by John Wiley & Sons Ltd.

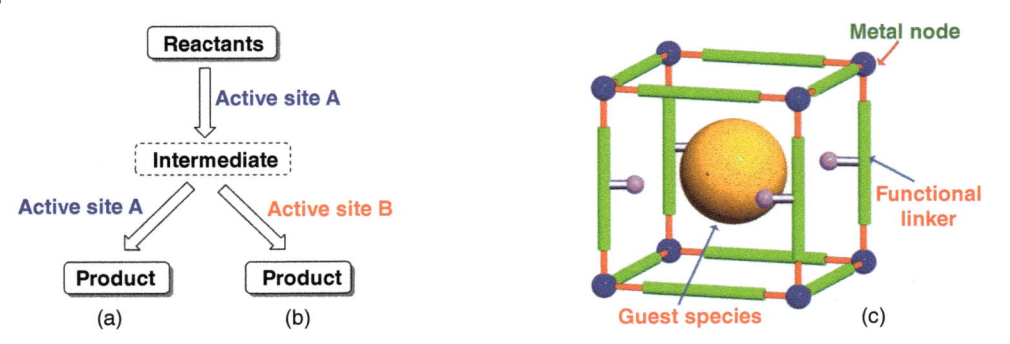

Figure 20.1 Tandem reactions catalysed by one active site (a) or two different active sites (b). (c) Various active sites on a metal-organic framework (MOF), namely metal nodes, catalytically active guest species, and functional organic linkers. Reproduced with permission from Ref 14 / Royal Society of Chemistry.

In this book chapter, we illustrate functionalized MOFs that have been used as catalysts for various liquid phase tandem reactions, thus contributing to understanding their significance as useful materials in tandem type catalysis. In the first part, favourable structural features of MOFs in this context are discussed. The main content of this chapter is systematized according to the various reported examples of particular tandem reactions catalysed by MOFs. The chapter does not intend to cover all the tandem reactions promoted by functionalized MOFs, but rather to demonstrate some important examples. Lastly, conclusions for future MOFs catalytic applications are specified.

20.2 MOFs as Catalysts for Tandem Reactions

MOFs, also known as porous coordination polymers, form a particular class of porous crystalline materials composed of metal ions or clusters and organic linkers [14–17]. Their relevance has been recognized in chemical and materials sciences due to their adaptable architectures, high surface areas, tunable pore sizes, as well as their applications in catalysis, gas storage and separation, sensing, drug delivery, proton conduction, and other techniques [18–20]. Among the applications, MOFs are gaining much consideration as heterogenous catalysts particularly in areas of tandem [21], asymmetric [22, 23], and photocatalysis [24, 25].

20.2.1 Active Sites in MOFs

The structures of MOFs can be tuned by changing the metal nodes and ligands, which provides a huge variety of possible compositions and structures [26], with a diversity of properties including concomitant acidity and basicity. Other important features of MOFs include their high surface areas and pore volumes, which permit active guest molecules to be introduced into the pores and allow substrates admittance to the internal active sites [27]. Functional MOFs can catalyse organic reactions via three different types of active sites: (i) open metal centres or clusters; (ii) organic ligands holding different functional organic groups (e.g. amine, amide, urea, thiol, and sulfonate); (iii) catalytically active guest molecules present in the MOF pores [13]. This blend of types of MOF active sites allows MOFs to behave as relevant multifunctional materials for tandem reactions (Figure 20.1b).

20.2.2 Advantages and Limitations of MOFs

MOFs own an important place in heterogeneous catalysis offering several advantages, including: (i) easy separation, recovery, and reuse of catalyst with preservation of catalytic activity; (ii) homogenously dispersed active sites which can improve the catalytic efficiency; (iii) tunable pore sizes that can promote regio-selective catalysis; (iv) low formation of by products; and (v) various types of stabilized catalytically guest species, such as NPs, metal complexes, or polyoxometallates, into their pores [28, 29].

The structural stability of MOFs during the catalytic process is an important concern in view of the limited thermal and chemical stability of the metal–ligand bonds. The solvents, high reaction temperature, or certain reagents can cause partial or total destruction of the crystal structures [30, 31]. Moreover, leaching of metal ions can result from a strong binding affinity of reagents or substrates. Another relevant point to be considered is the deactivation of MOF as a catalyst. It can undergo partial or total deactivation after a few recycling steps due to pore blocking, metal leaching, or structure collapse [32]. Thus, a careful survey of MOF stability under the reaction conditions is important. MOF stability after the catalytic process must be monitored by a combination of powder X-ray diffraction analysis and porosity measurements of the reused material. Moreover, the hot filtration test and chemical analysis of the liquid phase can provide significant information about leaching during the catalytic process. It is also recommended to find the reasons for deactivation and suggest a regeneration protocol.

20.3 Examples of MOFs Used as Catalysts for Tandem Reactions

MOFs have been used as heterogenous catalysts for various liquid phase organic transformations owing to characteristics such as their insolubility in common organic solvents, thermal and photostability, and good distribution of catalytically active sites [33]. MOFs having suitable metal and ligand systems can also catalyse several types of tandem reactions and examples are discussed in this reaction, addressing their activity, selectivity, and mechanistic roles.

20.3.1 Deacetalization-Knoevenagel Condensation

An interesting example of tandem reactions concerns the deacetalization-Knoevenagel condensation, in which the reaction of benzaldehyde dimethyl acetal and malononitrile directly produces benzylidene malononitrile. Different MOFs decorated with acidic and basic groups have effectively catalysed such reactions (Scheme 20.1) [34–36].

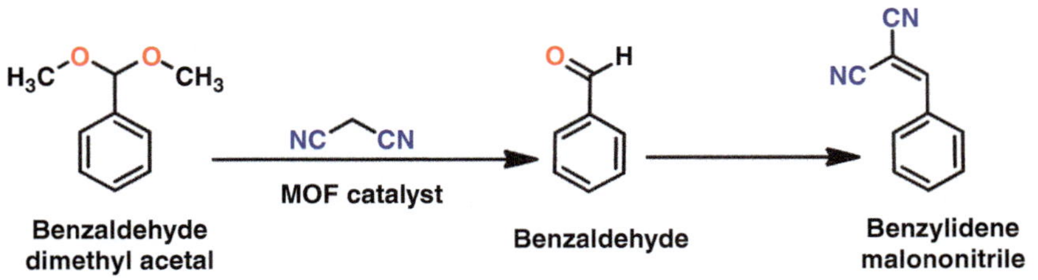

Scheme 20.1 Tandem deacetalization-Knoevenagel condensation reactions of benzaldehyde dimethyl acetal and malononitrile. Ref [35] / Royal Society of Chemistry.

For example, Zhou et al. reported a Cu(II) MOF formulated as $[Cu_2(L1)(H_2O)]_n$ (**1**) using a pyridine based amido carboxylate pro-ligand, 5,5'-[(pyridine-3,5-dicarbonyl)bis-(azanediyl)] diisophthalate (H_4L1) (Figure 20.2a), and this framework shows an efficient catalytic activity for the tandem one-pot deacetalization-Knoevenagel condensation reaction of benzaldehyde dimethyl acetal and malononitrile [34]. By using 0.5 mol% of **1** as catalyst for a such a reaction, 100% yield of benzylidene malononitrile was reached within 12 hours at 50 °C. Moreover, they have recycled the catalyst for three cycles and only a slight decrease in reaction yield (from 100% to 92%) was observed. Powder x-ray diffraction analysis before and after the catalysis did not show any alteration in the peaks pattern, confirming catalyst stability under the reaction conditions. The authors have also proved that the presence of both the Cu(II) sites and the amide groups from the ligands is crucial for this reaction. Addition of *p*-toluenesulfonic acid facilitates the deacetalization reaction but hampers the Knoevenagel condensation, whereas the addition of ethylenediamine (blocking the Cu(II) sites) only promotes the Knoevenagel condensation and hampers the deacetalization reaction, decreasing the overall yield of benzylidene malononitrile.

Recently, we have synthesized a two dimensional MOF, $[Zn_5(L2)_4(OH)_2(H_2O)_4]_n \cdot 8n(DMF) \cdot 4n(H_2O)$, in which DMF is dimethyl formamide, (**2**), by the solvothermal reactions between the amide functionalized dicarboxylic acid 4,4'-{(pyridine-2,6-dicarbonyl)bis(azanediyl)}dibenzoic acid (H_2L2) and zinc(II) nitrate (Figure 20.2b) [35]. This MOF heterogeneously catalyses the tandem deacetalization-Knoevenagel condensation reactions carried out under conventional heating, microwave, or ultrasonic irradiation. Comparative studies show that ultrasonic irradiation (yield of 98% after two hours) provides the most favourable method compared to microwave or normal heating (yield of 91–93% after three hours). Moreover, the catalysts can be reused at least for five consecutive cycles without losing activity significantly.

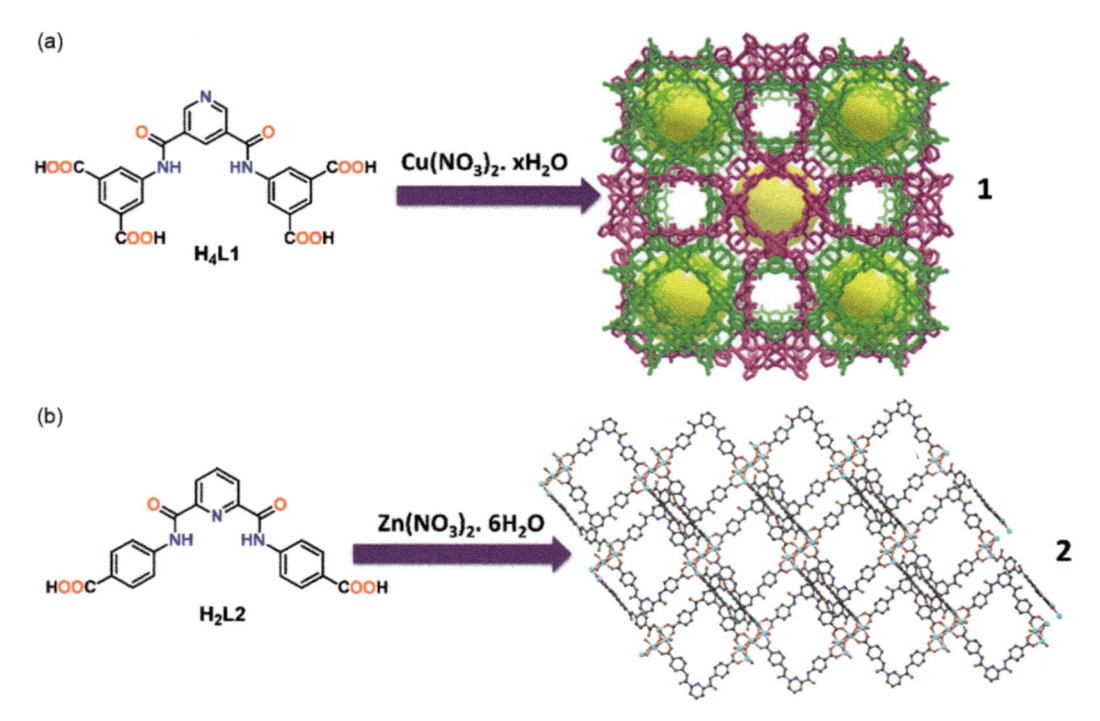

Figure 20.2 Synthesis and crystal structures of metal-organic frameworks (MOFs) **1** (a) and **2** (b). Adapted from Refs [34] and [35].

Figure 20.3 Synthesis and crystal structures of metal-organic frameworks (MOFs) **3** (A) and **4** (B). Ref [36] / Frontiers Media S.A.

In another work, we have reported the pyridine based amide functionalized tetracarboxylic acid 5,5′-{(pyridine-2,6-dicarbonyl)bis(azanediyl)}diisophthalic acid (H$_4$L3) and its coordination chemistry with Zn(II) and Cd(II) ions. The reactions of H$_4$L3 with Zn(II) and Cd(II) nitrates led to the formation of the two 2D MOFs [Zn$_2$(L3)(H$_2$O)$_4$]$_n$·4n(H$_2$O) (**3**) and [Cd$_3$(HL3)$_2$(DMF)$_4$]$_n$·4n(DMF) (**4**), respectively (Figure 20.3) [36]. On account of the presence of Lewis acid (Zn or Cd centers) and basic (uncoordinated pyridine and amide groups) sites, **3** effectively (yield of 99% after 3 h) catalyses the one-pot cascade deacetalization-Knoevenagel condensation reactions, under considerably mild conditions. These MOFs act as heterogeneous catalysts and can be recycled a few times without losing their activity. The stability of these MOFs after the catalytic reactions were proved by powder X-ray diffraction and FT-IR analysis.

20.3.2 Deacetalization-Henry Reaction

Deacetalization-Henry reactions are another important type of tandem reactions which involve the conversion of an acetal to aldehyde (first step) followed by the Henry or nitroaldol reaction between the intermediate aldehyde and nitromethane or nitroethane, leading to the formation of a nitroalkene as the final product (Scheme 20.2).

Scheme 20.2 Example of deacetalization-Henry tandem reactions of acetal and nitroalkane. Adapted from Ref [37].

In 2012, Li and co-workers prepared a Cr(III) MOF, formulated as [Cr$_3$(F)(O)(H$_2$O)$_2$(BDC)$_3$]$_n$ (MIL-101-Cr) (BDC = 1,4-benzene dicarboxylate), by solvothermal reaction between Cr(NO$_3$)$_3$·9H$_2$O and 1,4-benzene dicarboxylic acid. Later, they prepared a bifunctionalized MOF, namely MIL-101-SO$_3$H-NH$_2$ (**5**), by introducing sulfonic acid (SO$_3$H$^-$) and ethylenediamine (NH$_2$CH$_2$CH$_2$NH$_2$) groups into the MIL-101-Cr MOF via post-synthetic modification methods (Scheme 20.3) [37].

They tested the catalytic activity of MIL-101-SO$_3$H-NH$_2$ (**5**) towards the one-pot deacetalization-Henry tandem reactions of benzaldehyde dimethyl acetal and nitromethane, achieving 97% yield

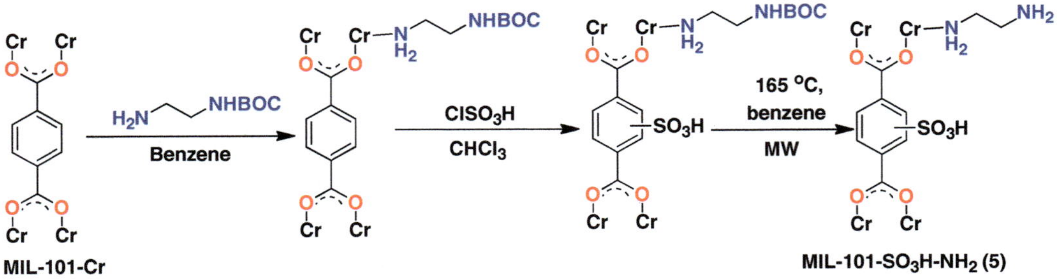

Scheme 20.3 Synthesis of MIL-101-SO$_3$H-NH$_2$ (**5**) via post-synthetic modification of MOF-101-Cr (BOC = *tert*-butyloxycarbonyl protecting group). Adapted from Ref [37].

of 2-nitrovinyl benzene as the final product after 24 h of reaction time at 90 °C. Upon performing the reaction with monofunctionalized MIL-101-SO$_3$H and MIL-101-NH$_2$ (only one of the two essential functional groups of the ligand is present), no such final product was detected. This experiment demonstrates that the presence of both sulfonic acid and amine groups is necessary to promote the abovementioned tandem catalytic process. Moreover, the use of catalyst **5** with an excess amount of acid (*p*-toluene sulfonic acid) or base (ethylamine) would inhibit the activity probably due to the formation of ion pairs. They have also reused the catalyst for three times and no changes in the framework crystallinity or structure were observed by means of powder X-ray diffraction analysis.

20.3.3 Meinwald rearrangement-Knoevenagel Condensation

Meinwald rearrangement concerns the conversion of epoxides into aldehydes or ketones by ring-opening and 1,2-shift of an hydride or alkyl group, whereas Knoevenagel condensation is the condensation reaction between aldehydes or ketones with nitriles to form benzylidene malononitriles (Scheme 20.4).

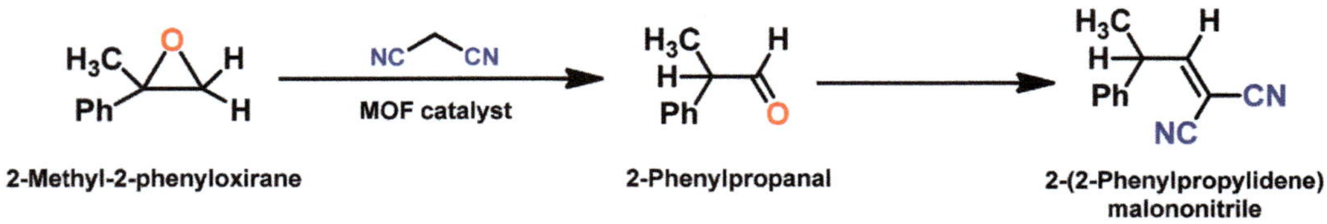

Scheme 20.4 Example of Meinwald rearrangement-Knoevenagel condensation tandem reactions. Adapted from Ref [38].

In 2012, Kim et al. reported a 3D Al(III) MOF, [NH$_2$-MIL-101(Al), **6**] formulated as [Al(OH)(BDC-NH$_2$)]$_n$ (BDC-NH$_2$ = 2-aminoterephthalate) (Figure 20.4), which effectively catalyses the tandem Meinwald rearrangement-Knoevenagel condensation reactions. The tandem reaction between 2-methyl-2-phenyloxirane and malononitrile in the presence of catalyst **6** at 60 °C resulted in 70% yield of the final product 2-(2-phenylpropylidene)malononitrile within 48 h of reaction time [38].

The Al(III) metal centres can act as Lewis acid catalytic sites and catalyse the first step (Meinwald rearrangement) of the tandem reactions, which is ring opening of 2-methyl-2-phenyloxirane to generate the intermediate 2-phenylpropanal [38]. The second step (Knoevenagel condensation) is

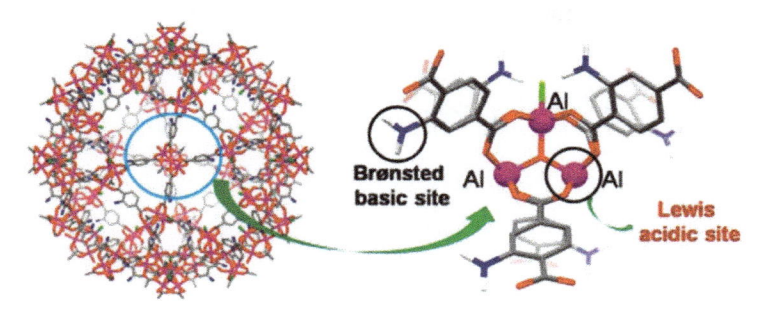

Figure 20.4 Crystal structure and trinuclear Al(III) secondary building unit (SBU) in NH$_2$-MIL-101(Al) (**6**). Adapted from Ref [38].

mostly promoted by free amino (-NH$_2$) moieties of the 2-amino terephthalate which act as Brønsted base. Upon performing the tandem reaction of 2-methyl-2-phenyloxirane and malononitrile in the presence of AlCl$_3$ or a mixture of AlCl$_3$ and 2-aminoterephthalate as a catalyst, the Meinwald rearrangement (first step) occurred with 85% of conversion, but the Knoevenagel condensation failed. This failure may be to the lack of free amine groups of 2-aminoterephthalate available for catalysis. However, the reaction of 2-phenylpropanal and malononitrile using 2-aminoterephthalate as catalyst produces 65% of 2-(2-phenylpropylidene)malononitrile. The results confirm that the presence of both Al(III) sites and free amine (-NH$_2$) group is essential to promote such tandem reactions. Moreover, hot filtration experiments established that the tandem epoxide ring opening and Knoevenagel condensation reaction happened in heterogeneous medium and powder X-ray diffraction analysis indicated that the structural integrity of the catalyst **6** remained unaltered after two reaction cycles.

20.3.4 Reductive Amination of Aldehydes with Nitroarenes

An interesting example of one-pot three step tandem reactions is the reductive amination of aldehydes with nitroarenes or hydrogenation-reductive amination. This overall reaction involves three steps, firstly the chemoselective reduction of the nitro compound to the corresponding amine in presence of H$_2$ and catalyst, followed by the condensation between the aromatic amine and the carbonyl group of aldehyde to produce an imine compound, and lastly the hydrogenation of the resulting imine to produce the corresponding secondary amine (Scheme 20.5).

Scheme 20.5 Example of tandem hydrogenation-reductive amination reactions. Ref [39] / Elsevier.

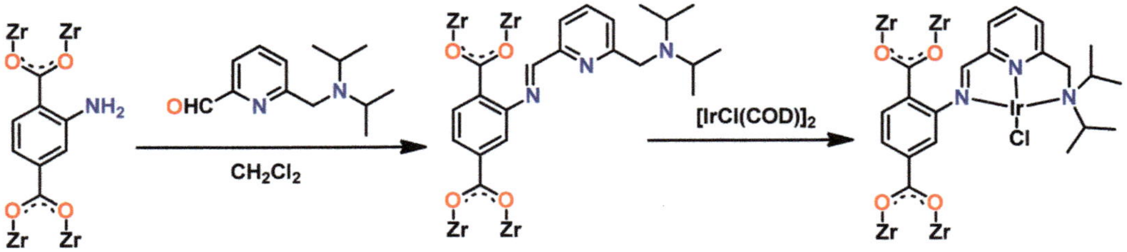

Scheme 20.6 Synthesis of the Ir(II) complex bearing metal-organic framework (MOF) **7a** via post-synthetic modification of the Zr(IV) MOF (**7**). Adapted from Ref [39].

Corma et al. prepared a Zr(IV) MOF, formulated as $[Zr_6O_4(OH)_4(BDC-NH_2)_6]_n$ (**7**) (BDC-NH$_2$ = 2-aminoterephthalate), by the solvothermal reaction of a Zr(IV) salt and 2-aminoterephthalic acid. Later, they incorporated an iridium complex into the MOF **7** via post-synthetic modification as shown in Scheme 20.6 to produce an Ir(II) complex bearing MOF (**7a**) able to catalyse hydrogenation-reductive amination tandem reaction [39].

Performing the reaction of benzaldehyde and nitrobenzene at 100 °C, in the presence of isopropanol, H$_2$ (6 bar) and MOF catalyst **7a**, resulted in 99% conversion of nitrobenzene to the secondary amine with a higher selectivity (99%) towards the amine compared to the imine product after 24 h of reaction time [39]. In this reaction, the nitrobenzene was reduced to the aniline in the presence of the iridium site and hydrogen. Then, the aniline and the benzaldehyde formed an imine via a condensation reaction which was mostly catalysed by the Zr(IV) acid site of **7a**. The final step, conversion of imine to the secondary amine by H$_2$ was also catalysed by Ir sites. Moreover, the authors have also used various substituted aromatic and aliphatic aldehydes, as well as substituted nitrobenzene, and achieved 90–100% conversion to the corresponding products. They have recycled the catalyst **7a** for four times and it maintained catalytic activity. Hot filtration test, powder x-ray diffraction and SEM analyses suggest that no leaching occurred during the catalytic process and the overall structure of the catalyst remained intact.

20.3.5 Epoxidation–Ring Opening of Epoxide

Farha et al. reported an Hf(IV) MOF, formulated as $[Hf_6(O)_4(OH)_4(FeTCP-Cl)_3]_n$ (**8**), prepared by the solvothermal reaction of HfOCl$_2$, *meso*-tetra(4-carboxyphenyl)-porphyrin-Fe(III) chloride (H$_4$FeTCP-Cl), and benzoic acid as a modulator (Figure 20.5). Later, to ensure every porphyrin was metalated the synthesized MOF **8** was suspended in a DMF solution containing anhydrous FeCl$_3$ and surprisingly a new material **8a** was formed [40].

In MOF **8a**, the Fe(III) ions from anhydrous FeCl$_3$ were coordinated to the hydroxy (–OH) and water ligands, which was confirmed by single-crystal X-ray diffraction analysis and diffuse reflectance infrared Fourier transform spectroscopy (DRIFTS). They have tested the catalytic activity of MOF **8a** towards the tandem styrene epoxidation–ring opening of epoxide reaction. The reaction of styrene with trimethylsilyl azide (TMSN$_3$) in presence of *tert*-butyraldehyde (to regenerate the catalyst), O$_2$ (5 atm) and MOF **8a** (as catalyst) in acetylnitrile (MeCN) medium at 60 °C led to formation of a protected 1,2-hydroxylamine within 10 h of reaction time (Scheme 20.7) [40].

The first step of this tandem reaction which is epoxidation of styrene to produce styrene oxide intermediate was mainly catalysed by the Fe(III) sites of **8a**, and the reaction of styrene epoxide with TMSN$_3$ was catalysed by the Lewis acid Hf(IV) nodes of **8a**. By attempting to perform the

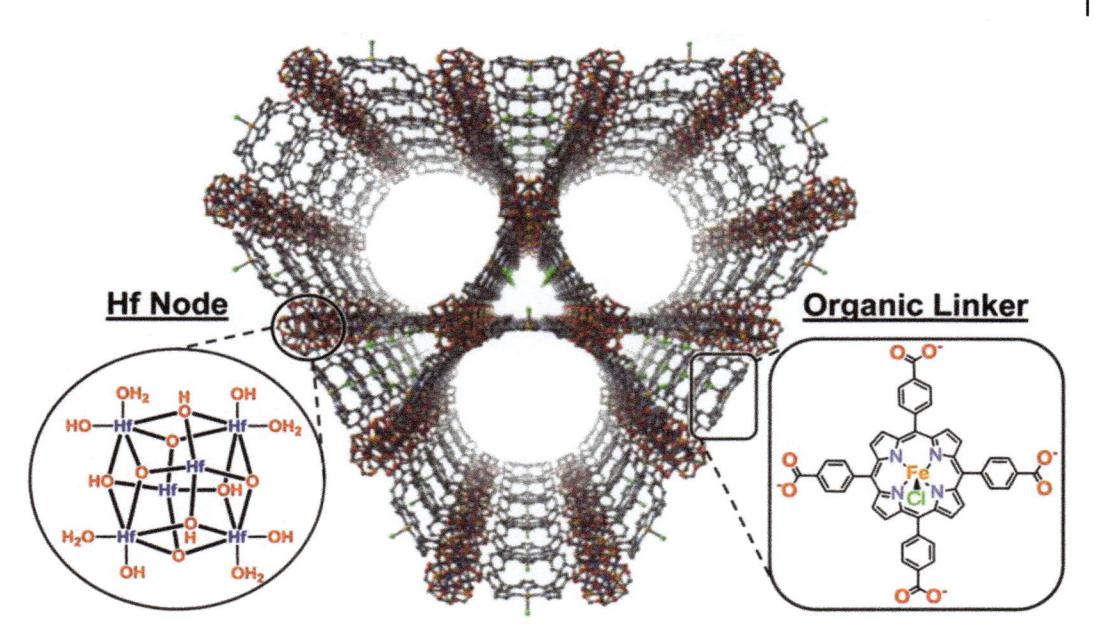

Figure 20.5 Structure of metal-organic framework (MOF) **8** and its [Hf$_6$(O)$_4$(OH)$_4$] secondary building unit (SBU) and FeTCP-Cl ligand. Ref [40] / American Chemical Society.

Scheme 20.7 Example of tandem epoxidation of styrene and ring opening of 2-phenyloxirane. Adapted from Ref [40].

reaction with MOF **8**, no final product was obtained. Moreover, no leaching of the heterogenous **8a** catalyst was observed during the catalytic reaction.

20.3.6 Oxidation-Esterification

Incorporation of metal NPs into MOFs pores can result in a catalytically more active system [41]. To study the catalytic activity of NPs incorporated MOFs, Fischer and co-workers prepared two zeolite imidazolate frameworks, namely [Zn(MeIM)$_2$]$_n$ (**9**) (MeIM = imidazolate-2-methyl) and [Zn(ICA)$_2$]$_n$ (**10**) (ICA = imidazolate-2-carboxyaldehyde), and embedded them with Au NPs to produce MOFs **9Au** and **10Au**, respectively [41].

They then studied the catalytic activities of these Au NPs encapsulated MOFs towards the tandem oxidation-esterification reactions of benzyl alcohol to methyl benzoate in the presence of methanol and O$_2$ (Scheme 20.8) [41]. The loading of Au NPs was varied between 5 and 30 wt.% and they were distributed throughout the MOFs with particle sizes ranging from 1 to 5 nm. Upon testing the catalytic activity of MOFs **9** and **10** (without incorporated Au NPs) towards the tandem

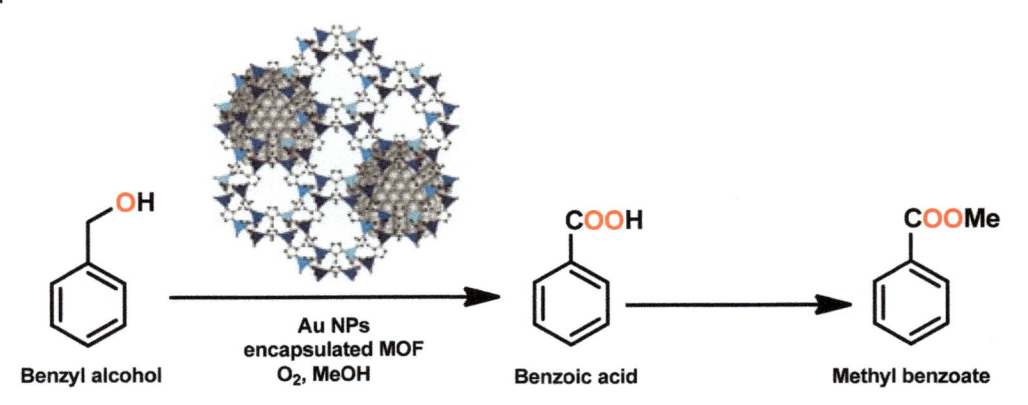

Scheme 20.8 Tandem oxidation-esterification reactions of benzyl alcohol to methyl benzoate using Au nanoparticles (NPs) supported metal-organic frameworks (MOFs) (**9** and **10**). Adapted from Ref [41].

oxidation-esterification reaction of benzyl alcohol to methyl benzoate, no activity was observed. However, by using the Au NPs incorporated MOFs **9Au** and **10Au** as catalysts, a conversion of 81 and 13% with a selectivity towards methyl benzoate of 98% and 50%, respectively, was obtained. Moreover, small amounts of benzaldehyde were formed during this tandem reaction, but no benzaldehyde dimethyl acetal was detected.

The same research group also reported metal oxide and nanoparticle encapsulated MOFs for the tandem oxidation-esterification reaction of benzyl alcohol to methyl benzoate [42]. Firstly they prepared a Zn(II) MOF, formulated as $[Zn_4O(BDC)_3]_n$ (**11**), and loaded it with ZnO and TiO_2, and prepared **ZnO@11** and **TiO$_2$@11**, respectively. The reaction of **11**, **ZnO@11** and **TiO$_2$@11** with [ClAuCO] produced intermediate materials denoted as **[ClAuCO]@11, [ClAuCO]/ZnO@11** and **[ClAuCO]/TiO$_2$@11**. These composites were decomposed at 100 °C under hydrogen and produced **Au@11**, **Au/ZnO@11** and **Au/TiO2@11**. The characterization of these materials revealed an homogeneous distribution of Au NPs (size range within 1–3 nm) over the MOF. The catalytic activity of these materials was then tested towards liquid-phase oxidation-esterification of benzyl alcohol to methyl benzoate in the presence of MeOH and O_2. Unlike the previous case, these materials did not show any activity without the presence of a base. However, upon performing the reaction in the presence of K_2CO_3 as base, **Au@11** displayed 50% conversion of benzyl alcohol with 48% yield of methyl benzoate in 30 min. The presence of metal oxide improved the benzyl alcohol conversion, as well as the selectivity to the methyl ester. For example, in the presence of K_2CO_3 base, **Au/ZnO@11** and **Au/TiO$_2$@11** resulted in 66 and 72% selectivity, respectively, towards methyl benzoate with trace amounts of benzaldehyde and no formation of benzoic acid.

20.3.7 Oxidation-Hemiacetal Reaction

Mixed NPs encapsulated MOFs have already emerged in heterogenous tandem catalysis [43]. For example, Luque et al. prepared a Au–Pd NPs supported MOF (**12**) by immersing a Cr(III) MOF, formulated as $[Cr_3(O)(F)(H_2O)_2(BDC)_3]_n \cdot 25nH_2O$ (BDC = 1,4-benzenedicarboxylate), into a solution containing Au(III) and Pd(II) NPs. They have tested its activity in the aerobic oxidation-hemiacetal tandem reactions of toluene and its derivatives to conversion into the corresponding esters [43]. The solvent-free oxidation-hemiacetal reaction of toluene was carried out in the presence of 1 wt% of catalyst **12** (Au:Pd ratio of 1:1.5) at 120 °C and under 1.0 MPa of O_2 and after 48 h of

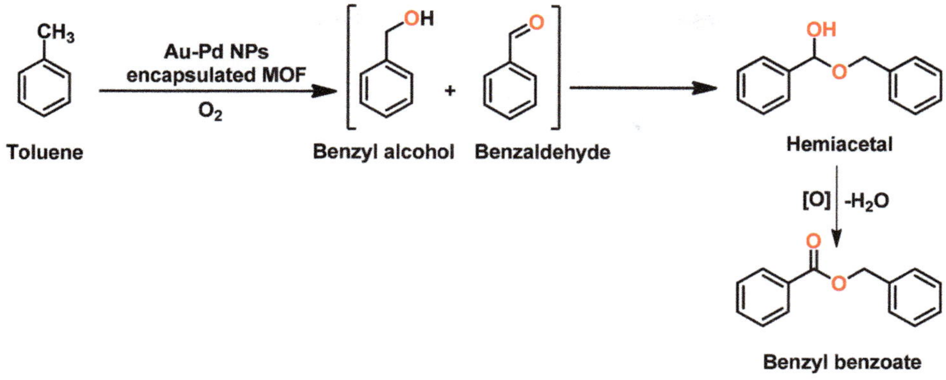

Scheme 20.9 Preparation of benzyl benzoate from toluene via tandem oxidation-aldol condensation catalysed by Au–Pd nanoparticles (NPs) supported metal-organic framework (MOF) (**12**). Adapted from Ref [43].

reaction time 98.6% of toluene conversion with 93% selectivity for benzyl benzoate was achieved (Scheme 20.9). Moreover, only 2 and 3% of benzaldehyde and phenyl benzoate were detected, respectively. They have also tested this reaction without any metal catalyst and with the parent Cr(III)-MOF as catalyst and almost no conversion was achieved in both cases.

The oxidation-hemiacetal tandem reactions of other substrates, such as *o*-, *m*-, and *p*-xylenes; 2-, 3-, 4-methoxytoluene; and 4-fluorotoluene, were also tested using **12** as a catalyst, and the system was also efficient [43]. For *p*-xylene, 40.2% conversion (TON value of 1206) with 65% selectivity for benzyl benzoate was obtained by performing the reaction at 120 °C for 48 h. The oxidation of substituted toluene resulted in the development of the corresponding esters as the major products with an overall selectivity within the range of 66–71% without formation of acid or CO_2. The catalyst **12** was stable under the reaction conditions and preserved its structure after three catalytic cycles as demonstrated by powder X-ray diffraction. This tandem reaction is mostly catalysed by the Au-Pd NPs and proceeds through the oxidative conversion of toluene to benzyl alcohol and benzaldehyde, followed by the formation of the hemiacetal [PhCH(OH)OCH$_2$Ph] via reaction of these intermediates, which finally is further oxidized to benzyl benzoate.

20.3.8 Asymmetric Tandem Reactions

In this context, the design and synthesis of MOF based catalysts for enantioselective tandem catalysis is a further challenging task. Homochiral MOFs containing both acid and basic sites have been engaged in multi-component tandem asymmetric synthesis [44].

For example, Duan et al. synthesized two homochiral enantiomorphs MOFs (**13a** and **13b**), formulated as [ZnW$_{12}$O$_{40}$].[Zn$_2$(NH2-BPY)$_2$(HPYI)$_2$(H$_2$O)(CH$_3$CN)], by the combination of the Keggin polyoxometalate [ZnW$_{12}$O$_{40}$]$^{6-}$ anion, NH$_2$-BPY (3-amino-4,4′-bipyridine), Zn(II) ions and L- or D-pyrrolidine-2-yl-imidazole (PYI) (Figure 20.6). By using these homochiral MOFs they have performed olefin epoxidation followed by the conversion of epoxide into cyclic carbonate using CO_2, and obtained good yield and selectivity [44].

The reaction of styrene, *tert*-butyl hydroperoxide (TBHP), and CO_2 in the presence catalyst **13a** at 50 °C led to the formation of (R)-phenyl(ethylene carbonate) with 92% yield and 80% enantiomeric excess (ee) within 96 h of reaction time (Scheme 20.10) [44]. By using **13b** as catalyst,

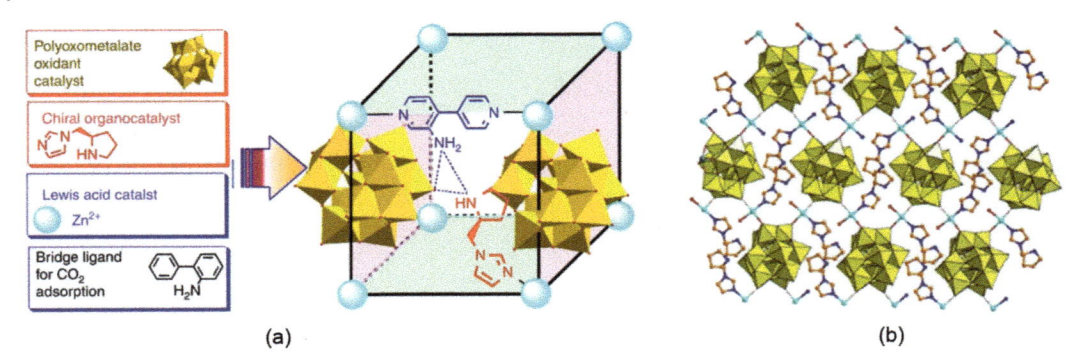

Figure 20.6 Synthesis (a) and crystal structure (b) of homochiral enantiomorphs metal-organic frameworks (MOFs) (**13a** and **13b**). Ref [44] / Springer Nature.

Scheme 20.10 Example of asymmetric tandem reactions catalysed by MOFs **13a** or **13b**. Adapted from Ref [44].

(S)-phenyl(ethylene carbonate) was obtained as the final product with similar yield and enantiomeric excess (ee). The asymmetric epoxidation of styrene (first step) is mostly catalysed by the $[ZnW_{12}O_{40}]^{6-}$ Keggin anion and the PYI ligand. But the second step of this tandem reaction (asymmetric cyclic carbonate formation) is primarily catalysed via Lewis acidic Zn(II) ions present in the MOF **13a** or **13b**, which interact synergistically and activate CO_2 insertion in view of the spatial location. Moreover, the amine (-NH_2) group not only acts as a basic site but also increases the CO_2 adsorption and promotes the formation of the cyclic carbonate product. The authors also tested the reactivity of their catalysts towards various substituted styrene derivatives and obtained good product yields (72–83%) and selectivities (55–70%). The removal of catalyst from the reaction mixture after 48 h stopped the reaction and no additional conversion was observed. The catalysts were reused at least three cycles with a moderate loss of activity (from 92 to 88% yield) and a small decrease in selectivity (from 80 to 77% ee). The powder X-ray diffraction analyses before and after catalysis indicate that the MOF catalysts remain stable during the catalytic process.

20.3.9 Photocatalytic Tandem Reactions

Photocatalysis is an important type of catalysis in which light can be used as an energy source to accomplish organic transformations under mild conditions. In this context the photocatalytic tandem reactions, in particular with MOF-based catalysts, deserve a particular attention [45, 46].

Amine-functionalized ligand-based MOFs coupled with photoactive metal nodes have been widely used for photocatalytic tandem reactions. For example, Matsuoka et al. have used the amine functionalized Zr(IV) MOF $[Zr_6O_4(OH)_4(BDC-NH_2)_6]_n$ (**7**) (BDC-NH_2 = 2-aminoterephthalate) for one-pot tandem reactions between benzyl alcohol and malononitrile to produce benzylidene malononitrile

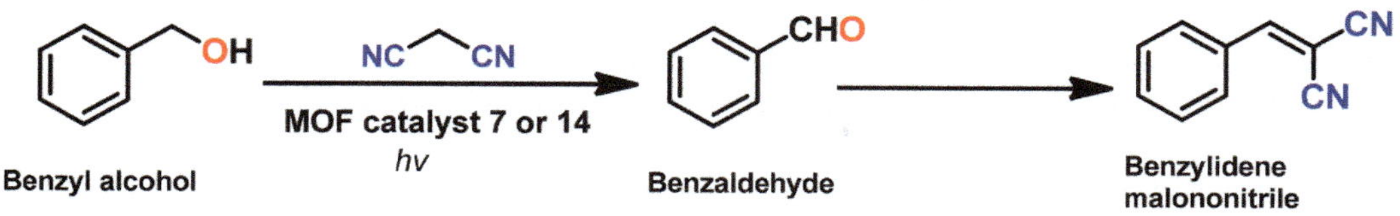

Scheme 20.11 Example of photocatalytic tandem reactions catalysed by metal-organic frameworks (MOFs) **7** and **14**. Adapted from Ref [45].

(Scheme 20.11) [45]. This overall conversion comprises the photocatalytic oxidation of benzyl alcohol to benzaldehyde, followed by the Knoevenagel condensation of the latter with malononitrile to produce an α,β-unsaturated benzylidene malononitrile. The reaction of benzyl alcohol with an excess amount of malononitrile in the presence of MOF **7** as catalyst under UV-light irradiation at 363 K yielded 91% benzylidene malononitrile after 48 h. The reaction did not procced without UV-irradiation and the use of a lower temperature (298 K) only promotes the first step of the tandem reaction (photocatalytic oxidation of benzyl alcohol) leading to the formation of benzaldehyde as major product (yield 89%). Moreover, the use of a Zr(IV)-benzenedicarboxylate MOF without the amine group or the test of ZrO_2 as catalyst for the above mentioned reaction only produces 28% or 2% of final product, respectively, suggesting a significant role of the amine (-NH_2) moiety in the reaction. The authors have also tested the activities of other amine containing MOFs e.g. the Zn-MOF-NH_2 [$Zn_4O(BDC-NH_2)_3$]$_n$ and the Ti-MOF-NH_2 [$Ti_8O_8(OH)_4(BDC-NH_2)_6$]$_n$ and their catalytic activities are lower than that of MOF **7**, suggesting a higher catalytic effect of Zr(IV) compared to the other metals. The recyclability of MOF **7** was tested and a slight decrease in catalytic activity was observed.

Li and co-workers employed an Fe-based MOF, [$Fe_3(O)(Cl)(BDC-NH_2)_3$]$_n$ (**14**) (BDC-NH_2 = 2-aminoterephthalate), as the catalyst for the tandem photooxidation-Knoevenagel condensation in the presence of O_2 and visible light at room temperature [46]. On account of extensive iron oxoclusters as well as the amine (-NH_2) group in MOF **14**, it shows intensive light absorption in the visible light region. In accord, MOF **14** efficiently catalyses the one-pot tandem reaction between benzyl alcohol and malononitrile via tandem photooxidation and base-catalysed Knoevenagel condensation, achieving 72% yield of benzylidene malononitrile after 40 h. As in the above earlier case, the reaction did not take place without catalyst or visible light irradiation. The authors also tested the catalytic activities of MOF **14** towards a variety of substituted aromatic alcohols and active methylene compounds and obtained product yields in the 20–76% range. The heterogeneous nature of the catalyst **14** was confirmed by a hot filtration experiment. Moreover, the catalyst can be recycled at least for three times without obvious reduction of the activity. Powder X-ray diffraction analyses, FT-IR and N_2 adsorption analyses after the catalytic reaction did not show significant changes, suggesting the stability of MOF **14** during the tandem reaction.

20.4 Conclusions

In recent decades, the use of MOFs as heterogeneous catalysts for different liquid-phase organic catalytic reactions, including tandem ones, has improved markedly, as illustrated in this chapter. MOFs can catalyse an organic reaction via different active sites. In most of the cases, the unsaturated metal nodes (Lewis acidic centres) or the functional active sites on the organic linkers (acid/base sites) are involved in the catalytic transformations. But in some cases the catalytically active NPs or metal complexes incorporated in the MOFs pores can also catalyse the organic reactions. Two main different methods have been used to incorporate the active sites into MOFs: (i) the direct method and (ii) the post-synthetic method. The direct method is simple and easy to incorporate different types of functional ligands into the MOF, but provides only a low control on the final structure and porosity. In contrast, the post-synthetic method can afford a good control on the structure and porosity of the MOFs, but its methodology is not easy. Due to the straightforwardness of the direct method, it has been extensively used for the synthesis of various catalytically active MOFs, such as Cu-MOF (**1**), Zn-MOFs (**2**, **3** and **13**), Cd-MOFs (**4**), Al-MOF (**6**), and Hf-MOF (**8**), systems. However, in some cases, the incorporation of catalytically active NPs and metal complexes into MOFs sites were done via post-synthetic modifications. For example, an Ir(II) complex was incorporated in the Zr-MOF (**7**) by the post-synthetic modification method, being hard to obtain using the direct method. Moreover, in the Cr-MOF (**5**), the basic sites (ethylene diamine) and acidic sites (sulfonic acid) were also introduced via a post-synthetic modification process.

These functionalized MOFs can act as effective catalysts for different types of tandem reactions including deacetalization-Knoevenagel condensation, deacetalization-Henry reaction, Meinwald rearrangement-Knoevenagel condensation, nitroarene reductive N-alkylation with aldehydes, epoxidation–epoxide ring-opening, and oxidation-esterification. High yields and selectivities can be achieved owing to the synergistic effect of metal sites, functional organic groups or encapsulated NPs or complex molecules within the structure. For example, MOF **1** catalyses the tandem deacetalization-Knoevenagel condensation where the deacetalization reaction is mostly catalysed by the Lewis acidic Cu(II) metal sites and the Knoevenagel condensation is promoted by a basic amide group present in the ligand. But in some tandem reactions the catalytic mechanisms are not clear and thus further characterization methods accompanied by theoretical calculations are needed for such a purpose.

Although important developments have been achieved in different tandem reactions catalysed by MOFs, their role in asymmetric and photocatalysis is still limited. A few of such MOFs (**7**, **13a**, **13b** and **14**) have been used for such a purpose, but, the product yield and selectivity (enantiomeric excess) are usually not remarkable. Thus, the use of MOF based catalysts in asymmetric and in photocatalysis deserves to be further explored.

Although hundreds of MOFs have already been reported as catalysts for a significant number of tandem reactions, the practical industrial applications of MOFs materials are still very limited. Some of the MOFs display high porosity and stability, but significant attention is required to produce high yielded and low-cost MOFs in an industrial scale. It is anticipated that this area will expand and MOFs will be used as robust catalysts for various tandem organic reactions of industrial significance.

Acknowledgements

This work was supported by the Fundação para a Ciência e Tecnologia (FCT), Portugal, projects UIDB/00100/2020 and UIDP/00100/2020 of Centro de Química Estrutural and LA/P/0056/2020. A. Karmakar expresses his gratitude to Instituto Superior Técnico and FCT for Scientific Employment contract (Contrato No: IST-ID/107/2018) under Decree-Law no. 57/2016, of August 29.

References

1 Climent, M.J., Corma, A., and Iborra, S. (2011). *Chem. Rev.* 111: 1072–1133.
2 Walsh, C.T. and Moore, B.S. (2019). *Angew. Chem. Int. Ed.* 58: 6846–6879.
3 Volla, C.M.R., Atodiresei, I., and Rueping, M. (2014). *Chem. Rev.* 114: 2390–2431.

4 Felpin, F.-X. and Fouquet, E. (2008). *ChemSusChem* 1: 718–724.
5 Wasilke, J.C., Obrey, S.J., Baker, R.T., and Bazan, G.C. (2005). *Chem. Rev.* 105: 1001–1020.
6 Chng, L.L., Erathodiyil, N., and Ying, J.Y. (2013). *Acc. Chem. Res.* 46: 1825–1837.
7 Liu, J., Chen, L., Cui, H. et al. (2014). *Chem. Soc. Rev.* 43: 6011–6061.
8 Islamoglu, T., Goswami, S., Li, Z. et al. (2017). *Acc. Chem. Res.* 50: 805–813.
9 Climent, M.J., Corma, A., Iborra, S., and Sabater, M.J. (2014). *ACS Catal* 4: 870–891.
10 Paul, A., Karmakar, A., Guedes da Silva, M.F.C., and Pombeiro, A.J.L. (2021). *Catalysts* 11: 90.
11 Dhakshinamoorthy, A. and Garcia, H. (2014). *ChemSusChem* 7: 2392–2410.
12 Zhang, Y., Huang, C., and Mi, L. (2020). *Dalton Trans.* 49: 14723–14730.
13 Huang, Y.-B., Liang, J., Wang, X.-S., and Cao, R. (2017). *Chem. Soc. Rev.* 46: 126–157.
14 Karmakar, A. and Pombeiro, A.J.L. (2019). *Coord. Chem. Rev.* 395: 86–129.
15 Karmakar, A., Guedes da Silva, M.F.C., and Pombeiro, A.J.L. (2014). *Dalton Trans.* 43: 7795–7810.
16 Tăbăcaru, A., Pettinari, C., and Galli, S. (2018). *Coord. Chem. Rev.* 372: 1–30.
17 Karmakar, A., Hazra, S., and Pombeiro, A.J.L. (2022). *Coord. Chem. Rev.* 453: 214314.
18 Pettinari, C., Pettinari, R., Nicola, C.D. et al. (2021). *Coord. Chem. Rev.* 446: 214121.
19 Karmakar, A., Paul, A., and Pombeiro, A.J.L. (2017). *CrystEngComm* 19: 4666–4695.
20 Kou, J., Lu, C., Wang, J. et al. (2017). *Chem. Rev.* 117: 1445–1514.
21 Karmakar, A., Soliman, M.M.A., Alegria, E.C.B.A. et al. (2022). *Catalysts* 12: 294.
22 Dybtsev, D.N. and Bryliakov, K.P. (2021). *Coord. Chem. Rev.* 437: 213845.
23 Mo, K., Yang, Y., and Cui, Y. (2014). *J. Am. Chem. Soc.* 136: 1746–1749.
24 Li, Y., Xu, H., Ouyang, S., and Ye, J. (2016). *Phys. Chem. Chem. Phys.* 18: 7563–7572.
25 Xiao, J.-D. and Jiang, H.-L. (2019). *Acc. Chem. Res.* 52: 356–366.
26 Karmakar, A., Titi, H.M., and Goldberg, I. (2011). *Cryst. Growth Des.* 11: 2621–2636.
27 Dhakshinamoorthy, A., Asiri, A.M., and Garcia, H. (2016). *Chem. Eur. J.* 22: 8012–8024.
28 Kang, Y.-S., Lu, Y., Chen, K. et al. (2019). *Coord. Chem. Rev.* 378: 262–280.
29 Gascon, J., Corma, A., Kapteijn, F., and Llabrés i Xamena, F.X. (2014). *ACS Catal.* 4: 361–378.
30 Karmakar, A., Martins, L.M.D.R.S., Hazra, S. et al. (2016). *Cryst. Growth Des.* 16: 1837–1849.
31 Karmakar, A., Rúbio, G.M.D.M., Guedes da Silva, M.F.C. et al. (2016). *RSC Adv.* 6: 89007–89018.
32 Conley, E.T. and Gates, B.C. (2022). *Chem. Mater.* 34: 3395–3408.
33 Wang, Q. and Astruc, D. (2020). *Chem. Rev.* 120: 1438–1511.
34 Park, J., Li, J.-R., Chen, Y.-P. et al. (2012). *Chem. Commun.* 48: 9995–9997.
35 Karmakar, A., Soliman, M.M.A., Rúbio, G.M.D.M. et al. (2020). *Dalton Trans.* 49: 8075–8085.
36 Karmakar, A., Paul, A., Rúbio, G.M.D.M. et al. (2019). *Front. Chem.* 7: 699.
37 Li, B., Zhang, Y., Ma, D. et al. (2012). *Chem. Commun.* 48: 6151–6153.
38 Srirambalaji, R., Hong, S., Natarajan, R. et al. (2012). *Chem. Commun.* 48: 11650–11652.
39 Pintado-Sierra, M., Rasero-Almansa, A.M., Corma, A. et al. (2013). *J. Catal.* 299: 137–145.
40 Beyzavi, H., Vermeulen, N.A., Howarth, A.J. et al. (2015). *J. Am. Chem. Soc.* 137: 13624–13631.
41 Esken, D., Turner, S., Lebedev, O.I. et al. (2010). *Chem. Mater.* 22: 6393–6401.
42 Müller, M., Turner, S., Lebedev, O.I. et al. (2011). *Eur. J. Inorg. Chem.* 1876–1887.
43 Liu, H., Li, Y., Jiang, H. et al. (2012). *Chem. Commun.* 48: 8431–8433.
44 Han, Q., Qi, B., Ren, W. et al. (2015). *Nat Commun* 6: 10007.
45 Toyao, T., Saito, M., Horiuchi, Y., and Matsuoka, M. (2014). *Catal. Sci. Technol.* 4: 625–628.
46 Wang, D. and Li, Z. (2015). *Catal. Sci. Technol.* 5: 1623–1628.

21

(Tetracarboxylate)bridged-di-transition Metal Complexes and Factors Impacting Their Carbene Transfer Reactivity

LiPing Xu[1,2], Adrian Varela-Alvarez[1], and Djamaladdin G. Musaev[1]

[1] *Cherry L. Emerson Center for Scientific Computation and Department of Chemistry, Emory University, 1515 Dickey Drive, Atlanta, Georgia, USA*
[2] *School of Chemistry and Chemical Engineering, Shandong University of Technology, Zibo, China*

21.1 Introduction

Multi-transition metal complexes have emerged as important thermal and photocatalysts in numerous vital synthetic and biological processes, as effective photosensitizers, and as vital multifunctional materials. For example, bourgeoning research has identified tetra-bridged paddlewheel-di-Rh, and di-Ru complexes as effective catalysts for C–H and C–C π-bond alkylation (by utilizing diazocarbenes $N_2CR^1R^2$ as a carbene source) [1–12] and amination (which operates best with sulfamate-derived nitrene species (NSO_3R)) (Figure 21.1) [13].

It has been demonstrated that metal centers stabilize reactive carbenes (and nitrenes) for consequent insertion into activated bonds (Figure 21.2) in a controllable and highly predictable manner during carbene (and nitrene) transfer reactions [14–18]. Some of the reported tetra-bridged paddlewheel-di-metallic complexes, such as the (tetracarboxylate)-di-Rh complexes, also have emerged as photosensitizers that adsorb visible light to initiate charge separation and charge transfer processes [19–21]. In many cases, these complexes act as the building blocks of multifunctional porous metal-organic frameworks (MOFs) [22]. These and other important features and applications of the tetra-bridged paddlewheel-di-metallic complexes have made them a target of numerous fundamental investigations. The existing extensive experimental and computational studies have emphasized the stability of these species, nature of the bridging ligands, and tunability of the redox potentials of metal centers as the success-limiting aspects [23–35]. To establish the factors impacting the stability and reactivity of these species under the catalytic conditions, the atomistic-level understanding of mechanisms of the targeted processes, roles of the transition metals (M) and their electronic structures, and roles of bridging and auxiliary ligands (L) have been shown to be important. Knowledge of these aspects of the tetra-bridged paddlewheel-di-metallic complexes is critical to improve their efficacy and broader utilization. Currently, the literature includes multiple high impact investigations of the tetra-bridged paddlewheel-$[Rh_2]$ and $-[Ru_2]$ complexes [7, 8, 12–18, 23–35]. However, the development of inexpensive analogs with the earth abundant first-row transition metals (such as Co, Ni, and Cu) requires additional research.

Catalysis for a Sustainable Environment: Reactions, Processes and Applied Technologies Volume 2, First Edition. Edited by Armando J. L. Pombeiro, Manas Sutradhar, and Elisabete C. B. A. Alegria.
© 2024 John Wiley & Sons Ltd. Published 2024 by John Wiley & Sons Ltd.

Figure 21.1 Schematic representation of the broadly used tetra-bridged paddlewheel-di-Rh catalysts in this study.

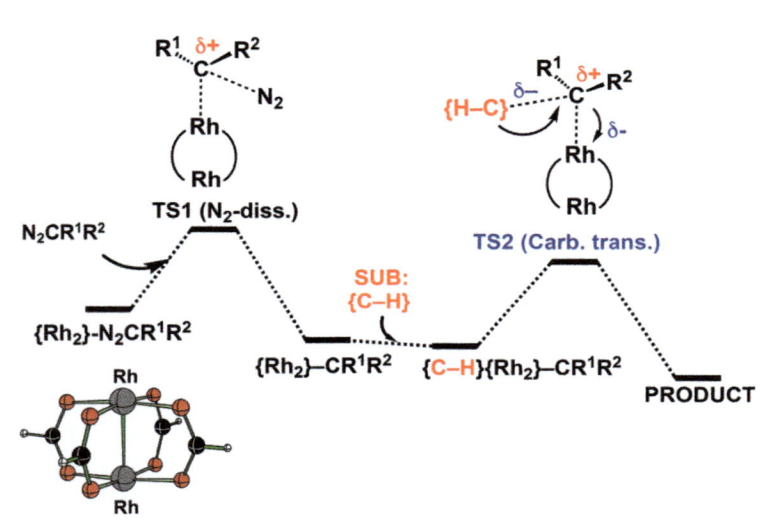

Figure 21.2 Schematic representation of the diazocarbene ($N_2CR^1R^2$) decomposition by the (tetracarboxylate)-bridged-di-Rh catalysts, and following carbene transfer to C–H bond.

Therefore, this chapter aims to discuss electronic structures of the $(RCOO)_4$-$[M_2]$ complexes, and their carbene derivatives $\{(RCOO)_4\text{-}[M_2]\}=CR^1R^2$, where M = Co, Ni, Cu, Rh, Pd, and Ag; R = Me, Ph and CF_3; $R^1 = p$-BrPh; and R^2 = COOMe. We anticipate that the comparison of the obtained data for different Ms and Rs will enable us to identify impact of the nature of the metal-centers and R-substituents of the bridging carboxylates on the stability and electronic structure of these complexes, as well as thermodynamic properties of diazocarbene decomposition (i.e. the reaction of $N_2=CR_2 + \{(RCOO)_4\text{-}[M_2]\} \rightarrow \{(RCOO)_4\text{-}[M_2]\}=CR^1R^2 + N_2$) by these complexes.

21.2 Computational Procedure

In this paper, we used the B3LYP density functional approximation [36–38] with Grimme's empirical dispersion-correction (D3) [39, 40] and the Becke-Johnson (BJ) damping-correction [41] in conjunction with the 6-31G(d,p) basis sets for C, H, N, and O atoms [42] and LANL2DZ basis sets (with their corresponding electrochemical potentials) for transition metal atoms and Br [43, 44]. Bulk solvent effects were incorporated for all calculations (including geometry optimizations and frequency calculations) using the self-consistent reaction field polarizable continuum model (IEF-PCM) [45–47]. We chose dichloromethane as the solvent. The reported thermodynamic data were computed at a temperature of 298.15 K and at 1 atm of pressure. Unless otherwise stated, energies are given as $\Delta H/\Delta G$ in kcal/mol, although only the Gibbs free energies will be discussed. We have calculated several lower-lying electronic states (with different spin multiplicities) for all reported species, including antiferromagnetically coupled (so called open-shell singlet) states. Regardless, below we will discuss only the energetically lowest representatives of the low-spin and high-spin states.

All reported structures were fully optimized without any geometry constraints. Previously, it was reported that the computational methods used in this paper accurately describe the energies and geometries of organometallic compounds [48–52]. Frequency calculations were carried out to verify the nature of the located stationary points. Graphical analysis of the imaginary vibrational normal modes was used to confirm the nature of the located equilibrium structures. All calculations were performed by the Gaussian-09 suite of programs [53]. NBO 3.1 program, which is included in the Gaussian-09 suite of programs, was used to obtain natural bond orbitals (NBOs), atomic net charges, densities of spin, and Wiberg bond indexes (WBI) at the optimized geometries.

21.3 Results and Discussion

21.3.1 Geometrical and Electronic Structures of the (RCOO)$_4$-[M$_2$] Complexes

The geometry and electronic structure of the (tetracarboxylate)-[M$_2$] complexes have been the subject of numerous studies [8, 30, 54–64]. These studies have identified the most critical two types of bonds that directly impact to structural stability of these complexes. They are: (i) the donor-acceptor M–O(carboxylate), and (ii) the M(d_{zz}) + M(d_{zz}) sigma bonds. The former requires empty or partially occupied d_{xy}-AOs of metals, whereas the latter is expected to be more effective if the M-centers will have a partially occupied d_{zz}-AO (see Figure 21.3).

M=Rh. The Rh(II)-centers with d^7-electron configuration is expected to be ideal for the formation of strong Rh-O(carboxylate) donor-acceptor bonds, as well as the M(d_{zz}) + M(d_{zz}) covalent bonds in the (tetracarboxylate)-[Rh$_2$] complex (see Figure 21.4).

Figure 21.3 Schematic representation of the important M−O(carboxylate) and metal-metal σ-bonds in the (RCOO)$_4$-di-metallic complexes.

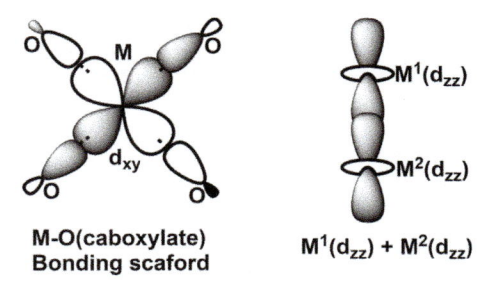

M-O(caboxylate) Bonding scaford

$M^1(d_{zz})$

$M^2(d_{zz})$

$M^1(d_{zz}) + M^2(d_{zz})$

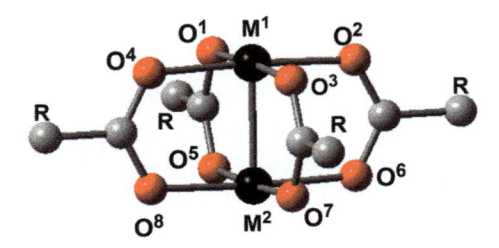

Figure 21.4 3D presentation of the lowest singlet and triple state (RCOO)$_4$-[Rh$_2$] and (RCOO)$_4$-[Co$_2$] complexes. For important geometry parameters of these species, see Table 21.1.

Consistently, previous extensive computational analyses have shown that, in its ground singlet $[\sigma(d_{zz}+d_{zz})]^2[\pi(d_{xz}+d_{xz})]^2[\pi(d_{yz}+d_{yz})]^2[\delta(d_{xy}+d_{xy})]^2[\delta^*(d_{xy}-d_{xy})]^2[\pi^*(d_{xz}-d_{xz})]^2[\pi^*(d_{yz}-d_{yz})]^2[\sigma^*(d_{zz}-d_{zz})]^0$ electronic state, this complex has a single σ[Rh(s)–Rh(s)] bond, along the Rh-O(carboxylate) donor-acceptor bonds (see Figure 21.5). Our NBO analyses of the (MeCOO)$_4$-[Rh$_2$] complex show that the Rh-O(carboxylate) donor-acceptor bonds are polarized toward the O-centers with 11.5% (Rh) + 88.5% (O) content. In (MeCOO)$_4$-[Rh$_2$], Rh-centers are reduced, and have only +0.72 |e|. In other words, (i) almost 2.56 |e| electron densities were transferred from four carboxylates to the Rh$_2$-core; and (ii) rhodium centers are in their +1 oxidation states. The calculated value of the Rh–Rh distance, 2.394–2.421 Å (Table 21.1), is consistent with the single bond character of the Rh–Rh bond. The calculated Wiberg bond index (WI) of 0.83–0.82 supports this assignment. The natural bond analyses shown that, the bonding σ-orbital has only 1.66 electrons, and its anti-bonding component, σ^* orbital, holds ca 0.27 electrons, in the (MeCOO)$_4$-[Rh$_2$] complex.

Interestingly, the energetically lowest excited triplet states of the (RCOO)$_4$-[Rh$_2$] complexes are the $(\sigma)^2(\pi)^4(\delta)^2(\delta^*)^2(\pi^*)^3(\sigma^*)^1$ triplet states, which can be formed from the singlet ground state by the $\pi^* \to \sigma^*$ one electron transition (which is symmetry-allowed only under C$_1$-symmetry [65]). These triplet states are calculated to be ~21–23 kcal/mol higher in free energy than the corresponding singlet states (Table 21.1). Spin density analyses show that in the excited triplet state, each Rh

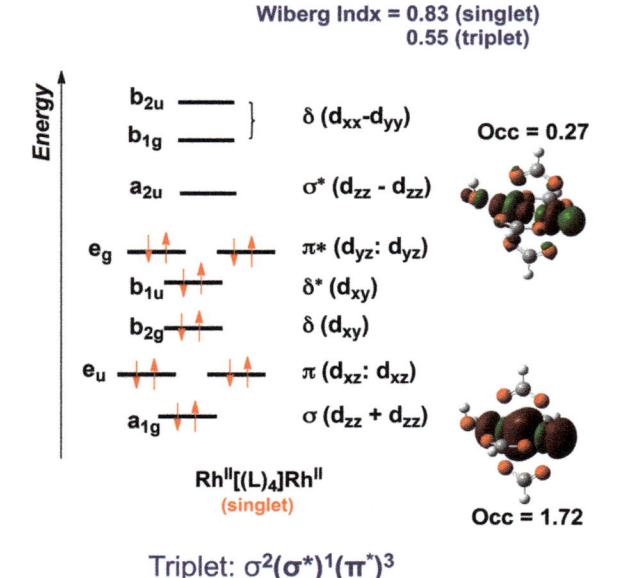

Figure 21.5 Electronic configuration of the ground singlet electronic state of the (MeCOO)$_4$-[Rh$_2$] complex.

Table 21.1 The calculated geometry parameters (in Å) of the (RCOO)-[M$_2$] complexes, at their energetically lowest low-spin (L) and high-spin (H) electronic states, as well as energy difference, Δ(L–H), in kcal/mol, between these electronic states (see Figures 21.4, 21.6, and 21.8 for the labeling used).

Complex	R	State	M^1-M^2	M^1-O^1	M^1-O^3	M^1-O^5	M^1-O^7	M^2-O^2	M^2-O^4	M^2-O^6	M^2-O^8	Δ(L–H)
M=Rh;	Me	^{1}A	2.394	2.065	2.064	2.065	2.064	2.064	2.065	2.064	2.065	0.00
		^{3}A	2.452	2.073	2.072	2.073	2.072	2.072	2.073	2.072	2.073	21.9
	Ph	^{1}A	2.395	2.060	2.060	2.060	2.060	2.060	2.060	2.060	2.060	0.00
		^{3}A	2.454	2.068	2.068	2.068	2.068	2.068	2.068	2.068	2.068	23.2
	CF$_3$	^{1}A	2.421	2.066	2.067	2.067	2.066	2.067	2.066	2.067	2.066	0.00
		^{3}A	2.486	2.079	2.080	2.077	2.075	2.080	2.079	2.075	2.077	21.2
M=Co;	Me	^{1}A$_{os}$	2.357	1.919	1.919	1.920	1.920	1.920	1.920	1.919	1.919	0.00
		^{3}A	2.452	1.912	1.910	1.911	1.908	1.941	1.923	1.941	1.924	3.5
	Ph	^{1}A$_{os}$	2.360	1.916	1.915	1.916	1.916	1.916	1.916	1.916	1.915	0.00
		^{3}A	2.438	1.929	1.923	1.923	1.930	1.925	1.920	1.919	1.925	5.4
	CF$_3$	^{1}A$_{os}$	2.439	1.924	1.923	1.926	1.925	1.926	1.926	1.924	1.925	0.00
		^{3}A	2.558	1.928	1.925	1.928	1.923	1.945	1.925	1.947	1.925	2.6
M=Pd;	Me	^{1}A	2.623	2.066	2.067	2.066	2.067	2.066	2.067	2.066	2.067	0.00
		^{3}A	2.532	2.232	2.233	2.234	2.235	2.042	2.041	2.041	2.041	2.4
	Ph	^{1}A	2.622	2.061	2.061	2.061	2.061	2.061	2.061	2.061	2.061	0.00
		^{3}A	2.534	2.230	2.228	2.230	2.230	2.036	2.037	2.036	2.036	2.3
	CF$_3$	^{1}A	2.675	2.071	2.072	2.072	2.072	2.070	2.070	2.070	2.070	1.7
		^{3}A	2.586	2.245	2.246	2.252	2.255	2.042	2.040	2.040	2.042	0.00
M=Ni;	Me	^{1}A	2.427	1.895	1.895	1.895	1.895	1.895	1.895	1.895	1.895	0.00
		^{3}A	2.340	1.958	1.957	1.957	1.957	1.957	1.957	1.957	1.957	2.6
	Ph	^{1}A	2.428	1.891	1.891	1.891	1.891	1.891	1.891	1.891	1.891	0.00
		^{3}A	2.338	1.952	1.952	1.952	1.952	1.952	1.951	1.951	1.951	3.5
	CF$_3$	^{1}A	2.498	1.900	1.899	1.899	1.900	1.900	1.900	1.900	1.900	0.00
		^{3}A	2.409	1.964	1.963	1.963	1.964	1.963	1.964	1.963	1.963	0.4
M=Ag;	Me	^{1}A	2.886	2.431	2.432	2.432	2.431	2.064	2.064	2.064	2.064	1.2
		^{3}A	2.727	2.221	2.204	2.217	2.205	2.203	2.219	2.207	2.219	0.00
	Ph	^{1}A	2.914	2.440	2.441	2.444	2.437	2.052	2.052	2.052	2.053	1.4
		^{3}A	2.730	2.208	2.205	2.201	2.204	2.201	2.204	2.207	2.205	0.00
	CF$_3$	^{1}A	2.847	2.226	2.226	2.223	2.223	2.233	2.234	2.236	2.237	1.6
		^{3}A	2.806	2.224	2.218	2.223	2.211	2.215	2.222	2.215	2.226	0.00

(Continued)

Table 21.1 (Continued)

Complex	R	State	M¹-M²	M¹-O¹	M¹-O³	M¹-O⁵	M¹-O⁷	M²-O²	M²-O⁴	M²-O⁶	M²-O⁸	Δ(L–H)
M=Cu;	Me	$^1A_{os}$	2.521	1.988	1.988	1.988	1.988	1.988	1.988	1.988	1.988	0.00
		^3A	2.507	1.973	1.972	1.973	1.973	1.973	1.973	1.973	1.973	0.3
	Ph	$^1A_{os}$	2.517	1.967	1.967	1.967	1.967	1.967	1.967	1.967	1.967	0.00
		^3A	2.512	1.967	1.967	1.967	1.967	1.967	1.967	1.967	1.967	0.1
	CF_3	$^1A_{os}$	2.633	1.990	1.992	1.992	1.990	1.992	1.990	1.990	1.992	0.00
		^3A	2.618	1.974	1.975	1.974	1.975	1.974	1.975	1.974	1.975	0.7

center has almost one (0.95 |e|) unpaired spin, and the Rh–Rh bond has slightly larger than half-bond character with 0.55 Wiberg bond indexes. NBO analysis also show the existence of two beta Rh–Rh bonding orbitals in the triplet state complex. The first one is a Rh–Rh σ-bond, whereas the second one is a Rh–Rh π-bond. Thus, the Rh–Rh bond with half σ- and half π-bonds has more than half-bond character. Bearing in mind that a one-electron π-bond is intrinsically weaker than one-electron σ-bond, one can explain the ~0.06Å elongation of the Rh–Rh bond distance in the triplet state complexes compared with the singlet state complexes.

The data presented in Tables 21.1 and 21.2 for the $(RCOO)_4$-$[Rh_2]$ complexes (where R = Me, Ph and CF_3) show only a slight impact of the R-groups on the calculated geometry of and singlet-triplet energy splitting in these complexes.

M= Co(II). As seen in Table 21.1, replacement of Rh(II) in $[RCOO]_4$–$[Rh_2]$ by its 3d-analog, Co(II), leads to formation of the high-spin $[RCOO]_4$–$[Co_2]$ complex with almost one unpaired

Table 21.2 The calculated important spin densities (in |e|) of the $(RCOO)_4$-$[M_2]$ complexes, at their energetically lowest low-spin and high-spin electronic states (see Figures 21.4, 21.6, and 21.8 for the labeling used).

Complex	R	State	M¹	M²	O¹	O²	O³	O⁴	O⁵	O⁶	O⁷	O⁸
M=Rh;	Me	^3A	0.95	0.95	0.01	0.01	0.01	0.01	0.01	0.01	0.01	0.01
	Ph	^3A	0.95	0.95	0.01	0.01	0.01	0.01	0.01	0.01	0.01	0.01
	CF_3	^3A	0.96	0.96	0.00	0.03	0.00	0.03	0.00	0.03	0.00	0.03
M=Co;	Me	$^1A_{os}$	-0.95	0.94	0.02	0.02	0.02	0.02	-0.02	-0.02	-0.02	-0.02
		^3A	1.00	1.01	0.00	-0.03	0.00	-0.03	0.00	0.00	0.01	0.00
	Ph	$^1A_{os}$	0.89	-0.89	-0.02	-0.02	-0.02	-0.02	0.02	0.02	0.02	0.02
		^3A	1.01	1.10	-0.02	0.00	0.00	-0.02	0.00	0.00	0.01	0.01
	CF_3	$^1A_{os}$	-0.94	0.94	0.02	0.02	0.02	0.02	-0.02	-0.02	-0.02	-0.02
		^3A	1.00	1.07	0.02	0.04	0.00	0.02	0.00	-0.01	-0.01	-0.01
M=Pd;	Me	^3A	1.27	0.31	0.02	0.02	0.02	0.02	0.09	0.09	0.09	0.09
	Ph	^3A	1.29	0.31	0.09	0.09	0.09	0.09	0.02	0.02	0.02	0.02
	CF_3	^3A	1.32	0.31	0.02	0.02	0.02	0.02	0.08	0.08	0.08	0.08

Table 21.2 (Continued)

Complex	R	State	M^1	M^2	O^1	O^2	O^3	O^4	O^5	O^6	O^7	O^8
M=Ni;	Me	^{3}A	1.05	1.05	0.01	0.01	0.01	0.01	0.01	0.01	0.01	0.01
	Ph	^{3}A	1.04	1.04	0.01	0.01	0.01	0.01	0.01	0.01	0.01	0.01
	CF$_3$	^{3}A	1.07	1.07	0.01	0.01	0.01	0.01	0.01	0.01	0.01	0.01
M=Ag;	Me	^{3}A	0.35	0.35	0.15	0.15	0.15	0.15	0.15	0.15	0.15	0.15
	Ph	^{3}A	0.36	0.36	0.15	0.15	0.15	0.15	0.15	0.15	0.15	0.15
	CF$_3$	^{3}A	0.38	0.38	0.14	0.14	0.14	0.14	0.14	0.14	0.14	0.14
M=Cu;	Me	^{1}A$_{os}$	0.57	−0.57	0.10	0.10	0.10	0.10	−0.10	−0.10	−0.10	−0.10
		^{3}A	0.57	0.57	0.10	0.10	0.10	0.10	0.10	0.10	0.10	0.10
	Ph	^{1}A$_{os}$	0.58	−0.58	0.10	0.10	0.10	0.10	−0.10	−0.10	−0.10	−0.10
		^{3}A	0.58	0.58	0.10	0.10	0.10	0.10	0.10	0.10	0.10	0.10
	CF$_3$	^{1}A$_{os}$	0.59	−0.59	0.10	0.10	0.10	0.10	−0.10	−0.10	−0.10	−0.10
		^{3}A	0.59	0.59	0.10	0.10	0.10	0.10	0.10	0.10	0.10	0.10

α-spin on each Co-centers. For the [RCOO]$_4$–[Co$_2$] complex, the antiferromagnetically coupled singlet ground electronic states of these complexes are lower in energy with the (s)$^{0.30}$(p)$^{0.34}$(d$_{xy}$)$^{0.61}$ (d$_{xz}$)$^{1.99}$(d$_{yx}$)$^{1.99}$(d$_{x2-y2}$)$^{1.97}$(d$_{zz}$)$^{1.05}$ valence electronic configurations of the Co-centers. The partial population of the Co(d$_{xy}$) orbital allows the formation of the strong Co-carboxylate interactions. The generated Co–O bonds have almost 11–12% contributions from the Co-centers and 89–88% contributions from the O-centers of carboxylate. Thus, these Co–O bonds are highly polarized toward oxygen centers. As a result of these interactions, some electron densities have transferred from carboxylates to the Co-centers: in [MeCOO]$_4$–[Co$_2$], Co-centers possess only +0.74 |e| positive charges, whereas each acetate ligand has a total of -0.42 |e| negative charge. As we could expect from the electronic configurations of the Co-centers in [RCOO]$_4$–[Co$_2$] reported here, the lowest singlet electronic states of these complexes may have a Co(sd$_{zz}$)–Co(sd$_{zz}$) σ-bond. Indeed, calculations show the presence of the Co–Co σ-bond with 1.77 |e| electron population. Because of the presence of such Co–Co bonding, the Co–Co distances are 2.357, 2.360, and 2.439 Å for R = Me, Ph, and CF$_3$, respectively. Because the calculated singlet-triplet energy differences are small and changes via 2.6 kcal/mol (R = CF$_3$)<3.5 kcal/mol (R = CH$_3$)<5.4 kcal/mol (R=Ph), one may conclude that the Co–Co interaction in the singlet state complexes is weak. As could be expected, the singlet-triplet excitation results in elongation of the Co–Co distance to 2.452, 2.438, and 2.558 Å for R = Me, Ph, and CF$_3$, respectively (Table 21.1).

M=Pd. NBO analyses show that Pd-ions in the singlet state [MeCOO]$_4$–[Pd$_2$] complexes are in their (s)$^{0.29}$(p)$^{0.29}$(d$_{xy}$)$^{0.83}$(d$_{xz}$)$^{2.00}$(d$_{yx}$)$^{2.00}$(d$_{x2-y2}$)$^{1.97}$(d$_{zz}$)$^{1.92}$ valence electron configurations, allowing them to participate in the Pd-carboxylate interactions. Existence of such Pd-O (carboxylate) interactions impacts electron distribution in the singlet state [MeCOO]$_4$–[Pd$_2$] complexes, where each Pd-center has +0.71 |e| positive charge (i.e. in the singlet electronic state of [RCOO]$_4$–[Pd$_2$], Pd-ions are in their +1 oxidation states). Because of the almost 2-electron population of the Pd(d$_{zz}$) atomic orbitals, there is no direct Pd–Pd interaction in the singlet electronic state of [RCOO]$_4$–[Pd$_2$]

M = Pd: Singlet
M = Ni: Singlet and Triplet

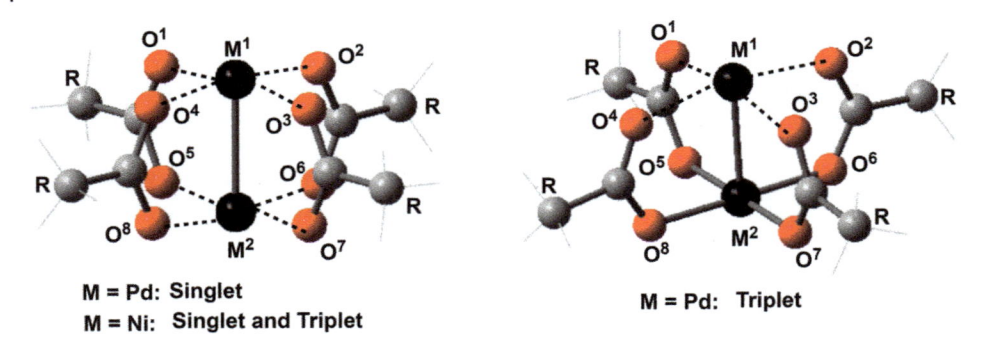

M = Pd: Triplet

Figure 21.6 3D presentation of the lowest singlet and triple state (RCOO)$_4$-[Pd$_2$] and (RCOO)$_4$-[Ni$_2$] complexes. For geometry parameters of these species, see Table 21.1.

complexes, while we cannot eliminate the bridging ligands mediated, as well as second order perturbation interactions between the Pd-centers. Calculations show that the singlet state [RCOO]$_4$–{Pd$_2$} complexes have a roughly symmetric geometry with the Pd–Pd distances of 2.623, 2.622, and 2.677 Å for R = Me, Ph, and CF$_3$, respectively (Figure 21.6).

Interestingly, in their triplet electron states, the [RCOO]$_4$–[Pd$_2$] complexes have a highly asymmetric structure (Figure 21.6), in which only one of the Pd-centers (Pd^2) is involved in a strong Pd^2-O(carboxylate) bonding. Meanwhile, the second Pd-center, Pd^1 in Figure 21.6, participates in weak Pd^1-O interactions (with the bond distance of ca 2.23 Å). Such an asymmetric structural motive of the triplet state [RCOO]$_4$–[Pd$_2$] complexes can be explained in two complementary ways, described as follows. Representative bonding molecular orbitals of the triple state [MeCOO]$_4$–[Pd$_2$] complex are given in Figure 21.7.

First, we can use MO-diagram presented in Figure 21.5. Compared with the [RCOO]$_4$–[Rh$_2$] complex MO diagram of the singlet state presented in Figure 21.5, [RCOO]$_4$–[Pd$_2$] has two more electrons. In the ground singlet state, these two electrons can occupy the σ^*(M–M) antibonding orbital and destabilize M–M bonding. However, in the triplet state [RCOO]$_4$–[Pd$_2$], these two additional electrons can be distributed between the σ^*(M–M) antibonding and $\delta(d_{xx}$-$d_{yy})$, which will lead to the (1-electron)-(2-center) σ(M+M) bonding and destabilization of Pd^1-carboxylate interactions. Secondly, we can use the concept in the difference in the electronic configuration of the Pd^1 and Pd^2-centers. Indeed, the Pd^1 and Pd^2-centers can be in two different electronic configurations originally: $(d_{xy})^{2.00}(d_{xz})^{2.00}(d_{yx})^{2.00}(d_{x2-y2})^{2.00}(d_{zz})^{0.00}$ and $(d_{xy})^{0.00}(d_{xz})^{2.00}(d_{yx})^{2.00}(d_{x2-y2})^{2.00}(d_{zz})^{2.00}$ for Pd^1 and Pd^2, respectively, allowing them to interact differently with the bridging carboxylates and each other.

Both of these (MO- and AO-based) concepts explain the calculated geometry and electronic properties of the triplet state [MCOO]$_4$–[Pd$_2$] complexes. Namely, (i) the presence of the 1.27 |e| and 0.31 |e| unpaired α-spin in Pd^1 and Pd^2, respectively; (ii) the formation of the Pd(sd$_{zz}$)+ Pd(sd$_{zz}$)

Pd-O (carboxylate)   **Pd-Pd bonding**

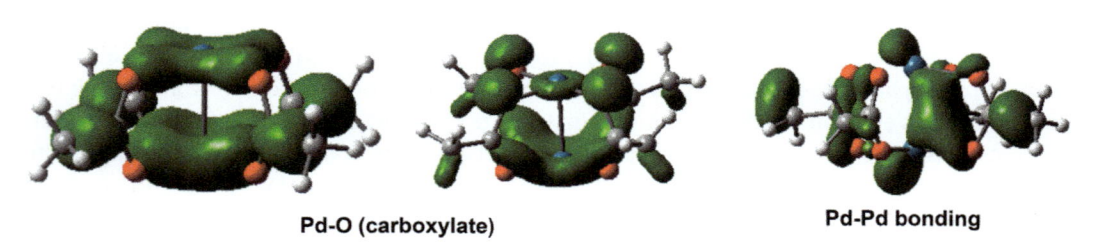

Figure 21.7 Representative bonding molecular orbitals of the triple state [MeCOO]$_4$–[Pd$_2$] complex.

bonding with 0.87 |e| electron population (interestingly, this bond is polarized toward the Pd^2 center and has 33.5% Pd^1 + 66.5% Pd^2 content); (iii) the shorter Pd–Pd bond distance (2.532, 2.534, and 2.586 Å for R = Me, Ph, and CF_3, respectively) compared with that in the corresponding singlet state complexes; (iv) the $(s)^{0.23}(p)^{0.26}(d_{xy})^{1.29}(d_{xz})^{2.00}(d_{yx})^{2.00}(d_{x2-y2})^{1.98}(d_{z2})^{1.39}$ and $(s)^{0.32}(p)^{0.33}(d_{xy})^{0.96}$ $(d_{xz})^{2.00}(d_{yx})^{2.00}(d_{x2-y2})^{1.97}(d_{z2})^{1.65}$ valence electron configurations of Pd^1 and Pd^2, respectively; (v) the fact that Pd^2-O(carboxylate) and Pd^1-O(carboxylate) bonds have 19.8% (Pd^2) + 80.2%(O) and 10.4% (Pd^1) + 89.6%(O) contents, respectively; and (vi) the +0.86 and +0.78 |e| total charges of the Pd^1 and Pd^2, respectively.

Because the existing Pd–Pd bond of the triplet electron state $[MCOO]_4$–$[Pd_2]$ complex is a weak (1-electron)–(2-center) type of bond, the calculated singlet and triplet electronic states of these complexes are close in energy. For R=Me and Ph, the singlet state is lower in energy than triplet state. However, for R = CF_3, the triplet state is lower than singlet state.

M=Ni. As could be anticipated, in general, electronic structures of the $[RCOO]_4$–$[Pd_2]$ and $[RCOO]_4$–$[Ni_2]$ complexes are very similar to each other. Indeed, for $[RCOO]_4$–$[Ni_2]$, as for their Pd-analogs, (i) ground electronic states are the singlet states, whereas their triplet states are only a few kcal/mol higher in free energy; and (ii) in the singlet state complexes there is no direct Ni–Ni bonding, whereas in the triplet state complexes there is one (1-electron)–(2-center) type of bond between the Ni-centers. However, compared with the corresponding Pd- carboxylate interaction, the weaker Ni-carboxylate interactionintroduces a few structural (both geometrical and electronic) differences between the $[RCOO]_4$–$[Ni_2]$ and $[RCOO]_4$–$[Pd_2]$ complexes (Figure 21.6; Tables 21.1 and 21.2): (i) in the $[RCOO]_4$–$[Ni_2]$, unlike the $[RCOO]_4$–$[Pd_2]$ complex, both metal centers have a symmetric ligand environment, regardless of the nature of electronic states, and, consequently; (ii) unlike the Pd–Pd bonding in $[MeCOO]_4$–$[Pd_2]$, the Ni–Ni bond $[MeCOO]_4$–$[Ni_2]$ is fully symmetrical with 50% contributions from the each metal centers. Consistent with these electronic structure analyses, $[RCOO]_4$–$[Ni_2]$ complexes in the singlet state have calculated Ni–Ni distances that are longer (2.427, 2.428, and 2.498 Å for R = Me, Ph, and CF_3, respectively) than those in the triplet state complexes (2.340, 2.338, and 2.409 Å for R = Me, Ph, and CF_3, respectively).

M=Ag. As already shown, there is critical importance of the (i) metal-O(carboxylate) donor-acceptor interaction involving d_{xy}-AO of metal and sp-hybrid orbitals of the oxygens; and (ii) the metal-metal interaction involving hybrid sd_{zz}-AOs of the metal center. These bonding schemes require availability of the (d_{xy}) and (d_{zz}) atomic orbitals of the M(II) ions, respectively, and may lead to roughly symmetric or asymmetric geometry structures with or without M–M sigma bonding depending on the strengths (and ratio) of these interactions. As seen in Figure 21.8 and Table 21.1, the ground singlet electronic states of the $[RCOO]_4$–$\{Ag_2\}$ complexes have asymmetric (with different ligand bonding motive to Ag^1 and Ag^2 centers) geometrical structures.

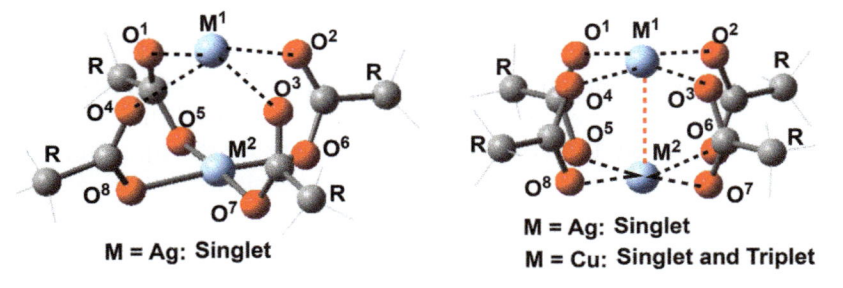

Figure 21.8 3D presentation of the lowest singlet and triple state $(RCOO)_4$-$[Ag_2]$ and $(RCOO)_4$-$[Cu_2]$ complexes. For important geometry parameters of these species, see Table 21.1.

The calculated Ag^1-O distances (of ~2.431Å) are ca 0.37Å longer than Ag^2-O distances (of ~2.064Å). Consistently, the Ag^2-ion in its $d^{9.92}$-electronic configuration, whereas the Ag^2-ion that involved in strong interactions with the carboxylate O-centers is in its $d^{9.33}$-electronic configuration. These findings are consistent with the presence of Ag^2–O bonding orbital that has 1.875 |e| population and with the 21.75% (Ag) + 78.25% (O) character. NBO analyses also shown that Ag^1 and Ag^2 centers have +0.96 |e| and +0.77 |e| positive charges, respectively, and each bridging carboxylate ligands has almost −0.57 |e| negative charge (instead of −1.0 |e|, each of the bridging carboxylates has donated overall almost 0.43 |e| electron density to the Ag-centers). Thus, in the ground singlet electronic states of these complexes, both Ag centers are in their +1 oxidation states. There is no direct bonding between the Ag-centers. The calculated Ag-Ag distances are 2.886 Å (for R = Me), 2.914 Å (for R = Ph), and 2.847 Å (for R = CF_3), for the singlet state $[RCOO]_4$–$[Ag_2]$ complexes.

In contrast, in their triplet exited states, the $[RCOO]_4$–$[Ag_2]$ complexes have symmetric (with very similar ligand bonding motive around the Ag^1 and Ag^2 centers) geometrical structures: the calculated Ag–O bond distances are in range of 2.20–2.22Å. Unpaired two α-spins are symmetrically distributed between the Ag-centers and bridging carboxylic ligands (Table 21.2). Interestingly, the Ag–Ag distance is 0.159 (for R = Me), 0.174 Å (for R = Ph), and 0.041Å (for R = CF_3) shorter in the triplet state complexes compared with the singlet state complexes. These geometry changes indicate the presence of the Ag–Ag interaction in the triplet state complexes. However, our analyses have revealed the presence of the (1-electron)-(2-center) type of Ag–O bonding orbitals with the 21.0% (Ag)+80.0% (O) character, but no direct Ag–Ag bond. The reported Ag–Ag distance reduction could be explained the presence of the second-order perturbation interactions between the Ag-center, either directly or via the bridging ligands. Consistent with the reported geometry and electronic structure parameters of the triplet state $[RCOO]_4$–$[Ag_2]$ complexes, each Ag-centers has +0.90 |e| positive charge and can be identified as an Ag(I)-centers. The energy differences between the ground singlet and excited triplet states of the $[RCOO]_4$–$[Ag_2]$ complexes are very small, regardless of the nature of the R-fragment.

M=Cu. The lower-lying antiferromagnetically coupled singlet and triplet states of the $[RCOO]_4$–$[Cu_2]$ complexes are energetically almost degenerate with similar geometry and electronic structures (Figure 21.8; Tables 21.1 and 21.2). Therefore, for simplicity, we discuss below only the triplet state complexes. As shown in Table 21.1, all studied Cu-complex Cu-centers have a symmetric ligand environment with a Cu-Cu distance of 2.521 and 2.507 Å (for R = Me), 2.517 and 2.512 Å (for R = Ph), and 2.633 and 2.618 Å (for R = CF_3), for the singlet and triplet state complexes, respectively. Data presented in Table 21.2 show that each Cu-center of these complexes bears 0.57–0.59 |e| unpaired α-spins; thus, almost 0.86–0.82 |e| spins are delocalized on the bridging carboxylates for all studied Rs. Based on the natural bond analyses: (i) each Cu-center of these complexes has +0.97 |e| positive charges and therefore the studied $[RCOO]_4$–$[Cu_2]$ complexes have Cu-centers that are in their +1 oxidation states; (ii) there is no direct Cu–Cu interactions, although we cannot eliminate the ligand-mediated and second-order perturbation interactions between the Cu-centers; and (iii) the Cu–O interactions are weaker than corresponding Ag–O interactions discussed above. Indeed, the reported (1-electron)–(2-center) type of M–O bonds have a 21.0% (M) + 80.0% (O) character for M = Ag, but 12.0% (M) + 88.0% (O) character for M = Cu complexes.

Thus, in the (tetracarboxylate)-$[M_2]$ complexes of late transition metals with d^7, d^8, and d^9 electron configurations, the most critical bonds that directly impact to structure and stability of these complexes are: (i) the donor-acceptor M–O(carboxylate); and (ii) the $M(d_{zz})+M(d_{zz})$ sigma bonds. To realize the M–O(carboxylate) bonding the availability of the empty or partially occupied d_{xy}-AOs of metals is critical, whereas the formation of the $M(d_{zz}) + M(d_{zz})$ sigma is expected to be more effective if M-centers have a partially occupied d_{zz}-AO. Rh(II) and Co(II) ions with d^7-electronic configurations are more suitable to form stable (tetracarboxylate)-[M_2] complexes. Calculations show that (tetracarboxylate)-[Rh_2] complexes with diamagnetic singlet electronic state are more stable than their Co-analogs, the ground electronic states of which are the ferromagnetically coupled singlet states. In the studied (tetracarboxylate)-[M_2] complexes, metal centers are in their +1 oxidation states. An increase in the number of electrons in the valence shell of the transition metal center (i.e. going to Pd and Ni with d^8-electron configurations, or to Ag and Cu with d^9-electron configurations) reduces the stabilizing M–O(carboxylate) and M(d_{zz})+M(d_{zz}) interactions in the corresponding (tetracarboxylate)-[M_2] complexes.

21.3.2 Diazocarbene Decomposition by the Reported [RCOO]$_4$–[M$_2$] Complexes

As mentioned above, ongoing experiments have identified some of the (tetra-bridged paddlewheel)-di-Rh complexes as effective catalysts for C–H and C–C π-bond alkylation. The first step of these processes is the decomposition of the diazocarbenes $N_2CR^1R^2$, leading to the formation of transient metal-carbene complexes (see Figure 21.9) that was the subject of numerous computational studies for the di-rhodium complexes [8, 14–18].

Unfortunately, neither the kinetics nor the thermodynamic properties of the decomposition of diazocarbenes by [RCOO]$_4$–[M$_2$] complexes for M = Co, Ni, Cu, Pd, and Ag were extensively studied. Here, we limit our study to report the thermodynamics of the decomposition of diazocarbenes, as well as the nature of the products. Because the nature of the M-carbene bonds was previously analyzed in details [8, 30, 66–68], we will not discuss it here. In our studies, we use the donor-acceptor diazocarbene $N_2C(COOMe)[(p\text{-}Br)\text{-}Ph]$.

The presented calculations show that the diazocarbene decomposition reaction:

$$N_2C(COOMe)[(p-Br)-Ph] + (RCOO)_4 - [M_2] \rightarrow \{(RCOO)_4 - [M_2]\} \\ -[C(COOMe)[(p-BrPh)] + N_2 \quad (21.1)$$

is exergonic for M = Ni, Cu, Rh, Pd, and Ag, but is slightly endergonic for M = Co (Table 21.3). In general, this reaction may lead to three types of products (Figure 21.9): (i) a carbene complex of the stable [RCOO]$_4$–[M$_2$] complex, called as a structure **I**; (ii) a metal-carbene complex with loosely coordination to the monometallic complex [RCOO]$_4$–[M] fragment, called as a structure **II**; and

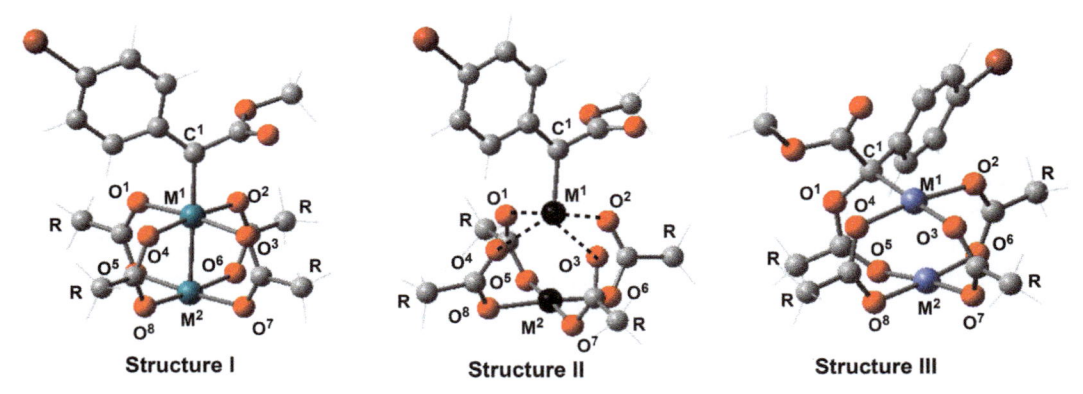

Figure 21.9 3D presentation of the possible products of the reaction of donor-acceptor diazocarbene $N_2C(COOMe)[(p\text{-}Br)\text{-}Ph]$ with the (AcO)$_4$-[M$_2$] complexes, where M = Co, Ni, Cu, Rh, Pd, and Ag.

Table 21.3 The calculated geometry parameters (in Å) of the products of diazocarbene decomposition by (AcO)$_4$–M[$_2$] complexes at their energetically lowest electronic states, and free energy difference (in kcal/mol) between them (see Figure 21.9 for the labeling used).

Comp.	Sta.	Str.	M^1-M^2	M^1-O^1	M^1-O^2	M^1-O^3	M^1-O^4	M^2-O^5	M^2-O^6	M^2-O^7	M^2-O^8	M–C	ΔG
M=Rh;	^{1}A	I	2.460	2.073	2.070	2.068	2.071	2.075	2.074	2.070	2.075	2.014	0.00
	^{3}A	I	2.572	2.079	2.062	2.070	2.062	2.072	2.032	2.064	2.033	2.013	19.2
M=Co;	^{1}A	I	2.283	1.934	1.928	1.931	1.935	1.931	1.910	1.912	1.914	2.009	0.00
	^{1}A$_{os}$	I	2.474	1.927	1.925	1.925	1.916	1.906	1.893	1.899	1.897	1.927	0.73
	^{3}A	I	2.475	1.928	1.925	1.925	1.917	1.906	1.895	1.899	1.897	1.926	0.15
M=Pd;	^{1}A	II	2.739	2.102	4.006	2.072	2.148	2.095	2.009	2.073	2.949	1.933	0.00
	^{3}A	I	2.625	2.257	2.291	2.280	2.302	2.046	2.045	2.042	2.050	2.087	4.16
M=Ni;	^{1}A	III	2.444	2.957	1.906	1.963	1.905	1.909	1.885	1.867	1.903	1.991	0.00
	^{1}A$_{os}$	I	2.418	2.044	2.038	2.033	2.038	1.875	1.878	1.873	1.876	2.122	19.2
	^{3}A	I	2.502	2.080	2.059	2.077	2.061	1.879	1.877	1.876	1.878	2.049	−0.69
M=Ag;	^{1}A	II	3.245	2.710	2.346	2.489	2.813	2.052	2.058	2.050	2.048	2.131	0.00
	^{3}A	III	2.921	3.221	2.405	2.381	2.356	2.447	2.162	2.200	2.159	3.669	−48.1
M=Cu;	^{1}A$_{os}$	I/II	2.697	2.112	1.972	2.212	1.969	1.946	2.000	1.939	1.999	2.079	0.00
	^{3}A	I/II	2.693	2.117	1.972	2.229	1.969	1.950	1.996	1.941	1.995	2.070	−0.15

(iii) complex **III**, in which a carbene fragment is inserted into the M–O bond of one of four carboxylate groups. As previously reported, structure **I** is a pre-reaction complex for the carbene insertion into both C–H and C–C π-bonds. Meanwhile, the formation of structure **III** is expected to be one of unproductive pathways of the reaction as it may lead to degradation of the [RCOO]$_4$–[M$_2$] complex. Stability of complex **II** can be critically dependent on the solvent used, and also can lead to the catalyst degradation.

Keeping these expectations in our minds, we report that the (AcO)$_4$-di-Rh and -di-Co complexes react with the diazocarbene and lead to carbene complex **I**. This reaction is exergonic by 15.2 kcal/mol for the di-Rh complex, but is endergonic by 6.7 kcal/mol for the di-Co complex. Therefore, one may expect the (AcO)$_4$-di-Rh complex to promote carbene generation and insertion into the C–H and C–C π-bonds, but its di-Co analog cannot because of the lack of a driving force for the transient metal-carbene intermediate formation from the donor-acceptor diazocarbene.

The reaction of the (AcO)$_4$-di-Ni complex with a donor-acceptor diazocarbene leads to formation of both triplet state complex **I**, and singlet state inserted complex **III**. Because the energy difference between these two isomeric forms of the (AcO)$_4$-di-Ni-C(MeCOO)(p-BrPh) species is very small (app. 0.7 kcal/mol), and the reaction (1) is highly, 21.7 kcal/mol, exergonic, one may expect formation of the stable (AcO)$_4$-di-Ni-carbene complex **I** in the specific reaction conditions. In contrast, the reaction of the di-Pd analog of the (AcO)$_4$-di-Ni complex with diazocarbene leads to the formation of the mono-Pd-carbene complex (i.e. structure **II**); reaction (1) for M = Pd is exergonic by 11.5 kcal/mol.

The reaction of the (AcO)$_4$-di-Cu with diazocarbene leads to formation of the structure **I**; it is exergonic only by 0.5 kcal/mol. However, fine tuning ligand environments, substituents of the diazocarbenes, and reaction conditions may make (AcO)$_4$-di-Cu applicable for the carbene transfer from diazocarbenes to the C–H and C–C π-bonds. In contrast, the reaction of the silver analog of the (AcO)$_4$-di-Cu with diazocarbene will lead exclusively to the Ag-O inserted complex **III**; this reaction is calculated to be ca 50.0 kcal/mol exergonic. Thus, it is unlikely that the (AcO)$_4$-di-Ag will promote carbene transfer from diazocarbenes to the C–H and C–C π-bonds.

To summarize, from the studied (AcO)$_4$-di-M complexes, only (AcO)$_4$-di-Rh could be an effective catalyst for the carbene transfer from diazocarbenes to the C–H and C–C π-bonds, which is consistent with the available experiments. The (AcO)$_4$-di-Ni and (AcO)$_4$-di-Cu complexes may be the next candidates, but the utilization of them as catalysts for the carbene transfer from diazocarbenes to the C–H and C–C π-bonds requires special experimental conditions, as well as suitable bridging ligands (such as esp or/and espn), and diazocarbenes.

Acknowledgement

This work was supported by NSF under the CCI Center for Selective C-H Functionalization (CHE-1700982). We gratefully acknowledge the NSF-MRI-R2 grant (CHE-0958205) and the use of the resources of the Cherry Emerson Center for Scientific Computation.

Note

The authors declare no competing financial interests.

References

1 Davies, H.M.L. and Beckwith, R.E.J. (2003). *Chem. Rev.* 103: 2861–2904.
2 Hansen, J. and Davies, H.M.L. (2008). *Coord. Chem. Rev.* 252: 545–555.
3 Davies, H.M.L. and Liao, K. (2019). *Nat. Rev. Chem.* 3: 347–360.
4 Davies, H.M.L. and Denton, J.R. (2009). *Chem. Soc. Rev.* 38: 3061–3071.
5 Davies, H.M.L. and Parr, B.T. (2014). Rhodium carbenes. In: *Contemporary Carbene Chemistry* (ed. R.A. Moss and M.P. Doyle), 363–403. Hoboken: John Wiley & Sons, Inc.
6 Davies, H.M.L. and Morton, D. (2011). *Chem. Soc. Rev.* 40: 1857–1869.
7 Liao, K., Negretti, S., Musaev, D.G. et al. (2016). *Nature* 533 (7602): 230–234.
8 Varela-Álvarez, A., Yang, T., Jennings, H. et al. (2016). *J. Am. Chem. Soc.* 138: 2327–2341.
9 Doyle, M.P. and Forbes, D.C. (1998). *Chem. Rev.* 98: 911–935.
10 Doyle, M.P., McKervey, M.A., and Ye, T. (1998). Modern catalytic methods for organic synthesis with diazo compounds: from cyclopropanes to ylides. In: *Modern Catalytic Methods for Organic Synthesis with Diazo Compounds: From Cyclopropanes to Ylides* (ed. M.P. Doyle, A. McKervey, and T. Ye), 112–162. Hoboken: John Wiley & Sons, Inc.
11 Doyle, M.P., Duffy, R., Ratnikov, M., and Zhou, L. (2010). *Chem. Rev.* 110: 704–724.
12 Espino, C.G. and Du Bois, J. (2001). *Angew. Chem. Int. Ed.* 40: 598–600.
13 Harvey, M.E., Musaev, D.G., and Du Bois, J. (2011). *J. Am. Chem. Soc.* 133: 17207–17216.
14 Nakamura, E., Yoshikai, N., and Yamanaka, M. (2002). *J. Am. Chem. Soc.* 124: 7181–7192.

15 Nowlan, D.T., III, Gregg, T.M., Davies, H.M.L., and Singleton, D.A. (2003). *J. Am. Chem. Soc.* 125: 15902–15911.
16 Lin, X., Zhao, C., Che, C.-M. et al. (2007). *Chem. Asian J.* 2: 1101–1108.
17 Hansen, J. and Davies, H.M.L. (2009). *J. Org. Chem.* 74: 6555–6563.
18 Li, Z., Boyarskikh, V., Hansen, J.H. et al. (2012). *J. Am. Chem. Soc.* 134: 15497–15504.
19 Millet, A., Xue, C., Turro, C., and Dunbar, K.R. (2021). *Chem. Commun.* 57: 2061–2064.
20 Sayre, H.J., Millet, A., Dunbar, K.R., and Turro, C. (2018). *Chem. Commun.* 54: 8332–8334.
21 Whittemore, T.J., Millet, A., Sayre, H.J. et al. (2018). *J. Am. Chem. Soc.* 140: 5161–5170.
22 Wang, C.-H., Gao, W.-Y., and Powers, D.C. (2019). *J. Am. Chem. Soc.* 141: 19203–19207.
23 Laconsay, C.J., Pla-Quintana, A., and Tantillo, D.J. (2021). *Organometallics* 40: 4120–4132, and references therein.
24 Wertz, B., Ren, Z., Bacsa, J. et al. (2020). *J. Org. Chem.* 85: 12199–12211.
25 McLarney, B.D., Hanna, S.R., Musaev, D.G., and France, S. (2019). *ACS Catal.* 9: 4526–4538.
26 Liao, K., Yang, Y.-F., Li, Y. et al. (2018). *Nature Chem.* 10: 1048–1055.
27 Liu, W., Ren, Z., Bosse, A.T. et al. (2018). *J. Am. Chem. Soc.* 140: 12247–12255.
28 Liao, K., Liu, W., Niemeyer, Z.L. et al. (2018). *ACS Catal.* 8: 678–682.
29 Ren, Z., Sunderland, T.L., Tortoreto, C. et al. (2018). *ACS Catal.* 8: 10676–10682.
30 Varela-Álvarez, A., Haines, B.E., and Musaev, D.G. (2018). *J. Organometallic Chem.* 867: 183–192.
31 Liao, K., Pickle, T., Boyarskikh, V. et al. (2017). *Nature* 551: 609–613.
32 McLarney, B.D., Cavitt, M.A., Donnell, T.M. et al. (2017). *Chem.-A Eur. J.* 23: 1129–1135.
33 Musaev, D.G., Figg, T.M., and Kaledin, A.L. (2014). *Chem. Soc. Rev.* 43: 5009–5031.
34 Wang, H., Guptill, D.M., Varela-Alvarez, A. et al. (2013). *Chem. Sci.* 4: 2844–2850.
35 Qin, C., Boyarskikh, V., Hansen, J.H. et al. (2011). *J. Am. Chem. Soc.* 133: 19198–19204.
36 Becke, A.D. (1988). *Phys. Rev. A* 38: 3098–3100.
37 Lee, C., Yang, W., and Parr, R.G. (1988). *Phys. Rev. B* 37: 785–789.
38 Becke, A.D. (1993). *J. Chem. Phys.* 98: 1372–1377.
39 Grimme, S., Antony, J., Ehrlich, S., and Krieg, H. (2010). *J. Chem. Phys.* 132: 154104.
40 Grimme, S., Hansen, A., Brandenburg, J.G., and Bannwarth, C. (2016). *Chem. Rev.* 116: 5105–5154.
41 Becke, A.D. and Johnson, E.R. (2005). *J. Chem. Phys.* 122: 154104.
42 Hehre, W.J., Radom, L., Schleyer, P.V.R., and Pople, J.A. (1986). *Ab Initio Molecular Orbital Theory*. Hoboken: John Wiley & Sons, Inc.
43 Hay, P.J. and Wadt, W.R. (1985). *J. Chem. Phys.* 82: 299–310.
44 Wadt, W.R. and Hay, P.J. (1985). *J. Chem. Phys.* 82: 284–298.
45 Cancès, E., Mennucci, B., and Tomasi, J. (1997). *J. Chem. Phys.* 107: 3032–3041.
46 Tomasi, J., Mennucci, B., and Cammi, R. (2005). *Chem. Rev.* 105: 2999–3094, and references cited therein.
47 Tomasi, J., Mennucci, B., and Cances, E. (1999). *J. Mol. Struct.: TheoChem* 464: 211–226.
48 Figg, F.M., Park, S., Park, J. et al. (2014). *Organometallics* 33: 4076–4085.
49 Xu, H., Muto, K., Yamaguchi, J. et al. (2014). *J. Am. Chem. Soc.* 136: 14834–14844.
50 Muto, K., Yamaguchi, J., Musaev, D.G., and Itami, K. (2015). *Nat. Commun.* 6: 7508.
51 Haines, B.E., Kawakami, T., Murakami, K. et al. (2017). *Chem. Sci.* 8: 988–1002.
52 Figg, T.M., Wasa, M., Yu, J.-Q., and Musaev, D.G. (2013). *J. Am. Chem. Soc.* 135: 14206–14214.
53 Frisch, M.J., Trucks, G.W., Schlegel, H.B. et al. (2013). *Gaussian 09, Revision D.01*. Wallingford, CT: Gaussian, Inc.
54 Cotton, F.A. and Walton, R.A. (1982). *Multiple Bonds between Metal Atoms*. Hoboken: John Wiley & Sons, Inc.

55 Chifotides, H.T. and Dunbar, K.R. (2005). Rhodium compounds. In: *Multiple Bonds Between Metal Atoms* (ed. F.A. Cotton, C.A. Murillo, and R.A. Walton), 465–589. New York: Springer-Science and Business Media, Inc.
56 Berry, J.F. (2012). *Dalton Trans.* 41: 700–713.
57 Berry, J.F. and Lu, C.C. (2017). *Inorg. Chem.* 56: 7577–7581.
58 Berry, J.F. (2021). *Comprehensive Coord. Chem.* 6: 4–42.
59 Wang, C.-H., Gao, W.-Y., Ma, Q., and Powers, D.C. (2019). *Chem. Sci.* 10: 1823–1830.
60 Pakula, A.J. and Berry, J.F. (2018). *Dalton Trans.* 47: 13887–13893.
61 Hrdina, R. (2021). *Eur. J. Inorg. Chem.* 2021 (6): 501–528.
62 Calo, F.P., Bistoni, G., Auer, A.A. et al. (2021). *J. Am. Chem. Soc.* 143: 12473–12479.
63 Singha, S., Buchsteiner, M., Bistoni, G. et al. (2021). *J. Am. Chem. Soc.* 143: 5666–5673.
64 Collins, L.R., van Gastel, M., Neese, F., and Fürstner, A. (2018). *J. Am. Chem. Soc.* 140: 13042–13055.
65 We are aware that singlet-triplet transitions are forbidden but it is also well-known that such processes take place and even some well-known reactions can only evolve through one of this forbidden surface-crossing. Spin-orbit coupling value for the system considered must be high in order to get an efficient singlet-triplet transition and metallic systems usually satisfy this requirement. Therefore, throughout the present work we will assume that singlet-triplet surface-crossing processes are fast enough to assure that those electronic states are as populated as their energetics dictate; that is, singlet and triplet states of each intermediate instantly reach their equilibrium population, and kinetic contributions of singlet and triplet states of the transition-state structures only depend on their respective free energy barrier heights.
66 Briones, J.F., Boyarskikh, V., Fullilove, F. et al. (2013). *Science* 342 (6156): 351–354.
67 Werlé, C., Goddard, R., and Fürstner, A. (2015). *Angew. Chem. Int. Ed.* 54: 15452–15456.
68 Werlé, C., Goddard, R., Philipps, P. et al. (2016). *J. Am. Chem. Soc.* 138: 3797–3805.

22

Sustainable Cu-based Methods for Valuable Organic Scaffolds

Argyro Dolla[1], Dimitrios Andreou[1], Ethan Essenfeld[2], Jonathan Farhi[2], Ioannis N. Lykakis[1], and George E. Kostakis[2]

[1] *Department of Chemistry, Aristotle University of Thessaloniki, University Campus, 1515 Dickey Drive, Thessaloniki, Greece*
[2] *Department of Chemistry, School of Life Sciences, University of Sussex, Brighton, UK*

22.1 Introduction

Sustainability is becoming increasingly important when people realise how valuable and limited resources are. One aspect of sustainable chemistry is the use of abundant resources to produce, in a catalytic manner, existing or new organic scaffolds in a way that eliminate wastes, ideally ones that replenish themselves at the same rate as expended [1]. In the past, a catalytic reaction often involved only two reactants, a single catalyst, and heat if needed. For example, the Shilov system, discovered in 1967, used Pt(IV) and Pt(II) salts in high loadings to catalyze carbon-hydrogen activation of alkynes at 120 °C [2]; for modern standards, this process is considered unsustainable. Currently, a catalytic reaction can involve more than two reactants and multiple catalysts in environmentally friendly conditions [3]. This concept is critical as it allows for minimal materials to be used in as few steps as possible while still achieving desirable yields of valuable organic scaffolds quickly. These reactions, also known as multi-component reactions (MCRs), follow complicated pathways, and there is considerable effort into understanding their mechanism(s) to identify what segments of the molecules are essential, which ones can be changed and what the role of the metal(s) is [4]. Incremental but systematic studies increase our understanding and pave the way for more sustainable methods to access valuable organic scaffolds; thus, it is vital to segment the role of each component, catalyst and reaction condition parameters.

Copper is an abundant element of low cost and redox-plural. A search of the inventory of a global chemical vendor identifies that the price (per gr) of common metal chloride salts in most of the countries in increasing order is: $CuCl_2$, $AgCl$, $PdCl_2$, $PtCl_4$ and $AuCl$. This explains the preference for economically viable catalytic in situ generated or well-characterised copper sources. Copper catalysis is frequently used for molecular transformations of bioactive molecules, agrochemicals, natural products, and organic functional materials [5–9]. Copper-based salts or compounds have been extensively used in the coupling, including MCR, reactions that form one or more new C_{sp}–X bonds (where X = C, O, or N) [7]. Several works study the reaction mechanism, whereas methodology optimisation is achieved either by fine-tuning the coordination environment of the catalyst or by varying co-ligands, solvents, and temperature. However, the oxidation state of the metal

Catalysis for a Sustainable Environment: Reactions, Processes and Applied Technologies Volume 2, First Edition. Edited by Armando J. L. Pombeiro, Manas Sutradhar, and Elisabete C. B. A. Alegria.
© 2024 John Wiley & Sons Ltd. Published 2024 by John Wiley & Sons Ltd.

center in the starting component may differ from those in the reaction intermediates [10, 11]. For example, a mixed-valent Cu(I/II) complex was crystallographically characterised from a Glaser coupling [12], whereas dicopper Cu(I/I) and Cu(I/II) complexes were used in click chemistry [13]. These studies signify the catalytic efficacy of the copper sources; however, more examples are needed to elucidate the role of the metal center in these molecular transformations.

In this chapter, we describe recently, and, in our opinion, sustainable methodologies that incorporate copper sources for the one-pot synthesis of valuable organic scaffolds. Part of this work discusses methods that our groups, individually or collectively, have developed in the last five years. Due to page limit restrictions, inevitably, some results have been neglected; therefore, we offer our sincere apologies to the authors of these works. Our effort is divided into three parts, starting with propargylamines (PAs), continuing with pyrroles (PYs), and finishing with dihydropyridines (DHPs). Each subsection debates the value of the organic scaffold and recent methods using well characterised or in situ prepared catalytic species. We envisage this effort to become a beacon and guide the development of more efficient catalytic protocols.

22.2 Synthetic Aspects

PA,s DHPs, and PYs are valuable organic scaffolds. They have significant importance as the core for synthesising multifunctional amino derivatives, natural products, and biologically active compounds [14–16]. MCRs, C–H activation, annulation, cyclization, and cascade transformations are the common routes to synthesising these N-based molecules. Among the several metal-free or metal-catalyzed, known procedures, the MCR of aldehydes, amines, and unsaturated C–C or C–N molecules dominate the synthesis of *N*-heterocycles [17, 18]. However, reactions involving C–C/C–N coupling and C–H activation processes can yield DHPs and PYs successfully [19]. The reported synthetic Cu-catalytic methodologies below are ordered and summarised based on the desired N-containing scaffold for valuable reader efforts.

22.2.1 Propargylamines

22.2.1.1 Seminal Work

The A^3 coupling reaction (Scheme 22.1) named for its three components (aldehyde-alkyne-amine) yields PAs; however, the production of chiral centered PAs is known as asymmetric A^3 coupling (AA^3). Two decades ago, the initial investigations of the A^3 coupling reaction incorporated Cu(I) salts or in situ generated catalytic species. In 2002, Li reported the first highly enantioselective Cu(I)-catalyzed direct alkyne-imine addition [20]. This pioneering work explored both Box and PyBox ligands and found that a PyBox ligand combined with Cu(OTf) provided the highest yields and enantioselectivities with a toluene solvent at a moderately elevated temperature of 40 °C. Li's pioneering work demonstrated that such ligands could be used to achieve high yields and enantioselectivity, but left open questions regarding ligand design, reaction conditions, and substrate scope. In the same year, another protocol that yielded PAs was reported involving CuBr as the metal source, *R*-(+)-Quinap as co-ligand, alkynes, and enamines as substrates, at room temperature for a 24 hour period with toluene as solvent [21]. A variety of alkynes were used, whereas the enamine substrate was kept relatively constant, derived from a secondary amines (dibenzylamine, diallylamine) and aliphatic aldehyde. The reaction proceeded via tautomerisation of the enamine to iminium, which subsequently reacted with the activated alkyne, yielding PAs in high yields and good to high % ee. These two inspirational works paved the way for the development of more

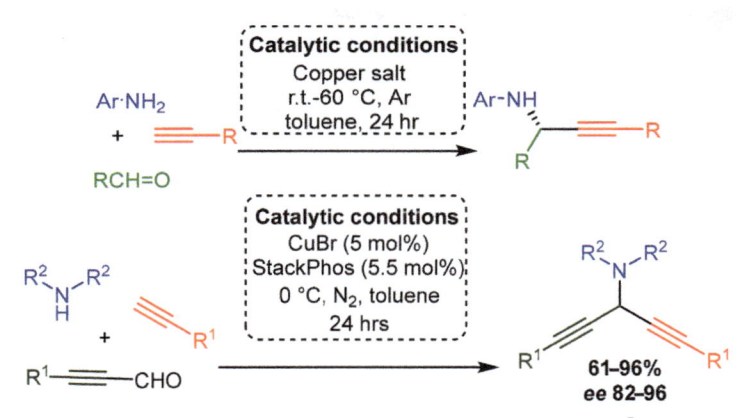

Scheme 22.1 (upper) The seminal aldehyde-alkyne-amine (A³) coupling reaction. (lower) A synthetic approach to access amino skipped diynes.

efficient and environmental friendly protocols. The recently reported total synthesis of (S)-(-)-N-acetylcolchinol (see 22.1.4), a tubulin polymerisation inhibitor [22]. Thus, the following sections will discuss current methods and their sustainable character.

22.2.1.2 Propargylamines from in Situ Generated Cu Catalytic Species

In 2012, Larsen et al. reported a protocol that used $Cu(OTf)_2$ as a catalyst in the A^3 coupling of the electron-deficient (tosylated) nitrogen sources with alkyl, aryl, and heteroaryl aldehydes [23]. Using a combination of CuOTf and $Cu(OTf)_2$ salts in different ratios highlighted the impact of the counterion in effective catalysis. Recently, Delpiccolo et al. employed a variety of substrates using solely $Cu(OTf)_2$ as a catalyst [24]. This work used aliphatic alkynes and cyclic secondary amines with differently substituted aromatic aldehydes. Variation in substitution at the *ortho* or *para* position increases yield compared to an unsubstituted aromatic aldehyde, whereas substitution at the *meta* position decreases the yield. Both protocols incorporated a metal salt loading of 10%, while the reactions were carried out in toluene at elevated temperatures and yielded PAs in very good to excellent yields.

In 2013, Aponick reported StackPhos, a chiral hybrid imidazole-phosphorous based ligand, which its use with CuBr (5% loading) yielded PAs in high yields and enantioselectivities at low temperatures (0°C) and short reaction times (24 hours) with aliphatic and aromatic aldehydes [25]. Three years later, the same catalytic system was successfully applied for the first enantioselective preparation of amino skipped diynes (Scheme 22.1). The protocol has a broad substrate scope, and the resulting scaffolds have found several applications in natural product synthesis such as in the chlorinated lipid family taveuniamides, for example [26]. In the same year, a different protocol that involved CuI, PPh_3 both in 10% loading yielded symmetrical and unsymmetrical 3-amino-1,4-diynes in excellent yields but at elevated temperatures [27].

In 2017, Guiry et al. reported the synthesis and application of StackPhim along with CuBr in the A^3 coupling. [28] The catalytic method was efficient even with 1% catalyst, although the reaction times were longer and the width of the protocol was limited with only nine reported examples, and the *ee*s varied from 65 to 98%. The scope of the StackPhim/CuBr protocol was further expanded in a later work, in which the more challenging propargylic alcohol was used with a wide range of aldehydes and amines, yielding the corresponding PAs in moderate to excellent yields and *ee*s [29]. In 2020, Khare reported an efficient three-component coupling protocol of aromatic aldehyde, deoxy sugar-based alkyne (α-2-deoxy propargyl glycoside), and heterocyclic amine yielded chiral PAs with good to excellent in a stereoselective manner [30]. The method uses CuI as the catalyst and a bifunctional ligand l-proline, in toluene, at elevated temperatures and applies to a wide range of substrates. In 2014, Naeimi et al. prepared a novel thiosalen ligand and its in situ generated complexes with copper(I) salts [31]. The new organometallic catalyst was used in 15% loading for imine's direct and enantioselective alkynylation. The reactions were performed in DCM or toluene and were completed in three to five days.

In 2014, Nakamura reported a variation of the well known PyBim ligands, which in combination with Cu(OTf) and a surfactant (sodium dodecyl sulfate), yielded an excellent library of PAs (20 examples) with high yields and *ee*s. The use of the surfactant promotes hydrophobic interactions and permits the reaction to proceed on tap or seawater water [32]. The reaction duration varies on the substrate (18–156 hours), whereas the catalyst loading is 10%.

22.2.1.3 Propargylamines from Well-Characterised Cu Catalytic Species

In 2015, the [Cu(I)$_2$(pip)$_2$] complex (where pip is the anion of (2-picolyliminomethyl)pyrrole) was found to promote the A^3 coupling with a very low loading (0.4%) [33]. The reaction is completed within two hours at 110 °C in toluene solvent. It applies to a wide range of PAs (31 examples) in very good to excellent yields. Notably, the scope of the protocol omits primary aromatic amines and propargylic alcohols. In 2017, Garcia and Ocando-Mavárez reported a tetrametallic Cu(I) complex based on a *tert*-butyldiallylphosphine ligand [34]. This polymetallic compound efficiently catalyzed (17 examples) the A^3 coupling reaction of aromatic and aliphatic aldehydes with cyclic amines and phenylacetylene, yielding the corresponding PAs under mild reaction conditions and absence of solvent. Later that year, the same team reported the use of a known mononuclear compound [Cu(I){1-phenyl-2,5-bis(2-thienyl)phosphole}$_2$Cl] [35] as a pre-catalyst in the A^3 coupling [36]. The protocol has a broad scope, including aromatic and aliphatic aldehydes with cyclic amines and phenylacetylene, low pre-catalyst loading, high stability, and no need to use purified reagents glovebox.

In 2017, our groups communicated their first effort in developing Cu(II) compounds as pre-catalysts for the A^3 coupling reaction [37]. Among a library of air-stable benzotriazole based one-dimensional coordination polymers, we identified that compound [Cu^{II}(L)$_2$(CF$_3$SO$_3$)$_2$]$_n$ was the most efficient pre-catalyst where L is 1,2-bis((1*H*-benzo[*d*][1,2,3]triazol-1-yl)methyl)benzene. The protocol uses relatively mild conditions and provides results for a good range of substrates, mainly when aliphatic aldehydes and secondary amines are employed. Furthermore, it eliminates the need for inert atmosphere and high loadings and uses 2-propanol, an environmentally friendly solvent [38]. We have also attempted to elucidate the reaction mechanism from an inorganic perspective; through a thorough synthesis and study of targeted coordination compounds, we evaluated how factors such as coordination geometry, anion, and ligand tuning affect the catalytic activity. In 2018, Pathak et al reported the synthesis of a copper salen complex build from the Schiff base ligand 6,6′-[(1*E*,1′*E*)-(cyclohexane-1,2-diylbis(azanylylidene))bis(methanylylidene)-bis(3-(diethylamino)phenol)] and Cu(OAc)$_2$·H$_2$O [39]. This compound promoted the A^3 coupling reaction at a low loading (0.9%) applied to various aldehydes, secondary amines and terminal alkynes. However, the organic transformation occurs at elevated temperatures and in toluene as solvent. In 2020, Peewasan and Powell reported a dimeric Cu(II) complex [Cu$_2$(H$_5$L)(NO$_3$)$_2$]NO$_3$ suitable to catalyze the A^3 coupling reaction where H$_6$L is bis(methylene)bis(5-bromo-2-hydroxylsalicyloylhydrazone) [40]. The method proceeds at room temperature for 24 h in *i*-PrOH and 1% catalyst loading with a good scope; however, no products were obtained when the aromatic and/or aliphatic primary amines were used.

In 2020, Kostakis et al. presented a Cu(II) based protocol that efficiently catalyzes the A^3 coupling reaction within 72 hours, in the open air and at room temperature using a salen type compound {Cu(II)L} where H_2L is the ligand derives from the condensation reaction of (1S,2S)-(−)-1,2-diaminocyclohexane and 3,5-di-*tert*-butyl-2-hydroxybenzaldehyde [41]. Notably, similar systems (see above) were reported before our work, but these protocols required elevated temperatures. The protocol has a good scope but is unsuitable for primary amines or aromatic aldehydes. Notably, the catalyst is recoverable and retains its structure. Mechanistic and control experiments studies suggested in situ generations of radical species and a transient Cu(I) active site via a Single Electron Transfer (SET) mechanism. In the same year, Singh et al. reported a family of Cu(II) complexes that efficiently catalyzed the A^3 coupling reaction at elevated temperatures, in toluene solvent, with very low catalyst loading (up to 0.9%) and an extended scope [42]. Among this family of complexes, compound [Cu(II)(HL)(ClO$_4$)] where HL is 6-(((methylpyridin-2-yl)ethylene)-1-((phenylhydrazinyl)pyridin-2-yl)-2-(phenylhydrazono)methyl)-4-(nitrophenol) achieves the highest yields.

Kostakis et al. incorporated compound [Cu(II)(OTf)$_2$(L)$_2$]·2CH$_3$CN where L is 1-(2-pyridyl)benzotriazole ligand [43] that enables the synthesis of a wide range of PAs at room temperature in the absence of additives. Control experiments and theoretical studies identify structural and electronic pre-catalyst control permitting alkyne binding with simultaneous activation of the C−H bond through an in situ catalytically active [Cu^I(OTf)(L)] species. In 2021, Kostakis et al. reported a dimeric Cu(II) complex [Cu(II)$_2$L$_2$(μ$_2$-Cl)Cl] where HL is (2-(((2-aminocyclohexyl)imino)methyl)-4,6-di-*tert*-butylphenol) which in dichloromethane solution exists in a monomeric [Cu(II)LCl] (85%)–dimeric (15%) equilibrium, and cyclic voltammetry (CV) and electron paramagnetic resonance (EPR) studies indicate structural stability and redox retention [44]. The addition of phenylacetylene to the DCM solution populates the monomer and leads to a transient radical species. Theoretical studies support this notion and show that the radical initiates an alkyne C–H bond activation process via a four-membered ring (Cu(II)–O···H–C$_{alkyne}$) intermediate. This unusual C–H activation method applies to the efficient synthesis of propargylamines, without additives, within 16 hours, at low loadings and in noncoordinating solvents, including late-stage functionalisation of important bioactive compounds molecules. Post-catalysis, single-crystal x-ray diffraction studies confirmed the framework's stability and showed that the metal center preserves its oxidation state.

22.2.1.4 Propargylamines as Intermediates

PAs are essential in synthesising various products, including isoindolines, oxazolidines, pyridones, and alkaloids and a few examples exhibit bioactivity (Scheme 22.2) [45]. In 2013, Periasamy reported a method that involved CuBr in 20% loading that yielded chiral PAs from aldehydes, alkynes, and chiral 2-dialkylamino methyl pyrrolidine at 25 °C, in toluene, in up to 96% yield and 99:1 dr. The PAs could be subsequently converted to the corresponding di-substituted chiral allenes in up to 81% yield and 99% ee upon reaction with CuI in dioxane at 100 °C [46]. In 2015, Seidel reported a protocol that involved CuI and an easily accessible hybrid ligand possessing both a carboxylic acid and a thiourea moiety [47]. PAs are obtained with up to 96% ee, and catalyst loadings can be as low as 1 mol%. In the absence of directing groups, pyrrolidine-derived PAs can be transformed to the corresponding allenes without loss of enantiopurity, which is a significant improvement compared to Periasamy's protocol [46].

The versatility of PAs in providing other scaffolds can be identified in Feng and Huang's highly chemoselective method, in which cyclic divalent PAs are accessed under microwave simple ambient conditions from diamines formaldehyde, and terminal alkynes, lead to *N,N′*-dipropargylimidazolidines or hexahydropyrimidines in moderate to excellent yields [48]. Dos Santos et al. demonstrated that different alkynols could be used as a substrate with a CuCl catalyst to produce the corresponding

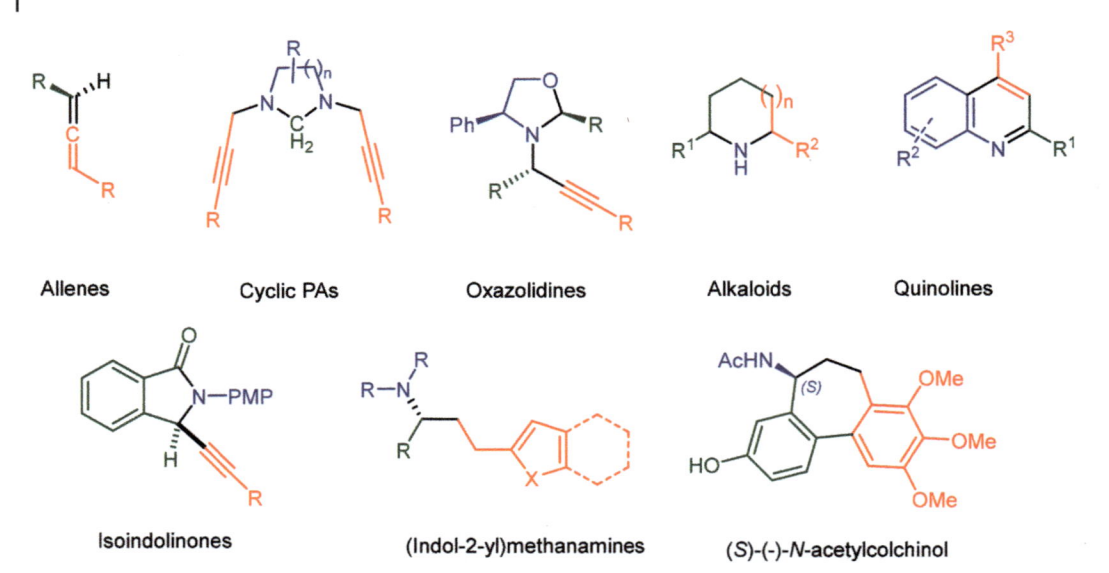

Scheme 22.2 A variety of organic scaffolds that propargylamines (PAs) can be transformed into. Color coding: alkyne (red), amine (blue), aldehyde (green), new-formed bonds (black).

hydroxy-PAs. The resulting hydroxy-PAs were achieved in moderate to high yields and could be converted in subsequent steps to alkaloids under a simple reduction process and intramolecular cyclization [49]. Iqbal et al. reported a procedure using CuCl to synthesise quinoline derivatives [50]. Arylamines, aromatic aldehydes, and alkynes yielded a PA intermediate that was subsequently converted to the quinoline derivative after a propargyl–allenyl isomerization cyclization process. The reaction for the PA synthesis was completed in a short period (10 hours); however, excess metal source, elevated temperatures (reflux) and an intermediate hazardous solvent was incorporated [38].

The combination of two different Cu(II) salts and amino alcohol as the amine component facilitated a cascade reaction that led to the formation of chiral oxazolidines [51]. In the optimised reaction, the combination of $CuBr_2$ and $CuCl_2$, each 10% loading, with the presence of the chiral plenylglycinol as the precursor and in solvent-free conditions, resulted in the target products in good yields (up to 90%) and excellent diastereomeric ratio (> 20:1). In 2017, Aponick reported a method incorporating an in situ generated catalytic species that yields alkynediols that could undergo gold-catalyzed cyclization [52]. The alkynylation/cyclization sequence is convergent, highly modular, and allows for a complementary scope to the heteroarylation of imines. Ma reports the most recent example that PAs are used as intermediates [22]. In this work, the combination of Pyrinap ligands and CuBr afforded optically active PAs with a catalyst loading as low as 0.1 mol% even in gram scale reactions. The protocol applies to the late-stage modification of drug molecules with highly sensitive functionalities and the asymmetric synthesis of the tubulin polymerisation inhibitor (S)-(-)-N-acetylcolchinol in four steps with moderate yields and high ee.

22.2.2 Pyrroles

22.2.2.1 General Synthetic Aspects of Pyrroles

Various methods are known to synthesise pyrrole (PY) scaffolds. The first synthetic methodologies, known as classical methods, are those reported by Knorr [53] and Hantzsch [54]. In recent years, variations of these methods with improved efficiency and refined experimental conditions have been developed. [55] In addition to these methods, new protocols to yield PYs have been established, including condensation, cyclization and cycloaddition processes, through multi-component [56, 57], metal-catalyzed [58–60], cycloisomerization [61] and cycloaddition [62, 63] reactions. In this section, we emphasise the Cu-mediated PY synthesis.

22.2.2.2 Cyclization/annulation Reactions of Enamino Compounds

Enamino based compounds have been extensively used to synthesise PYs (Scheme 22.3). CuI catalyzes the reaction of β-enamino ketones or esters and dialkyl ethylenedicarboxylates in the presence of O_2 to yield polysubstituted PYs. The reaction proceeds smoothly in moderate to good yields with various N-substituents, aromatic, alkyl substituents, and esters [64].

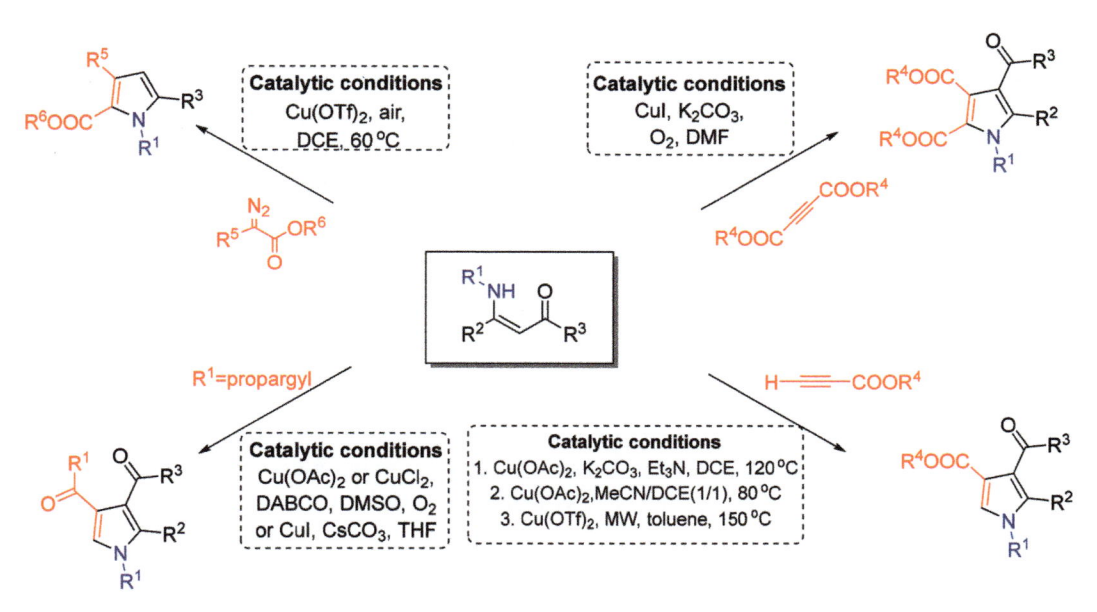

Scheme 22.3 Synthetic approaches that incorporate enaminones.

Cu(OAc)$_2$ facilitates the synthesis of 4-formylpyrroles and 4-benzoylpyrroles from N-allyl/propargyl enamine carboxylates under O_2 in 31–63% yield [65]. Also, Cu(OAc)$_2$ promotes the oxidative coupling of enamides with alkynes to synthesise polysubstituted PYs under mild conditions [66]. Recently, Y. Liu et al. reported the Cu(OAc)$_2$ catalyzed annulation of enaminones with alkynyl esters for the facile synthesis of 2,3,4,5-tetrasubstituted PYs; the scaffolds can be obtained in good yields varying between 46–68% [67]. In addition, the Cu(OTf)$_2$ catalyzes the microwave-assisted synthesis of pentasubstituted PYs through a tandem propargylation/alkyne azacyclization/isomerization sequence from β-enamino compounds and propargyl acetates. This approach provides access to α-arylpyrroles in moderate to good yields (up to 93%) [68]. In 2019, Liu et al. studied the CuI catalyzed annulation of alkyne-tethered enaminones to synthesise 2,3-ring fused PYs through a 5-exo-dig cyclization strategy [69]. The reaction proceeds smoothly under mild conditions and forms a variety of substituted PYs in good to excellent yields, within 59–99%. Also, Cu(OTf)$_2$ promotes the synthesis of polysubstituted PYs under mild conditions in moderate to-good yields through carbene insertion/ester migration/cyclization of enaminones and a-diazo compounds via 1,2-ester migration [70]; this method provides a series of 2,4-diaryl, N-Bn substituted PYs in 23–67% yields.

22.2.2.3 Condensation/Cyclization with Diazo Compounds or Oximes

Cu(hfacac)$_2$, where hfacac is hexafluoroacetylacetonate, has been used by Pal et al. to catalyze the stereoselective synthesis of α-diazo oxime ethers with 3-aminoalkenoates to form polysubstituted PYs in good to excellent yields. Mechanistically, an initial nucleophilic addition of the enamine to the electrophilic carbenoid yields a zwitterionic intermediate, which subsequently undergoes a plausible ring expansion to the desired PYs [71]. Cu(hfacac)$_2$ has also been used to facilitate the synthesis of polysubstituted PYs from silyl enol ethers and α-diazo-β-ketoesters in moderate to good yields [72]. Cu(OTf)$_2$ has been used to catalyze the synthesis of 2,4,5-trisubstituted PYs by coupling α-diazoketones with β-enaminoketones and esters in good yields. This catalytic protocol applies to synthesising a wide range of 2,3-disubstituted indole derivatives from α-diazoketones and 2-aminoaryl or alkyl ketones [73].

Copper (I) salts successfully catalyzed the radical cyclization procedure between oximes and C–C unsaturated compounds. For example, Jiang et al. reported the CuCl catalyzed coupling of oxime acetates with dialkyl acetylenedicarboxylates under aerobic conditions to synthesise polysubstituted PYs. The reaction proceeds smoothly with Na$_2$SO$_3$, as an additive, in DMSO as solvent at 120 °C and forms after a final dehydrogenation pathway the desired products in moderate to good yields [74]. Guan et al. used CuBr to catalyze the synthesis of 2-arylpyrroles via a 5-endo-trig cyclization of ketoxime carboxylates. The reaction proceeds smoothly in high yields in toluene as solvent, at 120 °C [75]. Recently, Jiang and Wei et al. reported the CuCl catalyzed synthesis of polysubstituted PYs via a cascade [3+2] spiroannulation/aromatisation of oximes and azadienes. The reaction exhibits broad substrate scope good functional group tolerance and occurs under milder conditions without any additive [76].

22.2.2.4 Condensation/cyclization Reactions with Alkynes or Isonitriles

Wan et al. reported the Cu(OAc) catalyzed ring-opening reaction of 2*H*-azirines with terminal alkynes to yield 3-alkynylated PYs. The reaction proceeds through a copper acetylide with the 2*H*-azirine that generates a copper-imine intermediate, which isomerizes to the corresponding copper-enamine intermediate. Finally, after intramolecular cyclization and aromatisation, the corresponding 3-alkynated PY derivatives are obtained in moderate to high yields (Scheme 22.4) [77]. CuCl$_2$ has been used by Sakai et al. to catalyze the synthesis of polysubstituted PYs via an [4+1] annulation of PAs with *N,O*-acetals. Three critical features of the *N,O*-acetal during the [4+1] annulation series via 5-endo-dig cyclization were described for the final synthesis of PY derivatives [78]. Multisubstituted PYs can be synthesised from a nucleophilic addition/cyclization/aromatisation cascade reaction from aromatic alkenes/alkynes, trimethylsilyl cyanide and *N,N*-disubstituted formamide (Scheme 22.4). The reaction, which is a new reaction mode for α-aminonitriles, is catalyzed by Cu(OTf)$_2$ and provides an efficient cyclization pattern for the PY synthesis in moderate to good yields, with high regioselectivities [79].

Isonitriles represent one of the highly active moieties for the [3+2] cycloaddition with alkynes leading to the desired five-membered *N*-heterocycles, i.e. PYs. The heterogeneous (Cu$_{nano}$/Al$_2$O$_3$) component catalyzes the synthesis of 2,4-disubstituted PYs from terminal alkynes and isocyanides through a [3+2] cycloaddition reaction in moderate yields [80]. The final products are obtained in high regioselectivity, while the authors propose that the presence of K$_2$CO$_3$ increases the catalytic yield (Scheme 22.4). On the other hand, Cu(OAC)$_2$ has been used by Tan and Xu et al. to catalyze the [3+1+1] cycloaddition reaction of nitrones and isocyanides to yield polysubstituted PYs in the presence of CsOAc, under N$_2$ atmosphere, *N*-methyl pyrrolidone (NMP) and 80 °C. This protocol gives a series of 2,4-dicarbonated, 3-aryl or alkyl-substituted PYs in 40–95% yields [81]. Recently, the groups of Zhou and Cai reported the CuI catalyzed PY-fused tetracyclic heterocycles synthesis

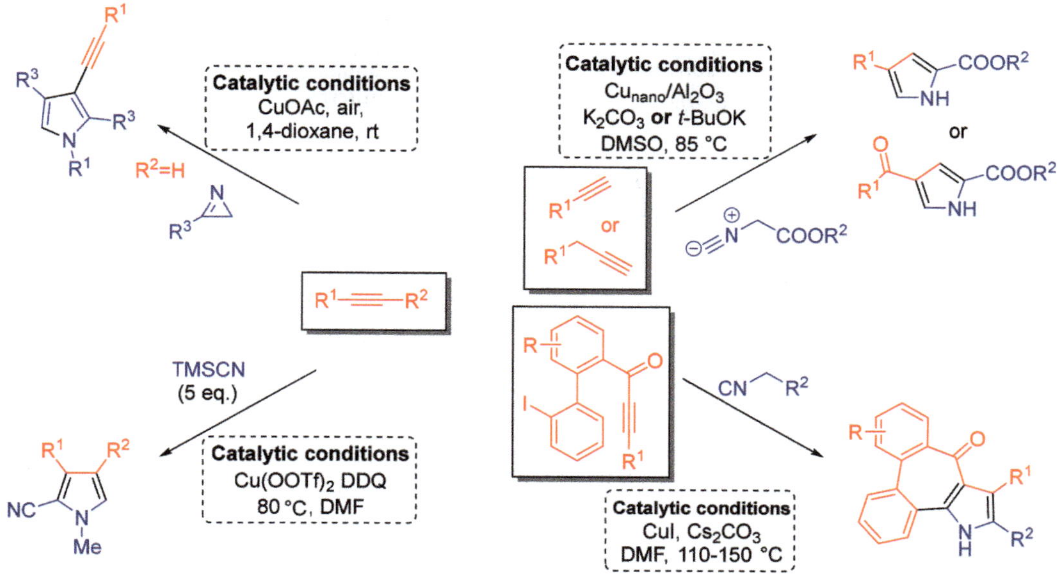

Scheme 22.4 Condensation/cyclization reactions with alkynes/isonitriles catalyzed by copper catalysts.

via a tandem [3+2] cycloaddition and C–C coupling reactions of 2-alkynoyl-2′-iodo-1,1′-biphenyls with isocyanoacetates. The reaction proceeds within a short time and with 2 equivalents of Cs_2CO_3 as a base, in DMF as solvent at 110–150 °C (Scheme 22.4) [82]. Recently, a CuI catalyzed synthetic protocol of trisubstituted PYs via the cycloaddition reaction between aldehydes, amines, and alkynes has been reported. The reaction proceeds in the presence of K_2CO_3, in DMF at 80 °C, with excellent yields and high regioselectivity [83].

22.2.2.5 Reaction of Nitroalkenes

The combination of nitroalkene, amine and aldehydes moieties is the most convenient approach for synthesising tri-, tetra- and penta-substituted PY derivatives. In this area, Wang et al. reported the $Cu(OTf)_2$ catalyzed synthesis of polysubstituted PYs from α-diazoketones, nitroalkenes, and amines under aerobic conditions through a cascade process, which involves oxidative dehydrogenation of the amine and a [3+2] cycloaddition reaction. In this study, PYs were formed in 43–70% yield under reflux conditions, in THF as the solvent and within 12 h (Scheme 22.5) [84]. Under milder conditions and using the same copper salt, Punniyamurthy et al. obtained polysubstituted PYs from amines and 3-nitro-1,3-enynes. The reaction proceeds through a cascade inter-/intramolecular cyclization pathway in THF and at room temperature, with a broad substrate scope and in a short time [85]. Moderate to good yields were also observed during the [3+2] annulation reaction between aziridines and nitroalkenes, catalyzed by $Cu(OTf)_2$, leading to the corresponding nitro-substituted PY derivatives (Scheme 22.5) [86]. Our collaborative work on the catalytic processes with Cu(II)-coordination polymers led to the synthesis of polysubstituted PYs using commercially available starting materials such as aldehydes, amines, and nitroalkenes. The reaction proceeds via a condensation/cyclization and 1,2-phenyl/alkyl migration processes, under mild catalytic conditions, low catalyst loading (0.3–1 mol%), with a broad substrate scope, excellent functional-group tolerance, and high yields (Scheme 22.5) [87]. The same synthetic protocol applies for the facile synthesis of seven-membered ring fused PYs using CuI as the catalyst and the

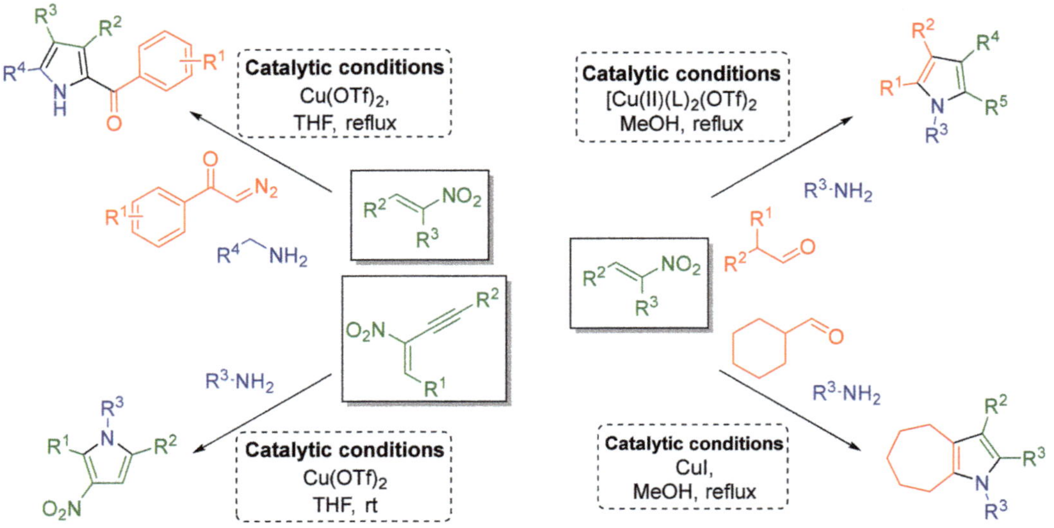

Scheme 22.5 Condensation/cyclization/migration reactions with nitroalkenes, aldehydes and amines catalyzed by copper catalysts.

cyclohexanecarbaldehyde as the aldehyde. The corresponding PY derivatives are formed via an in situ spiro-intermediate ring expansion, tolerating a library of previously inaccessible scaffolds with a broad range of functional groups in a simple step with tangible parameters and substrate adaptations. Selected examples of these moieties exhibited high antioxidant activity (Scheme 22.5) [88].

22.2.2.6 Reaction of Dicarbonyl Compounds

An efficient method for polysubstituted PY synthesis incorporates CuCl, as the catalyst, terminal alkenes, amines and β-keto esters [89]. The reaction proceeds under aerobic conditions, and the polysubstituted PYs are synthesised through a tandem cross-coupling/cyclization/oxidation in good yields in DMSO and at 80 °C. Besides, $Cu(OAc)_2$ mediates the synthesis of 2,3,5-trisubstituted via the one-pot condensation of 1,3-dicarbonyl compounds, acrylates and ammonium salts. This protocol is compatible with several functional groups and provides the polysubstituted PYs in moderate to good yields, within 48–75% and hexafluoroisopropanol (HFIP) as the solvent [90].

22.2.2.7 Miscellaneous Reactions

Other miscellaneous methods yield polysubstituted PYs efficiently. For example, Xi et al. reported the CuI catalyzed tandem vinylation of anilines with dienyl diiodides in the presence of *t*-BuONa. The reaction proceeds in good to excellent yields with electron-donating and -withdrawing substituted anilines and heteroaromatic amines [91]. Carretero et al. used sulfonyl alkenes as active species to synthesise 2,5-disubstituted PYs via a 1,3-dipolar cycloaddition with azomethine ylides. $[Cu(CH_3CN)_4][PF_4]$ catalyzes the reaction, which yields products with different regioselectivity when PPh$_3$ or DTBM-Segphos are incorporated as co-ligands. Depending on the structure of the sulfonyl alkene, the corresponding 4-acyl-2,5-disubstituted PYs or 3-acyl-2,5-disubstituted PYs are obtained [92]. Jiao et al. reported the $Cu(OAc)_2$ catalyzed synthesis of di-substituted PYs through a selective denitrogenation annulation of vinyl azides with aryl acetaldehydes. The reaction proceeds under mild and neutral conditions and the 2,4-disubstituted PYs are obtained in moderate to good yields [93]. $CuCl_2$ catalyzes the synthesis of 2-chloro-substituted PYs through aerobic oxidative annulation of *N*-furfuryl-β-enaminones, which are derived from furan derivatives [94]. CuCl$_2$ has also been applied in the regioselective synthesis of *N*-methoxy-polysubstituted PYs via cycloisomerization of 3-iminocyclopropenes. The protocol applies to the synthesis of steroidal PYs [95]. Opatz et al. investigated photochemical PY synthesis from isoxazoles. [96] They used Cu(II)-2-ethylhexanoate and aryl-substituted isoxazoles, which could be converted to highly substituted by the corresponding 2-acylazirine intermediates 2,4-diacylpyrroles. Adjustment of the temperature allowed to control the photoinduced pathway and the thermal step.

22.2.3 Dihydropyridines

Arthur Hantzsch described the preparation of 1,4-DHP (**1**) more than a century ago. He observed that in the process of synthesising pyridine by the one-pot three-component condensation reaction of acetoacetic ester (**2**), aldehyde (**3**), and ammonia (**4**), compound **1** could be isolated easily (Scheme 22.6).

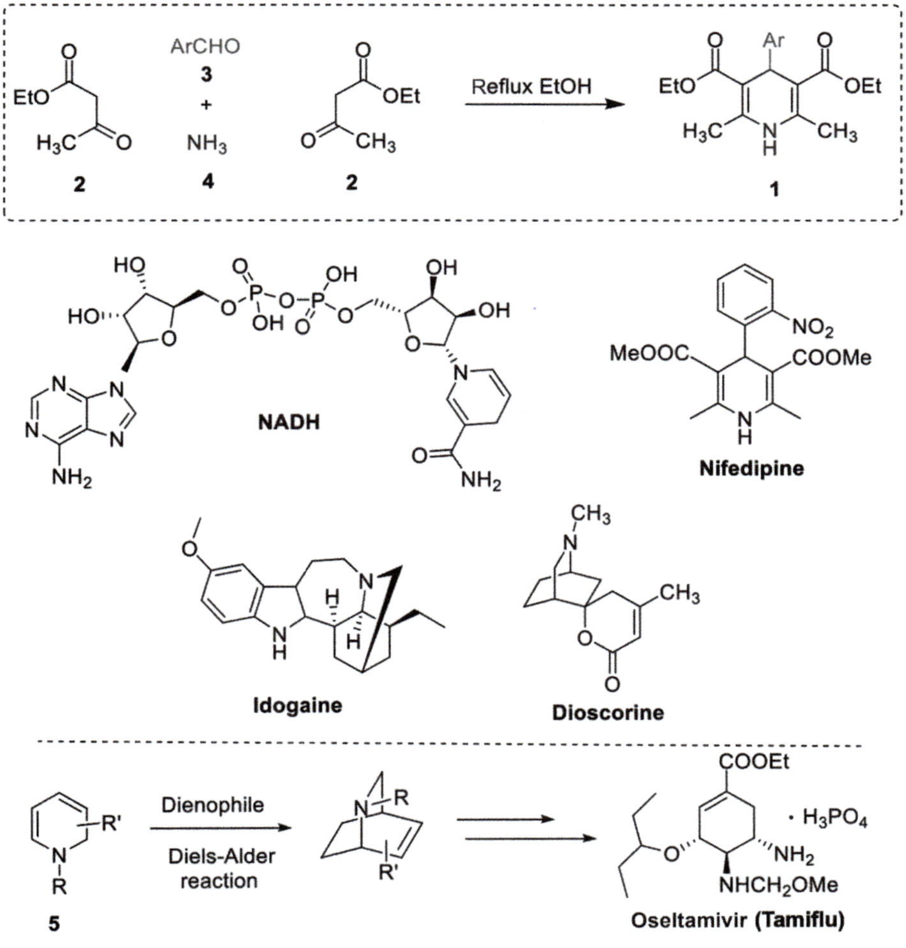

Scheme 22.6 Hantzsch synthesis and structure of biological compounds contain or derived from dihydropyridine (DHP).

Since then, this reaction has been successfully employed in synthesising 1,4-DHPs and bears his name known as Hantzsch dihydropyridine synthesis. [97] Progress in the Hantzsch DHP chemistry can be attributed to the resemblance of these molecules with NADH and their biological activity as anti-hypertensive agents. Hantzsch DHPs are a subset of the co-enzyme NADH, an established hydrogen transferring agent in biological processes. Besides, DHPs have been explored to possess the anti-tumour, anti-inflammatory, anti-oxidant and anti-tubercular activity and have been used as multidrug-resistance-reversing agents in cancer chemotherapy and as antimycobacterial and anticonvulsant agents. [98] Compounds such as nifedipine, a DHP derivative, are Ca^{2+} channel antagonists and are used in various cardiovascular disorders as anti-hypertensive agents to lower blood pressure [99]. The less studied 1,2-DHPs represent an essential scaffold to prepare 2-azabicyclo[2.2.2]octanes (isoquinuclidines). The isoquinuclidine ring system is widely found in natural products such as the alkaloids ibogaine and dioscorine, which have a large spectrum of interesting biological properties. Isoquinuclidines, and consequently the 1,2-DHPs (**5**), can be used as synthetic intermediates to synthesise oseltamivir phosphate (Tamiflu), which essentially is an anti-influenza drug. [100]

22.2.3.1 Synthesis of 1,4-DHPs with Copper Catalysts

Methodologies including Hantzsch, MCR, cycloaddition, C–C coupling reactions, nucleophilic addition to iminium salts, or the cycloaddition/coupling of azines are known to yield 1,4-DHPs (Scheme 22.7).

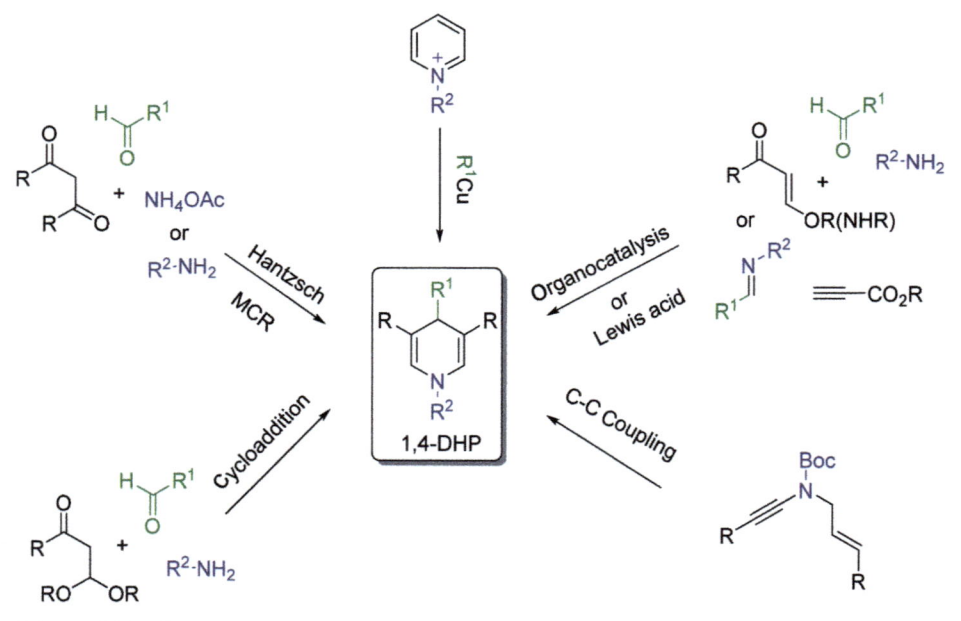

Scheme 22.7 Retrosynthesis of 1,4-dihydropyridines (1,4-DHP) derivatives.

22.2.3.1.1 Hantzsch-Type Reactions

Among several variations of the metal-catalyzed Hantzsch reaction, an efficient and straightforward method that generates a range of DHPs analogues is the microwave-assisted copper-catalyzed one-pot four-component reaction (Scheme 22.8a) [101]. In that case, $Cu(OTf)_2$ is incorporated as the catalyst in an ethanolic solution and the related products are formed in high yields, 85–96%.

Instead of 1,3-dicarbonyl compounds, malononitrile and acetylene dicarboxylates have successfully yielded polysubstituted 1,4-DHPs in very good to excellent yields. CuI catalyzes the ultrasound-mediated 1,4-DHPs synthesis in an aqueous medium (Scheme 22.8b) [102].

Besides the homogeneous conditions, the one-pot synthesis of two derivatives of 1,4-DHPs under heterogeneous conditions using copper iodide nanoparticles (CuI NPs) as a catalyst has also been described. This method demonstrated the four-component coupling reaction of aldehydes and ammonium acetate via two pathways with the corresponding desired products produced in high yields. The CuI NPs can be recovered and reused without significantly losing their activity [103]. Recently, a heterogeneous catalyst has been designed and synthesised based on the functionalisation of manganese ferrite nanoparticles encapsulated in a silica layer with Schiff base and subsequent incorporation of copper. The prepared organic-inorganic hybrid material was successfully used as an efficient and recoverable catalyst for synthesising 1,4-DHPs under mild conditions (Scheme 22.8c) [104].

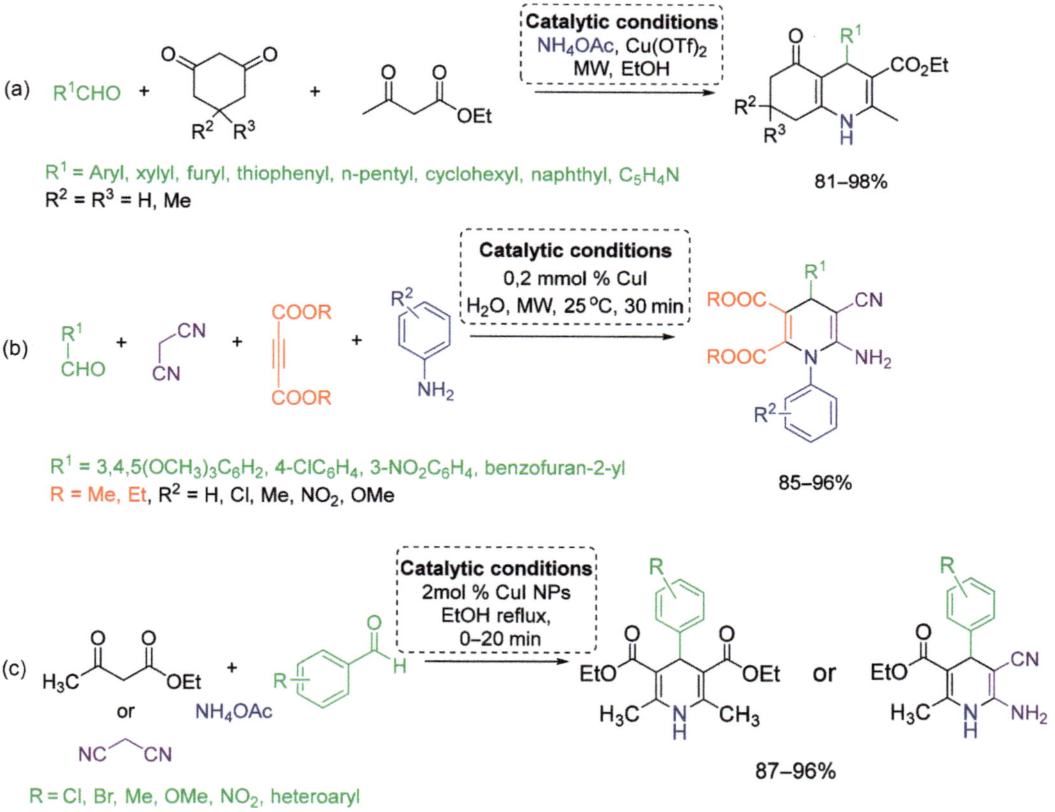

Scheme 22.8 Hantzsch-type synthesis of 1,4-dihydropyridines (1,4-DHPs) in the presence of copper catalysts.

22.2.3.1.2 Nucleophilic Addition to Iminium Salts

The nucleophilic addition to N-substituted pyridinium salts gives both 1,2- and 1,4-DHPs. Hard nucleophiles preferentially attack at C-2 while the softs ones prefer to attack at C-4 position of the pyridinium salts. Comins et al. reported the reaction of 1-acetylpyridinium chloride with alkyl Grignard reagent, which led to 1,4-DHPs. When a catalytic amount of CuI is present, the addition is regiospecific and results in the exclusive formation of 1,4-DHPs. [105] Stoichiometric organocopper reagents such as R$_2$CuLi [106], RCu, RCu•BF$_3$ [107], or copper hydrides [108] also give the 1,4-addition product (Scheme 22.9a).

Bannasar et al. reported the synthesis of 3,5-diacyl-4-phenyl-1,4-DHPs via a regioselective addition of heterocuprate Ph$_2$Cu(CN)Li$_2$ to N-substituted alkyl pyridinium salts, followed by acylation of the intermediate 1,4-DHPs with trichloroacetic anhydride (TCAA) and subsequent haloform type reaction (Scheme 22.9b). [109] Later, they focus on synthesising N-H 4-aryl-3,5-diacyl-1,4-DHPs with electron-withdrawing groups in benzene via chemo- and regioselective copper-mediated addition of functionalised aryl magnesium reagents to N-alkyl-3- acylpyridinium salts, followed by acylation [110].

Softer mixed types of organometallic reagents can be formed by the reaction of copper with alkyl and aryl-zinc reagents. Wang et al. have used alkyl zinc halide or dialkyl zinc base CuCN • 2LiBr to achieve high selective addition at the 4-position to synthesise para-substituted piperidine derivatives with 1,4-DHPs as intermediates. [111] Chia and Shiao use mixed organometallic copper-zinc reagents that regioselectively react with N-acyl pyridinium salts to synthesise 1,4-DHPs. They found the high γ-selectivity of the benzyl organometallics for the 3-substituted pyridines and substituted benzyl organometallics (Scheme 22.9c). [112]

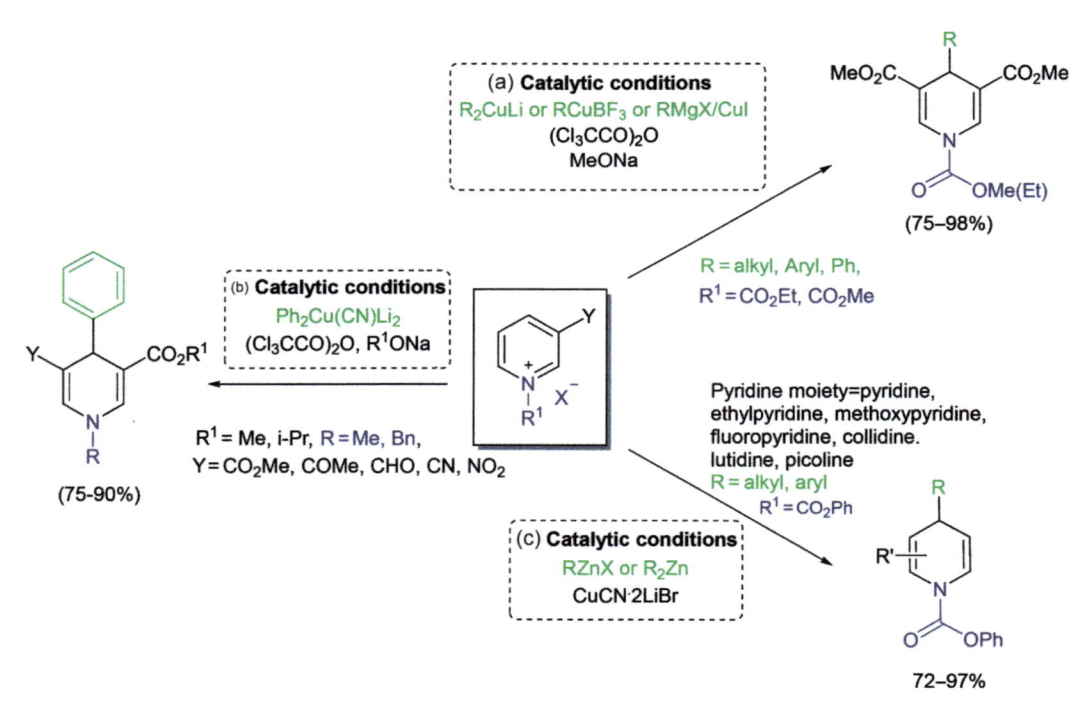

Scheme 22.9 C-4 Nucleophilic addition to N-acyl pyridinium salts.

22.2.3.1.3 Cycloaddition and C-C/C-N Coupling

Yan et al. have reported a mild three-component synthetic approach to 2-amino-1,4-DHPs via a hetero-Diels Alder reaction (HDA). It is a Cu-catalyzed reaction of terminal alkynes, sulfonyl azides, and N-sulfonyl-1-aza-1,3-butadiene. It relies on the in situ generations of metallated ynamide intermediates to achieve a formal inverse electron-demand hetero-Diels–Alder reaction.

Experimental results suggest that Li(I) and Cs(I) ions might play a critical role in this formal [4+2] cycloaddition (Scheme 22.10a). [113]

A recent study by Lyakakis et al. suggests the use of azines and alkynes for the synthesis of (N-substituted)-hydrazo-4-aryl-1,4-DHPs, via a one-pot reaction catalyzed by Cu(II) coordination polymers [Cu(II)(L)$_2$(MeCN)$_2$]•2(ClO$_4$)•2MeCN (Scheme 22.10b) [114]. The copper species consists of N-heterocyclic triazole groups, such as L = 1- {2-[(1Hbenzo [d] [1,2,3] triazol-1-yl) methyl] benzyl} -1Hbenzo [d] [1,2,3] triazole. The reaction proceeds in MeOH has good scope and yields the corresponding DHPs in moderate to high yields. Notably, fine-tuning the pre-catalysts provided useful insights regarding the plausible reaction mechanism.

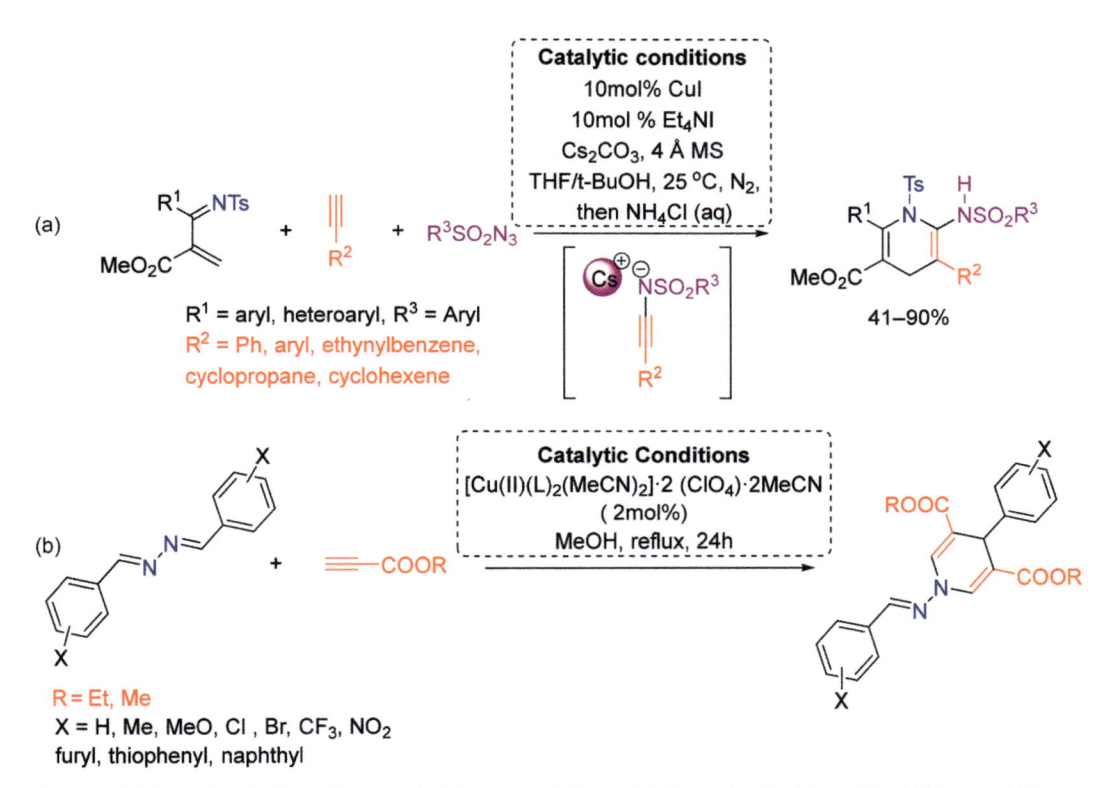

Scheme 22.10 (N-substituted)-4-aryl-1,4-dihydropyridines (DHPs) synthesised by a C-C/C-N/cycloaddtion copper-catalyzed reaction.

22.2.3.2 Synthesis of 1,2-DHPs with Copper Catalysts

1,2-DHPs can be mainly synthesised via nucleophilic addition to pyridines, pyridinium salts, and pericyclic reactions. Bennasar et al. considered the nucleophilic addition of a series of organocuprates to 3-acyl-N-alkylpyridinium salts followed by acylation of the intermediate 1,2-DHPs with TCAA. The alkenyl and alkynyl reagents gave good regioselectivity to α-position. This preferential attack at the α-position was observed for the allyl copper reagents [115]. Later on, Ma et al. reported that an asymmetric copper-catalyzed reaction of propiolates with 1-acylpyridinium could be performed with high enantioselectivity using chiral bis(oxazoline) ligands for the synthesis of 1,2-DHPs derivatives. This study signifies the impact of the carbonyl group in the 3-position of the 1-alkynes to the enantioselectivity of the reaction [116]. Pérez et al. described an elaborate transformation involving four consecutive catalytic cycles leading to the synthesis of N-substituted 1,2-DHPs [117]. The reaction is of two molecules of mono-substituted furans and one molecule of PhI = NTs and a Tp^{X}M complex (Tp^X = hydrotrispyrazolylborate ligand; M = Cu, Ag) as the catalyst.

22.3 Conclusions

Our contribution to this book describes synthetic methodologies that yield organic scaffolds such as PAs, PYs, and DHPs, for which our groups have dedicated, collectively or individually, their research efforts in the last five years. These methods use commercially available materials and Cu species but follow complicated mechanistic pathways. The use of abundant materials (substrates and Cu) is sustainable; however, in most of the described paradigms, the reactions are performed at elevated temperatures, with high Cu-species loading, thus prohibiting reaction, or catalyst, monitoring, which would be beneficial to elucidate the reaction mechanism and make these protocols more sustainable. From this discussion, it becomes evident that there is a plethora of combinations for starting materials, solvents, reaction conditions and Cu catalysts; thus, by incrementally altering each parameter, a variety of organic scaffolds can be efficiently synthesised in the future in a predictive, or not, manner. Our collaborative work, which involves well-characterised pre-catalysts, allowed us to obtain these organic scaffolds conveniently in low catalyst loadings and gain significant mechanistic evidence. Thus, the present approach paves the way for future investigations to conveniently ease optimisation and access to organic scaffolds, which is beneficial for the inorganic, organic, and catalytic communities.

References

1 Horváth, I.T. (2018). *Chem. Rev.* 118: 369–371.
2 Labinger, J.A. and Bercaw, J.E. (2015). *J. Organ. Chem.* 793: 47–53.
3 Sonogashira, K. (2002). *J. Organomet. Chem.* 653: 46–49.
4 Huang, G., Niu, Q., Zhang, J. et al. (2022). *Chem. Eng. J.* 427: 131018.
5 Trammell, R., Rajabimoghadam, K., and Garcia-Bosch, I. (2019). *Chem. Rev.* 119: 2954–3031.
6 Bhunia, S., Pawar, G.G., Kumar, S.V. et al. (2017). *Angew. Chem. Int. Ed.* 56: 16136–16179.
7 Guo, X.X., Gu, D.W., Wu, Z., and Zhang, W. (2015). *Chem. Rev.* 115: 1622–1651.
8 McCann, S.D. and Stahl, S.S. (2015). *Acc. Chem. Res.* 48: 1756–1766.
9 Allen, S.E., Walvoord, R.R., Padilla-Salinas, R., and Kozlowski, M.C. (2013). *Chem. Rev.* 113: 6234–6458.
10 Su, L., Dong, J., Liu, L. et al. (2016). *J. Am. Chem. Soc.* 138: 12348–12351.
11 Yao, B., Wang, D.X., Huang, Z.T., and Wang, M.X. (2009). *Chem. Commun.* 2899–2901.
12 Zhang, S. and Zhao, L. (2019). *Nat. Commun.* 10: 4848.
13 Ziegler, M.S., Lakshmi, K.V., and Tilley, T.D. (2017). *J. Am. Chem. Soc.* 139: 5378–5386.
14 Eicher, T., Hauptmann, S., and Speicher, A. (2012). *The Chemistry of Heterocycles : Structure, Reactions, Synthesis and Applications*. Wiley-VCH.
15 Joule, J.A. and Mills, K. (2010). *Heterocyclic Chemistry*, 5e. Wiley.
16 Majumdar, K.C. and Chattopadhyay, S.K. (2011). *Heterocycles in Natural Product Synthesis*. Wiley-VCH.
17 Van Der Eycken, E.V. and Sharma, U.K. (2021). *Multicomponent Reactions Towards Heterocycles*. Wiley.

18 Ameta, K.L. and Dandia, A. (2020). *Multicomponent Reactions: Synthesis of Bioactive Heterocycles*. Milton Park: Routledge.
19 Wu, X.F. (2016). *Transition Metal-Catalyzed Heterocycle Synthesis via C-H Activation*. Hoboken: Wiley.
20 Wei, C. and Li, C.J. (2002). *J. Am. Chem. Soc.* 2022 (124): 5638–5639.
21 Koradin, C., Polborn, K., and Knochel, P. (2002). *Angew. Chem. Int. Ed.* 41: 2535–2538.
22 Liu, Q., Xu, H., Li, Y. et al. (2021). *Nat. Commun.* 12: 1–10.
23 Meyet, C.E., Pierce, C.J., and Larsen, C.H. (2012). *Org. Lett.* 14: 964–967.
24 Martinez-Amezaga, M., Giordano, R.A., Prada Gori, D.N. et al. (2020). *Org. Biomol. Chem.* 18: 2475–2486.
25 Cardoso, F.S.P., Abboud, K.A., and Aponick, A. (2013). *J. Am. Chem. Soc.* 135: 14548–14551.
26 Paioti, P.H.S., Abboud, K.A., and Aponick, A. (2016). *J. Am. Chem. Soc.* 138: 2150–2153.
27 Choi, Y.J. and Jang, H.Y. (2016). *Eur. J. Org. Chem.* 2016: 3047–3050.
28 Rokade, B.V. and Guiry, P.J. (2017). *ACS Catal.* 7: 2334–2338.
29 Rokade, B.V. and Guiry, P.J. (2019). *J. Org. Chem.* 84: 5763–5772.
30 Thakur, K. and Khare, N.K. (2020). *Carbohydr. Res.* 494: 108053.
31 Naeimi, H. and Moradian, M. (2014). *Tetrahedron Asymmetry* 25: 429–434.
32 Ohara, M., Hara, Y., Ohnuki, T., and Nakamura, S. (2014). *Chem. Eur. J.* 20: 8848–8851.
33 Chen, H.-B., Zhao, Y., and Liao, Y. (2015). *RSC Adv.* 5: 37737–37741.
34 Rosales, J., Garcia, J.M., Ávila, E. et al. (2017). *Inorg. Chim. Acta* 467: 155–162.
35 Alfonso, S., González, S., Higuera-Padilla, A.R. et al. (2016). *Inorg. Chim. Acta* 453: 538–546.
36 Cammarata, J.R., Rivera, R., Fuentes, F. et al. (2017). *Tetrahedron Lett.* 58: 4078–4081.
37 Loukopoulos, E., Kallitsakis, M., Tsoureas, N. et al. (2017). *Inorg. Chem.* 56: 4898–4910.
38 Prat, D., Hayler, J., and Wells, A. (2014). *Green Chem.* 16: 4546–4551.
39 Agrahari, B., Layek, S., Ganguly, R., and Pathak, D.D. (2018). *New J. Chem.* 42: 13754–13762.
40 Peewasan, K., Merkel, M.P., Fuhr, O. et al. (2020). *RSC Adv.* 10: 40739–40744.
41 Sampani, S.I., Zdorichenko, V., Danopoulou, M. et al. (2020). *Dalton Trans.* 49: 289–299.
42 Singh, A., Maji, A., Mohanty, A., and Ghosh, K. (2020). *New J. Chem.* 44: 18399–18418.
43 Sampani, S.I., Zdorichenko, V., Devonport, J. et al. (2021). *Chem. Eur. J.* 27: 4394–4400.
44 Devonport, J., Sully, L., Boudalis, A.K. et al. (2021). *JACS Au.* 1: 1937–1948.
45 Lauder, K., Toscani, A., Scalacci, N., and Castagnolo, D. (2017). *Chem. Rev.* 117: 14091–14200.
46 Gurubrahamam, R. and Periasamy, M. (2013). *J. Org. Chem.* 78: 1463–1470.
47 Zhao, C. and Seidel, D. (2015). *J. Am. Chem. Soc.* 137: 4650–4653.
48 Zhang, Y., Feng, H., Liu, X., and Huang, L. (2018). *Eur. J. Org. Chem.* 2018: 2039–2046.
49 Carmona, R.C., Wendler, E.P., Sakae, G.H. et al. (2015). *J. Braz. Chem. Soc.* 26: 117–123.
50 Syeda Huma, H.Z., Halder, R., Singh Kalra, S. et al. (2002). *Tetrahedron Lett.* 43: 6485–6488.
51 Feng, H., Zhang, Y., Zhang, Z. et al. (2019). *Eur. J. Org. Chem.* 2019 (9): 1931–1939.
52 Paioti, P.H.S., Abboud, K.A., and Aponick, A. (2017). *ACS Catal.* 7: 2133–2138.
53 Knorr, L. (1884). *Chem. Ber.* 17: 1635–1642.
54 Hantzsch, A. (1890). *Ber. Dtsch. Chem. Ges.* 23: 1474–1476.
55 Balakrishna, A., Aguiar, A., Sobral, P.J.M. et al. (2019). *Catal. Rev. – Sci. Eng.* 61: 84–110.
56 Estévez, V., Villacampa, M., and Menéndez, J.C. (2014). *Chem. Soc. Rev.* 43: 4633–4657.
57 Rostami, H. and Shiri, L. (2020). *ChemistrySelect* 5: 11197–11220.
58 Gulevich, A.V., Dudnik, A.S., Chernyak, N., and Gevorgyan, V. (2013). *Chem. Rev.* 113: 3084–3213.
59 Zhou, N.N., Zhu, H.T., Yang, D.S., and Guan, Z.H. (2016). *Org. Biomol. Chem.* 14: 7136–7149.
60 Chelucci, G. (2017). *Coord. Chem. Rev.* 331: 37–53.
61 Donohoe, T.J., Bower, J.F., and Chan, L.K.M. (2012). *Org. Biomol. Chem.* 10: 1322–1328.

62 Neto, J.S.S. and Zeni, G. (2020). *ChemCatChem* 12: 3335–3408.
63 Wang, Y., Zhang, C., and Li, S. (2020). *ChemistrySelect* 5: 8656–8668.
64 Yan, R.L., Luo, J., Wang, C.X. et al. (2010). *J. Org. Chem.* 75: 5395–5397.
65 Toh, K.K., Wang, Y.-F., Ng, E.P.J., and Chiba, S. (2011). *J. Am. Chem. Soc.* 133: 13942–13945.
66 Zhao, M.-N., Ren, Z.-H., Wang, -Y.-Y., and Guan, Z.-H. (2014). *Chem. – A Eur. J.* 20: 1839–1842.
67 Fu, L., Wan, J.P., Zhou, L., and Liu, Y. (2022). *Chem. Commun.* 58: 1808–1811.
68 Zhang, X.Y., Yang, Z.W., Chen, Z. et al. (2016). *J. Org. Chem.* 81: 1778–1785.
69 Li, W., Usman, M., Wu, L.Y., and Liu, W.B. (2019). *J. Org. Chem.* 84: 15754–15763.
70 Li, M., Sun, Y., Xie, Y. et al. (2020). *Chem. Commun.* 56: 11050–11053.
71 Lourdusamy, E., Yao, L., and Park, C.M. (2010). *Angew. Chem. Int. Ed.* 49: 7963–7967.
72 Tan, W.W. and Yoshikai, N. (2016). *J. Org. Chem.* 81: 5566–5573.
73 Reddy, B.V.S., Reddy, M.R., Rao, Y.G. et al. (2013). *Org. Lett.* 15: 464–467.
74 Tang, X., Huang, L., Qi, C. et al. (2013). *Chem. Commun.* 49: 9597–9599.
75 Du, W., Zhao, M.N., Ren, Z.H. et al. (2014). *Chem. Commun.* 50: 7437–7439.
76 Lin, J., Zheng, T.Y., Fan, N.Q. et al. (2021). *Org. Chem. Front.* 8: 3776–3782.
77 Li, T., Xin, X., Wang, C. et al. (2014). *Org. Lett.* 16: 4806–4809.
78 Sakai, N., Hori, H., and Ogiwara, Y. (2015). *Eur. J. Org. Chem.* 2015: 1905–1909.
79 Mou, X.Q., Xu, Z.L., Xu, L. et al. (2016). *Org. Lett.* 18: 4032–4035.
80 Tiwari, D.K., Pogula, J., Sridhar, B. et al. (2015). *Chem. Commun.* 51: 13646–13649.
81 Tian, Z., Xu, J., Liu, B. et al. (2018). *Org. Lett.* 20: 2603–2606.
82 Ouyang, Y., Wu, K., Zhou, W., and Cai, Q. (2021). *Org. Chem. Front.* 8: 2456–2460.
83 Hsu, M.H., Kapoor, M., Pradhan, T.K. et al. (2021). *Synth.* 53: 2212–2218.
84 Hong, D., Zhu, Y., Li, Y. et al. (2011). *Org. Lett.* 13: 4668–4671.
85 Bharathiraja, G., Sengoden, M., Kannan, M., and Punniyamurthy, T. (2015). *Org. Biomol. Chem.* 13: 2786–2792.
86 Wang, S., Zhu, X., Chai, Z., and Wang, S. (2014). *Org. Biomol. Chem.* 12: 1351–1356.
87 Andreou, D., Kallitsakis, M., Loukopoulos, E. et al. (2018). *J. Org. Chem.* 83: 2104–2113.
88 Andreou, D., Essien, N.B., Pubill-Ulldemolins, C. et al. (2021). *Org. Lett.* 23: 6685–6690.
89 Liu, P., Liu, J.L., Wang, H.S. et al. (2014). *Chem. Commun.* 50: 4795–4798.
90 He, J.P., Zhan, Z.Z., Luo, N. et al. (2020). *Org. Biomol. Chem.* 18: 9831–9835.
91 Liao, Q., Zhang, L., Wang, F. et al. (2010). *Eur. J. Org. Chem.* 2010: 5426–5431.
92 Robles-Machín, R., López-Pérez, A., González-Esguevillas, M. et al. (2010). *Chem. Eur. J.* 16: 9864–9873.
93 Chen, F., Shen, T., Cui, Y., and Jiao, N. (2012). *Org. Lett.* 14: 4926–4929.
94 Liu, J., Zhang, X., Peng, H. et al. (2015). *Adv. Synth. Catal.* 357: 727–731.
95 Konishi, K., Takeda, N., Yasui, M. et al. (2019). *J. Org. Chem.* 84: 14320–14329.
96 Paternoga, J. and Opatz, T. (2019). *Eur. J. Org. Chem.* 2019: 7067–7078.
97 Sharma, V.K. and Singh, S.K. (2017). *RSC Adv.* 7: 2682–2732.
98 Saini, A., Kumar, S., and Sandhu, J.S. (2008). *J. Sci. Ind. Res. (India)* 67: 95–111.
99 Triggle, D.J. (1990). *Can. J. Physiol. Pharmacol.* 68: 1474–1481.
100 Silva, E.M.P., Varandas, P.A.M.M., and Silva, A.M.S. (2013). *Synth.* 45: 3053–3089.
101 Pasunooti, K.K., Nixon Jensen, C., Chai, H. et al. (2010). *J. Comb. Chem.* 12: 577–581.
102 Tabassum, S., Govindaraju, S., Khan, R.U.R., and Pasha, M.A. (2016). *RSC Adv.* 6: 29802–29810.
103 Safaei-Ghomi, J., Ziarati, A., and Teymuri, R. (2012). *Bull. Korean Chem. Soc.* 33: 2679–2682.
104 Ahadi, N., Mobinikhaledi, A., and Bodaghifard, M.A. (2020). *Appl. Organomet. Chem.* 34: 18–20.
105 Comins, D.L. and Abdullah, A.H. (1982). *J. Org. Chem.* 47: 4315–4319.
106 Piers, E. and Soucy, M. (1974). *Can. J. Chem.* 52: 3563–3564.

107 Akiba, K.Y., Iseki, Y., and Wada, M. (1982). *Tetrahedron Lett.* 23: 429–432.
108 Danner, D.A. (1984). *J. Org. Chem.* 49: 59.
109 Bennasar, M.-L., Juan, C., and Bosch, J. (1998). *Pergamon Tetrahedron Lett.* 39: 9275–9278.
110 Bennasar, M.L., Roca, T., Monerris, M. et al. (2002). *Tetrahedron* 58: 8099–8106.
111 Wang, X., Kauppi, A.M., Olsson, R., and Almqvist, F. (2003). *Eur. J. Org. Chem.* 2003 (23): 4586–4592.
112 Chia, W.L. and Shiao, M.J. (1991). *Tetrahedron Lett.* 32: 2033–2034.
113 Yan, X., Ling, F., Zhang, Y., and Ma, C. (2015). *Org. Lett.* 17: 3536–3539.
114 Kallitsakis, M., Loukopoulos, E., Abdul-Sada, A. et al. (2017). *Adv. Synth. Catal.* 359: 138–145.
115 Bennasar, M.-L., Juan, C., and Bosch, J. (2001). *Tetrahedron Lett.* 42: 585–588.
116 Sun, Z., Yu, S., Ding, Z., and Ma, D. (2007). *J. Am. Chem. Soc.* 129: 9300–9301.
117 Fructos, M.R., Álvarez, E., Díaz-Requejo, M.M., and Pérez, P.J. (2010). *J. Am. Chem. Soc.* 132: 4600–4607.

23

Environmental Catalysis by Gold Nanoparticles

Sónia Alexandra Correia Carabineiro

LAQV-REQUIMTE, Department of Chemistry, NOVA School of Science and Technology, Universidade NOVA de Lisboa, Caparica, Portugal

23.1 Introduction

Gold is found in nature in veins and alluvial deposits. It has atomic number of 79, being one of the naturally occurring elements with the highest atomic number.

Au, the chemical symbol for gold, derives from the Latin *Aurum*, which means "glowing dawn". Gold has a melting point of 1064 °C and a boiling point of 2700 °C [1]. This explains why the Au–Au bond is very strong, comparable to the Cu–Cu and Ag–Ag bonds [1]. Gold atoms located in the lattice are kept in place by powerful electrostatic attraction forces with the delocalised surrounding 6s electrons. Gold has no stable oxide, unlike other elements, including Ag [2]. Au also has several oxidation states [3], with Au^0 being the metallic state and Au^+ and Au^{3+} (aurous and auric compounds, respectively) being common. Its high electronegativity (2.039 eV) [3] also leads to an additional feature, the auride anion (Au^-). Its electron affinity is higher than oxygen [3]. Thus, Au compounds with other electronegative elements (like O_2 and S) are unstable, leading to the wrong idea of the past that it was a "noble" metal. Au is not affected by most acids and can only be dissolved in *aqua regia* (HCl and HNO_3, 3:1 v/v), forming a soluble tetrachloroaurate anion.

Gold is a good conductor of electricity and heat, being unaffected by water, air, and several reagents [3]. It is a good reflector of infrared light. Au has a high density (19.3 g/cm^3), which means that a gold cube with 37.27 cm sides long would weigh 1 ton [4]. Gold is also the most ductile and malleable metal known [3]. 1 gram of gold can be beaten to 3 m^2 (a semi-transparent thin leaf) and can be made into wires of 5 μm diameter [4] (much thinner than a human hair, which can have a mean diameter from ~50 to ~100 μm).

For a long time, the catalytic potential of gold remained unknown. Previous experience with platinum group metals (PGMs) that exhibited good chemisorption and good activity in many reactions lead to the incorrect belief that the low chemisorption capacity of gold also meant a low catalytic performance [2].

The first report of catalysis by gold was in 1906 for hydrogen oxidation [5]. In 1925, the first use of gold for CO oxidation came out [6]. Considerable work was performed in the following decades and was reviewed in 1972 by Bond, including references to the potential industrial applications of gold [7].

Catalysis for a Sustainable Environment: Reactions, Processes and Applied Technologies Volume 2, First Edition. Edited by Armando J. L. Pombeiro, Manas Sutradhar, and Elisabete C. B. A. Alegria.
© 2024 John Wiley & Sons Ltd. Published 2024 by John Wiley & Sons Ltd.

In 1996, Hutchings predicted that the standard electrode potential could be a good indicator of catalytic activity for hydrochlorination of ethyne [8]. The following work, reviewed by Hutchings in 2005 [9], confirmed this to be true.

Haruta and colleagues also found that gold could be very active for CO oxidation if present as nanoparticles (smaller than 10 nm) [10]. They also discovered that Au catalysts were active between −70 °C and +70 °C [11], whereas Pt catalysts only operated above 100 °C. Moreover, the presence of water is beneficial for this catalytic activity [12, 13], unlike from most Pt group metal catalysts.

After this work, considerable research was carried out as discussed in many reviews [1, 3, 11, 14–18]. Heterogeneous catalysis by gold is a hot topic due to potential applications in several reactions of environmental and industrial importance, namely, oxidation of CO, selective oxidation of CO in the presence of H_2 (preferential oxidation [PROX]); water-gas shift (WGS); total oxidation of volatile organic compounds (VOCs); selective oxidation of sugars, alcohols, and hydrocarbons; selective hydrogenation of unsaturated hydrocarbons, ketones, and aldehydes; removal of nitrogen oxides; and many others.

Several factors influence the catalytic activity of gold catalysts, including the preparation method, the kind of support, the calcination and pre-treatment, and, above all, the (low) gold nanoparticle size [1, 14, 16, 18]. To achieve active Au catalysts, careful preparation is needed to obtain very small nanoparticles (2–10 nm) well dispersed on the support. It is crucial to avoid allowing them to sinter, or the catalyst will become inactive. Sinterisation can be induced by the presence of chloride or high temperature [1, 14, 16, 19]. A particle of ~2 nm can melt at 300 °C, a much lower value than the melting point of gold in bulk (1064 °C) [20].

The cooperation between the low size and good thermal resistance is challenging and depends on the support used. The most common supports are titania [21–24], iron oxide [22–25], alumina [22–24], other oxides [22–24, 26, 27], and carbon materials [28–30].

23.2 Preparation Methods

Many procedures are described in the literature to produce small and highly dispersed Au nanoparticles. We will describe those that are the most common and effective.

23.2.1 Sol-immobilisation (COL)

Colloidal (COL) gold can be obtained in solution with an excess of stabilising (capping) agent/ ligand or surfactant (amines, thiols, phosphines, polymers, and others). This allows control of the shape and size of particles, inhibiting agglomeration. Colloids are obtained by reduction of $HAuCl_4$ by $NaBH_4$, citric acid, or another reducing agent, to achieve gold nanoparticles on a carbon or oxide support. This allows deposition from the colloid and a high dispersion of Au nanoparticles [16]. 1% Au/carbon, prepared by a similar procedure by Rossi and colleagues [15, 17], was supplied as a reference material by the World Gold Council. Other authors also utilised this procedure to prepare effective Au/carbon catalysts [28, 29, 31–34].

Polyvinylpyrrolidone (PVP) is frequently used as a stabilising agent [28–34], and its removal is very important. It can usually be eliminated by decomposition around 300 °C [28, 29, 31–34], solvothermal, ultraviolet light, or ozone treatments.

23.2.2 Impregnation (IMP) and Double Impregnation (DIM)

Impregnation is a classic procedure to obtain supported PGM catalysts. It involves a simple impregnation of the support with a metal salt solution, often with suspension of the support in a large volume, with the solvent removed later. Incipient wetness (IW) technique is a variation, consisting in filling the pores of the support with the metal solution.

Usual Au precursors are chloroauric acid (HAuCl$_4$) and auric chlorides (Au$_2$Cl$_6$ or AuCl$_3$). Complex salts, such as ethylenediamine complex [Au(en)$_2$]Cl$_3$ and potassium aurocyanide (KAu(CN)$_2$), can also be used. Common supports are alumina, silica, magnesia, alumina, boehmite (AlO(OH)), titania, magnesium hydroxide, and ferric oxide (α–Fe$_2$O$_3$) [3]. After drying, calcination at high temperature (800 °C) is necessary to decompose the precursor. Reduction is also needed by H$_2$ above 200 °C, a solution of oxalic acid at 40 °C, or a magnesium citrate solution [3].

However, IMP often results in large gold particles that can be inactive [3, 14, 16, 19, 35], and high dispersions of Au are hard to achieve. Particles can also agglomerate during calcination [14, 36, 37], as seen in Figure 23.1a. In this figure, a transmission electron microscopy (TEM) image shows a (large) gold nanoparticle with 400 nm diameter, supported on CeO$_2$.

Moreover, the existence of chloride is disadvantageous, as it enhances the movement of Au on the support and leads to sinterisation [1, 14, 16, 19]. Au and chloride come together to form bridges, which favors the growth of the particles with heating [38]. The first use of conventional IMP to prepare an Au catalyst explains why gold was first considered inactive for catalysis, compared with PGMs, such as Pt and Pd.

Bowker et al. used the double impregnation method (DIM) to remove chloride on Au/TiO$_2$ materials [40]. The support was impregnated with a solution of HAuCl$_4$ and later with a solution of Na$_2$CO$_3$, then washed and dried. This procedure was successfully employed for several carbon and metal oxide supports [25, 26, 28, 29, 39, 41–49]. Figure 23.1b shows an example, where small and well dispersed Au nanoparticles are observed on CeO$_2$. However, COL is better for carbon catalysts [28, 29], as stated previously.

23.2.3 Co-Precipitation (CP)

Co-precipitation is a very simple way to prepare Au catalysts [1, 14, 50]. It was a serendipitous discovery made in 1987 by Haruta's group that resulted in Au/Fe$_2$O$_3$ with low gold particle size by

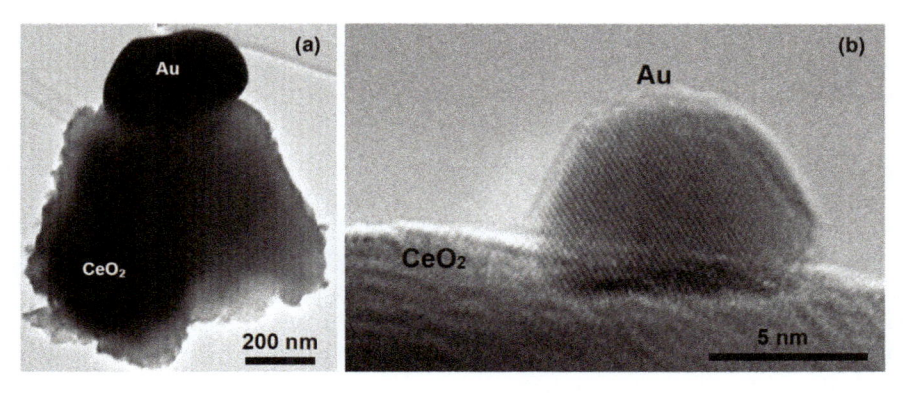

Figure 23.1 Transmission electron microscopy (TEM) images of Au/ceria prepared by impregnation (IMP) (a) and double impregnation method (DIM) (b). Reproduced with permission from Refs [19] and [39] / Elsevier.

mixing HAuCl$_4$, Fe nitrate, and Na carbonate [50]. Before that, IMP was the procedure commonly used to obtain PGM catalysts, and it was unsuitable for Au, as mentioned previously. Au/Fe$_2$O$_3$ prepared by co-precipitation (CP) showed remarkable activity in catalysing CO (and H$_2$) oxidation below room temperature, which had never been achieved for this reaction [50].

The method is still used with good results [51]. It comprises mixing aqueous solutions of HAuCl$_4$ and metal salts (e.g. nitrate) and pouring them into a solution of NH$_4$OH and/or Na$_2$CO$_3$ before stirring for some minutes. Both hydroxides and hydrated oxides are simultaneously precipitated. After 1 hour, the solid is washed, filtered, dried, calcined, and reduced [14, 52]. The final materials might have sodium and chloride ions, if metal chlorides are used as precursors. Both can be catalyst poisons [1, 3, 14, 16, 19, 36] (nevertheless, alkali can also enhance the activity of gold catalysts [30]). This procedure can only be used for metal compounds able to be co-precipitated with Au(OH)$_3$.

23.2.4 Deposition Precipitation (DP)

Deposition precipitation (DP) simple procedure is used to produce commercial Au supported catalysts [11, 14] and can be utilised in many supports. It was another discovery from Haruta's group and has been used ever since [21–24, 47, 48, 53]. The precursor is brought from the solution where the support is suspended, often by increasing the pH to cause precipitation of a hydroxide. The support surface behaves as nucleating agent, and the majority of the precursor is linked to the support. A solution of HAuCl$_4$ has the pH adjusted by addition of NaOH, to a certain value between 6–10, and the oxide support is submerged in the solution. [Au(OH)$_n$Cl$_{4-n}$]$^-$ (n = 1–3) species react with the support surface and after ageing for 1 hour, Au(OH)$_3$ is deposited on the support surface [3, 14, 36].

The pH value is important, as for pH > 6, AuCl$_4^-$ is transformed to [Au(OH)$_n$Cl$_{4-n}$]$^-$ (n = 1–3), and the particle sizes of gold nanoparticles are below 4 nm [36]. Bond and colleagues [21] showed that pH 9 is the best for Au/TiO$_2$, as most species in solution were anionic gold complexes, with residual chloride. Nevertheless, the best pH depends on the isoelectric point of the support.

Urea can also be used as an alternative to adjust the pH [3, 54], as it slowly decomposes in the solution, forming hydroxyl ions that are consumed as they are formed. Louis and colleagues realised that urea allowed them to obtain the same size of Au nanoparticles as NaOH (2–3 nm) [55]. Other authors also used this method, obtaining very active catalysts [56, 57].

However, DP is not adequate for some supports, such as activated carbon [11, 14, 16] or zeolites [58], as they have high isoelectric points. DP has advantages over CP as all of the active phase is on the support surface and not buried inside [3, 14]. It also produces narrow particle size distributions; however, it is preferrable that the support has a surface area above 50 m^2/g [3, 14]. Figure 23.2 shows some examples of gold nanoparticles prepared by DP.

23.2.5 Liquid-phase Reductive Deposition (LPRD)

The liquid-phase reductive deposition (LRPD) procedure was described by Sunagawa et al. and includes mixing aqueous solutions of HAuCl$_4$ and NaOH, aging for 24 hours at room temperature in the dark to complete hydroxylation, adding the support, dispersing ultrasonically for 30 minutes, and aging at 100 °C overnight [59]. The resulting solid is washed many times to remove chloride and dried. Adsorption of the gold ions takes place on the support surface, where reduction also happens. This method was effectively used to obtain Au nanoparticles on several supports [19, 28, 35, 46–48, 60, 61].

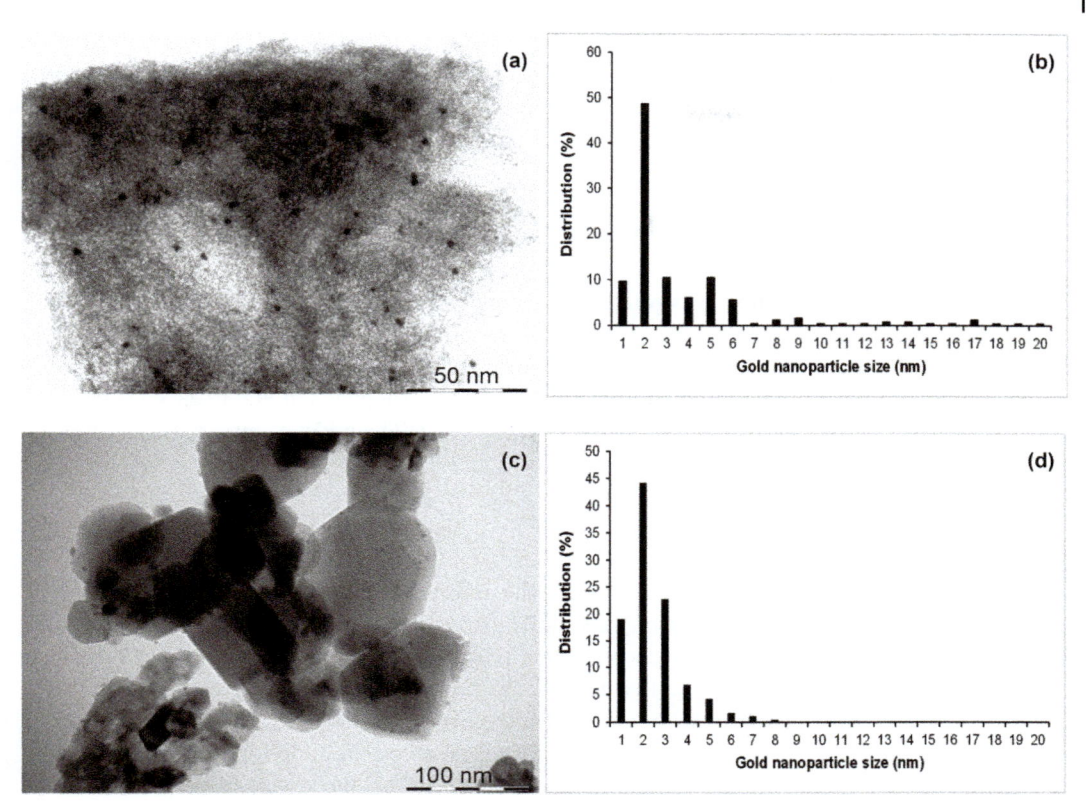

Figure 23.2 Transmission electron microscopy (TEM) images of Au/Al$_2$O$_3$ (a) and Au/CeO$_2$ (c) prepared by deposition precipitation (DP), with respective size gold nanoparticle size distributions (b, d). Reproduced with permission from Ref [22] / Elsevier.

23.2.6 Ion-Exchange

This method consists of replacing the ions of the support surface by gold ions. It is very efficient for zeolites, but adding active species inside the cavities of these materials, instead of having them on the surface, poses some obstacles, like the shortage of adequate cations or cationic complexes [3]. However, this method proved to be successful for several Au/zeolites [62, 63].

Pitchon and colleagues used direct anionic exchange (DAE) of the Au species with OH$^-$ groups of the support [64, 65]. HAuCl$_4$ (aqueous solution) is poured into the support, the mixture is heated to 70 °C and maintained there for 1 hour, then it is filtered, washed with warm water, dried during the night, and calcined in air at 300 °C. To assure a complete removal of chloride ions, a fraction of the dried catalyst is washed with a concentrated solution of ammonia. Nevertheless, caution must be taken, as gold and ammonia might produce fulminating gold that is explosive [1, 14, 16].

23.2.7 Photochemical Deposition (PD)

Photochemical deposition (PD) procedure ensures metal deposition on semiconductors, with metal ions being simultaneously reduced by conduction band electrons [66]. The method can be improved by the use of sacrificial electron donors (e.g. methanol, 2-propanol, or formaldehyde) that can provide a large quantity of electrons. The Au in aqueous solution is mixed with the sacrificial electron donor and the support, sonicated for 30 minutes, and photodeposited using an ultraviolet lamp. This procedure was used to prepare Au on TiO_2 [66–68], ZnO [60, 66, 69] and other supports.

23.2.8 Ultrasonication (US)

Ultrasonication (US) is analogous to PD but with no photodeposition; the sample is sonicated for 8 hours. It was a serendipitous discovery made by Carabineiro et al. while attempting to prepare Au/ZnO by PD [60]. The sample was going to be sonicated over 30 minutes (then photodeposited), but it was forgotten in the sonicator for 8 hours. As the mixture showed a deep purple colour, like those prepared by PD, it was filtered, washed, dried, tested in CO oxidation, and found to be a very active catalyst [60]. A TEM image and the size distribution are depicted in Figure 23.3. US was also used to prepare Au on Fe_2O_3 [25], MgO [44], CuO, La_2O_3, NiO, and Y_2O_3 materials [26], but the results were not so good as with Au/ZnO [60].

23.2.9 Vapor-Phase and Grafting

Vapor-phase and grafting procedures are similar; the difference is whether a solvent is used. In the vapor-phase (chemical vapor deposition), a volatile gold compound is carried to a support with a high area, with the help of an inert gas, and then reacts chemically with the support surface to form a precursor of the active species [14, 16, 36]. The most commonly used precursors are $AuCl_3$ or $HAuCl_4$, but other compounds without chloride have been used.

In physical vapor deposition, Au is vaporised and deposited on the support in high vacuum conditions [16]. According to 3M (Minnesota, USA) very active Au catalysts can be prepared like this for several supports, including water soluble materials, or those not suitable for DP, such as SiO_2 [70]. This procedure is cheap, reproducible, needs no washing or thermal treatments, and is not toxic. Very stable Au nanoparticles (up to 600 °C) were prepared on Al_2O_3 by this technique [71].

In grafting, a solution of a gold complex reacts with the support surface, and the formed species are converted to catalytically active forms. Gold phosphine complexes were grafted onto several

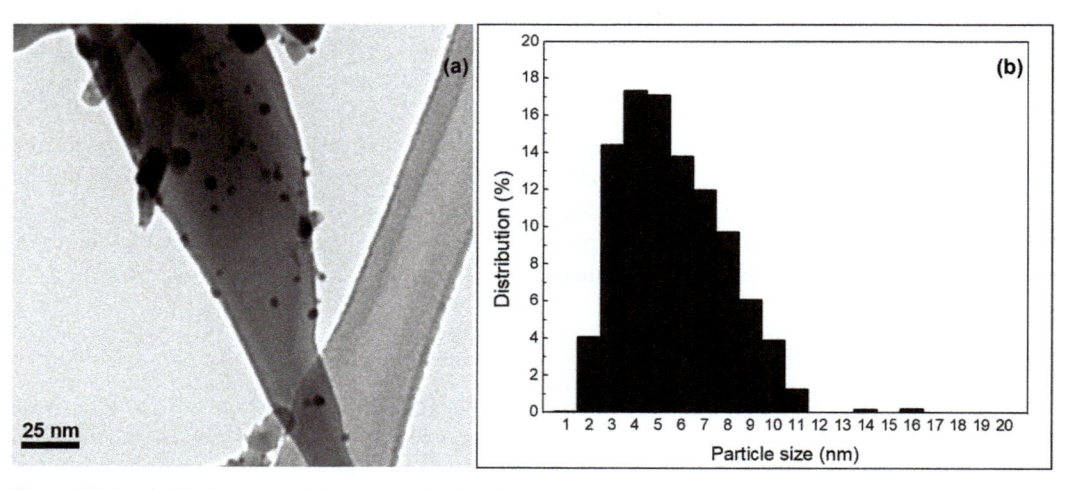

Figure 23.3 Au/ZnO prepared by ultrasonication (US): transmission electron microscopy (TEM) image (a) and Au nanoparticle size distribution (b). Reproduced with permission from Ref [60] / Elsevier.

hydroxides [3, 14, 72], as they have –OH groups that are able to react with the Au compounds. The procedure includes room temperature vacuum drying and air calcination, which leads to a simultaneous transformation of the precursors to oxides and Au particles. Au–phosphine complexes are great choices as they decompose by heat, forming metallic Au metal at a temperature similar to the value needed to converted metal hydroxides into oxides. These ligands are also expected to delay the growth of Au nanoparticles into larger ones. Gold can be supported on MCM–41, SiO_2, SiO_2–Al_2O_3, or activated carbon, as highly dispersed nanoparticles using a Au acetylacetonate complex, which is much more effective for these supports than liquid-phase procedures [73].

23.2.10 Bi- and Tri-metallic Au Catalysts

It is already well known that Au is an effective catalyst alone, but it is also effective when combined with other metals [14]. Gold-based bimetallic materials have a large potential for several reactions, such as selective oxidations and hydrogenations, C–C coupling, and photocatalysis, due to their high activities and selectivities in mild conditions [74]. Bimetallic Au catalysts can be prepared by many procedures, as described in many reviews [75].

23.2.11 Post-treatment and Storage

Several post-treatments can be used, such as calcination and reduction [1, 14, 16, 21, 46, 64, 72, 76–78]. However, several catalysts can be used without any further treatment. In fact, reduction or calcination can have negative effects in some cases [21, 79]. Also, the size of gold nanoparticles can increase with thermal treatments [1, 14, 16]. However, materials prepared by COL need a heat treatment for decomposition of the organic scaffold [28–34].

In terms of storage, freshly prepared samples should be kept at 0 °C and calcined materials should be kept in a cold place. Once dried, samples should be stored in a desiccator, under vacuum, in the dark, and be reduced immediately before use [16, 80].

23.3 Properties of Gold Nanoparticles

Gold nanoparticles have very different properties than bulk gold. The decrease in size of a gold particle leads to changes in the metal structure [81] and a large increase in the quantity of low coordinated surface atoms, such as edge and corner atoms [81]. It also produces changes in the electronic properties, as the metallic character is lost and particles become non-metallic or semi-conductive (caused by quantum size effects) [81]. These changes in structure and electronic properties produce alterations in the Au–Au bond distance [81]. For example, for ~1 nm sizes, the bond length is shorter (2.72 Å) than that of the bulk gold (2.88 Å) [81].

Moreover, if nanoparticles are supported, their behaviour will depend on the type of support, method of preparation, pre-treatment, calcination, and other factors [1, 3, 14, 16, 79]. Some of the more important properties of Au nanoparticles are succinctly discussed in the following.

23.3.1 Activity

It is widely known that the presence of small (usually <10 nm) gold nanoparticles well dispersed on the support is crucial for good catalytic activity [1, 3, 11, 14, 16, 48, 79, 82–89]. Some examples are shown in Figure 23.4. The improved activity of smaller nanoparticles was attributed to the larger amount of low-coordinated atoms of cubo-octahedral Au nanoparticles (corner sites) [90].

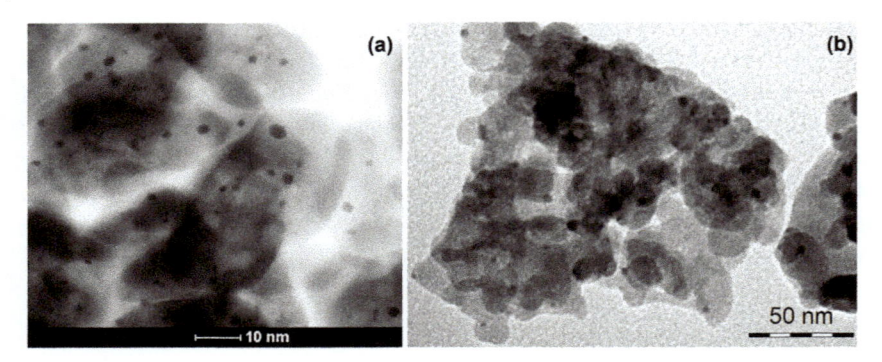

Figure 23.4 Transmission electron microscopy (TEM) images of gold nanoparticles supported on Fe_2O_3 (a) and ZnO (b). Reproduced with permission from Ref [22] / Elsevier.

Bond analysed the literature results of CO oxidation using Au/TiO_2 catalysts and found that the most active catalysts were non-metallic [91]. However, for other reactions and/or other catalysts, there is still no consensus, as some authors believe the cationic state is the preferable, while others claim it is the metallic state; and the coexistence of several oxidation states is also possible [1, 14, 16, 19, 22–25, 27, 29, 30, 39, 46, 53].

However, larger Au nanoparticles [11, 19, 84, 92], Au single crystals [37, 93, 94], bulk (unsupported powdered) Au [95–97], or naked Au [98] can also be active catalysts.

Nevertheless, the presence of small Au nanoparticles cannot fully explain the catalytic activity. There are studies with Au nanoparticles with similar sizes, but on different supports, that have different catalytic activities [1, 14]. This shows that the support also has a large effect [1, 3, 11, 14, 16, 22, 25, 39, 60, 84] based on its specific surface area and morphology [25, 60] as well as its reducibility [22, 23, 39, 42, 43].

The most common supports are transition metal oxides. Usually, two kinds of oxides are considered [1, 14, 16, 99]: the active, easily reducible supports (like Fe_2O_3, CeO_2, or TiO_2) [21–27, 39, 43, 47, 48, 67, 76] and the "inactive" or inert supports (like Al_2O_3, SiO_2, and MgO, which can also originate very good catalysts [22–24, 44, 88]). Mixed oxide or composite supports are also common [41, 46, 49, 61, 65, 69, 82, 100, 101], and might be more efficient that the single oxide components with Au. It is possible that synergistic effects exist between the Au nanoparticles and the support components [82].

A good interaction between Au and the support is also needed [11, 84], and that can be promoted by calcination. However, the use of high temperatures might lead to sinterisation of the nanoparticles [1, 14, 16], so calcinations at moderate temperatures (100–300 °C) are preferable. According to Haruta, the metal oxide precursors are converted to crystalline metal oxides during calcination, and the gold precursor is reduced to metallic Au [102]. However, some studies have proved that calcination can have detrimental effects [21].

23.3.2 Selectivity

Gold nanoparticles can be very selective in several reactions, such as selective oxidations and hydrogenations in mild conditions, as described in several reviews [1, 14–18]. Some examples include:

- The oxidation of glycerol to glyceric acid reported by Hutchings, with 100% selectivity for Au/charcoal or Au/graphite (1% wt.), for 3 hours, at 60 °C (with c. 55% conversion) by [103].

- The oxidation of D-glucose to D-gluconic acid, performed by Rossi, using a Au/C catalyst, with total selectivity under mild conditions [104].
- Hydroxymethylfurfural oxidation using gold on titania and nanoparticulate ceria, reported by Corma, with 100% selectivity to 2,5-furandicarboxylic acid, in alkaline conditions, in only 1 hour of reaction [105].
- The oxidation of propene to propene epoxide, where a remarkable selectivity of >99% was reported for Au/TiO$_2$ based catalysts, using H$_2$ and O$_2$ [106]. For Au nanoparticles above 2 nm, high selectivity to propene oxide is obtained [14, 107], but propane is obtained instead for sizes smaller than 2 nm [11, 14, 107]. Haruta used Au/TS-1-K-1 and found out that Au nanoparticles above 2 nm gave mostly acrolein, while sizes below 2 nm gave propene oxide [102].

The selective oxidation of CO in the presence of H$_2$, or PROX, discussed hereafter, is also a good example. Often, H$_2$ is obtained from a source containing carbon, like a natural gas, which leads to CO production. This gas can be detrimental, even in small quantities, as it can poison Pt-based proton exchange membrane (PEM) fuel cells (as CO tightly binds to Pt) [14, 108, 109]. Unlike PGMs, gold-based materials are more active for CO oxidation than for H$_2$ oxidation. Moreover, Au is nearly insensitive to CO$_2$, and its activity is enhanced by moisture [12, 110–113]. Those are products of CO oxidation (CO + O$_2$ $\rightleftarrows$ CO$_2$) and water-gas shift (CO + H$_2$O $\rightleftarrows$ CO$_2$ + H$_2$) reactions (also discussed later), common in the production and purification of H$_2$. This makes gold a good candidate for PROX, as shown in several articles [51, 65, 67, 69, 77, 82, 88, 101, 114].

23.3.3 Durability

If considering Tammann temperature, Au should sinter around 532 °C (half of the melting point of metallic Au: 1064 °C) [14]. However, small gold nanoparticles might sinter at a much lower temperature, as the melting values should significantly diminish with decreasing particle size (this means that a Au nanoparticle of ~2 nm should melt around 300 °C [14, 20], a much lower value than bulk gold). However, experience has been showing that gold catalysts have much higher durability than expected [1, 14].

Some examples can be found in literature for this, including:

- Au on cobalt oxide on zirconia-stabilised ceria, titania, and zirconia, that was able to resist at 500 °C for 157 hours, although with some deactivation [92].
- Adding transition metal oxides to obtain Au/MO$_x$/Al$_2$O$_3$ prevented Au sintering during methane oxidation up to 700 °C [115].
- Au/Al$_2$O$_3$ showed good activity in NO conversion, surviving pre-treatments of 600 °C in air for 24 hours, several cycles at 150–500 °C, and being kept overnight at 500 °C [52]. Another Au/Al$_2$O$_3$ sample was hydrothermally stable at 600 °C for 96 hours [116]. Au/Al$_2$O$_3$ can also be very efficient for the conversion of glucose to gluconic acid (>99% selectivity) and durability (no loss of activity or selectivity up to 110 days) [117], with 3.8 ton of product being produced per gram of Au in 70 days.
- Au/ZnO, used for PROX, resisted for 350 hours at 80 °C [118].
- A Au$_2$Sr$_5$O$_8$ material patented by Toyota Motor Company had improved durability (tested up to 800 °C for 5 hours) [119].
- Nanoaligned rutile rods limited the growth of Au up to 800 °C, and CO oxidation was very high even after exposure of the material to high temperature [120].
- Au/alumina with complete removal of chloride by washing with ammonia through DAE, were resistant to oxidation at 600 °C and to water [64]. Similar catalysts were also stable up to 600 °C for 100 hours [116].

- 1%Au/AC with YCl_3 showed 87.8% conversion for acetylene hydrochlorination to vinyl chloride monomer (VCM) and almost 100% selectivity [121]. The catalyst kept high activity for >2300 hours at 30 h^{-1} GHSV (C_2H_2) at 180 °C.
- A review refers several strategies to stabilise gold nanoparticles on solid supports, like developing gold core-shell or yolk-shell structures [122], which also increased catalyst durability. One example is gold nanoparticles within mesoporous frameworks, which enhanced assured thermal stability of Au nanoparticles up to 800 °C and a good durability (>130 h at 375 °C) for aerobic oxidation of benzyl alcohol and CO oxidation [123].
- Au/TiO_2 catalysts with 3D nanorod structures, inside a stable polymorph were able to maintain activity for CO oxidation below 115 °C, even after several cycles at 800 °C, being promising for industrial applications [120].
- A Keggin-type POM (polyoxometalate), $Cs_4[\alpha\text{-}SiW_{12}O_{40}]\cdot nH_2O$, was used as a support for <2 nm Au nanoparticles, being stable for at least one month, at 0 °C, with full CO conversion [124].

23.3.4 Poison Resistance

Au nanoparticles show low affinity for CO, and are poisoned by this gas below <200 °C, unlike PGMs [125]. Gold catalysts are also resistant to sulfur poisoning. In fact, Au on TiO_2 is much more active than pristine TiO_2 for Claus reaction and SO_2 reduction by CO [126] as follows:

$$SO_2 + 2H_2S \rightarrow 2H_2O + 3S_{solid}$$

$$SO_2 + 2CO \rightarrow 2CO_2 + S_{solid}$$

Gold on cobalt oxide on zirconia-ceria, titania and zirconia, remained active in a gaseous stream with 15 ppm SO_2 [92].

Haruta showed that low-temperature CO oxidation using Au/TiO_2 is inhibited by SO_2, but the effect on H_2 or propane oxidation is very small [127]. In contrast, Scurrel showed that the activity of Au/TiO_2 for room-temperature CO oxidation is very much enhanced by adding sulphate ions to the support, as they directly modify the active gold sites [128].

This shows that gold is not poisoned by sulfur and that S might even have a positive effect. In fact, a 2005 Chevron patent claimed that Au on sulphated ZrO_2 was efficient for WGS, and that sulfur improved the catalytic activity of gold [129].

23.4 Reactions Efficiently Catalyzed by Gold Nanoparticles

Because it was found that gold could be extremely active in the form of nanoparticles, extensive research occurred, as already referenced [1–3, 11, 14–17, 74, 75, 79, 83–87, 89, 102, 130–134]. Gold-based nanoclusters, with defined atom numbers, have also been used in catalysis [74].

Gold nanoparticles are good catalysts for several reactions, such as CO and VOC oxidation; WGS; PROX; nitrogen oxide removal; selective oxidation of hydrocarbons, sugars and alcohols; CO_2 abatement; selective hydrogenation of ketones, aldehydes and hydrocarbons, and others. Some of these reactions, with environmental interest, are addressed in the following.

23.4.1 CO Oxidation

Carbon monoxide is a poisonous toxic gas with no colour or odour. It can be formed during forest fires, volcanic activity and other combustions, or by photochemical reactions taking place in the troposphere. Only a few hundred ppm can lead to terrible human health damage or death, because CO is able to bind to haemoglobin, which is then less available for oxygen. Continued exposure, even at low levels, can lead to long term health issues. This makes CO one of the most hazardous pollutants that need to be detected and removed [1, 14, 16].

The oxidation ($CO + ½ O_2 \rightarrow CO_2$) is a very simple and widely studied model reaction for Au catalysts. CO oxidation at room temperature is important for indoor air purification appliances, gas masks, automotive exhaust cleaning, CO sensors, and fuel cells, where CO amount must be below 100 ppm in the H_2 rich gas [1, 3, 14, 19, 25, 26, 33, 39, 41–44, 46, 49, 60, 61, 79, 100, 109].

Au catalysts are highly advantageous for CO oxidation as they are *active at room temperature* down to −70 °C, as first surprisingly reported by Haruta [10, 11, 50, 76, 84–86, 89, 102, 135]. Since then, much more work was done on the topic [1–3, 10, 11, 50, 76, 79, 84–87, 89, 95, 102, 135], and Au proved to be active at temperatures much lower than PGMs [83]. The reason is that PGMs are able to easily dissociate O_2 at low temperatures and to strongly bind atomic oxygen and CO. Once strongly adsorbed, they require more energy to react and form CO_2, which only happens at higher temperatures. Although CO and oxygen loosely bound to gold, a higher CO binding energy on Au nanoparticles, compared to bulk Au, provides enough CO concentration on the surface to allow the oxidation reaction to take place with insignificant energy barriers [81].

For Au supported on metal oxides, the Bond and Thompson (BT) mechanism (Figure 23.5), first proposed by these authors in 2000 [79] and later confirmed by the results of many research groups [136], is accepted. It suggests that the lattice oxygen of the support reacts with CO and the supplied O_2 is only needed to replenish the support, as in the Mars-van Krevelen mechanism, where alternate reductions and oxidations of the oxide surface occur, with formation of oxygen vacancies at the surface, and their successive refill by the supplied oxygen [42, 43, 89, 135, 137, 138].

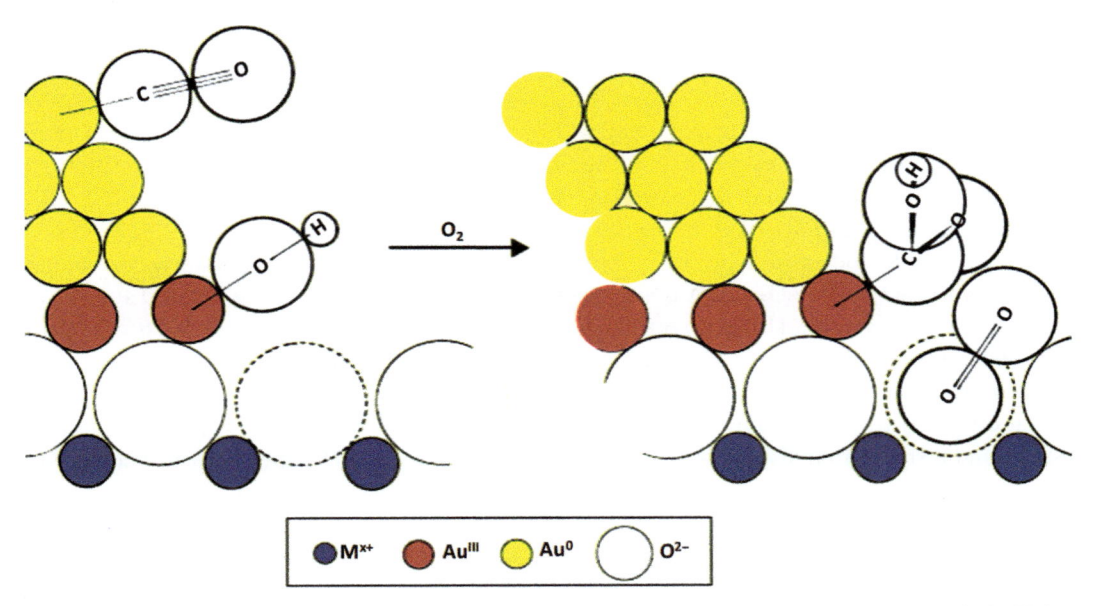

Figure 23.5 Bond and Thompson mechanism. Oxidation of CO at the periphery of a Au particle. Left: CO is chemisorbed on a low coordination Au atom, and a OH^- moves from the support to a Au^{3+} ion, forming an anion vacancy. Right: they react to form a carboxylate group, and O_2 occupies the anion vacancy as O_2^-, which oxidises the carboxylate group by extraction of a H atom, forming CO_2; the resulting HO_2^- oxidises carboxylate species, with formation of another CO_2 and restoration of two OH^- on the support surface, completing the catalytic cycle. Reproduced with permission from Ref [79] / Springer Nature.

This explains why the reaction depends so much on the *interaction of the support with Au*. TiO_2 is able to provide a strong metal-support interaction, which explains why it is such a good catalyst [21, 76, 138–144]. The activity is also related with the availability of surface oxygen, as shown on Au/exotemplated ceria, Au/exotemplated Mn_xO_y, Au/doped cryptomelane samples, by analysis of temperature programmed reduction (TPR) data (in particular, the temperature value at which the catalyst starts to be reduced) [42, 43, 49]. Experiments without oxygen in the feed for Au/exotemplated oxides [42, 43] showed that lattice oxygen can react with CO, confirming the BT mechanism. A comparison of Au on activated carbon with Au on C_3N_4 (inert to oxygen and stable in air up to 550 °C) showed that the material with no oxygen had little activity for CO oxidation, proving that Au only was not effective to activate O_2 [33]. ^{18}O isotope experiments showed that O_2 activation takes place on defects or oxygen vacancies [145].

It is accepted that CO oxidation takes place by a gold-assisted Mars-van Krevelen mechanism, in which the surface lattice oxygen (next to the Au nanoparticle) is removed by reacting with CO, partially reducing the TiO_2 surface, that is re-oxidised by O_2 [137]. Haruta proved that CO adsorption on Au/TiO_2 lead to partial reduction of the titania support [146]. Other authors also reported this to be the main reaction route for Au on reducible metal oxides [137].

There is also evidence that oxygen vacancies are involved for reaction at 80 °C and above [137, 143]. The removal of TiO_2 surface lattice oxygen by CO occurs at 120 °C. It is also possible at −20 °C, but at a slower rate and lower extent, being totally inhibited at −90 °C [140].

The active oxygen involved on Au/titania for CO oxidation at 80 °C and above, is a stable atomic species, formed at the perimeter of the Au/oxide interface [137]. At 80 °C, CO is able to remove this species. It is possible that the reaction only occurs on the Au/oxide perimeter, with CO adsorbing on Au and oxygen coming from the oxide [115, 147], or CO to be activated on gold and on the gold-support interface [1, 3, 54, 79, 136, 148].

However, for gold catalysts on non-reducible metal oxides, the mechanism must be different, probably only based on gold [137]. Bond and Thompson [79] also claimed that gold on inert supports, such as alumina and silica, were exceptions to the BT mechanism; that the mechanism would involve only Au nanoparticles, not the support; and the length of the periphery would be smaller, of Langmuir–Hinshelwood type, with oxygen (weakly) and carbon monoxide (more strongly) adjacently adsorbed on Au [79]. The supply of oxygen on those supports could be through hydroxyl groups, which could be renewed by water on the surface [14].

In fact, it is already known that the Au activity is improved by *moisture* [12, 110–113, 138]. This is very important for industrial applications, because CO streams often contain residual water. Haruta studied Au/Co_3O_4, Au/Fe_2O_3, and Au/NiO catalysts, showing that they could oxidise CO at 30 °C, at a relative humidity of 76% [10]. Also Au/Al_2O_3, Au/Fe_2O_3, Au/TiO_2, and Au/SiO_2 [12, 110, 112, 149] showed improvement up to two orders of magnitude for CO oxidation in the presence of moisture. Moreover, the reaction does not occur at low temperatures without some water on Au/SiO_2 [12].

Au on iron hydroxide [150], Au on boron nitride [113], and $Au/FeO_x/Al_2O_3$ [111] were also stable and resistant to moisture, which improved CO oxidation. COOH species easily activate O_2. Bond and Thompson [79] also claimed that perimeter interfaces contained hydroxyl groups, bound to Au atoms, for H_2O concentrations higher than 1 ppm.

Also the presence of *small* and *well dispersed Au nanoparticles* is crucial [1, 3, 11, 14, 16, 79, 81, 83–89, 138, 139]. A decrease in Au size leads to a large increase of the amount of low coordinated surface atoms, like edge and corner atoms [81]. The presence of these atoms is vital for catalytic activity, as they are needed at least for the adsorption of CO, and possibly for adsorption of O_2. The activity of Au/TiO_2 largely depends on the size of Au with best activity for ~3 nm, as reported by several researchers [11, 76, 84–86, 139]. However, larger Au nanoparticles also showed (surprising) good activity [11, 19, 39, 46, 84].

Unsupported gold can also be active for CO oxidation [95, 97], yet the majority of the studies were carried out on supported gold nanoparticles [1, 14, 16, 79, 87, 99]. Several authors mention the effect of the *support* [1, 3, 11, 14, 16, 25, 26, 37, 39, 42, 43, 60, 84, 151], in particular, its reducibility. The most common supports are transition metal oxides. As described previously, two types of oxides are defined [1, 14, 16, 99]: the active easily reducible supports, such as Fe_2O_3 [25, 78, 139], CeO_2 [19, 39, 43, 53], or TiO_2 [21, 35, 76, 138, 139, 143, 152], and the inert supports, such as SiO_2 [153] and Al_2O_3 [88, 138]. In terms of active supports, Fe_2O_3 has advantages, as it is cheaper, readily available, has good catalytic performance, is easily reducible and has large amounts of oxygen [1, 14, 25, 99]. But inert supports can also originate active catalysts, as mentioned earlier. Haruta showed that the activity of Au on Al_2O_3 and SiO_2 at room temperature was similar to that of Au on TiO_2 (prepared by gas-phase grafting), showing that the contribution of the support was similar, if gold was deposited as nanoparticles with good metal-support interactions [73]. Nevertheless, Au on silica–alumina and activated carbon, has much lower activity than Au on TiO_2 [73].

Other oxides, like CuO [26], Co_3O_4 [139, 154, 155], MgO [44, 156], Mn_xO_y [42], Nb_2O_5 [157], NiO [26, 154, 155, 158], silica-derived materials [159, 160], Y_2O_3 [26, 161], and ZnO [60, 162, 163] were also used. Y_2O_3 was considered as not adequate, but Corma et al. obtained a highly active catalyst using Au on nanocrystalline Y_2O_3 [161]. They also obtained better results with nanocrystalline ceria compared to regular CeO_2 [161, 164]. Carabineiro et al. [25] also obtained improved activity with Au on nanostructured Fe_2O_3 (prepared by calcination of the nitrate precursor at 300 °C for 1 hour, depicted in Figure 23.6a, showing higher surface area and larger amounts of hydroxylated iron species) than with a commercial Fe_2O_3 support (shown in Figure 23.6b) or with the reference World Gold Council (WGC) Au/Fe_2O_3 catalyst. The best result (0.94 $mol_{COg_{Au}}^{-1}h^{-1}$ at room temperature), showed an activity almost 10 times higher than the WGC material (0.10 $mol_{COg_{Au}}^{-1}h^{-1}$). Also, ZnO twin-brush-like mesocrystals with Zn/O-vacancy defects originated gold catalysts 153 times more active than defect-free ZnO nanorods [163].

Mixed oxide supports are also very efficient, given the synergistic effects between Au and the support [41, 46, 49, 61, 100, 165–167]. In particular, Au/perovskite [167, 168] is stable and durable for CO oxidation. Also, the addition of Ce to cryptomelane (KMn_8O_{16}) increased the number of defects that anchored gold nanoparticles and enhanced the charge transference between Au and cryptomelane, in addition to improving the stabilisation of polyhedron faceted nanoparticles below 3 nm, with larger catalytic activity for CO oxidation [61].

Composite Au/MO_x (M = Ba, Ce, Co, Fe, La, Li, Mg, Mn, Rb, Ti) on Al_2O_3 [46, 54, 82, 111, 115, 169–172] and SiO_2 [169, 173] were also used, with the latter being less active, due to larger Au particles [169]. The alkali (earth) metal oxides are structural promoters, stabilising the gold

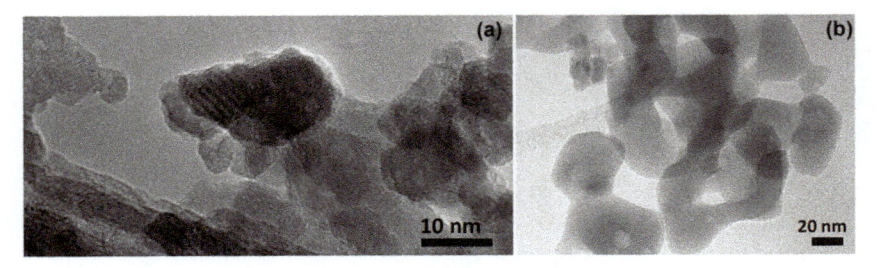

Figure 23.6 High resolution transmission electron microscopy (HRTEM) images of nanostructured iron oxide prepared by calcination of the nitrate precursor at 300 °C for 1 hour (a) and commercial iron oxide from Sigma Aldrich (b). Reproduced with permission from Ref [25] / Royal Society of Chemistry.

nanoparticles, preventing sintering in mild conditions; and accelerating oxygen activation, by oxygen vacancies or surface OH groups, and supplying active O, possibly by a Mars-van Krevelen redox cycle [54, 147, 148].

Nieuwenhuys tested Au/M^{I}O$_x$/M^{II}O$_x$/Al$_2$O$_3$ (M^I = transition metal; M^{II} = alkaline earth metal) catalysts [82, 148, 170]. They suggested that O$_2$ activation takes place on M^{I}O$_x$ through a Mars-van Krevelen mechanism, while M^{II}O$_x$ inhibits gold sintering. The most active catalyst was Au/MgO/MnO$_x$/Al$_2$O$_3$ with MgO being a stabiliser for gold and MnO$_x$ being a co-catalyst [82, 148, 170].

The *oxidation state* of gold might also influence catalytic activity. As said previously, Bond concluded from literature that the best Au/TiO$_2$ catalysts for CO oxidation were non-metallic [91]. However, Au$^+$ might also play a role [138]. Some authors reported better results for the cationic state [19, 25, 53, 139, 174]. For example, Carabineiro et al. [25] reported that Au/Fe$_2$O$_3$ catalysts prepared by LPRD had the lowest Au size (4.3 to 6.8 nm) compared to the analogues prepared by DIM (7.4 nm to 11.8 nm). Nevertheless, the DIM catalyst were up to almost 4 times more active (at room temperature) than those prepared by LPRD. A possible explanation is the oxidation state, which was Au$^+$ for DIM and Au^0 for LPRD samples. Jin et al. found better results for ceria supports with 20 nm size (CeO$_2$-20), compared to ceria with 5 nm (CeO$_2$-20) due to the strong CO adsorption of Au$^+$, present in the former [53].

However, other authors obtained better results with Au^0 [11, 25, 39, 42, 43, 46, 72]. For example, Wei et al. [142] tested Au/TiO$_2$ catalysts with Au^{n+} (1<n<3) and Au^0. Au^{n+}/TiO$_2$ was originally inactive, but CO transformed Au^{n+} to Au^0 and this improved CO adsorption ability and enhanced the catalytic activity. The formation of Au^0 favored the creation of oxygen vacancies in the support, which was beneficial for O$_2$ activation. There is no consensus as several oxidation states can coexist [168, 174], as already referenced.

The interaction/contact between Au and the support is also a crucial factor, as referenced by Haruta [135, 137]. Carabineiro et al. prepared for the first time Au/ZnO catalysts by DIM, PD, and US [60]. It was shown that the DIM sample had the lowest Au nanoparticle size (2.9 nm); however, it was the US sample, although with larger size (5.2 nm), that showed the best catalytic activity for CO oxidation (2 mol$_{CO}$ g$_{Au}$$^{-1}$h^{-1} at room temperature, which was among the best values reported at the time on Au/ZnO materials). A fast Fourier transform analysis (FFT) of the high resolution transmission electron microscopy (HRTEM) image is shown in Figure 23.7a, enlarged and indexed in Figure 23.7b. Several Au nanocrystals are epitaxially oriented on the ZnO rod, providing a unique interaction between catalyst and support, which improves catalytic activity [60].

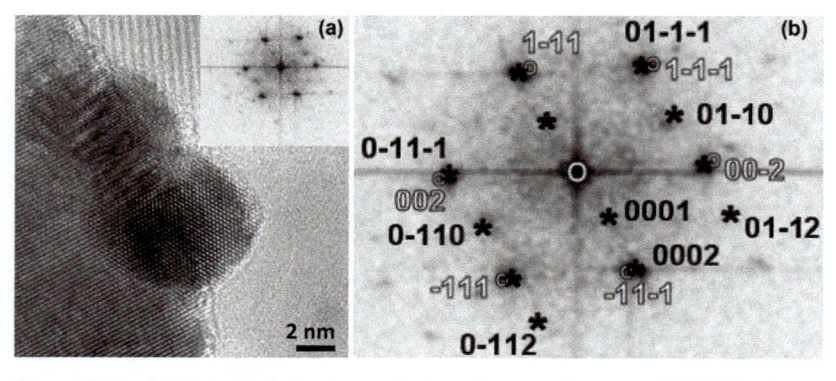

Figure 23.7 (a) High resolution transmission electron microscopy (HRTEM) micrograph of an epitaxially grown Au particle on ZnO support with fast Fourier transform analysis (FFT) as inset and (b) indexed diffraction spots in a FFT image (open circles and white numerals correspond to Au; black numerals and asterisks are used to label ZnO planes). Reproduced with permission from Ref [60] / Elsevier.

Haruta also indicated that *calcination* is able to increase the contact between Au and the support [11, 76, 84]. Calcinations up to 100–300 °C are better as sinterisation can occur at higher temperatures [1, 14, 16]. They also reported that Au/TiO$_2$ catalysts calcined at 200 and 300 °C had the best CO oxidation activity [76], as calcination at 600 °C originated larger particles, but with stronger contact with the support, allowing to achieve much higher catalytic activity at 300 °C [76]. A higher number of defects (edges, steps, and corners) in small Au nanoparticles improved the catalytic activity, and the contact with the support also improved the catalyst performance [95].

However, other authors reported that calcinations can have negative effects on the activity. Ayastuy et al. prepared Au/Fe$_2$O$_3$ catalysts by DP, non-calcined (only dried at 110 °C) and calcined at 150, 200, and 300 °C [78]. The Au/Fe$_2$O$_3$ interface perimeter was lower on the uncalcined material and also the oxygen storage and release ability were smaller, yet the catalyst showed a higher amount of lattice defects and surface OH groups, which resulted in the highest catalytic activity. Other authors also obtained good activities with uncalcined catalysts [3, 14, 16, 19, 21, 25, 26, 39, 42, 43, 78].

The CO activity is also related with the *preparation procedure*. As mentioned previously, the conventional IMP method that works well with PGMs, does not work for Au, as materials with large and inactive particles are formed. Haruta prepared Au/TiO$_2$ by IMP that were less active than Pt analogues; Au materials prepared by DP were quite active [175]. Haruta stated that DP is more effective as small hemispherical Au nanoparticles are in strong contact with support flat planes, whereas IMP results in larger spherical Au nanoparticle simply mixed with smaller particles of the support [36]. The activity of Au/TiO$_2$ was influenced by the preparation process and surpassed Pt/TiO$_2$ by one order of magnitude [175].

Haruta also showed that CO oxidation takes place on Pt surfaces of Pt/TiO$_2$, as the support is less involved in the process. This explains why Pt catalysts can be prepared by different methods without much difference in the activity [35, 175]. This was also proved by Carabineiro et al. [35], as they prepared Au, Ir, Pd, Pt, and Rh on TiO$_2$ by LPRD and IMP. Figure 23.8 shows that Au IMP was the worst catalyst for CO oxidation, whereas Au LPRD was the most active one (due to its lower particle size). LPRD also produced better results for other metals than IMP, but with difference not as high as for Au. The same authors also obtained improved results with Au/CeO$_2$ prepared by LPRD versus IMP [19].

Louis used a DP procedure with urea to obtain Au/TiO$_2$ catalysts, achieving good activities below room temperature [55]. Other authors also reported that Au/TiO$_2$ DP catalysts were much more active at room temperature than IMP analogues, due to chloride being present [40], which causes sinterisation of Au nanoparticles, as already stated [1, 14, 16, 19].

Carabineiro et al. prepared Au on CuO, La$_2$O$_3$, Y$_2$O$_3$, and NiO by DIM, LPRD, and US [26]. The best results were obtained for Au/NiO DIM (most likely related to its having the lowest Au nanoparticle size, 4.8 nm in average). All other DIM samples showed activities superior to those reported in literature for similar oxides (showing DIM to be more efficient than DP or CP) [26].

Haruta et al. proved that CO oxidation on Au/Co$_3$O$_4$, Au/Fe$_2$O$_3$, and Au/TiO$_2$ was independent of *CO concentration* and a little dependent on O$_2$ concentration [139]. Nevertheless, CO conversion increased when concentrations of CO and O$_2$ decreased [95]. Most reports in literature refer 1% CO or less [1, 3, 79, 81, 110, 138, 175, 176]. Only a few authors used values as high as 5% CO in the feed [19, 25, 26, 33, 35, 39, 41–44, 46, 49, 60, 61, 100]. That is the reason why full conversions are only achived at 100 °C in Figure 23.8, leading to the wrong idea that these catalysts are not so active. For example, the activity of Au/TiO$_2$ LPRD at room temperature was 2.05 mol$_{CO}$g$_{Au}^{-1}$h^{-1}, much larger than Au/Fe$_2$O$_3$ WGC reference material (0.10 mol$_{CO}$g$_{Au}^{-1}$h^{-1} [25]) under the same conditions).

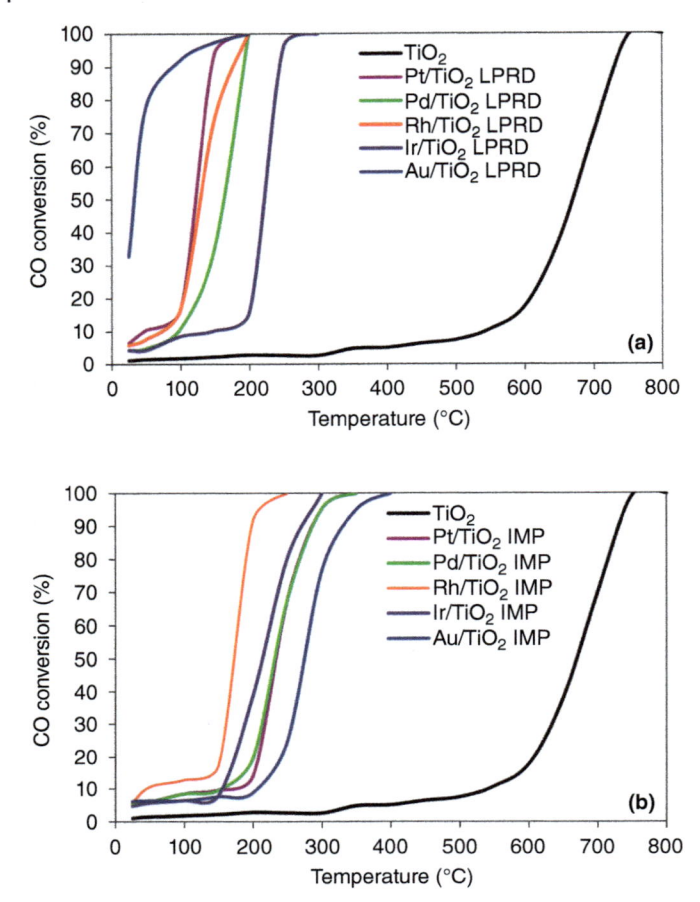

Figure 23.8 CO conversion for noble metal catalysts prepared by liquid-phase reductive deposition (LPRD) (a) and impregnation (IMP) (b) with 5%CO, 10%O_2 (balance He, total flow rate: 50 ml/min), 200 mg catalyst. Metal loading ~1% for Au loaded by IMP, 0.5% for Au by LPRD. Adapted from Ref [35].

Naturally, activities are also dependant on the *quantity of Au* on the catalyst (and the amount of catalyst used). Carabineiro et al. used 1% Au (or less) [19, 25, 26, 33, 35, 39, 41–44, 46, 49, 60, 61, 100], like other authors [12, 34, 76, 110, 157, 158, 168], but other researchers reported Au loadings up to 10% [12, 34, 54, 64, 115, 169–171, 175, 176].

A comparison of different results can be obtained if *specific activities* (or rates) are used, for example as converted amounts of CO per amount of Au, per time as in some examples given previously. One of largest activities for CO oxidation ever reported at room temperature was 8.41 $mol_{CO}g_{Au}^{-1}h^{-1}$ for Au on mesoporous iron oxide nanoflakes [177].

23.4.2 Preferential Oxidation of CO in the Presence of H_2 (PROX)

Gold catalysts also proved to be very active in the PROX of CO in a H_2-rich environment, which is of great interest in the purification of hydrogen to be used for fuel cells [51, 65, 67, 69, 77, 82, 88, 101, 114, 178–185]. A total CO conversion and catalyst stability in both CO_2 and water are crucial for an effective PROX catalyst to be used in proton exchange membrane (PEM) fuel cells. PEM fuel cells can improve efficiency and cleanliness of automobiles, given the low operation temperature, high power density, quick cold start, and high efficiency in energy conversion, and the electrolyte had excellent CO_2 tolerance [108, 109, 171, 178, 179, 186–193].

Hydrogen is a promising candidate for fuel cells, as it maximises the system efficacy, simplifies integration, only produces water during combustion (significantly reducing pollution and greenhouse effects), and is more powerful and more efficient than gasoline or other fossil fuels [187, 188, 191, 194, 195]. But, sadly, H_2 is not readily available, so it needs to be produced [195].

H_2 can be obtained from methanol and other fuels, by partial oxidation of hydrocarbon molecules, steam reforming, or water-gas shift (WGS) reactions, which produce a large quantity of CO_2 and some H_2O as by products, along with H_2 [148, 187, 188, 191, 192, 196–198]. The products of steam reforming contain around 10% CO, being passed inside a WGS system, where CO reacts with H_2O to form CO_2 and H_2, and the CO content is afterwards reduced to a few ppm in the PROX reactor [187], as H_2 can only be used in fuel cells after CO is below 10 ppm [109, 148, 178, 179, 187, 193, 198, 199].

Poisoning by CO_2 is possible but only significant if a low amount of CO is found in the reactant gas [109], and the presence of water diminishes the effect of CO_2 [200]. Thus, fuels cells are quite tolerant to CO_2 levels up to 25%. The presence of higher concentrations of CO_2 in the fuel, can largely decrease the cell performance, due to the reverse water-gas shift reaction (RWGS: $CO_2 + H_2 \rightleftharpoons CO + H_2O$) [201]. The dilution of H_2 with CO_2 leads to CO formation by RWGS. That CO is adsorbed on the Pt surface of PEM fuel cell, blocking the Pt active sites, with detrimental effects on the cell performance. This effect is smaller on Pt–Ru or Pt–Au alloys anode catalysts as Ru is CO-tolerant in nature and Au is able to oxidise the adsorbed CO, freeing the catalyst site. Ammonia (up to 20 ppm) can also harm the fuel cells [201].

Gold shows similar or even better activity, compared to PGMs, especially at low-temperature WGS. Au is more selective for CO oxidation than for H_2 oxidation, and its activity is improved by moisture and nearly insensitive to CO_2 [148, 187–189, 196–198, 202]. Compared to PGMs, Au has a low ability to chemisorb hydrogen, meaning lower activity in H_2 dissociation reactions [1, 114], which is advantageous since the oxidation of H_2 back to H_2O must be prevented, or at least kept lower. Dissociative adsorption of H_2 only happens on corners and edge sites of supported Au nanoparticles (such Au atoms dissociate H_2, but it does not spill-over to the face sites) [114].

Similarly to CO oxidation, the activity of Au for PROX also depends on the Au nanoparticle size, oxidation state, type of support (with fine results for both reducible and non-reducible oxides), interaction of the support with Au, preparation procedure, calcination conditions, existence of other metal oxide promoters or metal ions, temperature and gas composition, reaction conditions, and other factors, as shown in some reviews [114]. Gold catalysts allow complete CO conversion below 100 °C [114].

In 1997, Haruta et al. already reported >95% CO conversion for Au on Mn oxide between 50 and 80 °C [197]. This was absolutely promising, as normal operation temperatures of PEM cells are around 80 °C. Wang et al. [203] credited the better PROX activity of Au on Mn oxide to the amazing redox properties and the easy generation of activated oxygen species on the support surface.

Au/Fe_2O_3 are also very efficient [77, 78, 186, 192, 198, 204–209]. They have an improved CO oxidation activity at low temperatures (50 °C) and lower H_2 consumption. Also, Au on ferric hydroxide is very active for PROX at lower temperatures [210]. Au/TiO_2 catalysts are also among the most active and most selective for this reaction [67, 184, 206, 211–213]. Other Au catalysts like Au/alumina [172, 199, 211, 212, 214], Au/ceria [183, 188, 212, 214–216], Au/Co_3O_4 [207], Au/CuO [214], Au/MgO [172], Au/silica [217], Au/ThO_2 [218], Au/ZnO [69, 180, 190, 206], and Au/ZrO_2 [207, 211, 212] were also reported. Nanoporous (or unsupported) gold also showed interesting results [96, 114].

However, some points remain to be solved in PROX reaction, like some eventual, not desired, oxidation of H_2 and negative effects of CO_2 and/or H_2O [114]. The effect of excess H_2 is quite important as it can influence PROX positively or negatively. At lower temperatures, hydrogen accelerates CO oxidation, but, at higher temperatures, H_2 can compete with CO and O_2 for adsorption on Au. This means that CO conversion improves with temperature, but, above a certain value, H_2 oxidation is prevalent, which decreases selectivity to CO_2.

Grisel and Nieuwenhuys [148] suggest the presence of an inflection point in the temperature profiles, above which the energy for H_2 activation decreases, leading to a growth of the undesired H_2 oxidation to H_2O. This effect might be linked to the nature of the oxide support, and the presence of H_2O and CO_2 also play a role [114]. The addition of promoters also improves activity.

As an example, for $Au/MO_x/Al_2O_3$ (where M = Mg, Mn), the addition of MgO [147, 148] and MnO_x [148, 219] increased the CO oxidation activity and selectivity to CO_2 (Figure 23.9). The positive effect of MgO was ascribed to stabilisation of small gold nanoparticles, more active for CO than for H_2 oxidation. MnO_x could provide the active oxygen necessary for CO oxidation. The selectivity to CO_2 was larger than 90% for $Au/MgO/Al_2O_3$ and $Au/MnO_x/MgO/Al_2O_3$ at 100 °C and below. Above 50 °C, the thermal desorption rate of CO was high enough for the H_2 oxidation to start, which decreased the selectivity. Water has a good effect on the rate, even at room temperature, explained by the positive role of surface OH groups in CO oxidation [82, 147, 148].

Addition of MnO_x and FeO_x to $Au/MgO/Al_2O_3$ enhanced even more the low-temperature CO oxidation leading to better CO_2 selectivity [147]. $Au/MO_x/Al_2O_3$ (M = Ce, La and Mg) catalysts were also tested by Lakshmana et al. [101]. The catalytic activity diminished in the order: $Au/La_2O_3/Al_2O_3 > Au/CeO_2/Al_2O_3 > Au/MgO/Al_2O_3 > Au/Al_2O_3$. The $Au/La_2O_3/Al_2O_3$ catalyst showed a good performance, being stable in the presence of CO_2 [101, 185].

Au/MgO materials, modified with Mn and Fe, had higher stability, activity, and selectivity than unmodified Au/MgO [220]. The addition of these modifiers changed the type of metal ion–metal nanocluster active sites linked to CO activation, and the new sites were more active for temperatures higher than room temperature. MgO-supported Au catalysts modified by La, Pb, Sm, V, and

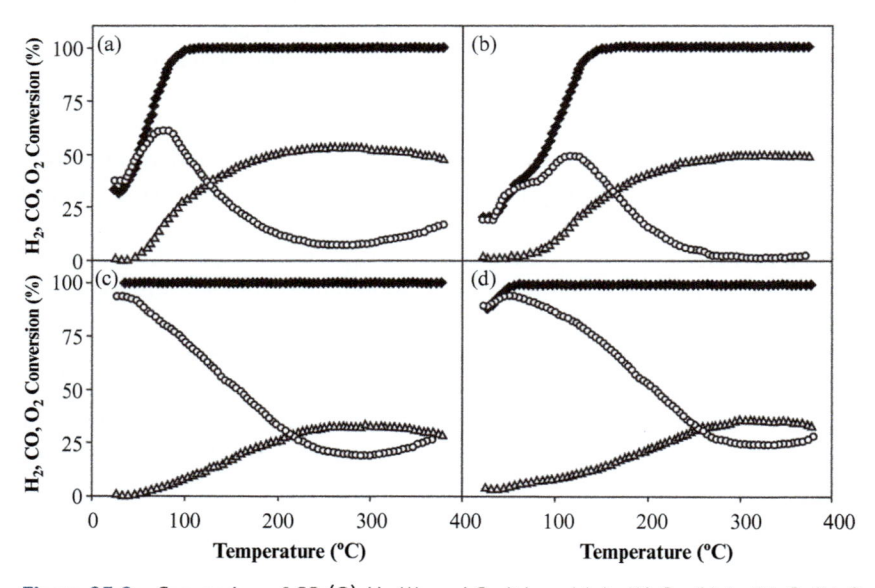

Figure 23.9 Conversion of CO (O), H_2 (Δ), and O_2 (◆) on (a) Au/Al_2O_3, (b) $Au/MnO_x/Al_2O_3$, (c) $Au/MgO/Al_2O_3$, and (d) $Au/MnO_x/MgO/Al_2O_3$ in the presence of H_2. Reproduced from Ref (148), with permission from Elsevier

Y were also tested [221]. The formed interfaces of the multi-component catalysts allowed a new type of active site ensembles able to adsorb and activate CO and O_2, but prevented activation of H_2.

Au/Fe_2O_3-TiO_2 (1:4) showed higher CO conversion than single Au/TiO_2 and Au/Fe_2O_3, especially below 100 °C, due to the presence of reducible Au species [181]. Au on mixed Ni-Fe supports were also studied [200].

Laguna et al. reported that adding 10% Fe to Au/ceria resulted in an improvement of catalytic activity, with the composite having a higher CO conversion than Au/ceria below 150 °C, and being scarcely affected by CO_2 and H_2O from the gas stream [222]. The strong interaction between Fe_2O_3 and CeO_2 contributed to larger oxygen mobility, leading to a higher catalytic activity of Au/Ce-Fe-O [65, 208]. Nice results were obtained for Au/La-Ce-O, which was stable, even with 10% water in the feed stream [223]. Also improved results were obtained when doping Au/ceria with Cu [182, 193], Mn [208], Ni [224], Sm [225], Y [51], Zn [225, 226], or Zr [207, 226, 227].

However, no improvement was found when Co_3O_4 was added to CeO_2 [208]. H_2 oxidation (accompanied by Co_3O_4 reduction to inactive CoO) occurred, together with methane formation (with a contribution from Co being reduced to the metallic state). However, addition of Co_3O_4 to TiO_2 had better results, as it increased the oxygen defects on TiO_2, forming more strongly bound Au atoms [213]. Co_3O_4 amorphous together with TiO_2 enhanced electronic interaction between Co_3O_4-TiO_2 and Au, stabilising the Au particles and improving the catalytic activity.

Au/CeO_2-MO_x/Al_2O_3 catalysts (M = Cr, Cu, Fe, La, Ni, Y) were also used [202]. Cu and Fe enhanced the catalytic performance of $Au/CeO_2/Al_2O_3$. The Cu-containing material was water resistant and CO oxidation efficient.

Better catalysts were obtained adding Pt to Au. For example, Au/ZnO had good stability at 80 °C, but its activity was slightly decreased above 350 hours [190]. But when small quantities of Pt were added, the stability improved, up to 500 hours [196]. Also, better results were found for Au–Pt/ceria than for Au/ceria [228]. Zhang et al. [196] tested Au–Pt/ZnO catalysts and found that 1% Pt improved the stability, but a larger amount of Pt had a detrimental effect on the CO_2 selectivity.

Hutchings et al. [204, 205] prepared a Au/Fe_2O_3 catalyst by CP of Au^{3+} and Fe_2O_3 and calcination in two stages. They obtained a commercially acceptable material to competitively oxidate diluted CO in the presence of moisture excess of H_2 and CO_2. Calcining the material at 400 and then at 550 °C yielded a catalyst able to remove >99.5% of the CO at 80 °C, the operating temperature of fuel cells [204, 205].

Concerning the reaction mechanism, the lattice oxygen of the oxide supports can also participate in the reaction (without Au or being enhanced by Au), and the Mars–van Krevelen mechanism is accepted, together with Au perimeter effects [114, 147, 148, 198, 209, 215, 216, 227].

23.4.3 Water-gas Shift

Water-gas, syngas or synthesis gas is a mixture of CO and H_2, first synthesised from the reaction of steam with coke [14]. Hydrogen can be obtained from biomass and fossil fuels (coke, coal, naphtha, oil, and natural gas) [194, 229, 230]. Steam reforming of methane from natural gas is a common and cheap process used to produce around half of the H_2 used globally [231]. However, the resulting gas contains a small amount of CO, which needs to be removed [187], as mentioned previously.

The water-gas shift (WGS) reaction ($CO+H_2O \rightarrow CO_2+H_2$) is often used to reduce CO and produce H_2 [231], and can be combined with steam reforming of methane (and other hydrocarbons) [232], allowing tuning of the H_2/CO ratio in the water gas [233] to produce methanol [234], ammonia [235], or liquid hydrocarbons (like kerosene, gasoline and lubricants; that is, the

Fischer-Tropsch synthesis) [231], and purification of H_2 for fuel cells [231]. Highly purified hydrogen is mandatory for fuel cells (as referenced previously), and also for the synthesis of ammonia, as CO acts as a poison, not just for the fuel cell Pt electrode anode (as already mentioned), but also for the Fe catalyst used in the production of ammonia [236, 237]. WGS is also a key step in automobile exhaust reactions, and H_2 is a very effective reductant to remove NO_x [14].

WGS is a relatively mild, equilibrium-limited, exothermic reaction, that is thermodynamically favored at low temperatures, although the kinetics is decreased under such conditions, reducing the yield of H_2 and increasing the quantity of catalyst needed to attain sufficient CO conversion [238–242]. The way to obtain a good reaction rate and a high CO conversion is to carry the WGS reaction in two steps: high-temperature (HT-WGS) and low-temperature (LT-WGS) [242–244]. HT-WGS operates around 350–500 °C and has a rapid reaction rate yet low CO conversion. LT-WGS, performed around 200–350 °C, is favored by thermodynamics to convert CO, tuning the CO:H_2 ratio of the gas coming out of the HT-WGS reactor, but has slower reaction rate, as previously mentioned [238–242]. A good part of the feed CO is converted in the first reactor due to the fast kinetics. The amount of CO is further reduced in the second reactor, at lower temperature, given the exothermal nature of the reaction.

Two different types of materials are industrially used for the WGS reaction: HT-WGS uses Fe–Cr oxides, whereas LT-WGS needs Co–Mo or Cu–Zn oxides [14, 229, 231, 242, 243]. Catalysis containing Cu has disadvantages, such as extended pre-reduction time, sensitivity to air and water, susceptibility to being poisoned by sulfur, slow kinetics at low temperature [231, 245]. Thus, they need precise reduction conditions, which makes them be very reactive to air after activation (pyrophoric), which is unsafe during operations. This makes them unsatisfactory for automotive applications [246].

Cu catalysts also operate from 200–300 °C, whereas gold is active from 100–200 °C. This is important because lowering the temperature causes the CO equilibrium concentration to be improved and the CO amount in the reformate gas to be lowered, avoiding the use of further CO removing systems that are still used in fuel processing commercial systems (as CO is poisonous, as already referred) [14]. Gold has an extra advantage, as it selectively catalyzes this oxidation (PROX reaction), as its selectivity for hydrogen/oxygen competing reaction is much lower at room temperature, as already mentioned.

Noble metals (like Pt and Au) on oxide supports efficiently catalyze LT-WGS [1, 14, 16, 47, 48, 84, 152, 229–231, 244, 246–262]. They are promising for fuel cells, even though they are expensive [229, 236]. There is the need for catalysts that are: (i) non-pyrophoric; (ii) usable without requiring pre-reduction treatments; and (iii) durable in quick heating and cooling cycles. Au catalysts are good candidates and have been successfully used in WGS [1, 14, 16, 47, 48, 152, 229–231, 244–246, 251, 252, 254, 256–260, 262–268]. When Au is added to metal oxides, there is an improvement in the activity of LT-WGS, explained by a synergistic effect between Au and the oxide [231]. Moreover, as described previously, a Chevron patent from 2005 already claimed that gold on sulphated ZrO_2 could be used for HT-WGS and LT-WGS [129], which is very advantageous.

In the 1990s, Andreeva et al. found that Au/Fe_2O_3 was unexpectedly highly active for WGS [247, 269]. CO conversion had started already at 120 °C, being more efficient than the WGS industrial catalysts of that time. Those authors kept studying gold and reported that Au/TiO_2 and Au/ZrO_2 were also active, even more than Au/Co_3O_4 [249, 251].

Shortly afterwards, Haruta also reported a high activity for Au/TiO_2, compared to Au/ZnO, Au/Al_2O_3, and Au/Fe_2O_3 [84, 248]. Andreeva et al. also tested mixed oxide supports (Fe_2O_3-ZnO and Fe_2O_3-ZrO_2) and amorphous TiO_2 and ZrO_2 [250]. Well-crystallised metal oxides were more efficient than amorphous oxides. Similar results were found for mixed oxides [250, 251].

Flytzani-Stephanopoulos reported the high activity of Au/ceria for WGS [260, 263, 266, 270, 271]. This is very important for automotive catalysts, as CeO_2 is widely used in automotive three-way catalysts of emission-control, given its ability to quickly change oxidation state if the redox potential of the exhaust gases change. Several studies reveal that CeO_2 is a very active support [1, 14, 16, 229–231, 246, 251, 252, 254, 256, 258–260, 262, 263, 265, 266, 270, 271].

It is known that ceria and ceria-containing materials show high capacities to transport oxygen and can easily shift between reduced and oxidised states (Ce^{3+} to Ce^{4+}), and this can be enhanced by adding transition metal ions into the ceria lattice [231]. Thus, adding rare earth metals (REMs) to ceria has been tested. REMs are highly active when added to ceria, leading to enhanced thermal stability and improved catalytic activity [267]. Au/ceria doped with La and Gd also displayed higher catalytic activity [271]. $Au/CeO_2–ZrO_2$ catalysts are also very active for WGS and allow H_2 to be obtained from H_2O at 100 °C [272]. $AuCe_{50}Zr_{50}$ showed high activity and stability due to high Au dispersion [272]. $Au/CeO_2/Al_2O_3$ was also investigated and transition metals like Cu, Fe and Zn were used as ceria dopants, improving the redox activity and, consequently, the activity [273].

Adding ZrO_2 to Au/Fe_2O_3 highly improved activity and stability, but the addition of CaO, Cr_2O_3, CuO, La_2O_3, MgO, or NiO had a smaller effect [274]. Some $Au/M_xO_y/TiO_2$ materials (M_xO_y = CaO, Al_2O_3, ZrO_2 Y_2O_3, and REM oxides) had higher activity than Au/TiO_2 if calcined at high temperature [275]. Au/carbide were also used [276]. Au/MoC_x also showed exceptional LT-WGS activity. Au was well dispersed on the carbide and interacted with the support by strong charge transfer.

Several reviews on Au catalysts for WGS showed that not just the nature of the support, but also the gold size, oxidation state, and preparation method are important factors [1, 14, 16, 229–231, 251, 252, 256, 258–260, 263, 265, 277]. DP and CP are the most common procedures used to prepare catalysts for LT-WGS reaction [14, 229–231, 246, 247, 251, 252, 256, 258–260, 265, 269, 278]. Carabineiro et al. used Au/Fe_2O_3 [47] and Au/TiO_2 [48] obtained by LPRD and DIM, but DP produced more active catalysts [47, 48].

Bimetallic Au-Ru catalysts on Fe_2O_3 were also used [279]. When compared to Au/Fe_2O_3 and Ru/Fe_2O_3 analogues, the bimetallic system had better activity at 100–300 °C. $Au-M/Fe_2O_3$ (M = Ag, Bi, Co, Cu, Mn, Ni, Pb, Ru, Sn, or Tl) were also used, with Ru and Ni having the best results [280]. Concerning $Au–Cu/TiO_2$ [281], the presence of Cu^+ enhanced the dispersion of reduced gold and also helped keep it highly dispersed during the reaction. Other bimetallic catalysts include $Au–Pt/CeO_2$ [282], $Au–Pd/CeO_2$ [282], and $Au–Re/CeO_2$ [245].

WGS, catalyzed by Au or other catalysts, is usually believed to progress by two pathways: a redox regenerative mechanism and an associate mechanism [259, 265, 277]. In the former, CO is reacted with a reducible oxide, forming CO_2 and partially reducing the support, which is then oxidised by water, replenishing oxygen and releasing hydrogen. This takes place with a reducible support and at high temperature. The second mechanism includes the development of surface CH_xO_y intermediates (formates and carboxyls), and their decomposition to H_2 and CO_2, with possible reduction and reoxidation of the metal oxide support (Figure 23.10). The mechanism depends on the type of catalyst, preparation method, pre-treatment, reaction temperature, and gas composition [259, 265].

Au catalysts might suffer WGS activity loss due to agglomeration of gold nanoparticles (this happens for Au/ceria-zirconia, which are among the most active materials for LT-WGS [261], and Au/ceria [264]). Au/ceria catalysts can deactivate due to poisoning or blockage of the active sites by hydrocarbons, carbonates, or formates [264], formed from CO and H_2, on oxygen-deficient sites of ceria [254]. Nevertheless, regeneration is possible by calcination and 95% of the initial activity can be restored by heating at 400 °C in air [254].

Au/Fe_2O_3 deactivated due to large reduction of the support surface area with time of use [283]. However, rapid deactivation of Au/Co_3O_4 was due to reduction of the support [251]. Gold can

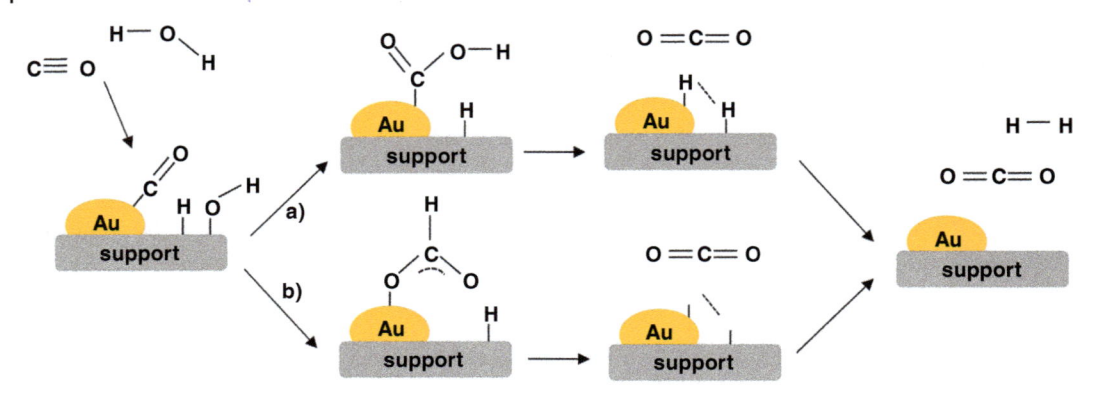

Figure 23.10 Carboxyl (a) and formate (b) mediated mechanism in the low-temerature water-gas shift (LT-WGS) reaction. The gold nanoparticle and support only show the position of surface sites. Adapted from Ref [277].

stimulate reduction of the supports, by activation of H_2 on Au and spill-over of the atomic hydrogen to the support [251].

Au/CeZrO$_4$ deactivated due to water that progressively removed the gold atoms from the support, diminishing the metal-support interaction [284]. The rate of exchange in Au^0–CO infrared bands was well related with deactivation, while Au^{d+}–CO species vanished more quickly on stream [278].

These considerations show that deactivation of Au catalysts in WGS is dependent on the type of catalyst and reaction conditions, such as temperature and gas composition [259]. Regeneration by calcination at high temperature is able to remove intermediates and strong adsorbates, but also re-disperse gold nanoparticles in the support and recover the surface oxygen [259].

23.4.4 Total Oxidation of Volatile Organic Compounds (VOCs)

Volatile organic compounds (VOCs) are present in printing, painting, petroleum refinery, motor vehicles and fuel storage, being harmful pollutants to humans and environment [14, 285]. VOCs are divided into several types:

- *Saturated* or *unsaturated aliphatic* (including methane, propane and propene, with alkanes being the less reactive).
- *Oxygenated* (alcohols, such as methanol, ethanol, and propanol; aldehydes, like formaldehyde; ketones, such as acetone; esters, like ethyl acetate). The order of reactivity is: alkanes<esters< ketones<aldehydes < alcohols [286]).
- *Aromatic hydrocarbons* (like benzene and toluene).
- *Chlorinated* (such as dichloromethane, methyl chloride, chlorobenzene, chlorophenol).
- *Nitrogenated* (like trimethylamine).
- *Sulfur-containing* (like dimethyldisulfide) [285].

Complete catalytic oxidation (preferably to CO_2 and H_2O) needs lower temperatures (250–500 °C) and leads to less NO_x formation, in comparison with uncatalyzed thermal oxidation that operates at higher temperature (650–1100 °C) [286, 287]. Metal oxides (like MnO_x, CuO, NiO, Fe_2O_3, and Co_3O_4) have been used for the total oxidation of several VOCs [288–292]. Mixed metal oxides, like perovskite and cryptomelane materials, were also extensively reported, with excellent results [290, 291, 293–308].

Metal oxides are more resistant to poisoning, but they are usually less active than noble metals for VOC oxidation [309]. Pd and Pt are usual noble metals, but Au shows excellent results [35, 309–315].

The first meaningful papers related to gold catalysts being used for VOC oxidation came out in the 1990s [316–318], and reported high activity of Au/Co_3O_4 for the total oxidation of methane [316] and chloromethane [318] and of Au/Fe_2O_3 for the combustion of formaldehyde, methanol, and formic acid [317] and 2-propanol, methanol, ethanol, acetone, and toluene [319, 320]. The good activity was related to the ability of well-dispersed Au to diminish the Fe-O bond and improve the mobility of lattice oxygen that participated in the oxidation, most likely through a Mars-van Krevelen mechanism [319].

Afterwards, more studies came out, as shown by many reviews on the topic [1, 14, 258, 285, 290, 321, 322]. Gold on cobalt oxides is one of the most active catalysts for the oxidation of light alkanes [299, 323]. The remarkable reducibility of Co_3O_4, compared to other oxides, might explain its activity [289].

Usual supports for Au include alumina, cobalt oxide, copper oxide, manganese oxide, iron oxide, silica, titania, zirconia, zeolites, and carbon based materials [27, 35, 45, 159, 310, 312, 319, 320, 323–331]. Au/ceria is also a good catalyst for the oxidation of several VOCs, like benzene, ethyl acetate, toluene, propanol, propene, and formaldehyde, among others [45, 322, 332–336]. The ability of Au to weaken the Ce-O surface bonds, next to Au atoms, improved the reactivity of the capping oxygen of the ceria surface [332], which participates in the oxidation through a Mars-van Krevelen mechanism [215, 319, 320, 332].

Au on ceria-mixed oxides [215, 311, 313, 315, 324, 333, 335–342] and on other mixed oxides [299, 314, 343–347] has also been used. Also bimetallic Au catalysts [348]. The contact between Au–Pd alloyed particles and α-MnO_2 nanotubes largely enhanced the reactivity of lattice oxygen, and the material was quite active in the oxidation of m-xylene, toluene, acetone and ethyl acetate, being also very stable, with good tolerance to CO_2 and water vapor [349].

The activity of Au catalysts for VOC oxidation, as with all other reactions, depends on many factors, like size and shape of Au nanoparticles, oxidation state of gold, type and characteristics of the support, method of preparation, and type and concentration of VOCs [1, 14, 285, 290, 321]. Only a smaller number of studies in literature compare the activity of Au on several supports for VOC oxidation [27, 337, 338], and propene oxidation is a common model [326–328, 350].

Hutchings compared Au on CoO_x, MnO_x, CuO, Fe_2O_3, CeO_2, and TiO_2 for the total oxidation of alkanes (methane, ethane, and propane) [323]. All of the supports were active for the reaction, but adding Au decreased the temperature at which the oxidation started. The most effective material was Au/CoO_x, prepared by CP, which maintained high activity during 48 hours.

Other authors investigated Au/Al_2O_3, Au/CeO_2, Au/Fe_2O_3, and Au/TiO_2 prepared by DP, for 2-propanol oxidation [325]. Au/ceria gave the best results. Andreeva also tested Au on ceria and titania, with molybdena or vanadia as promoters, in the total oxidation of benzene, and confirmed that ceria was the best support, and Mo was better than V as a promoter [337]. The catalytic activity and the catalyst reduction were well correlated.

Ilieva et al. also tested the total oxidation of benzene, using Au/ceria doped with Co, Fe, Mn, and Sn metal oxides, prepared by CP [338]. The average size and the distribution of Au particles were similar for all supports. In general, Au/ceria catalysts doped with other metal oxides were less active than Au/ceria alone, with the exception of $Au/ceria/CoO_x$ that showed better activity (and stability) than the Au/ceria material.

Au on alumina, ceria, and titania were also used for propene oxidation [326, 327]. Pristine alumina and titania were not able to convert the VOC up to 500 °C, but the addition of 1% wt. Au produced active catalysts, with Au/titania being better than Au/alumina. The ceria support had catalytic activity and loading with gold lead to the best catalyst of the study.

Nieuwenhuys also studied Au/alumina doped with alkali (earth) metal oxides (MO_x, M: Ce, Mn, Co, Fe, Li, Rb, Mg, or Ba) [350] for propene oxidation and found that ceria was the best dopant [350]. The extra metal oxides behave as co-catalysts and/or structural promoters, preventing gold nanoparticles from sintering, as also found in the PROX reaction studied in the same group (see previously).

Carabineiro et al. used Au catalysts supported on CuO, Fe_2O_3, La_2O_3, MgO, NiO, and Y_2O_3, prepared by DIM, in the oxidation of ethyl acetate and toluene [27]. Ethyl acetate was easier to oxidise than toluene. The reducibility of the support and the Au nanoparticle size were crucial factors influencing the activity, following a Mars-van-Krevelen mechanism. Figure 23.11a shows that the best activity was for Au/CuO (although pristine CuO already had good performance). Au/NiO was the second best. MgO was not active alone, as up to 400 °C, only a maximum conversion of ~75% was obtained, however when Au was added, Au/MgO was almost as active as Au/CuO (Figure 23.11b). Similar results (yet not so remarkable) were obtained for the weak catalyst Y_2O_3, which also showed much better activity after Au addition. Smaller Au sizes were obtained on Au/MgO and Au/Y_2O_3 (5.4 and 5.5 nm, respectively), in comparison with Au/CuO (5.8 nm). Au/NiO had even smaller size (4.8 nm), but CuO was more reducible, which explained its better results, confirming the striking effect of the support in the reaction.

The performance of Au on Mn_xO_y and CeO_2 supports prepared by exotemplating using carbon xerogel and activated carbon as exotemplates, removed by calcination [45] was also studied. When templates are inside, in the growing solid, and create a pore system after their removal, they are called endotemplates. If the templates are materials with structural pores, where other solids are created, thus providing scaffolds for the synthesis, they are referred as exotemplates. In both cases, the template has to be removed to generate a high surface area material. The advantages of carbon templates are their large surface area, high porosity, low cost and be easily removed by combustion.

The exotemplated Mn and Ce oxide materials, pristine or loaded with Au by DIM, were tested in the oxidation of ethanol, ethyl acetate, and toluene [45]. Adding Au to ceria greatly improved the catalytic activity, but adding Au to manganese oxides did not produce better results, due to the large size of gold nanoparticles [45]. Toluene was the most difficult molecule to oxidise, while ethanol was the easiest. Both CeO_2 and MnO_x catalysts, with or without gold, were found to be stable during VOCs oxidation.

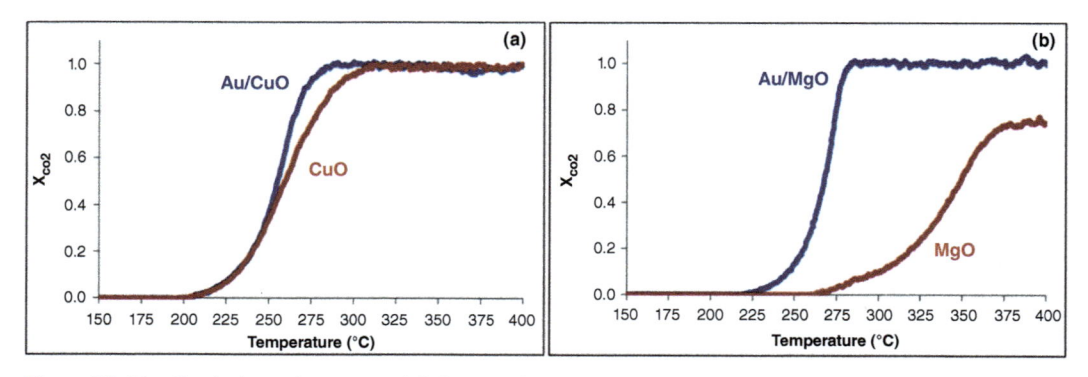

Figure 23.11 Catalytic performance of CuO (a) and MgO (b), with and without Au (1% wt.), prepared by double impregnantion (DIM), for the conversion of ethyl acetate. Adapted from Ref [27].

0.25% Au/α-MnO$_2$ with very small Au nanoparticles (downsized into approximate single atoms) was used for formaldehyde oxidation [331]. The material showed high tolerance to water and was very stable.

The mechanism of VOC total oxidation over gold on reducible oxides is accepted to be a redox or Mars–van Krevelen [1, 14, 27, 285, 290, 321]. Au enhances the reducibility and reactivity of the support and increases the exchange rate between surface and lattice oxygen.

23.5 Conclusions and Outlook

This chapter shows that Au nanoparticles are efficient catalysts for a several reactions, including environmentally related oxidation reactions. However, many more reactions can be catalyzed by gold, as shown in several reviews [1–3, 11, 14–17, 83–85, 87, 89, 98, 130–132, 134].

Gold catalysts are potential candidates for further research that might lead to even more environmental, commercial, and practical applications. The main conclusion is that catalysis by gold has a "golden" bright future!

Acknowledgements

This work was supported by national funds through FCT—Fundação para a Ciência e a Tecnologia, I.P., under the Scientific Employment Stimulus—Institutional Call (CEECINST/00102/2018) and partially supported by the Associate Laboratory for Green Chemistry—LAQV, financed by national funds from FCT/MCTES (UIDB/50006/2020 and UIDP/50006/2020).

References

1 Bond, G.C., Louis, C., and Thompson, D.T. (2006). *Catalysis by Gold* (ed. G.J. Hutchings). London, United Kingdom: Imperial College Press.
2 Bond, G.C. (2002). *Catal. Today* 72 (1–2): 5–9.
3 Bond, G. and Thompson, D. (1999). *Catal. Rev. Sci. Eng.* 41 (3&4): 319–388.
4 Stewart, D. (2018). Facts about gold. Available from: https://www.chemicool.com/elements/gold-facts.html (accessed 7 March 2020).
5 Bone, W.A. and Wheeler, R.V. (1906). *Philos. Transact. Roy. Soc. London Ser. A-Contain. Pap. Math. Phys. Char.* 206: 1–67.
6 Bone, W.A. and Andrew, G.W. (1925). *Proc. R Soc. London A Pap. Math. Phys. Charact.* 109 (751): 459–476.
7 Bond, G.C. (1972). *Gold Bull.* 5 (1): 11–13.
8 Hutchings, G.J. (1996). *Gold Bull.* 29 (4): 123–130.
9 Hutchings, G.J. (2005). *Catal. Today* 100 (1): 55–61.
10 Haruta, M., Yamada, N., Kobayashi, T., and Iijima, S. (1989). *J. Catal.* 115 (2): 301–309.
11 Haruta, A. (2003). *Chem. Rec.* 3 (2): 75–87.
12 Masakazu, D., Mitsutaka, O., Susumu, T., and Masatake, H. (2004). *Angew. Chem. Int. Ed.* 43 (16): 2129–2132.
13 Jin, C., Zhou, Y., Han, S., and Shen, W. (2021). *J. Phys. Chem. C* 125 (47): 26031–26038.

14 Carabineiro, S.A.C. and Thompson, D. (2007). Catalytic applications for gold nanotechnology. In: *Nanocatalysis. NanoScience and Technology* (eds. U. Heiz and U. Landman), 377–489. Berlin, Heidelberg: Springer.
15 Della Pina, C., Falletta, E., Prati, L., and Rossi, M. (2008). *Chem. Soc. Rev.* 37 (9): 2077–2095.
16 Carabineiro, S.A.C. and Thompson, D. (2010). Gold catalysis. In: *Gold: Science and Applications* (eds. C. Corti and R. Holliday), 89–122. Boca Raton, London, New York: CRC Press, Taylor & Francis Group.
17 Della Pina, C., Falletta, E., and Rossi, M. (2012). *Chem. Soc. Rev.* 41 (1): 350–369.
18 Carabineiro, S.A.C. (2019). *Front. Chem.* 7. https://doi.org/10.3389/fchem.2019.00702.
19 Carabineiro, S.A.C., Silva, A.M.T., Dražić, G. et al. (2010). *Catal. Today* 154 (3–4): 293–302.
20 Buffat, P. and Borel, J.P. (1976). *Phys. Rev. A* 13 (6): 2287–2298.
21 Moreau, F., Bond, G.C., and Taylor, A.O. (2005). *J. Catal.* 231 (1): 105–114.
22 Carabineiro, S.A.C., Papista, E., Marnellos, G.E. et al. (2017). *Mol. Catal.* 436 (Supplement C): 78–89.
23 Vourros, A., Garagounis, I., Kyriakou, V. et al. (2017). *J. CO2 Util.* 19 (Supplement C): 247–256.
24 Martins, L.M.D.R.S., Carabineiro, S.A.C., Wang, J. et al. (2017). *ChemCatChem.* 9 (7): 1211–1221.
25 Carabineiro, S.A.C., Bogdanchikova, N., Tavares, P.B., and Figueiredo, J.L. (2012). *RSC Adv.* 2 (7): 2957–2965.
26 Carabineiro, S.A.C., Bogdanchikova, N., Avalos-Borja, M. et al. (2011). *Nano Res.* 4 (2): 180–193.
27 Carabineiro, S.A.C., Chen, X., Martynyuk, O. et al. (2015). *Catal. Today* 244: 103–114.
28 Rodrigues, E.G., Carabineiro, S.A.C., Delgado, J.J. et al. (2012). *J. Catal.* 285 (1): 83–91.
29 Carabineiro, S.A.C., Martins, L.M.D.R.S., Buijnsters, J.G. et al. (2013). *Appl. Catal. A* 467: 279–290.
30 Ribeiro, A.P.C., Martins, L.M.D.R.S., Carabineiro, S.A.C. et al. (2017). *Appl. Catal. A* 547 (Supplement C): 124–131.
31 Onal, Y., Schimpf, S., and Claus, P. (2004). *J. Catal.* 223 (1): 122–133.
32 Demirel, S., Lehnert, K., Lucas, M., and Claus, P. (2007). *Appl. Catal. B* 70 (1–4): 637–643.
33 Zhu, J., Carabineiro, S.A.C., Shan, D. et al. (2010). *J. Catal.* 274 (2): 207–214.
34 Tofighi, G., Lichtenberg, H., Pesek, J. et al. (2017). *React. Chem. Eng.* 2 (6): 876–884.
35 Santos, V.P., Carabineiro, S.A.C., Tavares, P.B. et al. (2010). *Appl. Catal. B* 99 (1–2): 198–205.
36 Haruta, M. (1997). *Catal. Today* 36 (1): 153–166.
37 Meyer, R., Lemire, C., Shaikhutdinov, S.K., and Freund, H.-J. (2004). *Gold Bull.* 37 (1): 72–124.
38 Schulz, A. and Hargittai, M. (2001). *Chem. Eur. J.* 7 (17): 3657–3670.
39 Carabineiro, S.A.C., Silva, A.M.T., Draić, G. et al. (2010). *Catal. Today* 154 (1–2): 21–30.
40 Soares, J.M.C., Hall, M., Cristofolini, M., and Bowker, M. (2006). *Catal. Lett.* 109 (1–2): 103–108.
41 Carabineiro, S.A.C., Silva, A.M.T., Dražić, G., and Figueiredo, J.L. (2010). Preparation of Au nanoparticles on Ce-Ti-O supports. In: *Studies in Surface Science and Catalysis* (ed. E.M. Gaigneaux, M. Devillers, S. Hermans, et al.), 178: 457–461.
42 Carabineiro, S.A.C., Bastos, S.S.T., Órfão, J.J.M. et al. (2010). *Catal. Lett.* 134 (3–4): 217–227.
43 Carabineiro, S.A.C., Bastos, S.S.T., Órfão, J.J.M. et al. (2010). *Appl. Catal. A* 381 (1–2): 150–160.
44 Carabineiro, S.A.C., Bogdanchikova, N., Pestryakov, A. et al. (2011). *Nanoscale Res. Lett.* 6: 1–6.
45 Bastos, S.S.T., Carabineiro, S.A.C., Órfão, J.J.M. et al. (2012). *Catal. Today* 180 (1): 148–154.
46 Carabineiro, S.A.C., Tavares, P.B., and Figueiredo, J.L. (2012). *Appl. Nanosci.* 2 (1): 35–46.
47 Soria, M.A., Pérez, P., Carabineiro, S.A.C. et al. (2014). *Appl. Catal. A* 470: 45–55.
48 Pérez, P., Soria, M.A., Carabineiro, S.A.C. et al. (2016). *Int. J. Hydrogen. Energ.* 41 (8): 4670–4681.
49 Carabineiro, S.A.C., Santos, V.P., Pereira, M.F.R. et al. (2016). *J. Coll. Interf. Sci.* 480: 17–29.
50 Haruta, M., Kobayashi, T., Sano, H., and Yamada, N. (1987). *Chem. Lett.* 16 (2): 405–408.

51 Ilieva, L., Petrova, P., Pantaleo, G. et al. (2016). *Appl. Catal. B* 188: 154–168.
52 Seker, E. and Gulari, E. (2002). *Appl. Catal. A* 232 (1): 203–217.
53 Jin, Z., Song, Y.Y., Fu, X.P. et al. (2018). *Chin. J. Chem.* 36 (7): 639–643.
54 Gluhoi, A.C. and Nieuwenhuys, B.E. (2007). *Catal. Today* 122 (3–4): 226–232.
55 Zanella, R. and Louis, C. (2005). *Catal. Today* 107-08: 768–777.
56 Pakrieva, E., Ribeiro, A.P.C., Kolobova, E. et al. (2020). *Nanomater* 10: 151. (23 pages).
57 Kolobova, E., Mäki-Arvela, P., Grigoreva, A. et al. (2021). *Catal. Today* 367: 95–110.
58 Lin, J.-N. and Wan, B.-Z. (2003). *Appl. Catal. B: Environ.* 41 (1–2): 83–95.
59 Sunagawa, Y., Yamamoto, K., Takahashi, H., and Muramatsu, A. (2008). *Catal. Today* 132 (1): 81–87.
60 Carabineiro, S.A.C., Machado, B.F., Bacsa, R.R. et al. (2010). *J. Catal.* 273 (2): 191–198.
61 Santos, V.P., Carabineiro, S.A.C., Bakker, J.J.W. et al. (2014). *J. Catal.* 309: 58–65.
62 Chen, J.H., Lin, J.N., Kang, Y.M. et al. (2005). *Appl. Catal. A* 291 (1–2): 162–169.
63 Chen, B.B., Zhu, X.B., Wang, Y.D. et al. (2017). *Catal. Today* 281: 512–519.
64 Ivanova, S.I., Petit, C., and Pitchon, V. (2006). *Gold Bull.* 39 (1): 3–8.
65 Liao, X.M., Chu, W., Dai, X.Y., and Pitchon, V. (2012). *Appl. Catal. A* 449: 131–138.
66 Carabineiro, S.A.C., Machado, B.F., Drazic, G. et al. (2010). Photodeposition of Au and Pt on ZnO and TiO_2. In: *Scientific Bases for the Preparation of Heterogeneous Catalysts: Proceedings of the 10th International Symposium, Studies in Surface Science and Catalysis*, 175 (ed. E.M. Gaigneaux, M. Devillers, S. Hermans et al.), 629–633.
67 Yang, Y.F., Sangeetha, P., and Chen, Y.W. (2009). *Int. J. Hydrogen. Energ.* 34 (21): 8912–8920.
68 Kydd, R., Scott, J., Teoh, W.Y. et al. (2010). *Langmuir* 26 (3): 2099–2106.
69 Naknam, P., Luengnaruemitchai, A., and Wongkasemjit, S. (2009). *Int. J. Hydrogen. Energ.* 34 (24): 9838–9846.
70 Brady, J.T., Jones, M.E., Brey, L.A. et al. (2006). Heterogeneous, composite carbonaceous catalyst system and methods that use catalytically active gold, 3M Innovative Properties Co, WO Patent 2006/074126.
71 Smirnov, M.Y., Vovk, E.I., Kalinkin, A.V., and Bukhtiyarov, V.I. (2016). *Kinet. Catal.* 57 (6): 831–839.
72 Kozlov, A.I., Kozlova, A.P., Liu, H.C., and Iwasawa, Y. (1999). *Appl. Catal. A* 182 (1): 9–28.
73 Okumura, M., Tsubota, S., and Haruta, M. (2003). *J. Mol. Catal. A Chem.* 199 (1): 73–84.
74 Zhao, J.B. and Jin, R.C. (2018). *Nano Today* 18: 86–102.
75 Freakley, S.J., He, Q., Kiely, C.J., and Hutchings, G.J. (2015). *Catal. Lett.* 145 (1): 71–79.
76 Boccuzzi, F., Chiorino, A., Manzoli, M. et al. (2001). *J. Catal.* 202 (2): 256–267.
77 Ayastuy, J.L., Gurbani, A., and Gutierrez-Ortiz, M.A. (2016). *Int. J. Hydrogen. Energ.* 41 (43): 19546–19555.
78 Ayastuy, J.L., Iriarte-Velasco, U., Gurbani, A., and Gutierrez-Ortiz, M.A. (2017). *J. Taiwan Inst. Chem. Eng.* 75: 18–28.
79 Bond, G.C. and Thompson, D.T. (2000). *Gold Bull.* 33 (2): 41–50.
80 Wei, J.Y., Fan, G.F., Jiang, F. et al. (2010). *Chin. J. Catal.* 31 (12): 1489–1495.
81 Louis, C. (2008). Gold nanoparticles: recent advances in CO oxidation. In: *Nanoparticles and Catalysis* (ed. D. Astruc). Weinheim: Wiley-VCH Verlag GmbH & Co. KGaA.
82 Grisel, R., Weststrate, K.J., Gluhoi, A., and Nieuwenhuys, B.E. (2002). *Gold Bull.* 35 (2): 39–45.
83 Haruta, M. and Daté, M. (2001). *Appl. Catal. A* 222 (1–2): 427–437.
84 Haruta, M. (2002). *Cattech* 6 (3): 102–115.
85 Haruta, M. (2004). *Gold Bull.* 37 (1–2): 27–36.
86 Haruta, M. (2005). *Nature* 437 (7062): 1098–1099.

87 Hashmi, A.S.K. and Hutchings, G.J. (2006). *Angew. Chem. Int. Ed.* 45 (47): 7896–7936.
88 Gavril, D., Georgaka, A., Loukopoulos, V. et al. (2006). *Gold Bull.* 39 (4): 192–199.
89 Takei, T., Akita, T., Nakamura, I. et al. (2012). Heterogeneous catalysis by gold. In: *Advances in Catalysis*, 55 (eds. B.C. Gates and F.C. Jentoft), 1–126. San Diego: Elsevier Academic Press Inc. Advances in Catalysis. 55.
90 Janssens, T.V.W., Clausen, B.S., Hvolbæk, B. et al. (2007). *Top Catal.* 44 (1–2): 15–26.
91 Bond, G. (2010). *Gold Bull.* 43 (2): 88–93.
92 Mellor, J.R., Palazov, A.N., Grigorova, B.S. et al. (2002). *Catal. Today* 72 (1): 145–156.
93 Carabineiro, S.A.C. and Nieuwenhuys, B.E. (2009). *Gold Bull.* 42 (4): 288–301.
94 Carabineiro, S.A.C. and Nieuwenhuys, B.E. (2010). *Gold Bull.* 43 (4): 252–266.
95 Iizuka, Y., Tode, T., Takao, T. et al. (1999). *J. Catal.* 187 (1): 50–58.
96 Quinet, E., Piccolo, L., Daly, H. et al. (2008). *Catal. Today* 138 (1–2): 43–49.
97 Duan, H.M. and Xu, C.X. (2015). *J. Catal.* 332: 31–37.
98 Della Pina, C. and Falletta, E. (2011). *Catal. Sci. Technol.* 1 (9): 1564–1571.
99 Schubert, M.M., Hackenberg, S., van Veen, A.C. et al. (2001). *J. Catal.* 197 (1): 113–122.
100 Carabineiro, S.A.C., Silva, A.M.T., Dražić, G. et al. (2012). CO oxidation using gold supported on Ce-Mn-O composite materials. Carbon Monoxide: Sources, Uses and Hazards, 61–84.
101 Lakshmanan, P., Park, J.E., Kim, B., and Park, E.D. (2016). *Catal. Today* 265: 19–26.
102 Ishida, T., Koga, H., Okumura, M., and Haruta, M. (2016). *Chem. Rec.* 16 (5): 2278–2293.
103 Carrettin, S., McMorn, P., Johnston, P. et al. (2004). *Top Catal.* 27 (1): 131–136.
104 Biella, S., Castiglioni, G.L., Fumagalli, C. et al. (2002). *Catal. Today* 72 (1–2): 43–49.
105 Casanova, O., Iborra, S., and Corma, A. (2009). *ChemSusChem.* 2 (12): 1138–1144.
106 Zwijnenburg, A., Saleh, M., Makkee, M., and Moulijn, J.A. (2002). *Catal. Today* 72 (1–2): 59–62.
107 Haruta, M., Uphade, B.S., Tsubota, S., and Miyamoto, A. (1998). *Res. Chem. Intermed.* 24 (3): 329–336.
108 Das, S.K., Reis, A., and Berry, K.J. (2009). *J. Power Sourc.* 193 (2): 691–698.
109 Yan, W.-M., Chu, H.-S., Lu, M.-X. et al. (2009). *J. Power Sourc.* 188 (1): 141–147.
110 Daté, M. and Haruta, M. (2001). *J. Catal.* 201 (2): 221–224.
111 Wang, D.H., Dong, T.X., Shi, X.C., and Zhang, Z.L. (2007). *Chin. J. Catal.* 28 (7): 657–661.
112 Fujitani, T., Nakamura, I., and Haruta, M. (2014). *Catal. Lett.* 144 (9): 1475–1486.
113 Tran-Thuy, T.M., Chen, C.C., and Lin, S.D. (2017). *ACS Catal.* 7 (7): 4304–4312.
114 Lakshmanan, P., Park, J.E., and Park, E.D. (2014). *Catal. Surv. Asia* 18 (2–3): 75–88.
115 Grisel, R.J.H. and Nieuwenhuys, B.E. (2001). *Catal. Today* 64 (1): 69–81.
116 Xu, Q., Kharas, K.C.C., and Datye, A.K. (2003). *Catal. Lett.* 85 (3–4): 229–235.
117 Thielecke, N., Vorlop, K.-D., and Prüße, U. (2007). *Catal. Today* 122 (3): 266–269.
118 Wang, Y.H., Zhu, J.L., Zhang, J.C. et al. (2006). *J. Power Sourc.* 155 (2): 440–446.
119 Miyake, Y. and Tsuji, S. (2000). Catalyst for purifying an exhaust gas, Toyota JJK Patent, European Patent Application, EP1043059 A12000.
120 Barrett, D.H., Scurrell, M.S., Rodella, C.B. et al. (2016). *Chem. Sci.* 7 (11): 6815–6823.
121 Ke, J.H., Zhao, Y.X., Yin, Y. et al. (2017). *J. Rare Earths* 35 (11): 1083–1091.
122 Ma, Z. and Dai, S. (2014). Stabilizing gold nanoparticles by solid supports, Chapter 1. In: *Heterogeneous Gold Catalysts and Catalysis: The Royal Society of Chemistry* (ed. Z. Ma and S. Dai), 1–26.
123 Liu, B., Jiang, T., Zheng, H.Q. et al. (2017). *Nanoscale* 9 (19): 6380–6390.
124 Yoshida, T., Murayama, T., Sakaguchi, N. et al. (2018). *Angew. Chem. Int. Ed.* 57 (6): 1523–1527.
125 Vigneron, F. and Caps, V. (2016). *C R Chim.* 19 (1): 192–198.
126 Rodriguez, J.A., Liu, G., Jirsak, T. et al. (2002). *J. Am. Chem. Soc.* 124 (18): 5242–5250.
127 Ruth, K., Hayes, M., Burch, R. et al. (2000). *Appl. Catal. B* 24 (3–4): L133–L8.

128 Mohapatra, P., Moma, J., Parida, K.M. et al. (2007). *Chem. Commun.* 16 (10): 1044–1046.
129 Kuperman, A. and Moir, M. (2006). Method for making hydrogen using a gold containing water-gas shift catalyst. WO Patent 2005/0050322005.
130 Corma, A. and Garcia, H. (2008). *Chem. Soc. Rev.* 37 (9): 2096–2126.
131 Hashmi, A.S.K. and Rudolph, M. (2008). *Chem. Soc. Rev.* 37 (9): 1766–1775.
132 Rudolph, M. and Hashmi, A.S.K. (2012). *Chem. Soc. Rev.* 41 (6): 2448–2462.
133 Liu, X.Y., Wang, A.Q., Zhang, T., and Mou, C.Y. (2013). *Nano Today* 8 (4): 403–416.
134 Pflasterer, D. and Hashmi, A.S.K. (2016). *Chem. Soc. Rev.* 45 (5): 1331–1367.
135 Haruta, M. (2011). *Faraday Disc.* 152 (0): 11–32.
136 Bond, G. and Thompson, D. (2009). *Gold Bull.* 42 (4): 247–259.
137 Widmann, D. and Behm, R.J. (2014). *Acc. Chem. Res.* 47 (3): 740–749.
138 Saavedra, J., Pursell, C.J., and Chandler, B.D. (2018). *J. Am. Chem. Soc.* 140 (10): 3712–3723.
139 Haruta, M., Tsubota, S., Kobayashi, T. et al. (1993). *J. Catal.* 144 (1): 175–192.
140 Widmann, D., Krautsieder, A., Walter, P. et al. (2016). *ACS Catal.* 6 (8): 5005–5011.
141 Sandoval, A., Zanella, R., and Klimova, T.E. (2017). *Catal. Today* 282: 140–150.
142 Wei, S., Fu, X.P., Wang, W.W. et al. (2018). *J. Phys. Chem. C* 122 (9): 4928–4936.
143 Schlexer, P., Widmann, D., Behm, R.J., and Pacchioni, G. (2018). *ACS Catal.* 8 (7): 6513–6525.
144 Widmann, D. and Behm, R.J. (2018). *J. Catal.* 357: 263–273.
145 Zhang-Steenwinkel, Y., van der Zande, L.M., Castricum, H.L., and Bliek, A. (2004). *Appl. Catal. B-Environ.* 54 (2): 93–103.
146 Okumura, M., Coronado, J.M., Soria, J. et al. (2001). *J. Catal.* 203 (1): 168–174.
147 Grisel, R.J.H., Weststrate, C.J., Goossens, A. et al. (2002). *Catal. Today* 72 (1): 123–132.
148 Grisel, R.J.H. and Nieuwenhuys, B.E. (2001). *J. Catal.* 199 (1): 48–59.
149 Date, M., Ichihashi, Y., Yamashita, T. et al. (2002). *Catal. Today* 72 (1–2): 89–94.
150 Wu, K.C., Tung, Y.L., Chen, Y.L., and Chen, Y.W. (2004). *Appl. Catal. B* 53 (2): 111–116.
151 Mansley, Z.R., Paull, R.J., Savereide, L. et al. (2021). *ACS Catal.* 11 (19): 11921–11928.
152 Jozwiak, W.K., Kaczmarek, E., and Ignaczak, W. (2008). *Pol. J. Chem.* 82 (1–2): 213–222.
153 Zhu, H., Ma, Z., Clark, J.C. et al. (2007). *Appl. Catal. A* 326 (1): 89–99.
154 Liu, Y., Dai, H., Deng, J. et al. (2014). *J. Catal.* 309: 408–418.
155 Qiao, B., Lin, J., Wang, A. et al. (2015). *Chin. J. Catal.* 36 (9): 1505–1511.
156 Hao, Y., Mihaylov, M., Ivanova, E. et al. (2009). *J. Catal.* 261 (2): 137–149.
157 Murayama, T. and Haruta, M. (2016). *Chin. J. Catal.* 37 (10): 1694–1701.
158 Sreethawong, T., Sitthiwechvijit, N., Rattanachatchai, A. et al. (2011). *Mater. Chem. Phys.* 126 (1–2): 212–219.
159 Wu, H., Pantaleo, G., Venezia, A.M., and Liotta, L.F. (2013). *Catalysts* 3 (4): 774–793.
160 Moreno-Martell, A., Pawelec, B., Nava, R. et al. (2018). *Materials* 11 (6).
161 Guzman, J. and Corma, A. (2005). *Chem. Commun.* (6): 743–745.
162 Liu, J., Qiao, B., Song, Y. et al. (2016). *J. Energ. Chem.* 25 (3): 361–370.
163 Liu, M.-H., Chen, Y.-W., Lin, T.-S., and Mou, C.-Y. (2018). *ACS Catal.* 8 (8): 6862–6869.
164 Guzman, J., Carrettin, S., Fierro-Gonzalez, J.C. et al. (2005). *Angew. Chem. Int. Ed.* 44 (30): 4778–4781.
165 Solsona, B., Hutchings, G.J., Garcia, T., and Taylor, S.H. (2004). *New. J. Chem.* 28 (6): 708–711.
166 Del Rio, E., Hungria, A.B., Tinoco, M. et al. (2016). *Appl. Catal. B* 197: 86–94.
167 Tian, C.C., Zhu, X., Abney, C.W. et al. (2017). *ACS Catal.* 7 (5): 3388–3393.
168 Mokoena, L., Pattrick, G., and Scurrell, M.S. (2016). *Gold Bull.* 49 (1–2): 35–44.
169 Dekkers, M.A.P., Lippits, M.J., and Nieuwenhuys, B.E. (1999). *Catal. Today* 54 (4): 381–390.
170 Gluhoi, A.C., Dekkers, M.A.P., and Nieuwenhuys, B.E. (2003). *J. Catal.* 219 (1): 197–205.
171 Gluhoi, A.C., Lin, S.D., and Nieuwenhuys, B.E. (2004). *Catal. Today* 90 (3–4): 175–181.

172 Szabo, E.G., Tompos, A., Hegedus, M. et al. (2007). *Appl. Catal. A* 320: 114–121.
173 Zhang, H.L., Ren, L.H., Lu, A.H., and Li, W.C. (2012). *Chin. J. Catal.* 33 (7): 1125–1132.
174 Guzman, J. and Gates, B.C. (2004). *J. Am. Chem. Soc.* 126 (9): 2672–2673.
175 Bamwenda, G.R., Tsubota, S., Nakamura, T., and Haruta, M. (1997). *Catal. Lett.* 44 (1–2): 83–87.
176 Abd El-Moemen, A., Abdel-Mageed, A.M., Bansmann, J. et al. (2016). *J. Catal.* 341: 160–179.
177 Kaneti, Y.V., Tanaka, S., Jikihara, Y. et al. (2018). *Chem. Commun.* 54 (61): 8514–8517.
178 Bion, N., Epron, F., Moreno, M. et al. (2008). *Top Catal.* 51 (1–4): 76–88.
179 Park, E.D., Lee, D., and Lee, H.C. (2009). *Catal. Today* 139 (4): 280–290.
180 Dulnee, S., Luengnaruemitchai, A., and Wanchanthuek, R. (2014). *Int. J. Hydrogen. Energ.* 39 (12): 6443–6453.
181 Luengnaruemitchai, A., Srihamat, K., Pojanavaraphan, C., and Wanchanthuek, R. (2015). *Int. J. Hydrogen. Energ.* 40 (39): 13443–13455.
182 Reina, T.R., Ivanova, S., Laguna, O.H. et al. (2016). *Appl. Catal. B* 197: 62–72.
183 Carltonbird, M., Eaimsumang, S., Pongstabodee, S. et al. (2018). *Chem. Eng. J.* 344: 545–555.
184 Pereira, J.M.D., Ciotti, L., Vaz, J.M., and Spinace, E.V. (2018). *Mater. Res. Iber Am. J. Mater.* 21 (2).
185 Lakshmanan, P. and Park, E.D. (2018). *Catalysts* 8 (5).
186 Avgouropoulos, G., Ioannides, T., Papadopoulou, C. et al. (2002). *Catal. Today* 75 (1–4): 157–167.
187 Choudhary, T.V. and Goodman, D.W. (2002). *Catal. Today* 77 (1–2): 65–78.
188 Luengnaruemitchai, A., Osuwan, S., and Gulari, E. (2004). *Int. J. Hydrogen. Energ.* 29 (4): 429–435.
189 Kandoi, S., Gokhale, A.A., Grabow, L.C. et al. (2004). *Catal. Lett.* 93 (1–2): 93–100.
190 Manzoli, M., Chiorino, A., and Boccuzzi, F. (2004). *Appl. Catal. B* 52 (4): 259–266.
191 Farrauto, R.J. (2005). *Appl. Catal. B* 56 (1–2): 3–7.
192 Luengnaruemitchai, A., Thoa, D.T.K., Osuwan, S., and Gulari, E. (2005). *Int. J. Hydrogen. Energ.* 30 (9): 981–987.
193 Laguna, O.H., Hernandez, W.Y., Arzamendi, G. et al. (2014). *Fuel* 118: 176–185.
194 Arregi, A., Amutio, M., Lopez, G. et al. (2018). *Energ. Conv. Manag.* 165: 696–719.
195 Luo, M., Yi, Y., Wang, S.Z. et al. (2018). *Renew. Sust. Energ. Rev.* 81: 3186–3214.
196 Zhang, J., Wang, Y., Chen, B. et al. (2003). *Energ. Conv. Manag.* 44 (11): 1805–1815.
197 Torres Sanchez, R.M., Ueda, A., Tanaka, K., and Haruta, M. (1997). *J. Catal.* 168 (1): 125–127.
198 Kahlich, M.J., Gasteiger, H.A., and Behm, R.J. (1999). *J. Catal.* 182 (2): 430–440.
199 Bethke, G.K. and Kung, H.H. (2000). *Appl. Catal. A* 194–195: 43–53.
200 Qwabe, L.Q., Dasireddy, V., Singh, S., and Friedrich, H.B. (2016). *Int. J. Hydrogen. Energ.* 41 (4): 2144–2153.
201 Rajalakshmi, N., Jayanth, T.T., and Dhathathreyan, K.S. (2003). *Fuel Cells* 3 (4): 177–180.
202 Reina, T.R., Ivanova, S., Centeno, M.A., and Odriozola, J.A. (2015). *Int. J. Hydrogen. Energ.* 40 (4): 1782–1788.
203 Wang, L.-C., Huang, X.-S., Liu, Q. et al. (2008). *J. Catal.* 259 (1): 66–74.
204 Landon, P., Ferguson, J., Solsona, B.E. et al. (2005). *Chem. Commun.* (27): 3385–3387.
205 Landon, P., Ferguson, J., Solsona, B.E. et al. (2006). *J. Mater. Chem.* 16 (2): 199–208.
206 Imai, H., Date, M., and Tsubota, S. (2008). *Catal. Lett.* 124 (1–2): 68–73.
207 Zhang, Q.J., Qi, P., and Zhou, Y.C. (2008). *Chin. J. Catal.* 29 (4): 361–365.
208 Ilieva, L., Pantaleo, G., Ivanov, I. et al. (2010). *Catal. Today* 158 (1–2): 44–55.
209 Scirè, S., Crisafulli, C., Minicò, S. et al. (2008). *J. Mol. Catal. A Chem.* 284 (1): 24–32.
210 Qiao, B., Zhang, J., Liu, L., and Deng, Y. (2008). *Appl. Catal. A* 340 (2): 220–228.
211 Rossignol, C., Arrii, S., Morfin, F. et al. (2005). *J. Catal.* 230 (2): 476–483.
212 Ivanova, S., Pitchon, V., Petit, C., and Caps, V. (2010). *ChemCatChem.* 2 (5): 556–563.

213 Chen, Y.W., Chen, H.J., and Lee, D.S. (2012). *J. Mol. Catal. A Chem.* 363: 470–480.
214 Ko, E.Y., Park, E.D., Seo, K.W. et al. (2006). *Catal. Today* 116 (3): 377–383.
215 Scire, S., Riccobene, P.M., and Crisafulli, C. (2010). *Appl. Catal. B* 101 (1–2): 109–117.
216 Scire, S., Crisafulli, C., Riccobene, P.M. et al. (2012). *Appl. Catal. A* 417: 66–75.
217 Laveille, P., Guillois, K., Tuel, A. et al. (2016). Durable PROX catalyst based on gold nanoparticles and hydrophobic silica.
218 Tabakova, T., Idakiev, V., Tenchev, K. et al. (2006). *Appl. Catal. B* 63 (1–2): 94–103.
219 Miao, Y.X., Li, W.C., Sun, Q. et al. (2015). *Chem. Commun.* 51 (100): 17728–17731.
220 Margitfalvi, J.L., Hegedűs, M., Szegedi, Á., and Sajó, I. (2004). *Appl. Catal. A* 272 (1): 87–97.
221 Tompos, A., Margitfalvi, J.L., Szabo, E.G. et al. (2009). *J. Catal.* 266 (2): 207–217.
222 Laguna, O.H., Centeno, M.A., Arzamendi, G. et al. (2010). *Catal. Today* 157 (1–4): 155–159.
223 Luengnaruemitchai, A., Chawla, S., and Wanchanthuek, R. (2014). *Int. J. Hydrogen. Energ.* 39 (30): 16953–16963.
224 Li, S.N., Zhang, Y.G., Li, X.J. et al. (2018). *Catal. Lett.* 148 (1): 328–340.
225 Manzoli, M., Avgouropoulos, G., Tabakova, T. et al. (2008). *Catal. Today* 138 (3–4): 239–243.
226 Laguna, O.H., Sarria, F.R., Centeno, M.A., and Odriozola, J.A. (2010). *J. Catal.* 276 (2): 360–370.
227 Morfin, F., Ait-Chaou, A., Lomello, M., and Rousset, J.L. (2015). *J. Catal.* 331: 210–216.
228 Liu, Y., Liu, B., Liu, Y. et al. (2013). *Appl. Catal. B* 142-143: 615–625.
229 Ratnasamy, C. and Wagner, J.P. (2009). *Catal. Rev. Sci. Eng.* 51 (3): 325–440.
230 Gradisher, L., Dutcher, B., and Fan, M. (2015). *Appl. Ener.* 139: 335–349.
231 Pal, D.B., Chand, R., Upadhyay, S.N., and Mishra, P.K. (2018). *Renew. Sust. Energ. Rev.* 93: 549–565.
232 Haryanto, A., Fernando, S., Murali, N., and Adhikari, S. (2005). *Energy Fuels* 19 (5): 2098–2106.
233 Shen, J., Wang, Z.-Z., Yang, H.-W., and Yao, R.-S. (2007). *Energy Fuels* 21 (6): 3588–3592.
234 Galindo Cifre, P. and Badr, O. (2007). *Energy Conver. Manag.* 48 (2): 519–527.
235 Vojvodic, A., Medford, A.J., Studt, F. et al. (2014). *Chem. Phys. Lett.* 598 (0): 108–112.
236 Cameron, D., Holliday, R., and Thompson, D. (2003). *J. Power Sourc.* 118 (1): 298–303.
237 Abbas, H.F. and Wan Daud, W.M.A. (2010). *Int. J. Hydrogen Energ.* 35 (3): 1160–1190.
238 Lee, J.Y., Lee, D.-W., Lee, K.-Y., and Wang, Y. (2009). *Catal. Today* 146 (1–2): 260–264.
239 Maroño, M., Ruiz, E., Sánchez, J.M. et al. (2009). *Int. J. Hydrogen Energ.* 34 (21): 8921–8928.
240 Adams Ii, T.A. and Barton, P.I. (2009). *Int. J. Hydrogen Energ.* 34 (21): 8877–8891.
241 Gawade, P., Mirkelamoglu, B., and Ozkan, U.S. (2010). *J. Phys. Chem. C* 114 (42): 18173–18181.
242 Mendes, D., Mendes, A., Madeira, L.M. et al. (2010). *Asia-Pacific J. Chem. Eng.* 5 (1): 111–137.
243 Liu, K., Song, C., and Subramani, V. (2010 January). *Hydrogen and Syngas Production and Purification Technologies*. Wiley, Alche.
244 Lenite, B.A., Galletti, C., and Specchia, S. (2011). *Int. J. Hydrogen. Energ.* 36 (13): 7750–7758.
245 Çağlayan, B.S. and Aksoylu, A.E. (2011). *Catal. Commun.* 12 (13): 1206–1211.
246 Andreeva, D., Ivanov, I., Ilieva, L., and Abrashev, M.V. (2006). *Appl. Catal. A* 302 (1): 127–132.
247 Andreeva, D., Idakiev, V., Tabakova, T., and Andreev, A. (1996). *J. Catal.* 158 (1): 354–355.
248 Sakurai, H., Ueda, A., Kobayashi, T., and Haruta, M. (1997). *Chem. Commun.* (3): 271–272.
249 Boccuzzi, F., Chiorino, A., Manzoli, M. et al. (1999). *J. Catal.* 188 (1): 176–185.
250 Tabakova, T., Idakiev, V., Andreeva, D., and Mitov, I. (2000). *Appl. Catal. A: Gen.* 202 (1): 91–97.
251 Andreeva, D. (2002). Low temperature water gas shift over gold catalysts. *Gold Bull.* 35 (3): 82–88.
252 Pattrick, G., van der Lingen, E., Corti, C.W. et al. (2004). *Top Catal.* 30-1 (1–4): 273–279.
253 Idakiev, V., Tabakova, T., Yuan, Z.Y., and Su, B.L. (2004). *Appl. Catal. A* 270 (1): 135–141.
254 Kim, C.H. and Thompson, L.T. (2005). *J. Catal.* 230 (1): 66–74.
255 Mendes, D., Garcia, H., Silva, V.B. et al. (2008). *Ind. Eng. Chem. Res.* 48 (1): 430–439.

256 Bond, G. (2009). *Gold Bull.* 42 (4): 337–342.
257 Zhang, Y., Zhan, Y., Chen, C. et al. (2012). *Int. J. Hydrogen Energ.* 37 (17): 12292–12300.
258 Barakat, T., Rooke, J.C., Genty, E. et al. (2013). *Energ. Environ. Sci.* 6 (2): 371–391.
259 Tao, F. and Ma, Z. (2013). *Phys. Chem. Chem. Phys.* 15 (37): 15260–15270.
260 Flytzani-Stephanopoulos, M. (2014). *Acc. Chem. Res.* 47 (3): 783–792.
261 Carter, J.H., Liu, X., He, Q. et al. (2017). *Angew. Chem. Int. Ed.* 56 (50): 16037–16041.
262 Abdel-Mageed, A.M., Kucerova, G., Bansmann, J., and Behm, R.J. (2017). *ACS Catal.* 7 (10): 6471–6484.
263 Fu, Q., Weber, A., and Flytzani-Stephanopoulos, M. (2001). *Catal. Lett.* 77 (1): 87–95.
264 Luengnaruemitchai, A., Osuwan, S., and Gulari, E. (2003). *Catal. Commun.* 4 (5): 215–221.
265 Burch, R. (2006). *Phys. Chem. Chem. Phys.* 8 (47): 5483–5500.
266 Deng, W., Frenkel, A.I., Si, R., and Flytzani-Stephanopoulos, M. (2008). *J. Phys. Chem. C* 112 (33): 12834–12840.
267 LeValley, T.L., Richard, A.R., and Fan, M. (2014). *Int. J. Hydrogen. Energ.* 39 (30): 16983–17000.
268 Shekhar, M., Lee, W.-S., Akatay, M.C. et al. (2022). *J. Catal.* 405: 475–488.
269 Andreeva, D., Idakiev, V., Tabakova, T. et al. (1996). *Appl. Catal. A* 134 (2): 275–283.
270 Fu, Q., Kudriavtseva, S., Saltsburg, H., and Flytzani-Stephanopoulos, M. (2003). *Chem. Eng. J.* 93 (1): 41–53.
271 Fu, Q., Deng, W., Saltsburg, H., and Flytzani-Stephanopoulos, M. (2005). *Appl. Catal. B* 56 (1): 57–68.
272 Vindigni, F., Manzoli, M., Tabakova, T. et al. (2012). *Appl. Catal. B* 125: 507–515.
273 Reina, T.R., Ivanova, S., Centeno, M.A., and Odriozola, J.A. (2015). *Catal. Today* 253: 149–154.
274 Hua, J., Zheng, Q., Zheng, Y. et al. (2005). *Catal. Lett.* 102 (1): 99–108.
275 Ma, Z., Yin, H., and Dai, S. (2010). *Catal. Lett.* 136 (1): 83–91.
276 Dong, J., Fu, Q., Jiang, Z. et al. (2018). *J. Am. Chem. Soc.* 140 (42): 13808–13816.
277 Carter, J.H. and Hutchings, G.J. (2018). *Catalysts* 8 (12): 627.
278 Daly, H., Goguet, A., Hardacre, C. et al. (2010). *J. Catal.* 273 (2): 257–265.
279 Venugopal, A., Aluha, J., Mogano, D., and Scurrell, M.S. (2003). *Appl. Catal. A* 245 (1): 149–158.
280 Venugopal, A., Aluha, J., and Scurrell, M.S. (2003). The Au-M (M=Ag, Bi, Co, Cu, Mn, Ni, Pb, Ru, Sn, Tl) on iron(III) oxide system. *Catal. Lett.* 90 (1): 1–6.
281 Magadzu, T., Yang, J.H., Henao, J.D. et al. (2017). *J. Phys. Chem. C* 121 (16): 8812–8823.
282 Hurtado-Juan, M.-A., Yeung, C.M.Y., and Tsang, S.C. (2008). *Catal. Commun.* 9 (7): 1551–1557.
283 Silberova, B.A.A., Makkee, M., and Moulijn, J.A. (2007). *Top Catal.* 44 (1): 209–221.
284 Goguet, A., Burch, R., Chen, Y. et al. (2007). *J. Phys. Chem. C* 111 (45): 16927–16933.
285 Scirè, S. and Liotta, L.F. (2012). *Appl. Catal. B* 125: 222–246.
286 Tichenor, B.A. and Palazzolo, M.A. (1987). *Environ. Prog.* 6 (3): 172–176.
287 Moretti, E.C. (2002 June). *CEP Magazine* 2002: 30–40.
288 Bastos, S.S.T., Orfao, J.J.M., Freitas, M.M.A. et al. (2009). *Appl. Catal. B* 93 (1–2): 30–37.
289 Chen, X., Carabineiro, S.A.C., Bastos, S.S.T. et al. (2013). *J Environ Chem Eng* 1 (4): 795–804.
290 Kamal, M.S., Razzak, S.A., and Hossain, M.M. (2016). *Atmos. Environ.* 140: 117–134.
291 Li, J., Liu, H., Deng, Y. et al. (2016). *Nanotechnol. Rev.* 147.
292 Xie, S.H., Liu, Y.X., Deng, J.G. et al. (2017). *J. Catal.* 352: 282–292.
293 Merino, N.A., Barbero, B.P., Ruiz, P., and Cadús, L.E. (2006). *J. Catal.* 240 (2): 245–257.
294 Delimaris, D. and Ioannides, T. (2008). *Appl. Catal. B: Environ.* 84 (1–2): 303–312.
295 Delimaris, D. and Ioannides, T. (2009). *Appl. Catal. B: Environ.* 89 (1–2): 295–302.
296 Li, W.B., Wang, J.X., and Gong, H. (2009). *Catal. Today* 148 (1–2): 81–87.
297 Santos, V.P., Bastos, S.S.T., Pereira, M.F.R. et al. (2010). *Catal. Today* 154 (3–4): 308–311.

298 Santos, V.P., Pereira, M.F.R., Órfão, J.J.M., and Figueiredo, J.L. (2010). *Appl. Catal. B: Environ.* 99 (1–2): 353–363.
299 Solsona, B., Garcia, T., Aylón, E. et al. (2011). *Chem. Eng. J.* 175 (1): 271–278.
300 Konsolakis, M., Carabineiro, S.A.C., Tavares, P.B., and Figueiredo, J.L. (2013). *J. Hazard. Mater.* 261: 512–521.
301 Chen, X., Carabineiro, S.A.C., Bastos, S.S.T. et al. (2014). *Appl. Catal. A* 472: 101–112.
302 Chen, X., Carabineiro, S.A.C., Tavares, P.B. et al. (2014). *J. Environ. Chem. Eng.* 2 (1): 344–355.
303 Carabineiro, S.A.C., Chen, X., Konsolakis, M. et al. (2015). *Catal. Today* 244: 161–171.
304 Carabineiro, S.A.C., Konsolakis, M., Marnellos, G.E.-N. et al. (2016). *Molecules* 21: 644.
305 Konsolakis, M., Carabineiro, S.A.C., Marnellos, G.E. et al. (2017). *J. Coll. Interf. Sci.* 496 (SupplementC): 141–149.
306 Konsolakis, M., Carabineiro, S.A.C., Marnellos, G.E. et al. (2017). *Inorg. Chim. Acta* 455 (Part 2): 473–482.
307 Luo, Y.J., Zheng, Y.B., Zuo, J.C. et al. (2018). *J. Hazard. Mater.* 349: 119–127.
308 Xue, T.S., Li, R.N., Gao, W.L. et al. (2018). *J. Nanosci. Nanotechnol.* 18 (5): 3381–3386.
309 Liotta, L.F. (2010). *Appl. Catal. B* 100 (3): 403–412.
310 Tidahy, H.L., Barakat, T., Cousin, R. et al. (2010). Titanium oxide nanotubes as supports of Au or Pd nano-sized catalysts for total oxidation of VOCs. In: Scientific Bases for the Preparation of Heterogeneous Catalysts: Proceedings of the 10th International Symposium. *Studies in Surface Science and Catalysis*, 175 (eds. E.M. Gaigneaux, M. Devillers, S. Hermans et al.), 743–746.
311 Gaalova, J., Topka, P., Kaluza, L., and Solcova, O. (2011). *Catal. Today* 175 (1): 231–237.
312 Morales-Torres, S., Carrasco-Marín, F., Pérez-Cadenas, A., and Maldonado-Hódar, F. (2015). *Catalysts* 5 (2): 774.
313 Topka, P., Delaigle, R., Kaluza, L., and Gaigneaux, E.M. (2015). *Catal. Today* 253: 172–177.
314 Chlala, D., Labaki, M., Giraudon, J.M. et al. (2016). *C R Chim.* 19 (4): 525–537.
315 Yang, H.G., Deng, J.G., Liu, Y.X. et al. (2016). *J. Mol. Catal. A Chem.* 414: 9–18.
316 Waters, R.D., Weimer, J.J., and Smith, J.E. (1994). *Catal. Lett.* 30 (1): 181–188.
317 Haruta, M., Ueda, A., Tsubota, S., and Torres Sanchez, R.M. (1996). *Catal. Today* 29 (1–4): 443–447.
318 Chen, B., Bai, C., Cook, R. et al. (1996). *Catal. Today* 30 (1–3): 15–20.
319 Minico, S., Scire, S., Crisafulli, C. et al. (2000). *Appl. Catal. B* 28 (3–4): 245–251.
320 Minico, S., Scire, S., Crisafulli, C., and Galvagno, S. (2001). *Appl. Catal. B* 34 (4): 277–285.
321 Huang, H.B., Xu, Y., Feng, Q.Y., and Leung, D.Y.C. (2015). *Catal. Sci. Technol.* 5 (5): 2649–2669.
322 Gaalova, J. and Topka, P. (2021). *Catalysts* 11 (7).
323 Solsona, B.E., Garcia, T., Jones, C. et al. (2006). *Appl. Catal. A* 312: 67–76.
324 Centeno, M.A., Paulis, M., Montes, M., and Odriozola, J.A. (2002). *Appl. Catal. A* 234 (1–2): 65–78.
325 Liu, S.Y. and Yang, S.M. (2008). *Appl. Catal. A* 334 (1–2): 92–99.
326 Delannoy, L., Fajerwerg, K., Lakshmanan, P. et al. (2010). *Appl. Catal. B* 94 (1–2): 117–124.
327 Ousmane, M., Liotta, L.F., Di Carlo, G. et al. (2011). *Appl. Catal. B* 101 (3–4): 629–637.
328 Ousmane, M., Liotta, L.F., Pantaleo, G. et al. (2011). *Catal. Today* 176 (1): 7–13.
329 Xie, S.H., Deng, J.G., Liu, Y.X. et al. (2015). *Appl. Catal. A* 507: 82–90.
330 Tang, Z., Zhang, W.D., Li, Y. et al. (2016). *Appl. Surf. Sci.* 364: 75–80.
331 Chen, J., Yan, D.X., Xu, Z. et al. (2018). *Environ. Sci. Technol.* 52 (8): 4728–4737.
332 Scirè, S., Minicò, S., Crisafulli, C. et al. (2003). *Appl. Catal. B* 40 (1): 43–49.
333 Yang, S.M., Liu, D.M., and Liu, S.Y. (2008). *Top Catal.* 47 (3–4): 101–108.
334 Li, J.W. and Li, W.B. (2010). *J. Rare Earths* 28 (4): 547–551.

335 Lakshmanan, P., Delannoy, L., Richard, V. et al. (2010). *Appl. Catal. B* 96 (1–2): 117–125.
336 Jiang, X., Hua, J.F., Deng, H., and Wu, Z.B. (2014). *J. Mol. Catal. A Chem.* 383: 188–193.
337 Nedyalkova, R., Ilieva, L., Bernard, M.C. et al. (2009). *Mater. Chem. Phys.* 116 (1): 214–218.
338 Ilieva, L., Petrova, P., Tabakova, T. et al. (2012). *React. Kinet. Mechan, Catal.* 105 (1): 23–37.
339 Idakiev, V., Dimitrov, D., Tabakova, T. et al. (2015). *Chin. J. Catal.* 36 (4): 579–587.
340 Fiorenza, R., Bellardita, M., D'Urso, L. et al. (2016). *Catalysts* 6 (8).
341 Manzoli, M., Vindigni, F., Tabakova, T. et al. (2017). *J. Mater. Chem. A* 5 (5): 2083–2094.
342 Ilieva, L., Venezia, A.M., Petrova, P. et al. (2018). *Catalysts* 8 (7).
343 Solsona, B., Garcia, T., Agouram, S. et al. (2011). *Appl. Catal. B* 101 (3–4): 388–396.
344 Solsona, B., Perez-Cabero, M., Vazquez, I. et al. (2012). *Chem. Eng. J.* 187: 391–400.
345 Li, X.W., Dai, H.X., Deng, J.G. et al. (2013). *Chem. Eng. J.* 228: 965–975.
346 Genty, E., Cousin, R., Capelle, S., and Siffert, S. (2013). *Catalysts* 3 (4): 966–977.
347 Duplancic, M., Tomasic, V., and Gomzi, Z. (2018). *Environ. Technol.* 39 (15): 2004–2016.
348 Fiorenza, R. (2020). *Catalysts* 10 (6).
349 Xia, Y.S., Xia, L., Liu, Y.X. et al. (2018). *J. Environ. Sci.* 64: 276–288.
350 Gluhoi, A.C., Bogdanchikova, N., and Nieuwenhuys, B.E. (2005). *J. Catal.* 229 (1): 154–162.

24

Platinum Complexes for Selective Oxidations in Water

Alessandro Scarso[1], Paolo Sgarbossa[2], Roberta Bertani[2], and Giorgio Strukul[1]

[1] Dipartimento di Scienze Molecolari e Nanosistemi, Università Ca' Foscari Venezia, via Torino 155, Mestre Venezia, Italy
[2] Dipartimento di Ingegneria Industriale, Università di Padova, via Marzolo 9, Padova, Italy

24.1 Hydrogen Peroxide and Its Activation

With a current annual production of about 7 million tons, hydrogen peroxide has always been considered the dream oxidant for a myriad of industrial applications, ranging from bleaching (pulp, paper, and textiles) to water purification, use in the mining industry, and chemical production. However, its successful use in large scale chemical processes is relatively recent and, in fact, it has been only in the past decade that the BASF-Dow HPPO process for the production of propylene oxide [1] and the Enichem-Sumitomo process for the production of caprolactam [2] entered into operation, both made possible thanks to the use of the hydrophobic zeolite TS-1 as the catalyst.

The idea of activating hydrogen peroxide with transition metal complexes in solution is a long story with many failures and only moderate success, complicated by several undesired H_2O_2 properties, that can be summarized as follows:

- It is sold as a water solution and produces water as the reduction product. This is the main reason to prefer hydrogen peroxide in oxidation reactions, but, when looking for possible catalysts, this property also excludes Lewis acidic high valent early transition metal centers such as Ti(IV), V(V), Cr(VI), and Mo(VI), otherwise known to be excellent epoxidation catalysts in the presence of alkylhydroperoxides as oxidants, because they deactivate with water.
- It is thermally unstable, thereby discouraging the use of reaction temperatures above 70–90 °C if necessary to increase reaction rates because of unacceptably high losses of oxidant due to decomposition.
- It easily undergoes radical decomposition, triggered by one-electron redox pairs. Again, this leads to loss of oxidant and the involvement of poorly selective radical oxidations. This obviously excludes for instance redox pairs such as Fe(II)/Fe(III), Cu(I)/Cu(II), Mn(II)/Mn(III), Co(II)/Co(III), and their complexes, but to some extent also species such as Ru, Os, Rh, Ir, and Ni as potential choices for H_2O_2 activation.

When approaching hydrogen peroxide activation, an important property that the candidate transition metal catalyst should possess is what we may call soft Lewis acidity. That means a moderate electron deficiency induced by the ligands present in the coordination sphere of the

Catalysis for a Sustainable Environment: Reactions, Processes and Applied Technologies Volume 2, First Edition. Edited by Armando J. L. Pombeiro, Manas Sutradhar, and Elisabete C. B. A. Alegria.
© 2024 John Wiley & Sons Ltd. Published 2024 by John Wiley & Sons Ltd.

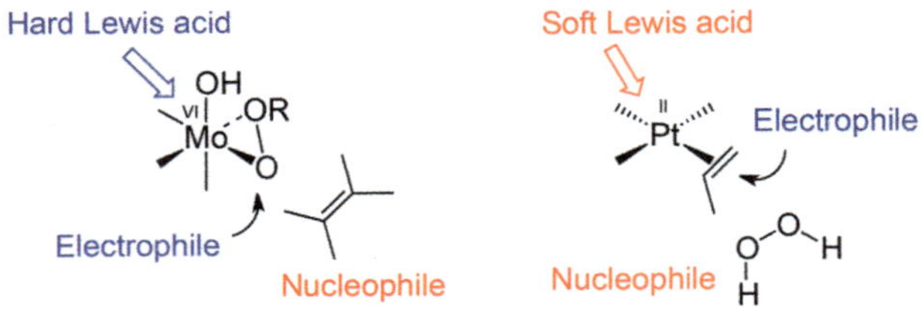

Figure 24.1 The concepts of electrophilic (left) and nucleophilic (right) oxidation promoted by hard and soft Lewis acids respectively.

metal and/or a positive charge. At the same time, the metal must be able to increase the electrophilicity of the substrate to be oxidized, generally a soft molecule (e.g. olefins, aromatics, organic sulfides) to exploit the moderate nucleophilic properties of hydrogen peroxide.

With respect to the well-known concept of electrophilic oxidation typical of Halcon chemistry with early transition metals in the epoxidation of olefins with alkylhydroperoxides, the features mentioned will reverse the roles of substrate and oxidant during the oxygen-transfer step so that we can speak of nucleophilic oxidation as shown in Figure 24.1.

In the former case, the hard Lewis acidic Mo(VI) center activates the alkylhydroperoxide, increasing its electrophilicity and making it susceptible to attack by the free electron-rich olefin (as in the Halcon process for the production of propylene epoxide). Systems of this type are intrinsically better suited for electron-rich substrates such as (highly) substituted olefins and allylic alcohols. In the latter case, a Lewis acidic but soft d^8 metal center [in this case Pt(II)] increases the electrophilicity of the olefin, making it susceptible to nucleophilic attack by hydrogen peroxide. Systems of this type will be more favorable to relatively electron-poor alkenes such as terminal ones and hardly oxidized by high valent transition metals. Metal centers with these characteristics are less oxophilic and more tolerant to water and hydrogen peroxide with respect to hard Lewis acids. Finally, because no monoelectron redox pairs are involved, these features will result in less oxidant decomposition.

24.2 Platinum Complexes

When we started studying epoxidation with hydrogen peroxide using cationic Pt(II) complexes as catalysts some forty years ago, these driving concepts were rather blurry in our minds, becoming clear only later. Everything started because one of us, who was interested in catalytic oxidation, came across some hydroxo complexes of Pt(II), stabilized by phosphines and a trifluoromethyl ligand, that had been just synthesized by our late colleague and friend Rino A. Michelin. These species allowed us to synthesize the first example of stable hydroperoxo and alkyperoxo complexes of Pt(II) and to accomplish the first example of catalytic epoxidation of terminal alkenes using hydrogen peroxide as an oxidant [3]. Catalytic studies made immediately clear that cationic complexes of the type $[(P-P)Pt(CF_3)(solv)]^+$ were much better catalysts than the corresponding hydroxo or hydroperoxo species, thus supporting the idea that reinforcing the Lewis acid character of the complex was essential.

The mechanism of the reaction (Figure 24.2) provided evidence for the twofold role of Pt, which both increases the electrophilicity of the substrate through coordination and activates the oxidant via hydrolysis [4]. It is also a clear experimental evidence for the concept of soft Lewis acidity as it has been defined previously.

In this reaction, we tested also the homologous Pd complexes, but they yielded only small amounts of methylketones as had been observed previously by Roussel and Mimoun [5] using

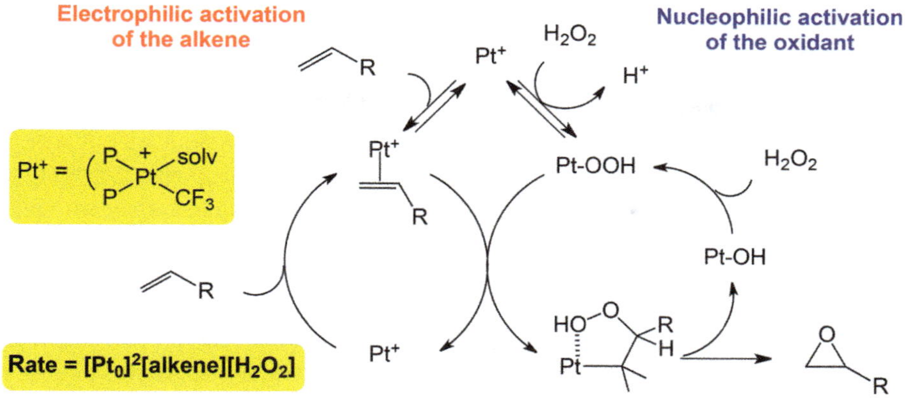

Figure 24.2 Mechanism of the epoxidation of terminal olefins with hydrogen peroxide catalyzed by cationic complexes of Pt(II).

Pd(OAc)$_2$, despite the similar structural features between Pd and Pt such as the atomic radii [6], the bond lengths in their homologous compounds [7], and the same coordination geometry. Similarly, some early attempts to extend this chemistry to other noble metals such as Rh or Ir were completely unsuccessful.

So, why only platinum? In coordination chemistry, platinum is normally considered as the lazy brother of palladium, but, in this specific case, some features in Pt chemistry that are normally considered as flaws for catalytic purposes turn out to be advantages. In fact:

- With respect to Pd, ligand lability is lower [8]; this means that Pt complexes, particularly those containing phosphines (easily oxidized by free hydrogen peroxide), are more stable under strong oxidizing conditions.
- Phosphine Pt^{II} complexes are inert to oxidative addition, at variance with N-containing ligands that promote the oxidative addition of hydrogen peroxide to yield *bis*-hydroxo Pt^{IV} complexes [9]: this reduces the waste of oxidant as well as the number of oxidant activation steps, favoring selectivity.
- It does not undergo 1-electron redox pathways, resulting in a better oxidant stability.
- It is slow in β-hydride elimination/addition, a landmark in Pd chemistry, thus avoiding the involvement of Wacker chemistry that would lead to ketones.

Together, these properties lead to reactions with no waste of oxidant, high product yields, >99% product selectivity, and mild experimental conditions. On the other hand, the use of phosphine ligands, while resulting in a better tuning of the steric and electronic properties of the catalyst, is also the main reason for catalyst decomposition in the long term.

The synthesis of the this class of catalysts was rather tedious and time consuming because the introduction of the strongly electron withdrawing –CF$_3$ ligand required several steps, among which the oxidative addition of CF$_3$Br on tetrakis phosphine Pt^0 species, a very slow and poorly selective reaction based on the use of a reactant that is no longer available on the market. This is why –CF$_3$ was successfully replaced with –C$_6$F$_5$ in the past 20 years. In this case, the starting material is commercially available bromo-pentafluorobenzene, which by lithiation can be used to substitute the chloride ligands on Pt$_{II}$ centres. The synthesis of the dinucelar precursor [Pt(C$_6$F$_5$)(μ-Cl)(tht)]$_2$ (tht = tetrahydrothiophene) as proposed by Uson et al. [10] can be accomplished in two steps: (i) the double pentafluorophenylation of [PtCl$_2$(tht)$_2$] to give the intermediate [Pt(C$_6$F$_5$)$_2$(tht)$_2$]; and (ii) its subsequent metathesis with one equivalent of PtCl$_2$ to afford the final product in high yield and purity. The main advantage of the dimeric Pt–C$_6$F$_5$ precursor is its good stability in air and the possibility to easily create two coordination sites on the metal centre by exchange of the labile tht and μ-Cl ligands. Addition of the appropriate diphosphine leads to the desired catalyst.

The epoxidation of terminal olefins was the first catalytic oxidation studied with the Pt/H$_2$O$_2$ system, but, on the basis of the aforementioned principles, other types of substrates could be tested, proving the generality of the nucleophilic oxidation and soft Lewis acidity concepts. In this respect, the Baeyer-Villiger (BV) oxidation of ketones to yield lactones and esters is an ideal case study. In fact, in organic synthesis this reaction is accomplished using organic peroxycarboxylic acids as oxidants under acid catalysis. Mechanistic studies [11] demonstrated that the reaction proceeds via protonation of the ketone carbonyl oxygen making the carbonyl carbon susceptible to nucleophilic attack by the peroxyacid to yield the so-called Criegee intermediate from which the ester is formed *via* a rearrangement.

The Pt/H$_2$O$_2$ system allowed us to achieve the first unequivocal example of transition metal catalysis of BV oxidation in the case of cyclic ketones to yield lactones [12]. As we observed, the original trifluoromethyl platinum complexes proved relatively slow in this reaction, albeit effective and selective, and could be conveniently replaced by a similar class of [(P–P)Pt(μ-OH)]$_2$X$_2$ (P–P = diphosphine; X = BF$_4$, ClO$_4$, triflate) complexes. These are the most efficient class of soluble catalysts for this reaction ever reported and can be synthesized by simple chloride abstraction with the desired silver salt (AgX with X = BF$_4$, triflate) on the corresponding diphosphino precursor complex [(P–P)PtCl$_2$] in wet dichloromethane at room temperature [13].

The mechanism proposed for the BV reaction is shown in Figure 24.3. As can be seen, the bridged hydroxy ligands present in the starting complex are split in the presence of the ketone and hydrogen peroxide to yield a monocationic species containing both the substrate and the activated oxidant. This increases the electrophilicity of the carbonyl carbon and, at the same time, the nucleophilicity of hydrogen peroxide generating a coordinated, formally OOH$^-$ ligand. The structure of the *quasi*-peroxymetallacycle involved as a reactive intermediate in the subsequent step closely resembles the Criegee intermediate involved in the organic reaction using organic peroxycarboxylic acids (Figure 24.3), where Pt$^+$ performs the same role as H$^+$.

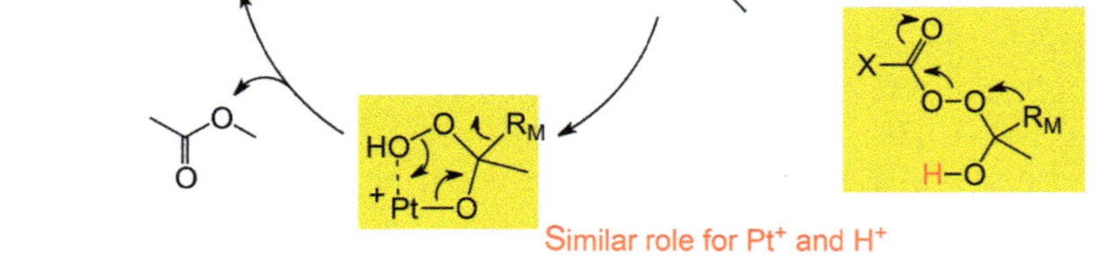

Figure 24.3 Mechanism of the Baeyer-Villiger oxidation of cyclic ketones with hydrogen peroxide catalyzed by cationic Pt(II) complexes. R$_M$ is the ketone migrating group.

But how does the acidity of the Pt center influence the activity in the epoxidation and BV oxidation? Is it really so important? We tried to answer these questions in epoxidation by using 2,6-dimethyl-phenylisocianide as a molecular probe. In fact, it is known that the value of the wavenumber shift Δv ($\Delta v = v(C\equiv N)_{coord} - v(C\equiv N)_{free}$) for the $C\equiv N$ stretching mode of 2,6-dimethyl-phenylisocyanide provides valuable information about the electrophilicity of the isocyanide carbon atom which in turn correlates well to the Lewis acidity of the metal complex to which it is coordinated [14].

On this basis, we prepared a series of homologous complexes of general formula $[(P-P)Pt(C_6F_5)(H_2O)](OTf)$ (P–P = **a-h** as in Figure 24.4 left) to which stoichiometric amounts of 2,6-dimethyl-phenylisocyanide were added. The observed Δv value for each complex was then correlated with the initial rate observed in the epoxidation of 1-octene with hydrogen peroxide catalyzed by the aquo complex [15]. The results are shown in Figure 24.4 (left) and resemble the volcano plots often observed in heterogeneous catalysis when plotting catalytic activity vs. the strength of

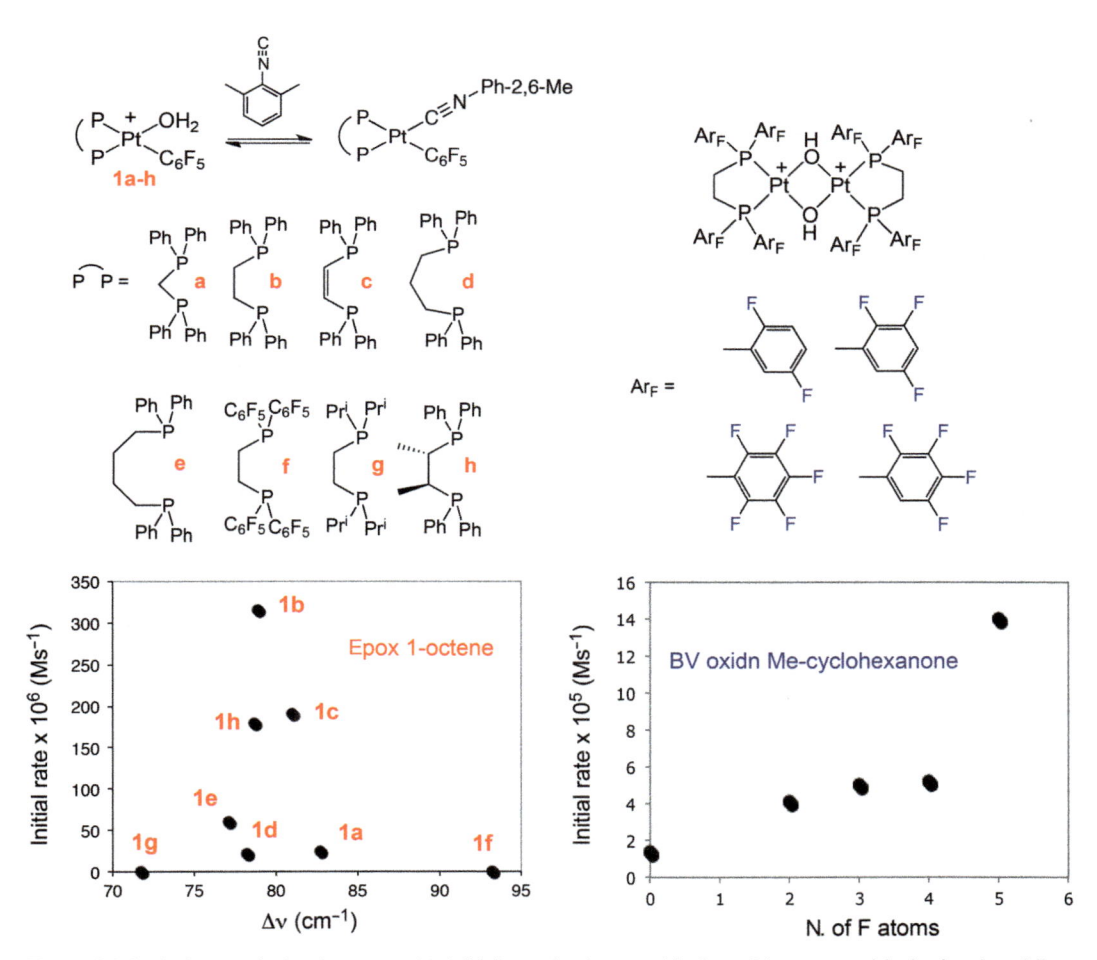

Figure 24.4 Left, correlation between the initial rate in the epoxidation of 1-octene with the Lewis acidity of the catalyst measured with 2,6-dimethyl-phenylisocyanide as a molecular probe. Right, correlation of the initial rate in the Baeyer-Villiger oxidation of 2-methyl-cyclohexanone with the number of fluorine atoms present in the phenyl rings of the diphosphine ligand.

chemisorption [16]. They seem to indicate that the Lewis acidity character of the complex must be of intermediate intensity to maximize activity, but, in fact, unreactive complexes such as those containing diphosphines **g** and **f** are respectively the least and the most acidic of the series. However, this evidence is not conclusive because the large differences in activity observed for complexes with similar acidity, such as those containing **b** and **d**, indicate that other factors play an important role. For instance, steric requirements such as ring size and bite angle have also a significant influence [15].

A clearer answer to the questions came from the BV oxidation of Me-cyclohexanone catalyzed by the bridging hydroxo complexes of Pt(II) shown in Figure 24.4 right [17]. These contain diphosphine ligands with a variable number of fluorine substituents in the phenyl rings and were prepared from 1,2-bis(dichlorophosphino)ethane by substitution with four equivalent of the corresponding lithium fluorophenyl derivative (from the brominated compounds treated with butyl lithium in anhydrous diethyl ether at low temperature) [18]. The dimeric µ-OH precatalyst is then obtained in a two-step procedure upon coordination of the diphosphine to Pt^{II} in $[PtCl_2(COD)]$ (COD = cyclooctadiene) followed by chloride abstraction with two equivalents of $AgBF_4$ in wet CH_2Cl_2 at room temperature.

An increasing number of electron-withdrawing fluorine atoms in the substituents implies a decrease in the donor capacity of the phosphines and consequently a metal center with a stronger Lewis acid character. A plot of the initial rate observed in the BV reaction vs. the number of F atoms present in each Ph substituent is shown in Figure 24.4 right. Here the differences in steric requirements among the different complexes are conceivably negligible, hence confirming the correlation between activity and acidity of the catalyst.

The significant Lewis acidity of this class of complexes was exploited to catalyze also other reactions such as the acetalization and thioacetalization of aldehydes and ketones [19], the Diels-Alder reaction [20] and the isomerization of simple olefins [21].

24.3 Enantioselective Oxidations

One of the main goals underlying these studies was to find catalysts able to perform oxidations with hydrogen peroxide in an enantioselective manner. Enantioselective oxidations are key synthetic steps in the synthesis of large selling pharmaceuticals such as esomeprazole, a chiral sulfoxide and a proton pump inhibitor used in the treatment of gastric ulcer, Taxol, a chiral aminoalcohol obtained by ring opening of the corresponding epoxide used as an anti-cancer drug, and indinavir, another chiral aminoalcohol used as antiviral drug in the treatment of AIDS. The advantage of the class of Pt complexes here discussed lies in the presence of diphosphines in the coordination sphere and chiral diphosphines are perhaps the largest family of chiral ligands synthesized for enantioselective transformations, many of which are commercially available. Thus, the transformation of ordinary catalysts into chiral catalysts using the synthetic approaches previously outlined was quite straightforward.

The $[(dppe)Pt(C_6F_5)(solv)]^+$ catalyst was quite specific in the epoxidation of terminal alkenes. In fact in the epoxidation of 1,4-hexadiene a comparison between the Pt^+/H_2O_2 system with *m*-chloroperbenzoic acid (MCPBA), a typical organic epoxidation reagent (Figure 24.5), showed >99% regioselectivity for the terminal epoxide with the former system, whereas 97% regioselectivity toward the internal epoxide was observed with the latter [22]. The same catalyst modified with chiraphos allowed the enantioselective terminal epoxidation of a variety of dienes with ee in the range 63–98% [23] (Figure 24.5). Previously, using the chiraphos, $-CF_3$ derivative, we had been

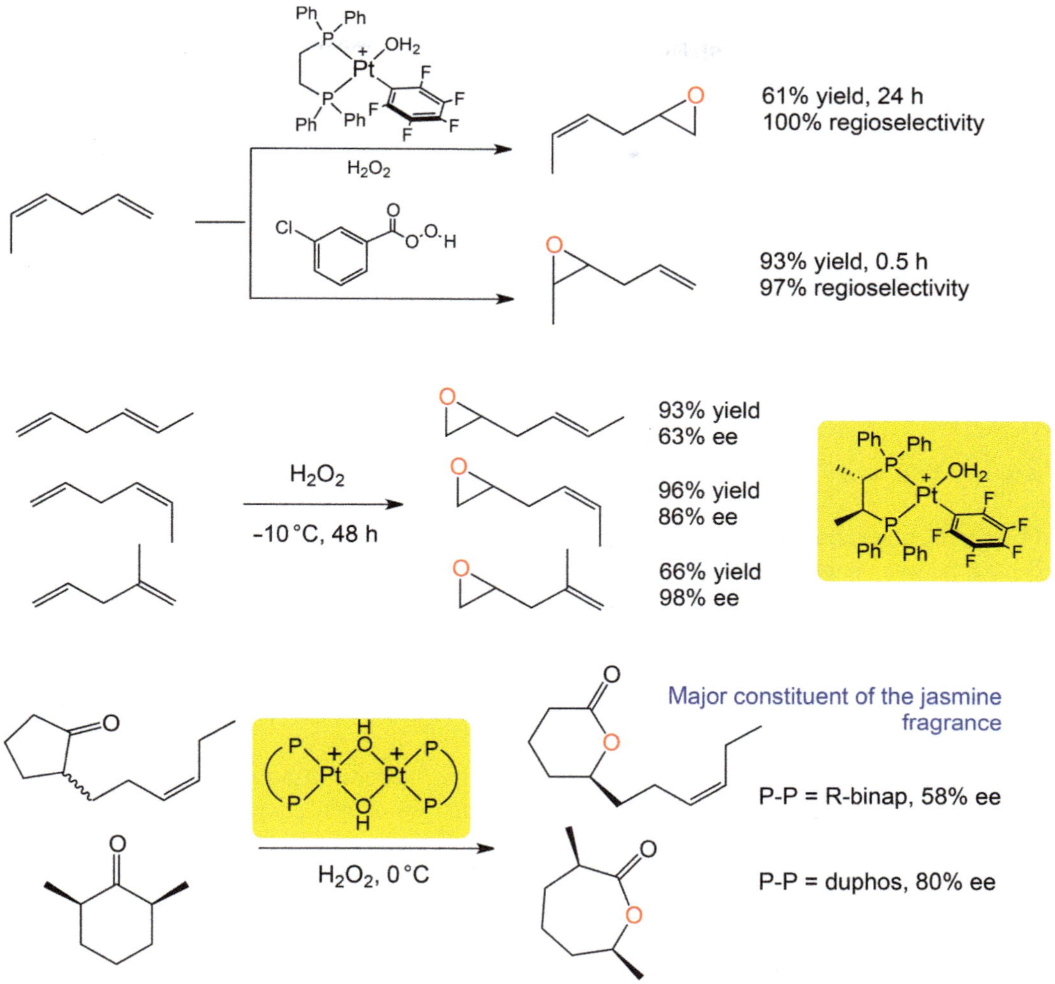

Figure 24.5 Enantioselective epoxidation of terminal alkenes and BV oxidation of cyclic ketones with hydrogen peroxide using chiral Pt(II) complexes as catalysts.

able to asymmetrically epoxidize also a very simple molecule like propene obtaining propene epoxide with 41% ee [24]. Although these enantioselectivities may not be exceptional in some cases, it must be borne in mind that simple alkenes, possessing no extra functional groups in the molecule, lack secondary binding to the metal, a factor that is recognized to play a fundamental role in increasing enantioselectivity.

Similarly, the desymmetrization of a series of racemic or *meso* ketones to yield the corresponding chiral lactones was successfully achieved using the hydroxo-bridged [(P–P)Pt(μ-OH)]$_2^{2+}$ complexes again modified with chiral diphosphines [25]. Some examples are reported in Figure 24.5.

The use of hydrogen peroxide in these reactions, the possibility of exploiting the wide library of phosphine ligands (chiral and non-chiral) available to tune the steric and electronic properties of the catalysts, the high product yields observed in most cases, the high product selectivities (even enantio-), the absence of oxidant waste in undesired side reactions, and the mild conditions applied, combine to make these oxidation reactions some good examples of *ante litteram* Green

Chemistry, since most of them were discovered long before the twelve principles of Green Chemistry were established [26]. However, an important step forward in this respect would be the possibility to easily separate the catalyst from the reaction mixture and recycle it, and to remove the organic solvent (usually dichloromethane or dichloroethane) used to dissolve all reaction ingredients (except hydrogen peroxide).

In fact, the solvent constitutes about 80% of the total mass of a reaction and in the industrial practice this sets two major problems: (i) the environmental acceptability of the solvent; and (ii) its recovery and recycling. With respect to the former point, methanol, acetone, and toluene are by far the most widely accepted solvents in industry for obvious reasons, while recovery and recycling have an efficiency that rarely exceeds 70%. In the pharmaceutical industry, where the standards of purity required are more stringent, the amount of solvents recycled having a sufficient purity is no more than 30%, the remaining part being simply incinerated to produce heat [27].

24.4 Water as the Reaction Medium

In a chemical reaction the solvent is required to play several roles at the same time: (i) to ensure contact between substrates of different polarity; (ii) to control heat transfer; and (iii) to favor the interaction that leads to the final transformation. In this framework, among all possible liquids, water is certainly the one with the smallest impact on the environment. Water as a possible reaction medium has many advantages over organic solvents [28]:

- It is environmentally safe, non flammable, non toxic, cheap, readily available anywhere and requires no synthetic procedure to obtain.
- It does not contribute to greenhouse emissions and its E factor is assumed equal to zero [29].
- It has a tunable acidity, and a large heat capacity and heat of evaporation that make it an ideal choice for exothermic reactions.
- It has a high polarity and H-bond donor and acceptor functionalities coexist.
- It is immiscible with most organic solvents making catalyst separation possible (at least in principle) in water/organic biphasic reactions.

On the other hand, it also has a few but crucial disadvantages that have precluded its widespread use in organic synthesis, first of all its poor solubility properties with respect to most organic substrates and catalysts (this is the major problem); furthermore, some chemical functionalities (especially organometallics) are not stable in the presence of water.

The hydrophobic effect that makes most organics insoluble in water is essentially an entropic effect that is due to the existence in liquid water of an extensive hydrogen bonding network that causes a large cohesive energy. When a hydrophobic organic molecule is added to water, dissolution (i.e. molecular dispersion into the aqueous medium) would require extensive hydrogen bond breaking. Conversely, aggregation (i.e. two-phase formation) will minimize hydrogen bond disruption and will restrict the organic molecule motion, thus resulting in an increase in order and a decrease in entropy.

In recent years, the interest in catalysis in water was stimulated by the Ruhrchemie-Rhone Poulenc process for the hydroformylation of propene and butene catalyzed by rhodium complexes modified with sulfonate-containing phosphine ligands in order to make them soluble in water [30]. Historically, it has been known since the 1980s that the hydrophobic effect can greatly accelerate reactions between poorly water-soluble substrates. Rideout and Breslow [31] found that the Diels–Alder cycloaddition reaction in water as the solvent was hundreds of times faster than in organic protic and aprotic solvents, an observation that excluded simple polarity or H-bonding effects on catalysis. It is commonly accepted [32] that the hydrophobic effect forces molecules or apolar groups to aggregate in order to minimize their contact with water. This leads to a compacting of the reagents in the transition state thus increasing reaction rate. Concepts like the hydrophobic effect and the donor-acceptor hydrogen bonding ability of water have allowed rationalization of the enhanced productivity as well as regio-, diastereo-, and enantioselectivity in several catalytic reactions [33].

Mother Nature has always handled reactions between hydrophobic substrates using water as the sole solvent. How does this happen? Because of the hydrophobic effect, enzymes, the natural catalysts, fold in water and, thanks to a complex network of hydrogen bonding and dipole-dipole interactions, they assume their characteristic structure where the hydrophilic groups of the macromolecule are exposed on the surface in contact with water, while the hydrophobic portions lie inside the structure where the active site is generally located. This complex strategy, based on an amphiphilic nature, allows a hydrophobic phase to be well dispersed in water as well as to be a suitable reaction medium for organic transformations. These processes are pivoted by weak supramolecular interactions (again dipole-dipole, hydrogen-bonding, and others) that drive not only the reaction rate and the extent of the catalytic phenomenon, but also product as well as substrate selectivity.

The simplest way to mimic, at least in part, enzymes behavior is to use a surfactant. Soaps have been used by humans for about 2000 years to remove fats. News on their preparation is reported by Pliny the Elder in his *Naturalis Historia* [34], in which he describes a strange practice by Gauls, who used to melt together tallow and wood ashes obtaining by cooling a solid material called *sapo*. Surfactants, the active ingredients of soaps, are amphiphilic molecules with a polar hydrophilic head and a hydrophobic tail that above a certain concentration (the so-called critical micellar concentration or cmc) self-assemble in solution to give nanometric aggregates (micelles) where hydrophilic heads are concentrated on the surface in contact with water, while hydrophobic tails point inside. Clearly, they are much smaller than enzymes, and, unlike enzymes, do not possess that large variety of internal functional groups and chemical complexity capable of driving exceptional activity and selectivity. However, similarly to enzymes, they constitute an organic nanophase dispersed in water capable of dissolving hydrophobic molecules. Surfactants are a large class of cheap chemicals, commercially available as cationic, anionic, and neutral molecules.

Micellar media are a good choice to perform homogeneous catalysis in water for several reasons:

- They enhance catalyst and reagents solubility in water because their inside is an organic phase.
- They behave like an organic solvent and therefore transition metal complexes already optimized to be used as catalysts in solution need no ligand modification, unlike with catalysts made soluble in water for which polar groups must be introduced into the ligands, possibly altering their electronic properties.
- Catalyst local concentration inside the micelle is about two orders of magnitude higher than in a regular solvent, thus a beneficial effect on reactivity is possible.
- Product isolation is feasible via extraction with apolar solvents immiscible with water, followed by phase separation.
- With appropriate extraction solvents the catalyst remains confined in the micellar medium and in principle can be recycled.

From a certain point of view, the behavior of micelles resembles that of enzymes, isolating species from the bulk solvent (water) and playing several roles at a time (such as improving solubilization of organic reagents in water, favoring compartmentalization of reagents, and imparting unique chemo-, regio- and stereoselectivities). A limitation of micellar catalysis is related to the amounts of substrate that can be loaded into the micelles, usually lower than in common organic solvents, although the higher activities and selectivities often observed with micelles can partially compensate the disadvantage of working in diluted media. The distribution of organic species between bulk water and micelles depends on their polarity, charge, and dimension. Apolar substrates are almost exclusively hosted inside micelles. On the other hand, charged micelles tend to concentrate species of opposite charge on their surface, hence in cationic micelles the surface local pH is slightly more basic than in the bulk solution, and the opposite is observed for anionic micelles [35]. Depending on the nature of the micelle, upon dissolution a transition metal complex can experience a second coordination sphere due to non-covalent interactions with the micelle, and as in metallo-enzymes [36] this can play a significant role in driving the activity and selectivity.

The interactions between catalyst and surfactants can also influence catalyst separation. Under the best conditions it could be possible to simply efficiently extract the product, leaving the catalyst and the surfactant in the micellar medium. This occurs when the metal catalyst and surfactant are oppositely charged and the organic reagents and reaction products are rather apolar and easily removable from the micellar medium. In other cases, product isolation is more difficult because extraction with a solvent removes partially the surfactant and the catalyst from the aqueous phase. This often occurs with neutral surfactants, neutral metal complexes, or organocatalysts.

24.5 Catalytic Oxidation Reactions in Water

On the basis of these principles, the catalytic enantioselective oxidation reactions reported previously seem to be the ideal choice to test homogeneous catalysis in micellar media. In fact, (i) water is already present both as the solvent for H_2O_2 and as the reaction product in the original system; (ii) with the solvent dichloromethane (DCM) or 1,2-dichloroethane (DCE), the reaction medium is already biphasic; and (iii) the Pt catalysts are quite stable in water, a condition that does not apply to many transition metal and organometallic complexes.

Our first attempt was performed in the oxidation of a series of thioanisoles. Selectivity issues were twofold: one concerned the sulfoxide/sulfone balance and the other the extent of enantioselectivity of the sulfoxide. Reactions were performed using the R-binap derivative of the hydroxo-bridged $[(P-P)Pt(\mu-OH)]_2^{2+}$ complexes as catalysts and the results obtained in the micellar medium were compared to those observed with a regular solvent. Some representative data are reported in Figure 24.6 and show the superior performance of the micellar medium both with respect to the sulfoxide/sulfone selectivity and in the enantioselectivity of the sulfoxide [37]. The catalyst being cationic, the best option as surfactant turned out to be an anionic one like sodium dodecyl sulfate (SDS). A cationic surfactant like cetrimonium bromide (CTABr) was the worst, whereas a neutral one was only slightly better than the latter. The micellar medium could not only improve the chemoselectivity of the process (sulfoxide/sulfone 80:1 in CH_2Cl_2 and >200:1 in water/SDS), but also significantly increase the enantioselectivity (up to 88%).

Figure 24.6 reports also some data on the scope of the reactions in water/SDS as the reaction medium. As shown an electron-releasing substituent in the aromatic ring increases the activity but decreases the enantioselectivity, while the opposite occurs with electron-withdrawing substituents. Optimization of the reaction conditions requires an investigation on the effect of the surfactant concentration. Figure 24.6 shows that both the catalytic activity and the enantioselectivity of

24.5 Catalytic Oxidation Reactions in Water

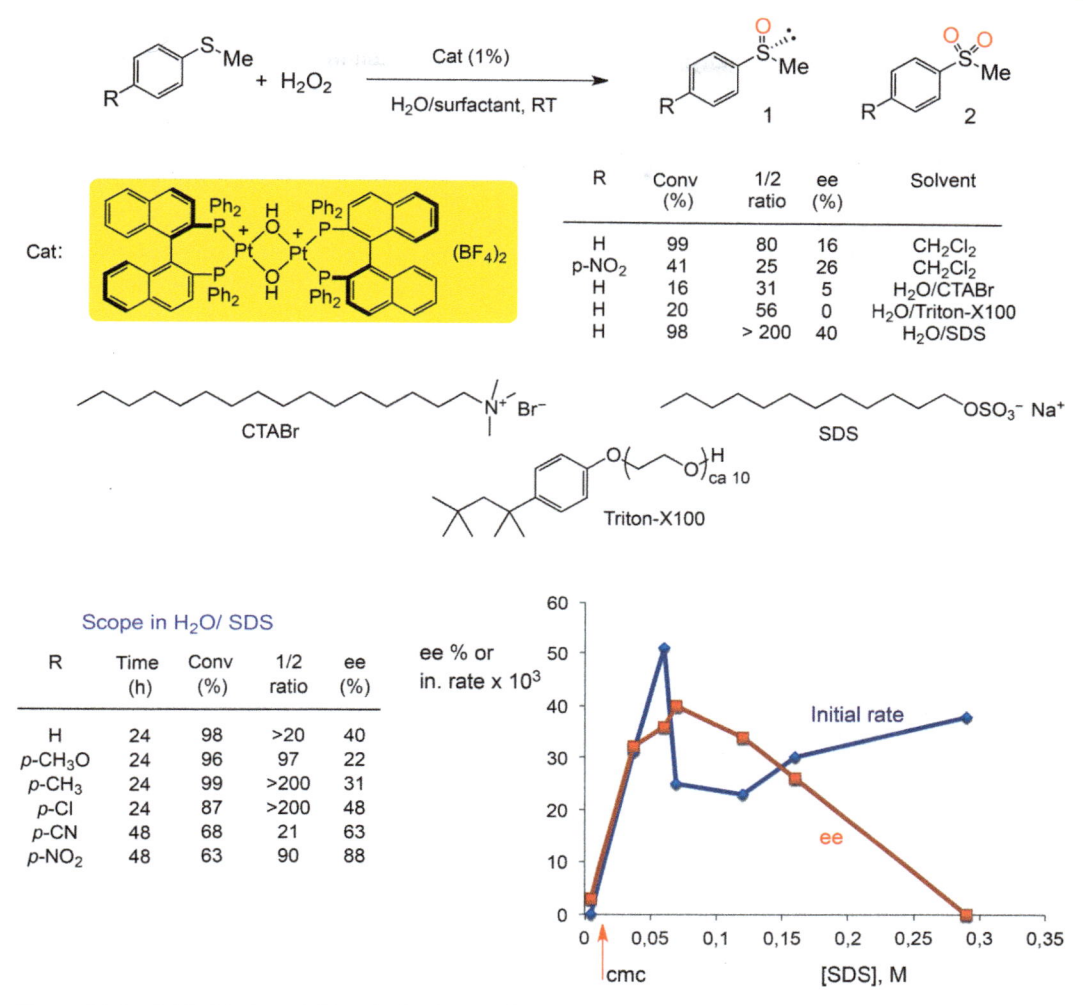

Figure 24.6 Sulfoxidation of thioanisoles in CH_2Cl_2 vs. micellar media and effect of the surfactant concentration on the activity and selectivity.

Reaction scheme: $R-C_6H_4-S-Me + H_2O_2 \xrightarrow{\text{Cat (1\%), } H_2O/\text{surfactant, RT}}$ sulfoxide **1** + sulfone **2**

Catalyst: [(Ph₂P-binaphthyl-PPh₂)Pt(μ-H)(μ-O)Pt(Ph₂P-binaphthyl-PPh₂)](BF₄)₂

R	Conv (%)	1/2 ratio	ee (%)	Solvent
H	99	80	16	CH_2Cl_2
p-NO₂	41	25	26	CH_2Cl_2
H	16	31	5	H_2O/CTABr
H	20	56	0	H_2O/Triton-X100
H	98	> 200	40	H_2O/SDS

Surfactants: CTABr, SDS (OSO_3^- Na^+), Triton-X100

Scope in H_2O/ SDS

R	Time (h)	Conv (%)	1/2 ratio	ee (%)
H	24	98	>20	40
p-CH₃O	24	96	97	22
p-CH₃	24	99	>200	31
p-Cl	24	87	>200	48
p-CN	48	68	21	63
p-NO₂	48	63	90	88

the reaction are strongly influenced by the latter parameter, showing a maximum. This test must be done for each reaction investigated because the outcome can change dramatically.

A study of the enantioselective epoxidation of terminal alkenes was also performed in micellar media using the same catalyst and the same conditions reported previously in Figure 24.7a summarizes some data using $[(\text{chiraphos})\text{Pt}(C_6F_5)(H_2O)]^+$ as catalyst in H_2O/Triton-X100 as the reaction medium that turned out to be the best choice [38]. The use of neutral surfactants was crucial to ensure sufficient catalyst solubilization as demonstrated by 2D-nuclear magnetic resonance (NMR) experiments. In the case of 4-methyl-2-pentene the epoxide was produced with 84% ee, while in dichloroethane under the same conditions, the enantioselectivity observed was 58% ee with very similar product yields. In this epoxidation system catalyst separation and recycling was possible by extraction of the reaction products with *n*-hexane. The micellar medium containing the catalyst was recycled at least three times with negligible differences in terms of activity and enantioselectivity [38].

The asymmetric Baeyer–Villiger oxidation of cyclic ketones with hydrogen peroxide catalyzed by chiral diphosphine-Pt(II) catalysts was extensively investigated in water in the presence of different surfactants (Figure 24.7b) [39]. The reaction resulted extremely sensitive to the combination of substrate, catalyst and surfactant employed and only careful optimization of all three parameters enabled good results in terms of asymmetric induction and yield of lactone. It was also found that the oxidation of *meso*-cyclobutanones in micellar media allowed the reaction to proceed in high yields and better enantiomeric excess (ee up to 56%) with respect to organic media. Extension to *meso*-cyclohexanones resulted in a general decrease in yields but an enhancement of enantioselectivity (ee up to 92%) among the best reported in the literature and second only to enzymatic catalysis.

Some striking effects on the extent of stereocontrol exerted by micelles in enantioselective reactions was observed in the Baeyer-Villiger oxidation of some *meso*-cyclobutanones catalyzed by the Co(salen) complex shown in Figure 24.7c [40]. This was completely inactive when employed in several organic solvents. The same complex turned out to induce significant asymmetric induction when employed in water, with best results using SDS as surfactant. Particularly interesting was the oxidation of chiral racemic cyclobutanones as shown in Figure 24.7c. Here the catalyst is called to control both the regioselectivity (lactone 1 vs. lactone 2) and the enantioselectivity. In the specific case reported in Figure 24.7c, in DCE as the solvent the Co catalyst is poorly active and non enantioselective (ee = 0%), while in the micellar medium the yield in stereoisomer 1 jumps to 80% and its ee to 88%. This striking effect is only due to the use of Triton-X114 that was found to be the best surfactant. In the latter examples, careful surfactant optimization was necessary since, as observed in other oxidation reactions, every combination of substrate and catalyst behave in a different way when using different surfactants and this needs optimization.

Over the years we have investigated also other (non oxidation) reactions in micellar media using the same class of cationic Pt(II) complexes, like the hydration of terminal alkynes to yield methylketones [41] and the hydroformylation of alkenes to give exclusively linear aldehydes [42]. In both cases the use of H_2O/SDS allowed both the increase of product yields and the recycle of the catalyst several times. Other reactions studied were the hydration of nitriles using Ru catalysts in H_2O/Triton-X114, again with improved yields and catalyst recycling [43] and the Diels-Alder reaction between cyclopentadiene and methacrolein catalyzed by a Cr(salen) complex in H_2O/SDS with better yields with respect to organic solvents [44].

Over the past 15 years the general subject of micellar catalysis using soluble metal complexes as catalysts in hydrogenation, C–C and C–heteroatom bond formation, olefin metathesis, cross-coupling, cycloaddition reactions, has been extensively investigated and reviewed [45] leading in some cases to potential large-scale applications [46]. Moreover, in order to extend the scope of these reactions, a decade ago Lipshutz introduced the concept of designer surfactants [47] (i.e. some new amphiphilic molecules specifically designed and developed for catalytic applications in water) [48] based for example on natural products like α-tocopherol or β-sitosterol as the hydrophobic part, often coupled with hydrophilic poly(ethylene-glycol) methyl ether chains of different length. In the recent years the field of designer surfactants has flourished with many more surfactants that further widen the applicability of micellar catalysis to an even larger range of reactions [49]. Alternatively, metallo-surfactants [50] have been proposed in which the hydrophilic head-group of the surfactant is a ligand for metal centers connected to a long alkyl chain forming metallo-micelles adorned on the surface with many catalytically active metal centers. This all in one approach has the advantage to provide often very good catalytic properties at the expense of a much more complicated synthesis.

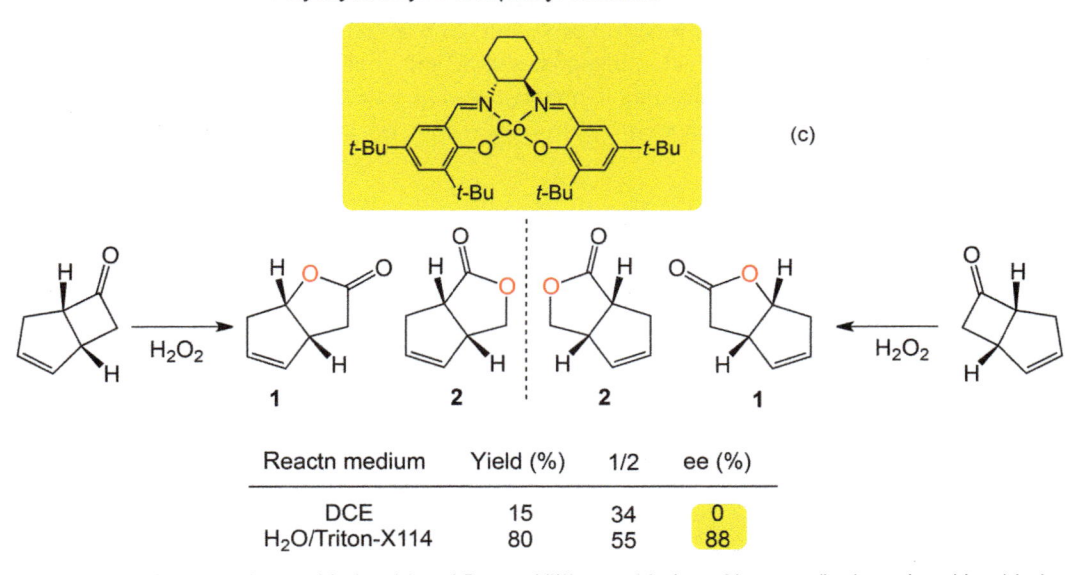

Figure 24.7 Asymmetric epoxidation (a) and Baeyer-Villiger oxidation of ketones (b, c) catalyzed by chiral Pt (a, b) and Co (c) complexes in different micellar media.

24.6 The Catalyst/micelle Interaction

The improvement in conversion and selectivity very often observed when moving from a traditional organic solvent to a micellar medium stems from the interactions between the micelle and the catalyst "dissolved" in it. Even if surfactants such as SDS or CTABr have a very simple internal structure (alkyl chains piled essentially side by side) there will always be a gradient of polarity on going from the micelle surface, where the polar groups are located, toward the core where the hydrophobic part is concentrated. Catalyst positioning in the micelle will depend on a reciprocal affinity basis with portions of different polarity.

To examine this issue, we chose three similar platinum catalysts, all cationic and containing diphosphines, already successfully tested in oxidation reactions (see previously), and investigated their positioning with respect to SDS micelles using 2D nuclear Overhauser effect (NOESY) NMR experiments [37–39], by checking the interactions between the diphosphine aryl groups and different portions of SDS alkyl chain (Figure 24.8).

As shown in the figure, the bridging hydroxo complex used for sulfoxidation (Figure 24.8 left) was found to interact with the $-CH_2-$ group adjacent to the sulfate groups, indicating that the catalyst is essentially sitting on the micelle surface in contact with the anionic groups. Less expected was the catalyst/micelle interaction observed between the Pt pentafluoro derivative used in epoxidation (Figure 24.8 center). Here the NOESY cross peaks indicated an interaction between the phosphine phenyl groups and the terminal $-CH_3$ of the SDS alkyl chain, indicating that the complex is deeply buried in the apolar core of the micelle, somehow surprisingly given its cationic nature. Finally, the bis-aquo complex used for the Baeyer–Villiger oxidation shows clear interactions between the phosphine phenyl groups and the $-CH_2-$ groups present along the surfactant chain, thus indicating that it is roughly halfway between the micelle's surface and core.

These experiments demonstrate that even simple supramolecular aggregates like micelles can strongly influence catalyst positioning by mutual interactions and even apparently small differences in the catalyst nature can lead to quite different situations. Hence the constrain of reaction partners in a confined anisotropic space (at variance with solutions) will exert a tighter supramolecular control on reactions and will influence the steric and electronic demand of the transition state in catalyzed reactions, thus leading to improvement in activity and selectivity. For example, in the case of the experiment reported Figure 24.7c, from the extent of the enantioselectivites

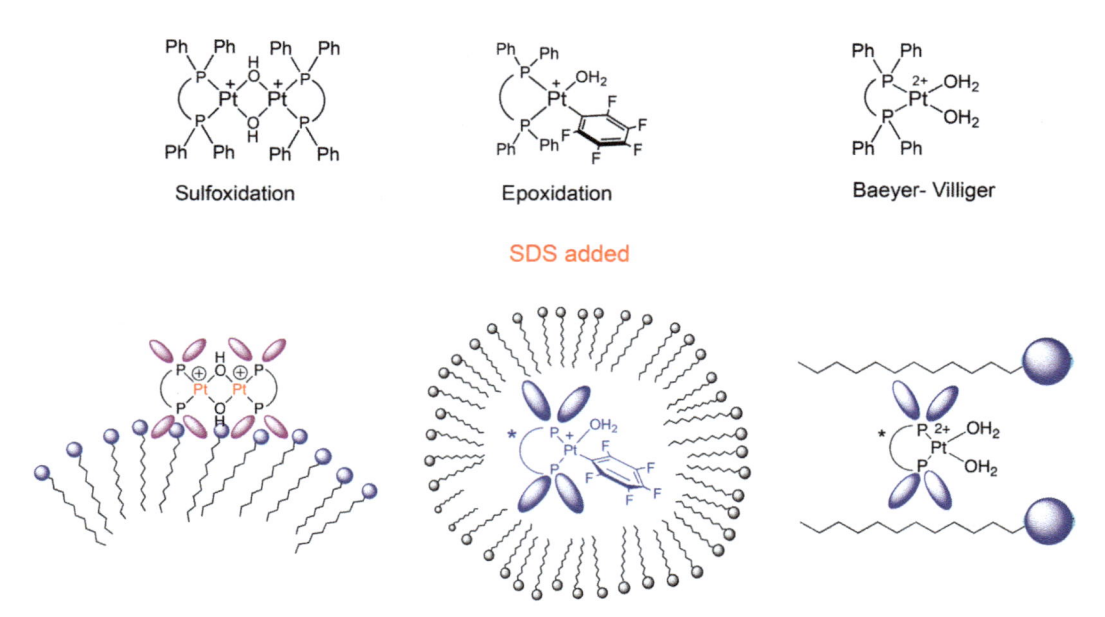

Figure 24.8 A graphical representation of the interactions between the complexes indicated on top with SDS micelles. Ellipses on phosphorus represent the phenyl rings.

observed it can be calculated from the Eyring equation that the $\Delta\Delta G^{\neq}$ values of the diastereomeric transition states when the reaction is carried out in DCE and the same reaction performed in H_2O/Triton-X114 differ by about −6.8 kJ/mol. To exploit sucessfully these effects in micellar catalysis it is clear that the nature and the concentration of the surfactant must be critically selected and carefully optimized.

The feature just discussed is typical of enzyme behavior where activity and selectivity issues are amplified by the high variety and complexity of supramolecular interactions occurring inside the enzyme structure. These interactions are capable of driving certain molecules to the active site, lowering the energy of the transition state, and producing specifically only one product enantiomer. They are also the basis for the ability of enzymes to pick only one specific molecule among many others present in biological fluids and lead to its transformation, a property known as substrate selectivity. Can micelles behave similarly?

We tested the Diels-Alder reaction between cyclopentadiene and a series of α,β-unsaturated aldehydes differing by the length of the alkyl chain (C_1 to C_7) adjacent to the double bond, that were fed all together into the reaction mixture. The reaction was catalyzed by a Cr(salen) complex and was tested both in chloroform and in water/SDS (Figure 24.9a). By analyzing the yields of the

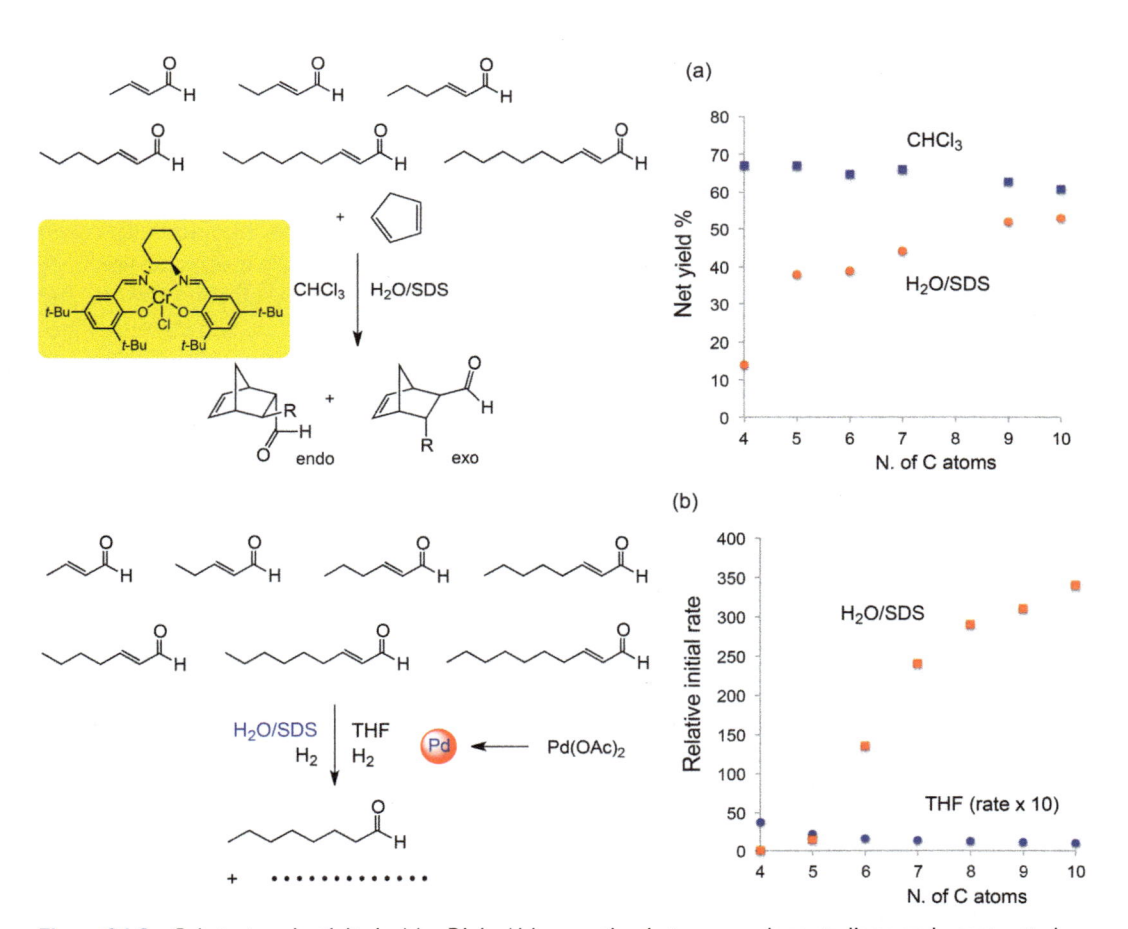

Figure 24.9 Substrate selectivity in (a) a Diels-Alder reaction between cyclopentadiene and unsaturated aldehydes of different length and (b) hydrogenation of the same aldehydes with Pd metal. In both cases, the graphs report a comparison between the use of an ordinary solvent and a micellar medium.

various cycloaddition products it was observed that in chloroform differences related to the alkyl chain length of the aldehyde were negligible. On the contrary in water/SDS an increase in yields was observed. Here the cycloaddition product bearing the longest alkyl chain was found to be favored over the one having the shortest alkyl chain by a factor of about 3.5.

More striking evidence for substrate selectivity effect was observed in the hydrogenation of the α,β-unsaturated aldehydes with Pd metal particles generated in situ either in tetrahydrofuran (THF) or in water/sodium dodecylbenzenesulfonate (SDBS) micellar medium (Figure 24.9b) [51]. Pd particles produced in either medium had the same size and morphology as was demonstrated by TEM measurements. When the mixture of aldehydes was fed into THF the initial hydrogenation rates for the individual aldehydes decreased on going from the shortest to the longest by a factor of 3.6. The contrary occurred in water/SDBS, with the longest aldehyde being preferred by a factor of 330. As in the Diels-Alder reaction the substrate selectivity rule seems to rely on the higher lipophilicity of longer aldehydes and their higher affinity with the micelle. It should be emphasized that in both cases we performed really competitive experiments because substrates were fed all together into the reaction mixture.

24.7 Environmental Acceptability Evaluation and Possible Industrial Applications

The green character of micellar catalysis as well as the advantages with respect to traditional catalysis in solution were evaluated in several reactions by comparing the E-factor of the regular reaction carried out in organic solution and the same reaction performed in a micellar medium. In these evaluations also the final work-up to isolate the products was taken into account. As an example, we consider three C–C bond forming reactions reported by Lipshutz et al. some years ago [52], all catalyzed by Pd complexes (Figure 24.10): (i) a Heck coupling between ethyl acrylate and 3-bromoquinoline; (ii) a Suzuki-Miyaura reaction between phenylboronic acid and 2-iodo-5-bromotoluene; and (iii) a Sonogashira coupling between 3-butynol and a 4-bromobenzoate ester. In all three cases the use of the micellar medium allowed to increase the yields, simplify the reaction procedure and lower the transition metal content. Notably in all three cases the E-factors including work-up were drastically reduced using micellar media based on neutral surfactants instead of the traditional procedure.

From what we have reported, it is clear that the industrial application potential of micellar catalysis is quite high. Indeed, the pharmaceutical industry has already adopted this technology either in the total synthesis of some active pharmaceutical ingredients or in the preparation of important intermediates. In particular, research groups at Novartis [53] and Abbvie [54] tested the use of micellar media in a series of reactions of industrial interest, addressing crucial issues such as increasing yields, making product isolation easier, and reducing the amount of catalyst. In 2016 [46b], they reported the total synthesis of an active pharmaceutical ingredient, showing only the functional groups involved in the chemical transformations and hiding the sensitive information of the molecules utilized. The final product synthesis consisted of five different steps and for each of them the product yield, the time necessary to accomplish it and the cost difference were evaluated (Table 24.1) comparing the traditional methodology in organic solvents with the one based on micellar media, both applied on industrial scale.

The surfactant technology, based on the nonionic designer surfactant TPGS-750-M in water proved more straightforward and highly advantageous all across the entire synthetic route with environmental as well as economic and productivity benefits. As can be seen from Table 24.1, this resulted in reduction of organic solvent use, reduction of water use, improved yields, milder

24.7 Environmental Acceptability Evaluation and Possible Industrial Applications

Heck

DMF, 90 °C, Pd catalyst / 5% PTS/H$_2$O 60 °C

E-factor		
Based on	Organic	Micellar
Total org solvent	40.3	2.2
Incl aqueous work-up	137	7.6

Suzuki–Miyaura

H$_2$O/acetone, Pd catalyst / 2% TPGS-750-M/H$_2$O

E-factor		
Based on	Organic	Micellar
Total org solvent	42	3.9
Incl aqueous work-up	83	8.3

Sonogashira

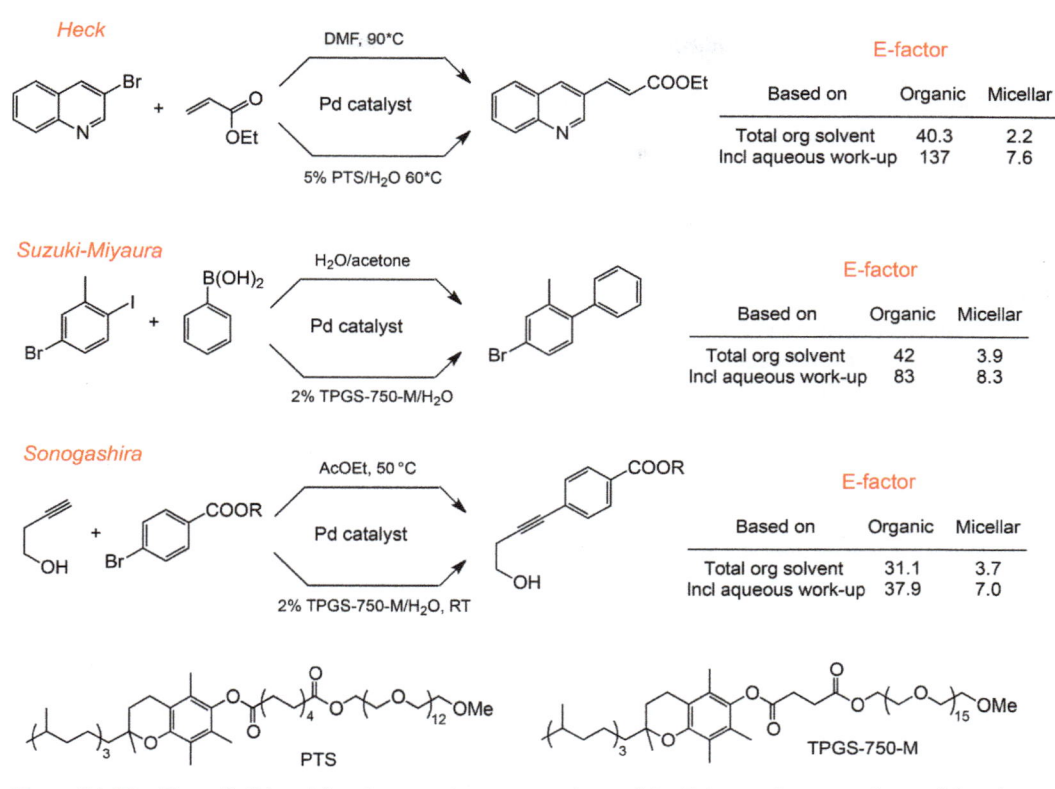

AcOEt, 50 °C, Pd catalyst / 2% TPGS-750-M/H$_2$O, RT

E-factor		
Based on	Organic	Micellar
Total org solvent	31.1	3.7
Incl aqueous work-up	37.9	7.0

Figure 24.10 Three C–C bond forming reactions: comparison of the E-factors between the traditional procedure based on organic solvents and the alternative use of aqueous micellar media.

Table 24.1 Synthesis of an active pharmaceutical ingredient: comparison between the traditional organic methodology and the use of micellar media (highlighted in yellow).

Step	Conditions		Yield (%)	Cycle time (hr)	Cost diff
1 S$_N$Ar	reactant 1.2 eq, i-PrOH/toluene, 80 °C, 5 hours		87	104	+4%
	reactant 1.0 eq, TPGS-750-M/water, RT, 12–16 hours		75	61	
2 Suzuki Miyaura	reactant 1.4 eq, Pd, t-AmOH, 85 °C, six hours		=	61	−38%
	reactant 1.0 eq, Pd, TPGS-750-M/water, RT, four to six hours			24	
3 hydrolysis	NaOH, MeOH/water, RT, eight hours, MeTHF		70	137	
	NaOH, RT, four to six hours		87	53	
4 amide bond formation	reactant 1.2 eq, acetonitrile, 50 °C, 10 hours		76	105	−7%
	reactant 1.0 eq, TPGS-750-M/water, RT, 10–12 h		80	76	
5 final deprotection	HCl, MeOH/water, 0 °C to RT, 16 hours		92	62	0%
	HCl, MeOH/water, 0 °C to RT, 16 hours		92	62	
	Overall	organic traditional	42.5	469	−17%
		micellar	48	276	

reaction conditions, reduction of cost, and reduction of cycle time, with the latter being an important issue in the pharmaceutical industry that typically operates in discontinuous, making use of multi-purpose facilities already present in the infrastructure. To quote the authors, "Quantitatively, the differences for some of these virtues approached 50% in favor of surfactant technology".

24.8 Conclusions

Despite the moderate applications of platinum complexes in homogeneous catalysis, the class of very similar electron-poor cationic complexes containing diphosphines that we have investigated over the past forty years has proven useful to elucidate and give experimental support to a series of concepts that can be used as guidelines to identify potential catalysts for oxidation reactions using hydrogen peroxide as oxidants. These concepts are: (i) soft Lewis acidity that makes possible the (ii) electrophilic oxidation, thus opening the way to the oxidation of electron-poor olefins and ketones with unprecedented activity, selectivity (including enantio-), product yields, without any waste of oxidant, and under very mild conditions (room temperature, in air). Additional advantages of these Pt complexes are: (iii) the possibility to tune the steric and electronic properties of the metal center through an appropriate choice of commercially available diphosphine ligands, including chiral ones to perform enantioselective transformations; and (iv) full compatibility with water necessary to operate with hydrogen peroxide. The latter property allowed these Pt complexes (v) to be used in aqueous media where they proved to be ideal to test the different features of micellar catalysis, thus removing organic solvents at least as reaction media.

Commercially available surfactants can be used as simple auxiliaries to create a separate phase in water, made of self-assembled devices (micelles) that can be seen as dispersed nanoreactors, in which to dissolve catalyst, reactants, products. Micellar media can improve activity and selectivity and, in many cases, make it possible catalyst separation and recycling. As we have shown, the gradient of polarity present in micelles can result in different catalyst and substrate positioning. This internal anisotropy is typical of enzymes, involving not only areas of different polarity but also hydrogen bonding and other weak interactions. The consequence of that is an amplification of the free energy differences of the possible transition states of a reaction, thus influencing its outcome and driving selectivity at any level. The different chemical affinity of different substrates for micelles, mainly driven by differences in hydrophobic properties, enable traditional metal catalysts to display unusual substrate selectivity [55], an important property in common with enzymes and also with zeolites, although on a different selection basis.

Practical examples have shown that the use of micellar media can significantly reduce the E-factor in organic transformations. However this is not a simple green chemistry advance, but rather a new technological approach that can lead to significant benefits across the entire synthetic route, thus paving the way to industrial applications.

References

1 BASF. (2009). New BASF and Dow HPPO plant in Antwerp completes start-up phase. https://www.basf.com/ru/ru/media/news-releases/2009/03/p-09-154.html (accessed 19 April 2023).
2 Chemical Online. (2000). Sumitomo to commericalize new caprolactam process. https://www.chemicalonline.com/doc/sumitomo-to-commercialize-new-caprolactam-pro-0001 (accessed 19 April 2023).

3 (a) Strukul, G., Ros, R., and Michelin, R.A. (1982), *Inorg. Chem.* 21: 495–500; (b) Strukul, G., Michelin, R.A., Orbell, J.D., and Randaccio, L. (1983). *Inorg. Chem.* 22: 3706–3713; (c) Strukul, G. and Michelin, R.A. (1984). *JCS Chem. Commun.* 1538–1539; (d) Strukul, G. and Michelin, R.A. (1985). *J. Am. Chem. Soc.* 107: 7563–7569.

4 Zanardo, A., Pinna, F., Michelin, R.A., and Strukul, G. (1988). *Inorg. Chem.* 27: 1966–1973.

5 Roussel, M. and Mimoun, H. (1980). *J. Org. Chem.* 45: 5381–5383.

6 (a) Suresh, C.H. and Koga, N. (2001). *J. Phys. Chem. A* 105: 5940–5944; (b) Bond, G.C. (2000). *J. Mol. Catal. A-Chem.* 156: 1–20.

7 For some example of homologous diphosphine complexes structural data, see: (a) Maddox, A.F., Rheingold, A.L., Golen, J.A. et al. (2008). *Inorg. Chim. Acta* 361: 3283–3293; (b) Ghent, B.L., Martinak, S.L., Sites, L.A. et al. (2007). *J. Organom. Chem.* 692: 2365–2374.

8 (a) Langford, C.H. and Gray, H.B. (1965). Chapter 2. Square-Planar Substitutions, pag. 18, In: *Ligand Substitution Processes (1966)* W.A. Benjamin, Inc. Ed., New York (USA). In: *Ligand Substitution Processes*. (ed. C.H. Langford and H.B. Gray), 18. New York: W.A. Benjamin, Inc; (b) Anderson, G.K. and Cross, R.J. (1980). *Chem. Soc. Rev.* 9: 185–215; (c) Appleton, T.G., Clark, H.C., and Manzer, L.E. (1973). *Coord. Chem. Rev.* 10: 335–422; (d) Chval, Z., Sip, M., and Burda, J.V. (2008). *J. Comput. Chem.* 29: 2370–2381.

9 (a) Aye, K.T., Vittal, J.J., and Puddephatt, R.J. (1993). *J. Chem. Soc., Dalton Trans.* 1835–1839; (b) Deacon, G.B., Lawarenz, E.T., Hambley, T.W. et al. (1995). *J. Organomet. Chem.* 493: 205–213; (c) Rashidi, M., Nabavizadeh, M., Hakimelahi, R., and Jamali, S. (2001). *J. Chem. Soc., Dalton Trans.* 3430–3434.

10 (a) Uson, R., Fornies, J., Espinet, P., and Alfranca, G. (1980). *Synth. React. Inorg. Met.-Org. Chem.* 10: 579–50-. (b) Uson, R., Fornies, J., Martinez, F., and Tomas, M. (1980). *J. Chem. Soc., Dalton Trans.* 888–894.

11 Criegee, R. (1948). *Justus Liebigs Ann. Chem.* 560: 127–135.

12 Del Todesco Frisone, M., Pinna, F., and Strukul, G. (1983). *Organometallics* 12: 148–156.

13 (a) Scarcia, V., Furlani, A., Longato, B. et al. (1988). *Inorg. Chim. Acta* 153: 67–70; (b) Gavagnin, R., Cataldo, M., Pinna, F., and Strukul, G. (1998). *Organometallics* 17: 661–667; (c) Sgarbossa, P., Scarso, A., Michelin, R.A., and Strukul, G. (2006). *Organometallics* 26: 2714–2719.

14 (a) Michelin, R.A., da Silva, M.F.C.G., and Pombeiro, A.J.L., (2001). *Coord. Chem. Rev.* 218: 75–112; (b) Belluco, U., Michelin, R.A., Uguagliati, P., and Crociani, B. (1983). *J. Organomet. Chem.* 250: 565–587.

15 Pizzo, E., Sgarbossa, P., Scarso, A. et al. (2006). *Organometallics* 25: 3056–3962.

16 Bond, G.C. (1987). *Heterogeneous Catalysis: Principles and Applications*, 2e. Oxford: Clarendon Press.

17 Michelin, R.A., Pizzo, E., Sgarbossa, P. et al. (2005). *Organometallics* 24: 1012–1017.

18 Wursche, R., Debaerdemaeker, T., Klinga, M., and Rieger, B. (2000). *Eur. J. Inorg. Chem.* 2063–2070.

19 (a) Cataldo, M., Nieddu, E., Gavagnin, R. et al. (1999). *J. Mol. Catal. A Chemical* 142: 305–316; (b) Nieddu, E., Cataldo, M., Pinna, F., and Strukul, G. (1999). *Tetrahedron Lett.* 40: 6987–6990; (c) Battaglia, L., Pinna, F., and Strukul, G. (2001). *Can. J. Chem.* 79: 621–625.

20 Pignat, K., Vallotto, J., Pinna, F., and Strukul, G. (2000). *Organometallics* 19: 5160–5167.

21 Scarso, A., Colladon, M., Sgarbossa, P. et al. (2010). *Organometallics* 29: 1487–1497.

22 Colladon, M., Scarso, A., Sgarbossa, P. et al. (2007). *J. Am Chem. Soc.* 129: 7680–7689.

23 Colladon, M., Scarso, A., Sgarbossa, P. et al. (2006). *J. Am Chem. Soc.* 128: 14006–14007.

24 Sinigalia, R., Michelin, R.A., Pinna, F., and Strukul, G. (1987). *Organometallics* 6: 728–734.

25 (a) Gusso, A., Baccin, C., Pinna, F., and Strukul, G. (1994). *Organometallics* 13: 3442–3451; (b) Paneghetti, C., Gavagnin, R., Pinna, F., and Strukul, G. (1999). *Organometallics* 18: 5057–5065.

26 (a) Anastas, P.T. and Warner, J.C. (1998). *Green Chemistry: Theory and Practice*. New York: Oxford University Press; (b) Anastas, P.T. and Kirchhoff, M. M. (2002). *Acc. Chem. Res.* 35: 686–695; (c) Bolm, C., Beckmann, O., and Dabard, O.G.A. (1999). *Angew. Chem. Int. Ed.* 37: 1198–1209.
27 (a) Seyler, C., Capello, C., Hellweg, S. et al. (2006). *Ind. Eng. Chem. Res.* 45: 7700–7709; (b) Capello, C., Fischer, U., and Hungerbühler, K. (2007). *Green Chem.* 9: 927–934.
28 (a) Cortes-Clerget, M., Yu, J., Kincaid, J.R.A. et al. (2021). *Chem. Sci.* 12: 4237–4266; (b) Lipshutz, B.H., Gallou, F., and Handa, S. (2016). *ACS Sust. Chem. Eng.* 4: 5838–5849.
29 Sheldon, R.A. (2005). *Green Chem.* 7: 267–278.
30 Joó, F. (2001). *Aqueous Organometallic Catalysis*. Dordrecht: Kluwer.
31 Rideout, D.C. and Breslow, R. (1980). *J. Am. Chem. Soc.* 102: 7816–7817.
32 (a) Gawande, M.B., Bonifácio, V.D.B., Luque, R. et al. (2013). *Chem. Soc. Rev.* 42: 5522–5551; (b) Simon, M.-O. and Li, C.-J. (2012). *Chem. Soc. Rev.* 41: 1415–1427.
33 (a) Lindström, U.M., (2006). *Angew. Chem., Int. Ed.* 45: 548–551; (b) Adams, D., Dyson, P., and Tavener, S. (2004). *Chemistry in Alternative Reaction Media*. Chichester: Wiley; (c) Li, C.J. (1993). *Chem. Rev.* 93: 2023–2035; (d) Li, C.J. (2005). *Chem. Rev.* 105: 3095–3166; (e) Pirrung, M.C. (2006). *Chem. Eur. J.* 12: 1312–1317; (f) Narayan, S., Muldoon, J., Finn, M.G. et al. (2005). *Angew. Chem., Int. Ed.* 44: 3275–3279.
34 Pliny the Elder. *Naturalis Historia*, 28, ch. 47.
35 (a) Breslow, R., Maitra, U., and Rideout, D., (1983). *Tetrahedron Lett.* 24: 1901–1904; (b) Grieco, P.A., Garner, P., and He, Z. (1983). *Tetrahedron Lett.* 24: 1897–1900; (c) Dey, A. and Patwari, G.N. (2011). *J. Chem. Sci.* 123: 909–918.
36 Zhao, M., Wang, H.-B., Ji, L.-N., and Mao, Z.-W. (2013). *Chem. Soc. Rev.* 42: 8360–8375.
37 Scarso, A. and Strukul, G. (2005). *Adv. Synth. Catal.* 347: 1227–1234.
38 Colladon, M., Scarso, A., and Strukul, G. (2007). *Adv. Synth. Catal.* 349: 797–801.
39 Cavarzan, A., Bianchini, G., Sgarbossa, P. et al. (2009). *Chem. Eur. J.* 15: 7930–7939.
40 Bianchini, G., Cavarzan, A., Scarso, A., and Strukul, G. (2009). *Green Chem.* 11: 1517–1520.
41 Trentin, F., Chapman, A.M., Scarso, A. et al. (2012). *Adv. Synth. Catal.* 354: 1095–1104.
42 Gottardo, M., Scarso, A., Paganelli, S., and Strukul, G. (2010). *Adv. Synth. Catal.* 352: 2251–2262.
43 Cavarzan, A., Scarso, A., and Strukul, G. (2010). *Green Chem.* 12: 790–794.
44 Trentin, F., Scarso, A., and Strukul, G. (2011). *Tetrahedron Lett.* 52: 6978–6981.
45 (a) Sorella, G.L., Strukul, G., and Scarso, A. (2015). *Green Chem.* 17: 644–683; (b) Gallou, F. (2020). *Chimia* 74: 538–548; (c) Sar, P., Ghosh, A., Scarso, A., and Saha, B. (2019). *Res. Chem. Int.* 45: 6021–6041; (d) Scarso, A. and Strukul, G. (2019). In: *Green Synthetic Processes and Procedures* (ed. R. Ballini), 268–288. London: The Royal Society of Chemistry.
46 (a) Kar, S., Sanderson, H., Roy, K. et al. (2022). *Chem. Rev.* 122: 3637–3710; (b) Gallou, F., Isley, N.A., Ganic, A. et al. (2016). *Green Chem.* 18: 14–19.
47 Lipshutz, B.H. and Ghorai, S. (2012). *Aldrichimica Acta* 45: 3–16.
48 (a) Lipshutz, B.H. (2018). *Curr. Opin. Green Sust. Chem.* 11: 1–8; (b) Lipshutz, B.H., Ghorai, S., and Cortes-Clerget, M. (2018). *Chem. Eur. J.* 24: 6672–6695; (c) Lipshutz, B.H. (2017). *J. Org. Chem.* 82: 2806–2816.
49 Lorenzetto, T., Berton, G., Fabris, F., and Scarso, A. (2020). *Catal. Sci. Technol.* 10: 4492–4502.
50 Mehta, S. and Kaur, R. (2022). *Metallosurfactants: From Fundamentals to Catalytic and Biomedical Applications*. Weinheim: Wiley-VCH.
51 La Sorella, G., Canton, P., Strukul, G., and Scarso, A. (2014). *ChemCatChem.* 6: 1575–1578.
52 Lipshutz, B.H., Isley, N.A., Fennewald, J.C., and Slack, E.D. (2013). *Angew. Chem. Int. Ed.* 52: 10952–10968.

53 (a) Li, X., Thakore, R.R., Takale, B.S. et al. (2021). *Organ. Lett.* 23: 8114–8118; (c) Yu, T.Y., Pang, H., Cao, Y. et al. (2021). *Angew. Chem. Int. Ed.* 60: 3708–3613.
54 (a) Petkova, D., Borlinghaus, N., Sharma, S. et al. (2020). *ACS Sustainable Chem. Eng.* 8: 12612–1267; (b) Sharma, S., Ansari, T.N., and Handa, S. (2021). *ACS Sustainable Chem. Eng.* 9: 12719–12628.
55 (a) Lindbäck, E., Dawaigher, S., and Wärnmark, K. (2014). *Chem. Eur. J.* 20: 13432–13481; (b) Otte, M. (2016). *ACS Catal.* 6: 6491–6510.

25

The Role of Water in Reactions Catalyzed by Transition Metals

A.W. Augustyniak and A.M. Trzeciak

University of Chemistry, Faculty of Chemistry, 14 F. Joliot-Curie, Wrocław, Poland

25.1 Water as a Solvent in Organic Reactions

Water is the most environmentally friendly solvent, cheap and easily available. In addition, because water is non-toxic and non-flammable, it is highly recommended as a medium for laboratory and industrial syntheses [1]. However, the high polarity of water complicates the performance of organic reactions because most organic compounds are non-polar and insoluble in water. In this context, the mediating effect of water in the course of organic reactions should be mentioned [2, 3]. In 1980, Rideout and Breslow reported an acceleration of the Diels-Alder reaction between non-polar substrates in the presence of water [4, 5]. A similar beneficial effect was confirmed next by other authors for Claisen rearrangement [6].

In 2005, Sharpless introduced the name on water for the reaction in which insoluble reactants are stirred in aqueous suspension [7]. He also mentioned the key role of the phase boundary in the acceleration of organic reactions in these conditions. Organic reactions that benefit from performance on water were reviewed by Fokin et. al [8].

Further interpretations were presented by Jung and Marcus, who focused on the hydrophobic interfacial structure of water and proposed models of on-water and aqueous catalysis [9]. In this kinetic approach, the highest values of rate constants were noted for reactions occurring on the catalyst surface. When reaction takes place at an oil-water interface, the formation of H-bonds with the reactant (B) and with the transition state (AB) facilitated the reaction course (Figure 25.1).

In contrast, only a moderate shortening of the reaction time was observed for aqueous homogeneous reactions (Figure 25.1). This could be related to the presence of the H-bond network formed by water molecules around the reactants.

Kühne et al. supported some conclusions of Jung and Marcus for the Diels-Alder reaction under on water conditions using the molecular dynamics of the Car-Parrinello method [10]. Calculations of the free energy profiles indicated the role of the dangling OH-bonds at a water-oil interface, whereas the stabilization of the transition state by H-bonding only had a small influence on the free energy barrier.

Recently Kitanosono and Kobayashi reviewed different aspects of on water organic reactions as well as reactions performed with and without a catalyst [11]. They proposed a new on water model

---

Catalysis for a Sustainable Environment: Reactions, Processes and Applied Technologies Volume 2, First Edition. Edited by Armando J. L. Pombeiro, Manas Sutradhar, and Elisabete C. B. A. Alegria.
© 2024 John Wiley & Sons Ltd. Published 2024 by John Wiley & Sons Ltd.

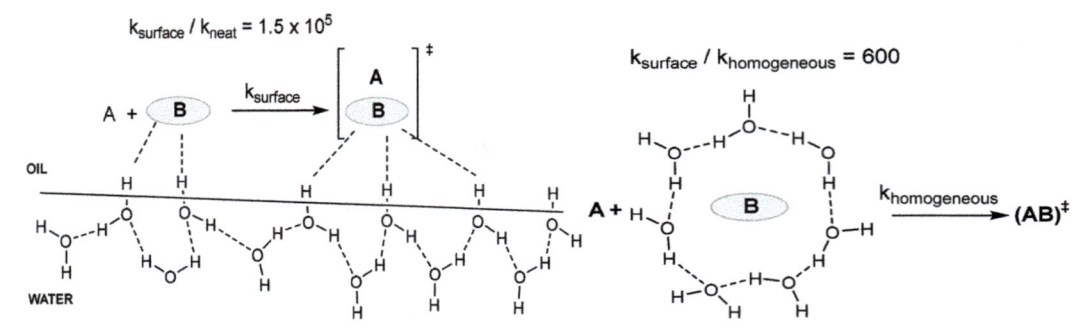

Figure 25.1 The kinetic data for the on water and homogeneous aqueous reactions.

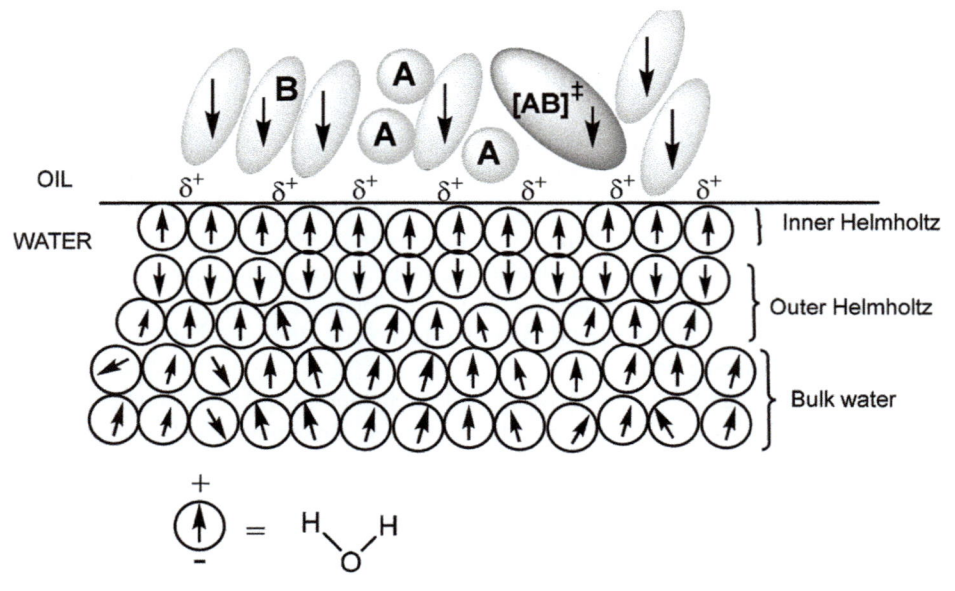

Figure 25.2 Orientation of water molecules on the acidic surface.

for acidic surfaces considering the partial polarization of water molecules. In this model, three water layers were distinguished according to the orientation of water dipoles. Accordingly, the most important for the reaction course was the orientation of molecules at an interface and the formation of hydrogen bonds (Figure 25.2). The authors mentioned that their model can better explain a decrease in the energy of the transition state as a result of the formation of H– bonds between the catalyst and reactants.

The promotional role of water in heterogeneous catalysis was discussed by Li et al., who considered experimental and theoretical aspects of such systems and distinguished different situations describing water assistance [12]. Thus, water molecules accelerate the catalytic reaction due to the stabilization of reactants and intermediates on the catalyst surface via hydrogen bonds. Water can also be involved in H-transfer processes, initiated by the formation of hydrogen bonds with donor or acceptor molecules. In this case, the dehydrogenation of the HA reactant is facilitated by water, which mediates the H-atom transfer to the surface.

Davies discussed the importance of water using examples of the heterogeneously catalyzed oxidation of CO and alcohols or Fischer-Tropsch synthesis [13]. Following some ideas of Roberts [14], he pointed out the contribution of water in the creation of hydroxyls, proton transfer and the stabilization of reactive intermediates.

Micek-Ilnicka reviewed the physicochemical properties of heteropolyacid (HPA)-water systems [15]. HPAs are strong mineral acids and are often used as acid-base or redox catalysts, also in industry. HPAs are typically used in hydrated form and water molecules present in the structure can interact with the terminal oxygen of the HPA anion or they can form hydrogen bonds to OH groups. The formation of $H_5O_2^+$ protonated clusters can also be considered. It is worth noting that water influences acid strength and the number of proton centers in HPAs, which is important for their catalytic activity. For example, the rate of acid-base reactions is related to the concentration of non-solvated protons involved in hydrogen bonds. Moreover, water influences the secondary structure of HPAs and competes with other polar molecules.

It can be concluded that the understanding of the accelerating effect of water molecules in catalytic reactions is quite well developed for heterogeneous systems [16]. Thus, water molecules can stabilize reactants and intermediates on the catalyst surface via the H-bond (a solvation-like effect). In this case, water can also be involved in H-transfer processes, but it is not consumed. Alternatively, water may be decomposed on the catalyst surface with the formation of OH and O species, which contribute next in the catalytic reaction, for example in the oxidation of CO to COOH. The formation of ˙OOH and ˙OH radicals may occur in the reaction of $O_2 + H_2O$. Water can contribute to other specific processes such as the removal of carbon from the catalytic surface, the blocking of active sites or surface reconstruction. In homogeneous systems similar processes can occur with the contribution of water coordinated as a ligand to the active catalyst. Moreover, similarly as in the case of HPAs, the second coordination sphere created by hydrogen bonds can have an additional influence on the access of substrates to the active center and reaction efficiency.

In the nextsections, selected examples, which present water as an active component of catalytic systems, including both heterogeneous and homogeneous ones, are discussed.

25.2 The Role of Water in Heterogeneous Catalytic Systems

25.2.1 The Transformations of Furfuryl Derivatives

The decarbonylation of 5-hydroxymethylfurfural to furfuryl alcohol was performed with a Pd/Al_2O_3 catalyst in a mixed solvent system containing water (Figure 25.3) [17]. The presence of 28 wt% of water caused an increase of selectivity to furfural to 97% due to the suppression of side reactions, such as hydrolysis or etherification. However, at a higher amount of water, 40 wt%, conversion decreased while the contribution of dehydrogenation/polymerization increased. The reaction mechanism was studied using the density functional theory (DFT) method and it was found that the formation of hydrogen bonds between furfuryl alcohol and water hinders further reaction. Moreover, the formation of hydrogen bonds between water and OH groups on the Al_2O_3 surface decreased the activity of Pd/Al_2O_3 and retarded side reactions of furfuryl alcohol.

The effect of water was disclosed in the ring opening of furfuryl alcohol to 1,2-pentanediol catalyzed by immobilized Pd catalysts, Pd/CeO_2, Pt/MgO, and Pd/La-Al_2O_3 (Figure 25.4) [18]. This reaction presents an important step during the transformation of biomass to valuable chemicals.

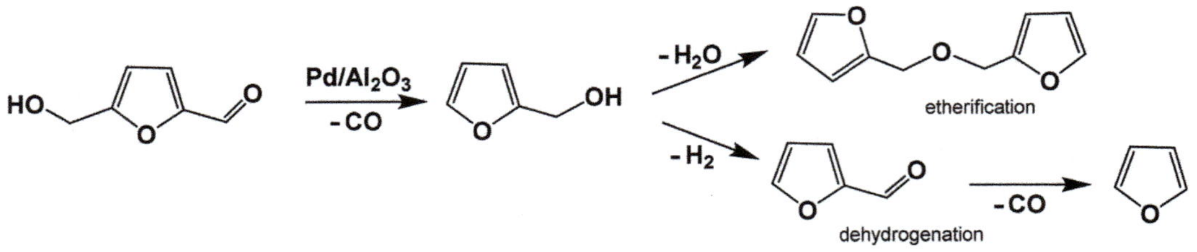

Figure 25.3 The decarbonylation of 5-methoxymethylfurfural.

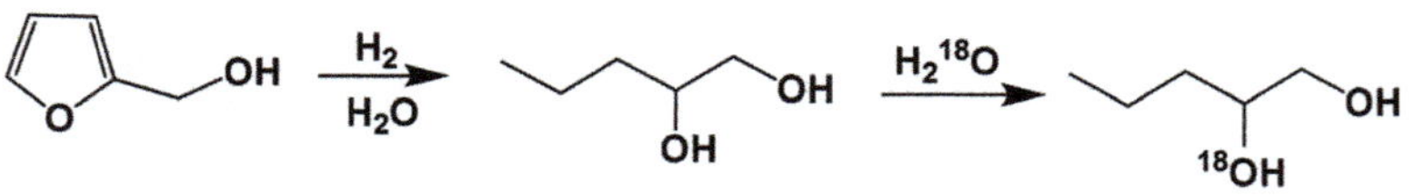

Figure 25.4 The transformations of furfuryl alcohol to 1,2-pentanediol.

By using DFT calculations, several interactions of water molecules with a support and an active metal phase were identified. An experiment performed with $H_2^{18}O$ directly evidenced the presence of ^{18}O in $CH_3CH_2CH_2CH(^{18}OH)$ ion originated from 1,2-pentanediol. This is direct proof for the involvement of water in this reaction.

25.2.2 Oxidation and Deoxygenation

The activity of palladium catalysts in methane combustion is inhibited by water and different explanations of this effect were proposed [19] One of the recent interpretations is based on limited methane adsorption on the active catalyst surface due to the formation of Pd(OH). Moreover, in the presence of water, the activation of C–H bonds in the methyl group was inhibited. The deprotonation of water on oxygen vacancies also has a negative influence on the formation of PdO and, in consequence, on the combustion of methane.

Water is formed by the condensation of surface hydroxyls, while their desorption generates oxygen vacancies (O_V). In these conditions the formation of Pd-OH species is preferred.

$$Pd-O + H_2O + Pd-O_V \rightleftarrows 2Pd-OH$$

Another example of water contribution was reported for the Fe-catalyzed hydrodeoxygenation of phenol to benzene (Figure 25.5) [20]. It was evidenced that water decreased the activation barrier due to the formation of Bronsted acid sites on the Fe(110) surface during water splitting. The formed acid centers are involved in proton transfer and promote the formation of benzene.

The role of water in the catalytic oxidation of hydrocarbons was recently reviewed by Ovchinnikova et al. [21]. Some important aspects related to the presence of water in the catalytic systems were pointed out. One of them is the formation of new Bronsted acid sites on the catalyst surface. These acid sites influenced the reaction course and, in particular, reaction selectivity. This is because they suppressed the formation of some intermediates. Water can also retard the reoxidation process by substituting oxygenate precursors on the catalyst surface. Moreover, water influences the oxidation of hydrocarbons because it competes with reaction products and intermediates for catalytically active sites. It can also facilitate the desorption of the product. As a source of OH groups, water moderates reaction kinetics.

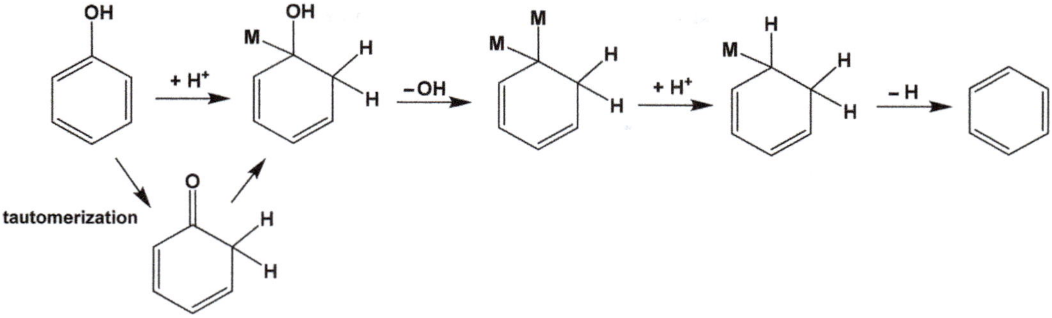

Figure 25.5 The hydrodeoxygenation of phenol.

An interesting example of the enhancement of the reaction rate in the presence of water was reported by Qiu et al. [22]. They studied the oxidation of different aliphatic and aromatic alcohols containing a sulfur or nitrogen atom catalyzed by an Ru catalyst supported on a carbon nanotube (Ru/CNT) in an oil/water emulsion. The beneficial effect of water was explained by the formation of emulsion droplets with an Ru catalyst reacting at the interfaces. An additional advantage of this system is its excellent selectivity to desired aldehydes and the easy recycling of the catalyst.

Mullen et al. reviewed the effect of surface water present on the catalytic activity of gold [23]. They found that the interaction of water with a single crystal gold surface is different in the presence and absence of oxygen. When the surface was covered with atomic oxygen, dissociative adsorption occurred and ˙OOH and ˙OH radicals were formed. They enhanced the selectivity of propene epoxidation, increased the efficiency of CO oxidation and the water-gas shift reaction.

The solvation effect of water can result in the acceleration of the catalytic reaction due to the stabilization of reactants and intermediates on the surface [12]. The formation of hydrogen bonds, without the splitting of O–H, plays the main role here. Theoretical calculations performed for the reaction of O_2 with H_2O on the surface of Au_{10} and Au_{38} clusters show that O_2 is activated and formed the HOO˙ species in reaction with hydrogen from the H_2O molecule.

$$O_2 + H_2O \rightarrow \text{˙OOH} + \text{˙OH}$$

The activation of water on the Au cluster occurred in several steps involving the elongation of the O–O bond in O_2 (from 1.27 Å to 1.38 Å) and an O–H bond in H_2O (from 1.0 Å to 1.42 Å) [24]. As a result HOO˙ and ˙OH radicals were formed. This result can be helpful in explaining the activating role of water in propene epoxidation with O_2 and H_2O on the $Au_7/\alpha\text{-}Al_2O_3$ catalyst. The overall equation for propene epoxidation with water being a catalytic promoter was:

$$2C_3H_6 + O_2 + H_2O \rightarrow 2PO + H_2O \text{ (PO = propene oxide)}$$

Interesting DFT calculations were performed for the conversion of H_2O to H_2O_2 in the presence of $Fe^{IV}O^{2+}$, which occurred according to the scheme:

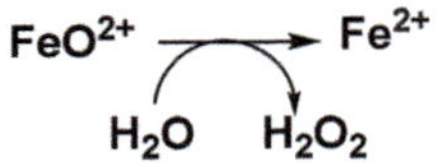

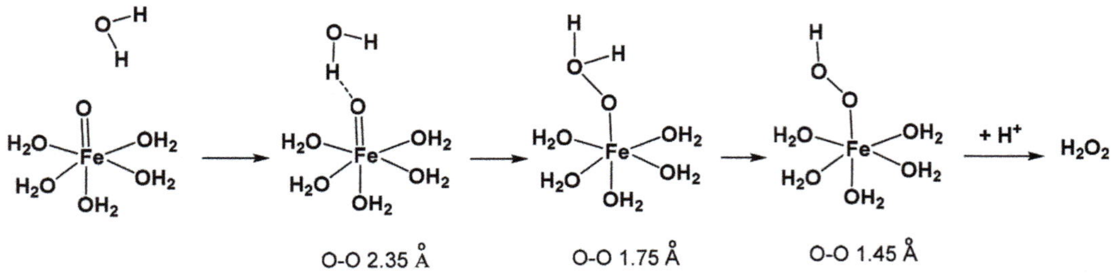

Figure 25.6 Oxidation of water with high-spin Fe(IV)-oxo species.

The reaction proceeded in two steps, namely the formation of the O–O bond between the O_{H2O} (water oxygen) and O_{oxo} ligand followed by the generation of H_2O_2 (Figure 25.6). In the first step contributed $[(H_2O)_5Fe(IV)O]^{2+}$ which coordinated the additional water molecule to the O_{oxo} ligand. Next, the transition state was formed with O–O at a distance of 1.76 Å. Two electrons were transferred from O_{oxo} to Fe and, consequently, Fe(IV) was reduced to Fe(II) and HOO⁻ was generated. Shortening the O–O distance to 1.45 Å indicated the formation of the peroxide fragment (in H_2O_2 the O–O distance is 1.47 Å). In the final step of the reaction, the H_2O_2 molecule was formed due to the H transfer from water to the HOO⁻ group with the O–O bond shortened to 1.35 Å. At this stage, Fe^{2+} formed the $[Fe(H_2O)_5]^{2+}$ complex. These studies explained the main steps of water oxidation and indicated the special role of a second solvation sphere in stabilizing the Fe–OOH intermediate as well as in proton transfer.

25.2.3 Arylcyanation and C–C Cross-Coupling

The arylcyanation of aryl bromides is a method used in the synthesis of aromatic cyanides, including biologically active compounds. It is generally assumed that this reaction is catalyzed by soluble palladium compounds in a homogeneous system. However, recently Fairlamb et al. have shown that water originating from $K_4[Fe(CN)_6]\cdot 3H_2O$ used as a cyanating agent, changed the reaction mechanism to a heterogeneous one [25]. A heterogeneous pathway is based on the activity of Pd(0) in the form of bimetallic species $[L_nPd]_2$ or Pd_n aggregates (Figure 25.7). An excess of H_2O influenced the equilibrium between these forms and the catalytic cycle based on L_nPd species leached from Pd_n, which became dominating. In addition, the contribution of Pd_n aggregates in the reaction course was confirmed by the Hg(0) test. The amount of water present in the system influenced arylcyanation efficiency and higher turnover frequency (TOF) and turnover number (TON) values were obtained for $K_4[Fe(CN)_6]$ containing less than 220 ppm of water. A higher amount of water, >4000 ppm, introduced with $K_4[Fe(CN)_6]\cdot 3H_2O$ caused a decrease in the reaction yield.

A solvent-free Suzuki-Miyaura reaction was successfully performed under conventional heating using PdNPs/MWCNT (multi-walled carbon nanotubes) [26]. In this original solid-state system, a small amount of water formed during the trimerization of phenyl boronic acid plays an important role (Figure 25.8). In agreement with this, calcinated phenylboronic acid as well as $KBPh_4$ or $PhBF_3K$, which cannot release water, provided a much lower conversion. The presence of water in the catalytic system facilitates the diffusion of substrates to the catalyst and mass transfer. Water can also be the source of basic nucleophiles, such as an OH⁻ anion, which can substitute the X ligand in the coordination sphere of Pd to form an R-Pd-OH intermediate. More data concerning the reaction course were collected from a SEM examination. In the reaction mixture, particles of a diameter of c. 100 nm formed while grinding reagents were identified. Interestingly, the coupling product present inside these particles was easily isolated by sublimation. At this stage, some

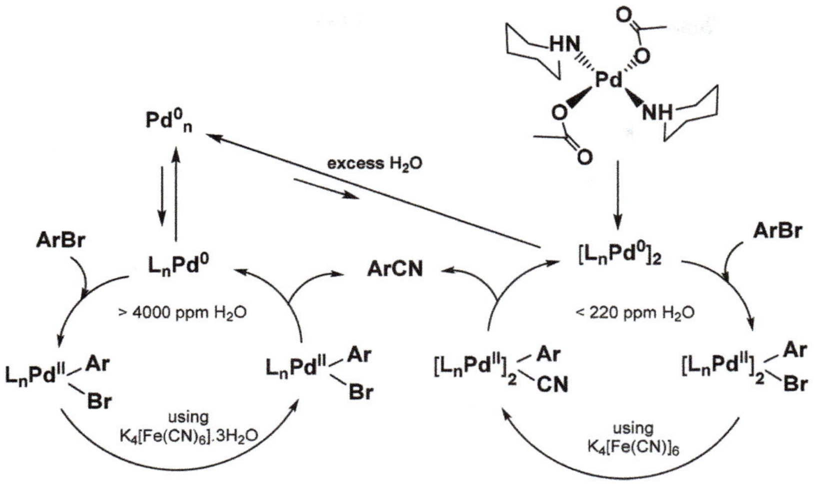

25.2 The Role of Water in Heterogeneous Catalytic Systems

Figure 25.7 Mechanism of arylcyanation.

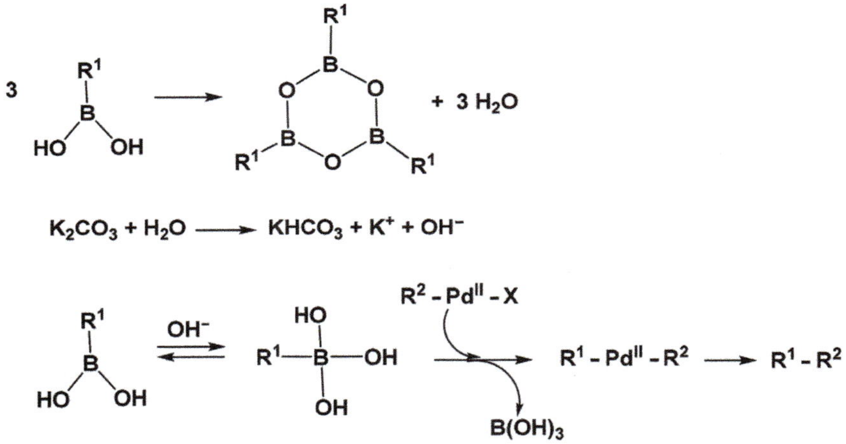

Figure 25.8 Contribution of water in the solid-phase Suzuki-Miyaura reaction.

morphology changes occurred and new pores were formed due to the removal of the product from particles. However, the catalyst, Pd/MWCNT, was reused with the same activity without Pd loss. The presented procedure enables the minimization of Pd leaching due to solvent elimination.

A direct H-atom transfer from H_2O to alkenes and alkynes was reported for transfer hydrogenation with the $B_2(OH)_4$ and Pd/C catalyst [27]. According to the proposed mechanism, supported by experimental data, in the first step, $B_2(OH)_4$ coordinates to Pd via an oxidative addition pathway with the splitting of the B–B bond. Next, water bonded to $B(OH)_2$ is split and acts as a source of the H ligand. Further steps of the catalytic cycle were analogous to other hydrogenation cycles and led to the Pd-alkyl intermediate. In the end the water molecule reacted with $B(OH)_2$ producing the H ligand and $B(OH)_3$ (Figure 25.9). In this reaction, water acted as the only source of hydrogen.

Kamal et. al., discussed the role of the hydrophobic effect on the catalyst surface for water mediated Heck and Ulmann coupling with Pd nanoparticles supported on amphiphilic carbon spheres

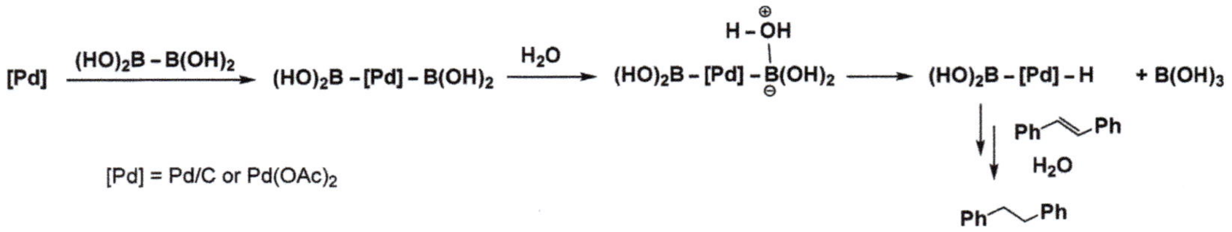

Figure 25.9 Hydrogenation of trans-stilbene by water and $B_2(OH)_4$.

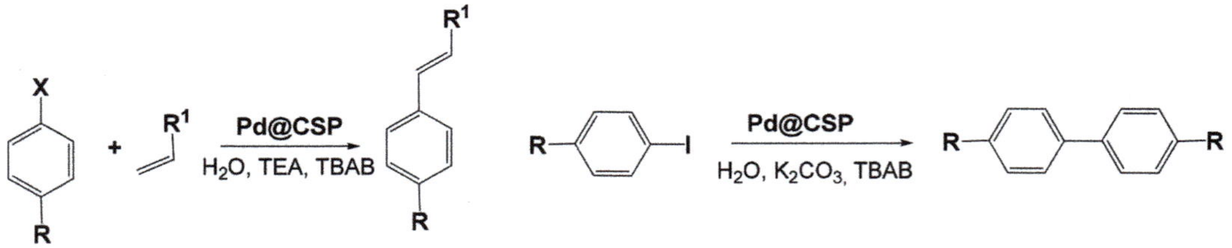

Figure 25.10 The Heck (left) and the Ullmann (right) coupling over Pd@CSP in water.

(CPS) prepared from glucose (Figure 25.10) [28]. Catalytic tests performed for Heck coupling in different solvents revealed the advantages of water, which was the best solvent considering catalyst activity, selectivity and stability. These advantages were related to the Breslow effect based on the amphiphilic structure of carbon support. The presence of carboxylic groups on its surface additionally limits the aggregation of Pd nanoparticles.

25.2.4 Hydrogenation

Experimental and theoretical studies on the mechanism of the hydrogenation of cinnamylaldehyde catalyzed by a Pt_3Fe nanocatalyst supported on a carbon nanotube (CNT) showed the involvement of water in this process [29]. It was found that the water molecule forms a bridge enabling hydrogen exchange between aldehyde and Pt sites (Figure 25.11). Synergistic interactions between water and the catalyst during the hydrogenation process resulted in an increase of both aldehyde conversion and selectivity to alcohol. In particular, the water-mediated pathway was energetically favorable, over the water-free one, due to a lower energy barrier. Thus, water mediates a hydrogen exchange and acts as a promoter of selective hydrogenation.

The hydrogenation of aromatic acids catalyzed by Pd supported on a carbon nanofiber in water occurred selectively in the aromatic ring [30]. The carboxylic acid groups were not affected and cyclohexanecarboxylic acids were formed as the only products as final products. This high selectivity of hydrogenation was explained by the interaction between water, the substrate and the metal surface. It was assumed that polar OH groups of acid form hydrogen bonds with water molecules blocking their interaction with the metal center. In such a case, the interaction of the aromatic ring with the active metal was preferred and hydrogenation occurred in the ring.

An interesting effect of water was observed during the hydrogenation of acetophenone catalyzed by Pd/S-DVB (S-DVB = styrene/divinylbenzene copolymer) in $MeOH/H_2O$ and $2\text{-PrOH}/H_2O$ mixtures [31]. The reaction proceeded efficiently in pure water or in pure alcohol, while in a mixture of solvents the yield was significantly lower. For example, 91% of acetophenone was converted in pure MeOH, while in the mixture 1.5 MeOH/1.5 H_2O conversion decreased to 24% only (Figure 25.12). It was assumed that hydrated clusters of alcohol formed in the mixed solutions interacted with the

Figure 25.11 The hydrogenation of cinnamaldehyde over $Pt_3Fe@CNT$ catalyst.

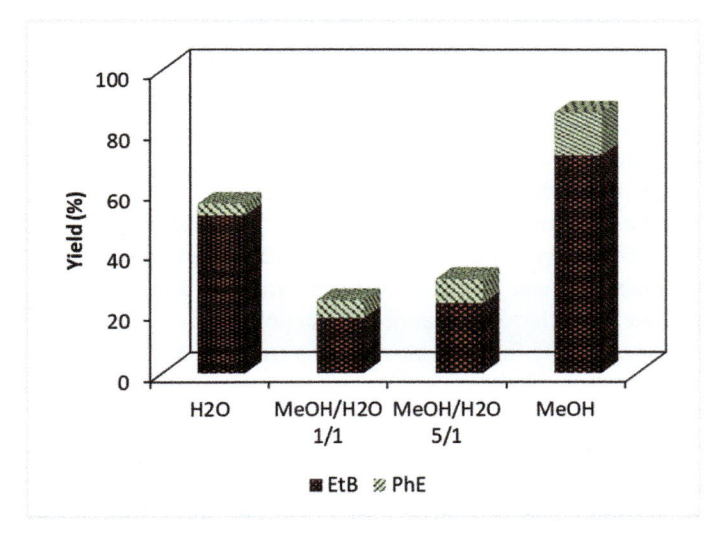

Figure 25.12 The hydrogenation of acetophenone in different solvents.

porous catalyst blocking the access of substrates to Pd active centers. As a result, catalytic efficiency decreased. This inhibiting effect was evident in $MeOH/H_2O$ and 2-PrOH solutions, while in $BuOH/H_2O$ it was significantly weaker due to a weaker hydration of bigger BuOH clusters.

25.2.5 Hydroformylation

The asymmetric hydroformylation of vinyl acetate catalyzed by the Rh/DNA catalyst in the presence of chiral diphosphine in water proceeded with significantly higher enantioselectivity when compared to the same reaction with the $Rh(acac)(CO)_2$ catalyst [32]. For instance, enantioselectivity (ee) achieved the value of 49% with Rh/DNA + (R)-BINAP and 11% with $Rh(acac)(CO)_2$ + (R)-BINAP (Figure 25.13). Importantly, the addition of free DNA to the reaction catalyzed by $Rh(acac)(CO)_2$ caused an increase of ee to 35%. It was concluded that the synergistic interaction of DNA support with chiral phosphine has an improving effect on the reaction course. The contribution of water in the stabilization of the active species should also be considered. This catalytic system also works well for the asymmetric hydroformylation of styrene.

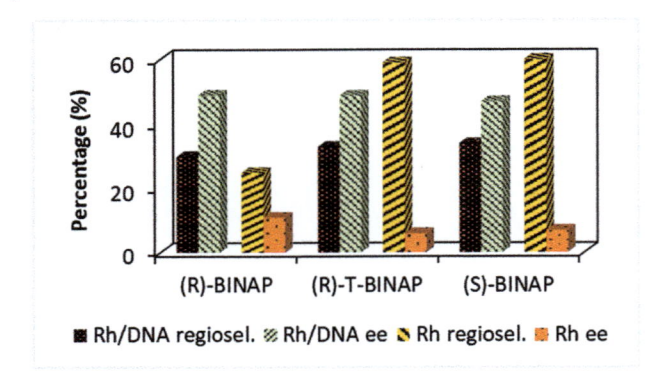

Figure 25.13 The asymmetric hydrogenation of vinyl acetate catalyzed by Rh/DNA in water (regioselectivity to 2-acetoxypropanal).

25.2.6 Catalytic Reactions with MOF-based Catalysts in an Aqueous Medium

For more than two decades, metal-organic frameworks (MOFs) and covalent organic frameworks (COFs) have been considered exciting platforms for applications in heterogeneous catalysis [33, 34]. Their well-defined structures, high surface areas, and tunable pores create a unique environment around catalytic centers. In the context of catalysis performed in an aqueous medium, it is important to emphasize that the hydrophobic/hydrophilic character of MOFs can accumulate selected substrates around catalytic centers and accordingly influence reaction selectivity.

A number of catalytic reactions have been developed using MOFs and COFs in heterogeneous catalysis, including reactions in an aqueous phase [35, 36]. However, still less work has been devoted to specific interactions with water, which can influence the reaction course.

In this chapter, special attention was paid to the function of water as a reactant or co-catalyst.

25.2.6.1 Cross-Coupling Reactions

Table 25.1 compiles selected reports concerning the use of Pd-MOFs as catalysts for C–C coupling reactions such as Suzuki–Miyaura, Heck, and Sonogashira, in which water was used as a solvent. The use of water in these reactions is strongly recommended, not only from an environmental point of view but also because of the application of their products in the pharmaceutical industry. However, in most cases, the influence of water on the reaction has not clearly been stated.

One of the first reports on a Suzuki reaction in a water medium was presented for Pd@MIL-101-Cr as a catalyst (Figure 25.14). Water in this system was primarily a green solvent; however, there was also some information about its other role. It was noticed that the combination of water and a NaOMe base led to the generation of a small amount of methanol and NaOH. This system was able to achieve up to 78% conversion of 4-chloroanisole, while much less, 63% of the product was obtained in a reaction performed in a methanol/water solution in the presence of NaOH. These results confirmed that the formation of MeOH and NaOH in situ increased the efficiency of this reaction [37].

Another MOF material, Pd@UiO-66-NH$_2$, used in the Suzuki reaction in water formed 53% of 4-methoxybiphenyl when bromobenzene was used as a substrate [39]. The proposed reaction mechanism indicates that water participated in the reaction of aryl boronic acid with K$_2$CO$_3$. An anionic species, R-B(OH)$_3^-$, formed in this reaction, is considered the most reactive species toward the transmetalation step according to previous works [48].

Martín-Matute et al. used the catalyst Pd@MIL-101-NH$_2$-Cr for Suzuki–Miyaura cross-coupling in water at room temperature (Figure 25.15) [38]. In this case water affected the reactivity, and electron-rich substrates reached full conversion while the electron-poor compound was converted

25.2 The Role of Water in Heterogeneous Catalytic Systems

Table 25.1 C–C coupling reactions catalyzed by MOFs-based Pd catalysts in water.

Entry	Catalysts	Reaction type	Ref.
1	Pd/MIL-101-Cr	Suzuki-Miyaura	[37]
2	Pd@MIL-101(Cr)-NH$_2$	Suzuki-Miyaura	[38]
3	Pd/UiO-66-NH$_2$	Suzuki-Miyaura	[39]
4	Pd(0)/MCoS-1	Suzuki-Miyaura, Sonogashira	[40]
5	Pd–NHC–MIL-101(Cr)	Suzuki-Miyaura	[41]
6	Pd@[Ni(H$_2$BDP-SO$_3$)$_2$]	Suzuki-Miyaura	[42]
7	Pd/Y-MOF	Suzuki-Miyaura, Sonogashira	[43]
8	Pd^{II}@Cu(BDC)/2-Py-SI	Heck	[44]
9	Pd@MIL-101	Heck Suzuki	[45]
10	Pd(II)-UiO-67	Heck Suzuki	[46]
11	Pd@MIL-101-NH2 Pd@MIL-88B-NH$_2$	Heck	[47]

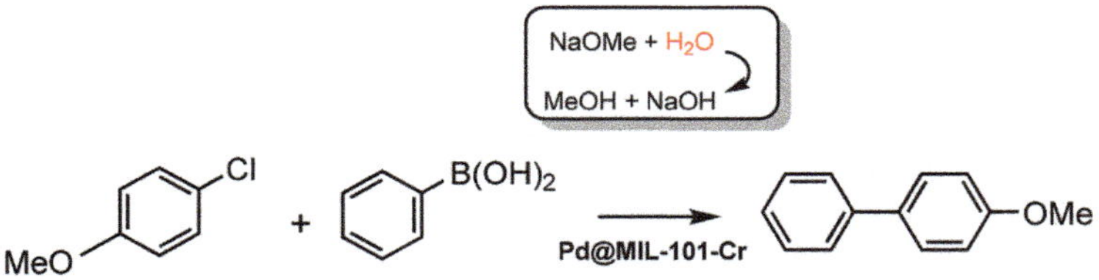

Figure 25.14 The Suzuki-Miyaura coupling of 4-chloroanisole catalyzed by Pd@MIL-101-Cr.

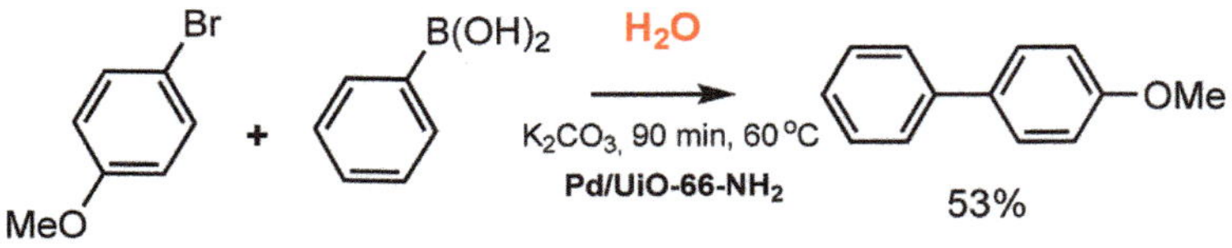

Figure 25.15 The Suzuki-Miyaura coupling of bromobenzene derivatives catalyzed by Pd@MIL-101-Cr.

into the corresponding biphenyl in only 84%. These differences were explained by the different solubility of these compounds in water.

Zhou investigated the mechanism of the Heck C–C coupling catalyzed by Pd@MIL-101-NH$_2$ and Pd@MIL-88B-NH$_2$ in the water medium (Figure 25.16) [47]. They demonstrated that in the absence of H$_2$O, Pd remained in an oxidation state +2. However, in the presence of water and olefin, Pd was reduced to Pd(0), which represents the active species that initiates the catalytic cycle.

COFs materials possess well-defined porous skeletons which can bind metal nanoparticles. Additionally, COFs are insoluble in organic solvents and in water, and are attractive supports for heterogeneous catalysts. Surprisingly, they have only been scarcely used for C-C coupling thus far. The first example of using COF material in the Suzuki-Miyaura reaction in an aqueous medium was presented by Esteves et al. (Figure 25.17) [49]. They used Pd(OAc)$_2$@COF-300 as a catalyst in

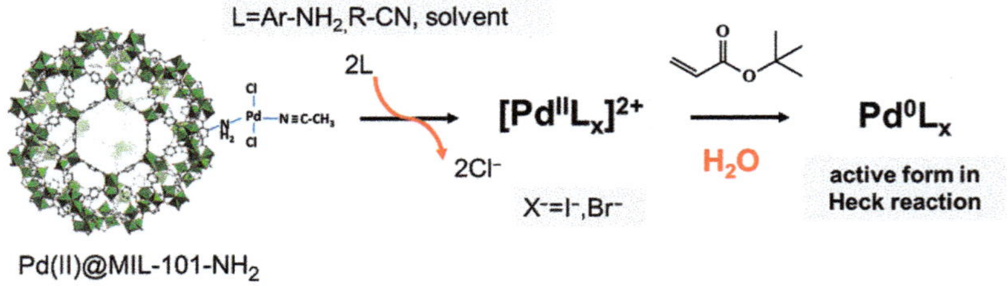

Figure 25.16 Formation of the catalytically active form of Pd catalyst.

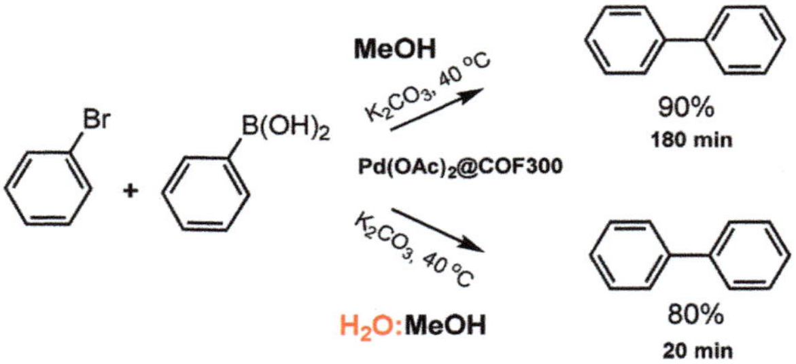

Figure 25.17 Influence of water on the Suzuki–Miyaura reaction rate.

reaction between bromobenzene and phenylboronic acid in methanol and obtained 90% yield during 180 minutes. The use of the MeOH:H_2O mixture (1:1) led to an 80% yield already after 20 minutes. Unfortunately, the authors did not perform the reaction in water alone, however, clearly, the reaction was improved by the addition of water.

In another work, the COF-based material Pd@COF-QA was used for the first time in Suzuki–Miyaura cross-coupling in only water [50]. In the reaction between iodobenzene and phenylboronic acid, 99% yield of biphenyl was obtained after six hours. Under the same conditions much less reactive chlorobenzene was converted in 95%, indicating the high efficiency of this catalyst in water.

25.2.6.2 Hydrogenation Reactions

Jiang et al. developed an efficient heterogeneous catalyst system based on Pd nanoparticles supported on MIL-101-Cr and nanoparticles (Figure 25.18) [51]. It was used for phenol hydrogenation in water and excellent selectivity >99.9% to cyclohexanone was obtained in remarkably mild reaction conditions, including low pressure (0.1 MPa) and temperature as low as 25 °C. It was emphasized that cyclohexanone could easily be separated from water, which is attractive for eventual practical applications.

An interesting example of the fundamental role of water in the transfer hydrogenation of phenylacetylene was reported by Trzeciak and Augustyniak (Figure 25.19) [52]. They used as [Pd(2-pymo)$_2$]$_n$ as a catalyst, which is a unique MOFs containing Pd(II) nodes. Mechanistic studies indicated that water is a donor of one hydrogen atom, and the second hydrogen is generated from the hydrolysis of NH_3BH_3. The same catalyst is active in the dehydrogenation of NH_3BH_3. It is worth mentioning that the presence of hydrophobic cages in [Pd(2-pymo)$_2$]$_n$ limits the contact of NH_3BH_3 with Pd in a

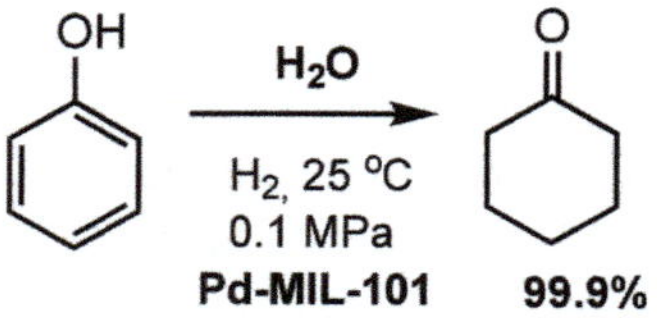

Figure 25.18 The hydrogenation of phenol to cyclohexanone.

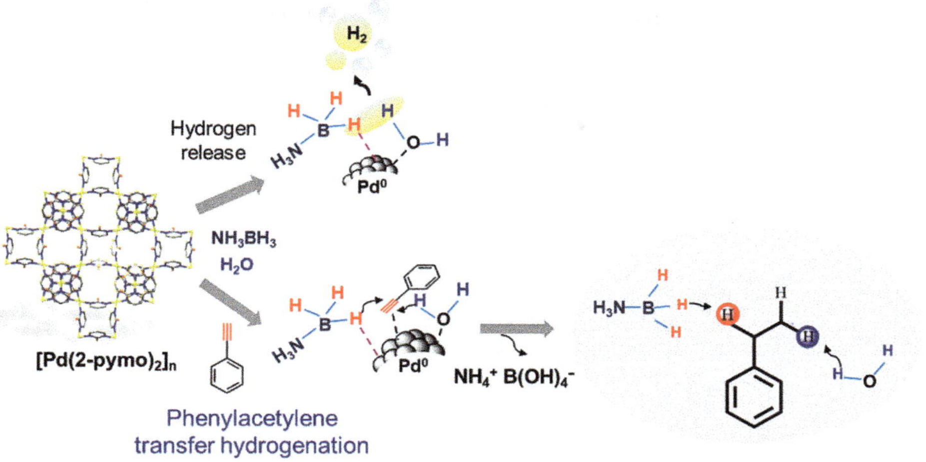

Figure 25.19 Reactions of NH_3BH_3, hydrogen release and transfer hydrogenation, catalyzed by $[Pd(2\text{-pymo})_2]_n$.

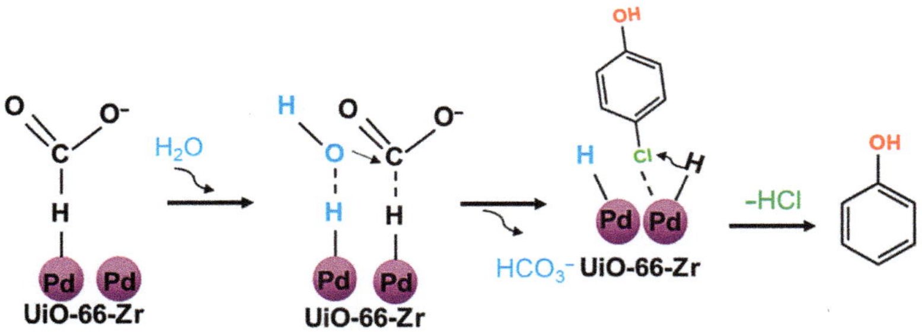

Figure 25.20 Activation of formate on UiO-66-Zr in hydrodehalogenation catalyzed by Pd@UiO-66.

water solution. This only resulted in a partial reduction of Pd(II) to Pd(0) and lowered the yield of H_2 formed from the hydrolysis of NH_3BH_3.

Recently, Olsbye et al. showed the positive influence of H_2O on CO_2 hydrogenation to methanol catalyzed by Pt@UiO-67 [53]. The obtained results indicated that water increased the selectivity to methanol because it promoted methanol desorption from the catalyst surface and inhibited the formation of CH_4.

Li et al. [54] evaluated the hydrodehalogenation of aryl and heteroaryl halides over Pd supported on the UiO-66 (Figure 25.20). In addition to ammonium formate being a hydrogen donor, the presence of water was required to trigger the reaction. The proposed catalytic cycle supposes that the first carboxyl from ammonium formate is activated on Pd active sites, and then the nucleophilic addition of water to the carbonyl groups with the H–O bond cleavage. As a result, two Pd–H active sites for hydrodehalogenation reaction are obtained.

25.2.6.3 Hydroamination

Cirujano et al. reported the synthesis of a defective nickel pyrazolate porous framework with cationic $[Pd(NH_3)_4]^{2+}$ sites. This material was employed as a catalyst in hydroamination reactions (Figure 25.21) [55] Interestingly, in case there is no water in the catalytic system, the reaction goes through by the coordination of palladium sites to the triple bond of the substrate, and, finally,

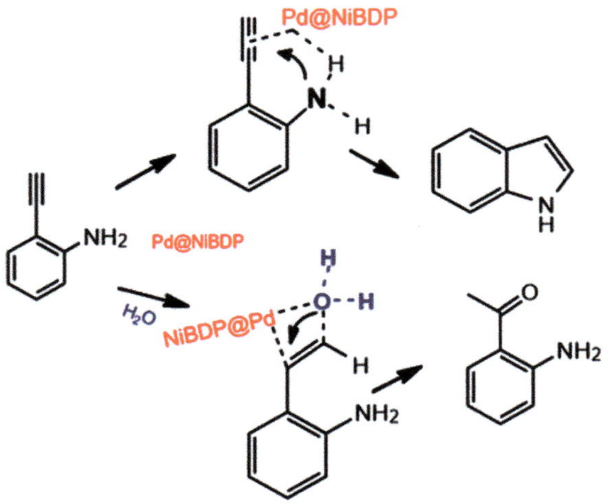

Figure 25.21 Mechanism of hydroamination and hydration over Pd@NiBDP catalyst.

indole was generated. However, water introduced into the system favored the hydration of the terminal alkyne, and 2-aminoacetophenone was obtained.

25.3 The Contribution of Water in Homogeneous Catalytic Systems

25.3.1 Oxidation and Epoxidation

The contribution of water in the oxidation of alkanes was studied experimentally and theoretically for the [MeReO$_3$]/H$_2$O$_2$/H$_2$O–CH$_3$CN system. According to this scheme, water is the H-transfer promoter and oxidation occurs with the contribution of HOO˙ and HO˙ radicals, formed in the water assisted process [56].

Thus, the Re(VII) complex, MeReO$_2$(OO)(H$_2$O), formed in the reaction of MeReO$_3$ with H$_2$O$_2$, reacted with the second H$_2$O$_2$ molecule to give MeReO$_2$(OO)(H$_2$O)(H$_2$O$_2$) (Figure 25.22). Next, the reaction of this complex with H$_2$O resulted in a cleavage of the Re-OOH bond and the formation of the HOO˙ radical and further reactions with H$_2$O$_2$ and H$_2$O generate HO˙ radicals involved in the oxidation process. It was found, that the coordination of water to the Re catalyst decreased the activation energy of radical formation and, in consequence, the oxidation reaction can occur under mild conditions, at room temperature.

The epoxidation of olefins by hydrogen peroxide catalyzed by the same complex, MeReO$_3$, was accelerated in the presence of H$_2$O, which acted as a co-catalyst. This improving effect was explained by the contribution of water in all steps of the catalytic process in agreement with results of DFT calculations [57]. In particular, the formation of cyclic transition states facilitates the proton transfer from H$_2$O$_2$ to the oxo or hydroxo ligand. As a result, peroxo complexes are formed and they react next with cyclohexene to form epoxide (Figure 25.23).

25.3.2 The Hydrogenation of Carbonyl Compounds and CO$_2$

Ab initio molecular dynamic calculations performed for the model Ru-catalysts in the transfer hydrogenation of formaldehyde, show the difference between reaction mechanisms operating in an aqueous solution, in methanol or in the gas phase. The formation of hydrogen bonds by water

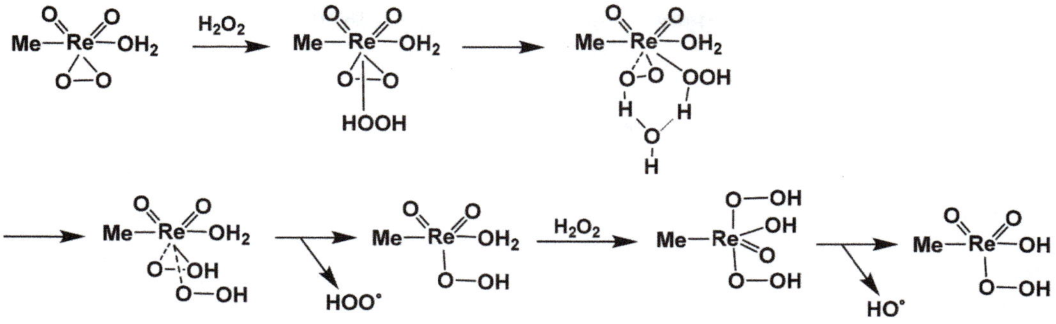

25.3 The Contribution of Water in Homogeneous Catalytic Systems

Figure 25.22 Formation of HOO· and HO· radicals in reaction with MeReO$_2$(OO)(H$_2$O) complex.

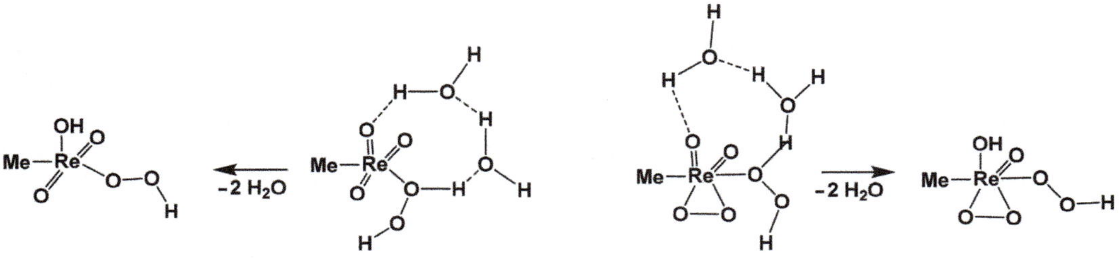

Figure 25.23 Contribution of water in olefin epoxidation with Re catalyst.

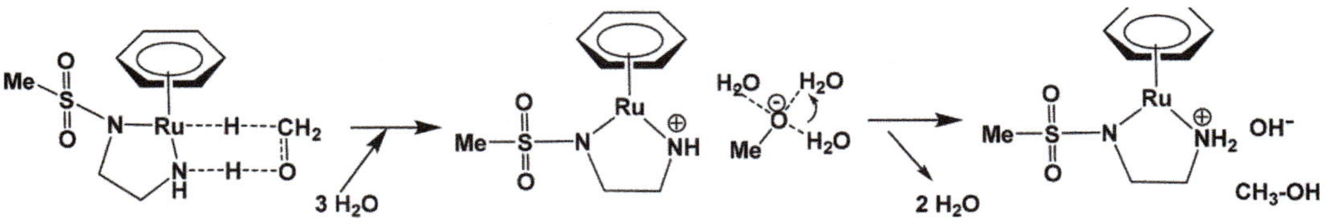

Figure 25.24 Ru-catalyzed transfer hydrogenation of formaldehyde in water.

and formaldehyde facilitated a proton transfer from a water molecule to substrate. A Ru-methoxide intermediate created at this stage was stabilized by three water molecules (Figure 25.24). In contrast, methanol interacted mainly with the catalyst [58].

Two energetic profiles were calculated for acetone hydrogenation catalyzed by [RhH$_2$(PR$_3$)$_2$S$_2$]$^+$ in the presence of water [59]. The first case was based on a reductive elimination pathway and the second, on outer sphere hydrogenation. It was found that the energetic span for the first path is lower and, consequently, it can be considered the energetically preferred route. The presence of two water molecules significantly reduced the relative energy of the Rh-intermediate from 27.1 kcal mol^{-1} to 10.9 kcal mol^{-1} (Figure 25.25).

The accelerating influence of water on the Ru-catalyzed hydrogenation of CO$_2$ was studied using the DFT method. It was found that the formation of hydrogen bonds between water and CO$_2$ facilitated the nucleophilic attack of the H ligand on CO$_2$ by lowering the activation barrier [60]. The water modulated interaction of CO$_2$ with the Ru-H species lead to the insertion product. It is worth noting that in the reaction carried out in the presence of water, the active species is cis-Ru(H)$_2$(PMe$_3$)$_3$(H$_2$O)$_2$, whereas cis-Ru(H)$_2$(PMe$_3$)$_3$ is catalytically active in the reaction performed without water. Interestingly, when the reaction was carried out in water, CO$_2$ interacted with hydride and aqua ligands but not with the Ru center.

25.3.3 The Cyclotrimerization of Alkynes

Water played an important role in the Pt(II) catalyzed cycloisomerization of enynes [61]. This reaction is mainly catalyzed by Pt and Au complexes and different mechanisms were proposed for these two types of catalysts. For Au, a linear coordination of alkyne to Au(I) is expected rather than the formation of metallacycle. In contrast, metallacycles are typical intermediates in the Pt catalyzed cycloisomerization of enynes. It was found that in this reaction water facilitated product formation and to explain this effect Echavarren et al. proposed that water prevented the chelating coordination of the 1,6-enyne to the Pt catalyst involving the coordination of 1,6-enyne via its double and triple bonds. This assumption was next supported by electrospray ionization-mass spectrometry (ESI-MS) studies and theoretical calculations. The Pt(II) anionic complex with the diphosphine ligand, used as a precatalyst, formed an adduct with the cyclic reaction product identified by ESI MS (m/z 394.6) using CAR/CID sequences (CAR = collision-activated reaction, CID = collision-induced dissociation). What was important, the amount of this adduct significantly increased (by factor c. 2.6), in the presence of H$_2$O compared to CHCl$_3$. In conclusion, the presence of the aqua ligand favored the coordination of 1,6-enyne to Pt via a triple bond and limited the formation of chelate and directed reaction selectivity toward cycloisomerization (Figure 25.26).

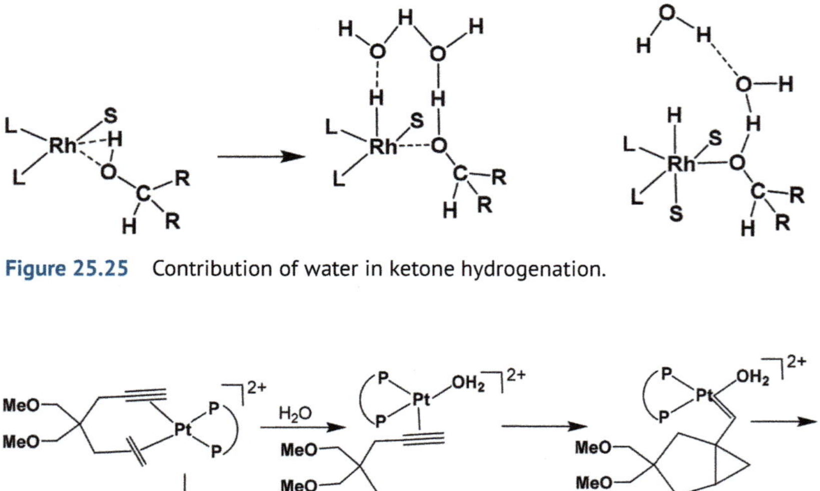

Figure 25.25 Contribution of water in ketone hydrogenation.

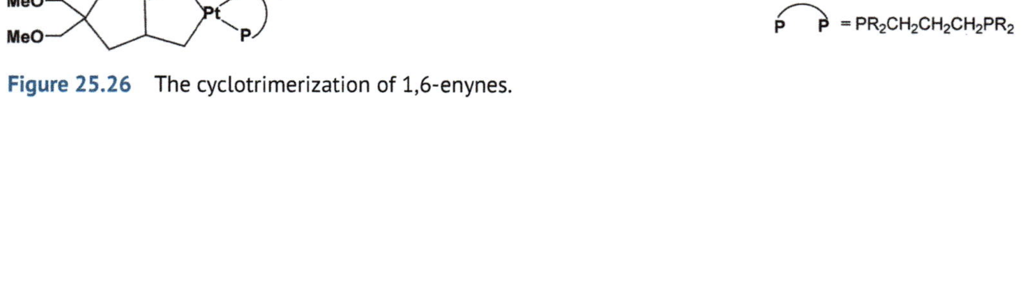

Figure 25.26 The cyclotrimerization of 1,6-enynes.

25.3.4 The Isomerization of Allylic Alcohols

The isomerization of allylic alcohols presents a simple atom-economical procedure for the synthesis of aldehydes or ketones which can be performed in water. In the review article published by Romerosa et al., different aspects of this reaction were discussed including the presentation of catalysts and mechanistic approaches. Considering catalysts of isomerization, ruthenium is the most frequently used and several versions of the reaction mechanism for Ru complexes were proposed [62, 63].

One of these mechanisms, proposed for the neutral Ru-Cl complex, is based on the π-allyl-hydride scheme. In the first step, alcohol coordinates to the Ru catalyst via a double bond. Next, the β-proton from the π-allylic fragment is transferred to the metal, forming the Ru-H species. The rearrangement of protons in this intermediate produces enol transformed next to aldehyde. In this classic mechanism water was not involved (Figure 25.27).

In 2011 Valera-Alvares et al. presented studies on the isomerization of allylic alcohols catalyzed by Ru complexes [{Ru(η^3:η^3-C$_{10}$H$_{16}$)Cl(μ-Cl)}$_2$], [Ru(η^3:η^3-C$_{10}$H$_{16}$)Cl$_2$(L)], (L=CO, PR$_3$, CNR, NCR, and [Ru(η^3:η^3-C$_{10}$H$_{16}$)(NCMe)$_2$]SbF$_6$ (C$_{10}$H$_{16}$ = 2,7-dimethylocta-2,6-diene-1,8-diyl) in water and in THF solutions [64]. For both reactions, the Gibbs free energy profiles were constructed and it was found that the coordination of water to Ru lowered the energy barrier by c. 7 kcal mol^{-1}. It was also characteristic that in all intermediates included in the catalytic process, Cl and H$_2$O ligands remained coordinated to Ru (Figure 25.28).

Another mechanism was proposed for the same reaction in a basic medium in which catalytic efficiency was significantly higher than in neutral conditions (Figure 25.29). In this case, water

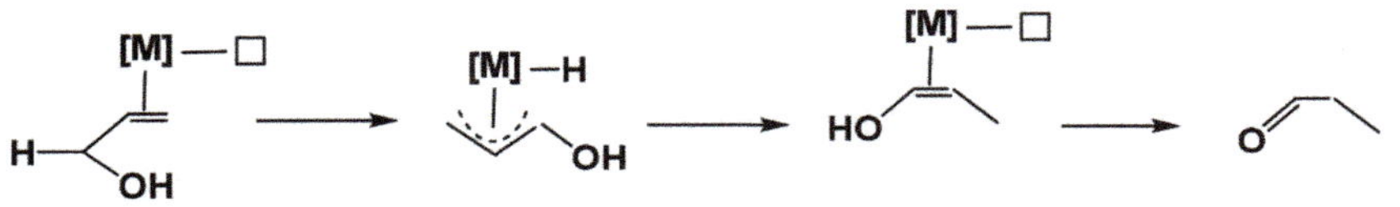

Figure 25.27 The isomerization of allylic alcohols.

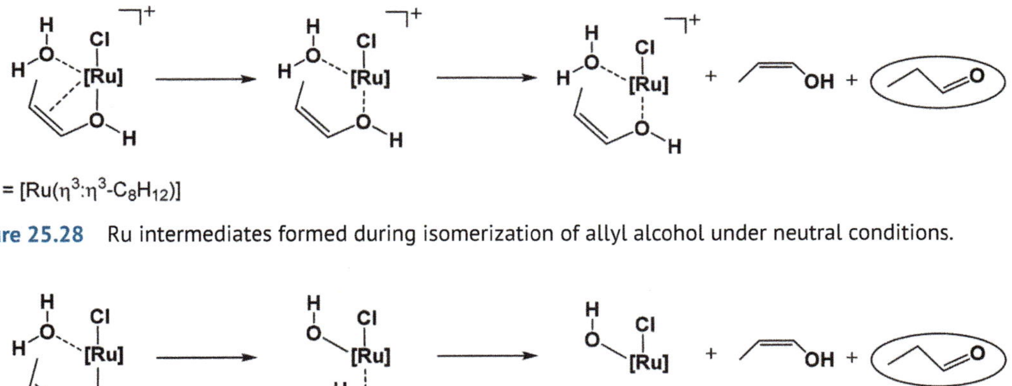

[Ru] = [Ru(η^3:η^3-C$_8$H$_{12}$)]

Figure 25.28 Ru intermediates formed during isomerization of allyl alcohol under neutral conditions.

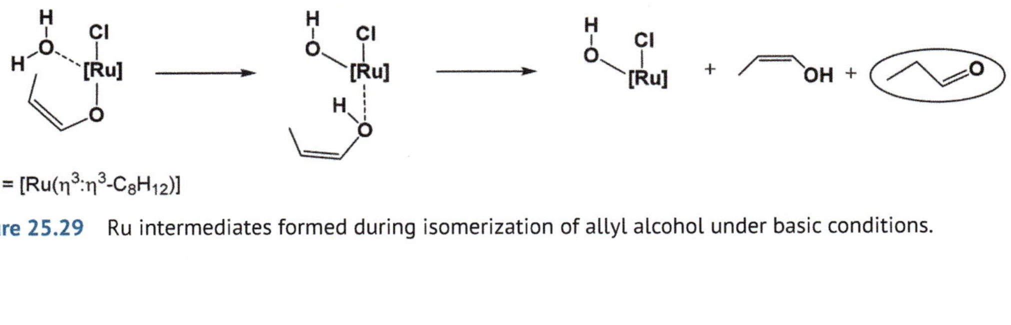

[Ru] = [Ru(η^3:η^3-C$_8$H$_{12}$)]

Figure 25.29 Ru intermediates formed during isomerization of allyl alcohol under basic conditions.

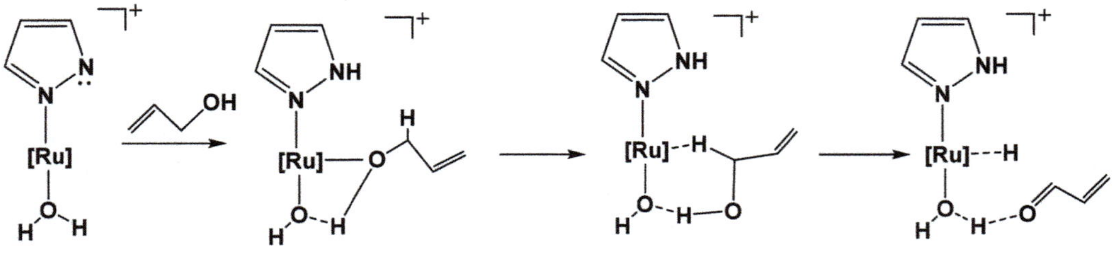

Figure 25.30 The isomerization of allyl alcohol in water with Ru-pyrazole catalyst.

contributed directly in the reaction as a donor of a proton transferred to an enolate ligand. Ru species involved in the catalytic cycle contained aqua or hydroxo ligands.

Gimeno et al. employed a Ru(IV) catalyst bearing the pyrazole-type ligand, [Ru(η^3:η^3-$C_{10}H_{16}$)Cl_2L] ($C_{10}H_{16}$ = 2,7-dimethylocta-2,6-diene-1,8-diyl, L = pyrazole ligand) in the isomerization of allylic alcohols in water [65]. They proposed the mechanism of this reaction with the contribution of water and pyrazole ligands, while the Cl ligand was not involved in an active intermediate (Figure 25.30). The studies performed for the most active complex show that water easily substituted the Cl ligand and formed Ru-OH species after the proton transfer to an N-atom of pyrazole. The substrate coordinated to Ru via an oxygen atom and formed the hydrogen bond with an OH ligand. Next, a bifunctional concerted mechanism occurred with a hydride transfer to Ru and a proton transfer to the OH ligand. The as formed Ru intermediate contained aqua and acrolein ligands. The transfer of two hydrogens from Ru-H and N-H groups to the C=C bond of acrolein was the final step of this reaction leading to the expected ketone.

The hydrogen transfer step was investigated in detail considering two pathways with a different role of water. The authors concluded that the presented mechanism of isomerization with the active contribution of water can be competitive to the chloride route in the classic mechanism. The new mechanism also explains that reaction in water is faster than in tetrahydrofuran (THF).

The effect of water was mentioned in the studies of the isomerization of allylic alcohols catalyzed by the water soluble [Rucp(H_2O)(PTA)$_2$](CF$_3$SO$_3$) complex (PTA = 1,3,5-triaza-7-phosphaadamantane) [66]. An increase of the reaction efficiency, observed in this reaction, was explained using the results of neutron scattering and theoretical calculations. The most interesting finding was the formation of a three-membered water chain connecting hydrogen atoms of the OH group of allylic alcohol with the N-atom of PTA. A similar water chain was also found for the second isomer of an Ru- intermediate, [Rucp(η^2-CH_2=CH-CH_2OH)(PTA)$_2$]$^+$. The interaction with the water chain resulted in the increased stabilization of the isomer involved in the further reaction. The improving effect of water was additionally evidenced by the fact that catalytic results were much worse in MeOH than in water. This is because in MeOH hydrogen bonds do not stabilize the most favorable conformation of substrate, which can rotate around the C–C bond. In summary, the specific solvation of catalytic intermediates was responsible for the high efficiency of allyl alcohol isomerization in water.

25.3.5 Hydroarylation with Boron Compounds

Very promising synthetic pathways can be realized due to the reactivity of diboron compounds with water. For example, diboron compounds mediate a palladium-catalyzed transfer of hydride from the water molecule to unsaturated substrates, such as alkenes and alkynes. In these reactions, the O–H bond in water is split on the highly oxophilic boron center and acts as a hydrogen donor [67].

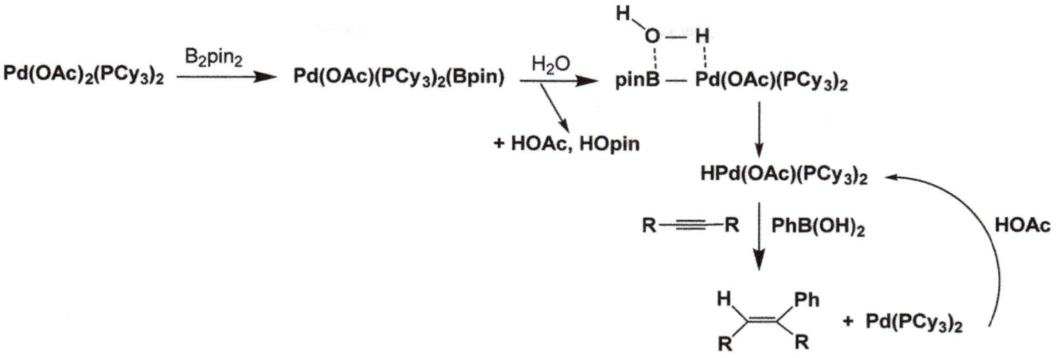

Figure 25.31 Plausible mechanism of hydroarylation catalyzed by Pd(OAc)$_2$(PCy$_3$)$_2$.

Figure 25.32 The hydroarylation of alkynes with D$_2$O.

The hydroarylation of alkynes presents another reaction which can be successfully performed with water as the hydride source. In this reaction, an internal alkyne is transformed to the olefin with the addition of aryl and hydride to the triple bond. Prabhu et. al., reported an example of such a process catalyzed by Pd(OAc)$_2$/Pcy$_3$ with the diboron compound B$_2$pin$_2$ [B$_2$pin$_2$ = bis(pinacolato) diboron] used for water activation and PhB(OH)$_2$ served as a donor of the phenyl group [68]. The reaction mechanism involved in the formation of the hydrido species of the H-Pd-OAc type in the reaction of Pd(OAc)$_2$(PCy$_3$)$_2$ with B$_2$pin$_2$ and H$_2$O (Figure 25.31). It is worth noting that Pd(OAc)$_2$(PCy$_3$)$_2$(Bpin) reacted with H$_2$O forming the catalytically active species, HPd(OAc)(PCy$_3$)$_2$. Water also contributed in another step of the catalytic process, namely the transmetalation of the Pd precursor with B$_2$pin$_2$ leading to pinB-[Pd]-OAc and AcO-Bpin. The as formed AcO-Bpin reacted with H$_2$O forming HOAc, which is employed in the end of the catalytic cycle to recover catalytically active H-[Pd]-OAc from [Pd(0)(PCy$_3$)$_2$].

The function of water as a hydrogen donor was also evidenced in the hydroarylation of alkynes performed in water with the arylating agent Na[BPh$_4$], catalyzed by PdCl$_2$(im)$_2$ (im = substituted imidazole) catalysts with different imidazole ligands [69]. The test experiment, performed with D$_2$O instead of H$_2$O showed the presence of D exclusively in the vinyl position of arylated olefin despite the presence of non-deuterated HOAc in this reaction (Figure 25.32).

25.3.6 Hydroformylation

An unexpected increase of hydroformylation selectivity was achieved using the Rh(acac)(CO)$_2$ and π-acceptor P(NC$_4$H$_4$)$_3$ ligand in the presence of water. The hydroformylation of 1-butene carried out in a toluene/water mixed solvent system provided n/iso as high as 46 (at 8 bar H$_2$/CO) or 51 (at 6 bar H$_2$/CO). For comparison, in the absence of water the highest n/iso value was 22. Similarly, in the hydroformylation of propene with the P(NC$_4$H$_4$)$_3$ ligand n/iso was 27.1 while without water it was 18.2.

It was supposed that water interacted with the Rh-H catalytic intermediate and influenced a migratory insertion step to form a linear alkyl complex and next a linear aldehyde (Figure 25.33).

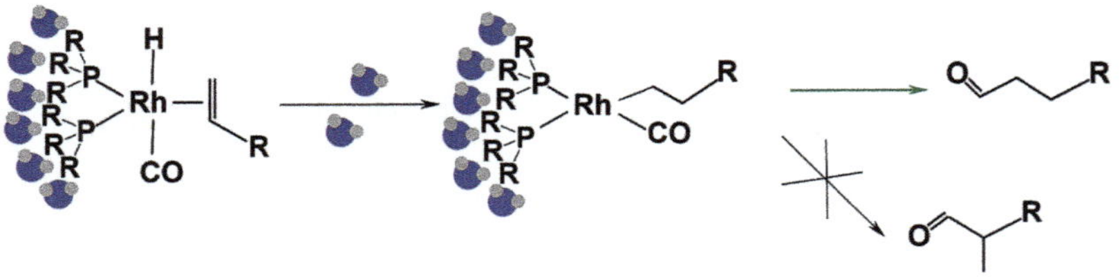

Figure 25.33 The proposed interactions of water with Rh-H and Rh-alkyl intermediates during selective hydroformylation.

Not only selectivity but also the reaction rate increased in the presence of water [70]. This can be demonstrated by a comparison of TON values increasing from 127 to 255 after the addition of water to toluene in the hydroformylation of propene with $Rh(acac)(CO)_2/P(NC_4H_4)_3$. An increase of TON values was also observed for PPh_3, PCy_3 and $PPh(NC_4H_4)_2$ in the same conditions.

25.4 Conclusions

In conclusion, it is clear that water used as the solvent in the catalytic reactions is not an innocent component of these systems. On the contrary, by forming hydrogen bonds with the catalyst and reactants, water can significantly affect the activation stage and increase the rate of the reaction. Theoretical models developed for the interaction between water and catalyst, developed for heterogeneous systems, explain well the phenomenon of acceleration of the reaction. In homogeneous systems, water often becomes a ligand coordinated to the metal center. It can undergo decomposition (e.g. with the formation of radicals that participate in the subsequent steps of the reaction). The course of these processes depends on the type of metal and the coordinated ligands and it would therefore be more difficult to elaborate a uniform description of these reactions. There is no doubt, however, that it is possible to model the efficiency and selectivity of catalytic reactions in a very wide range by conducting them in water.

References

1 Anastas, P.T. (ed.) (2010). Reactions in water. In: *Handbook of Green Chemistry*, vol. 5 (ed. C.-J. Li), Wiley VCH Verlag GmbH & Co. KGaA.
2 Gawande, M.B., Bonifacio, V.D.B., Luque, R. et al. (2013). *Chem. Soc. Rev.* 42: 5522–5551.
3 Butler, R.N. and Coyne, A.G. (2010). *Chem. Rev.* 110: 6302–6337.
4 Rideout, D.C. and Breslow, R. (1980). *J. Am. Chem. Soc.* 102: 7816–7817.
5 Breslow, R. (1991). *Acc. Chem. Res.* 24: 159.
6 Gajewski, J.J. (1997). *Acc. Chem. Res.* 30: 219–225.
7 Narayan, S., Muldoon, J., Finn, M.G. et al. (2005). *Angew. Chem. Int. Ed.* 44: 3275–3279.
8 Chandra, A. and Fokin, V.V. (2009). *Chem. Rev.* 109: 725–748.
9 Jung, Y. and Marcus, R.A. (2007). *J. Am. Chem. Soc.* 129: 5492–5502.
10 Karhan, K., Khaliullin, R.Z., and Kuhne, T.D. (2014). *J. Chem. Phys.* 141: 22D528.
11 Kitanosono, T. and Kobayashi, S. (2020). *Chem. Eur. J.* 26: 9408–9429.
12 Chang, C.-R., Huang, Z.-Q., and Li, J. (2016). *WIREs Comput. Mol. Sci.* 6: 679–693.

13 Davies, P.R. (2016). *Top. Catal.* 59: 671–677.
14 Roberts, M.W. (2014). *Catal. Lett.* 144: 767–776.
15 Micek-Ilnicka, A. (2009). *J. Mol. Catal. A: Chem.* 308: 1–14.
16 Li, G., Wang, B., and Resasco, D.E. (2020). *ACS Catal.* 10: 1294–1309.
17 Meng, Q., Cao, D., Zhao, G. et al. (2017). *Appl.Catal.B: Env.* 212: 15–22.
18 Ma, R., Wu, X.P., Tong, T. et al. (2017). *ACS Catal.* 7 (1): 333–337.
19 Li, X., Wang, X., Roy, K. et al. (2020). *ACS Catal.* 10 (10): 5783–5792.
20 Hensley, A.J., Wang, Y., Mei, D., and McEwen, J.S. (2018). *ACS Catal.* 8 (3): 2200–2208.
21 Andrushkevich, T.V. and Ovchinnikova, E.V. (2020). *Mol. Catal.* 484: 110734.
22 Yang, X., Wang, X., and Qiu, J. (2010). *Appl. Catal. A: General* 382: 131–137.
23 Mullen, G.M., Gong, J., Yan, T. et al. (2013). *Top. Catal.* 56 (15–17): 1499–1511.
24 Bernasconi, L., Kazaryan, A., Belanzoni, P., and Baerends, E.J. (2017). *ACS Catal.* 7 (6): 4018–4025.
25 Bray, J.T., Ford, M.J., Karadakov, P.B. et al. (2019). *React. Chem.Eng.* 4 (1): 122–130.
26 Pentsak, E.O. and Ananikov, V.P. (2019). *Eur.J. Org. Chem.* 26 (2019): 4239–4247.
27 Cummings, S.P., Le, T.N., Fernandez, G.E. et al. (2016). *J. Am. Chem. Soc.* 138 (19): 6107–6110.
28 Kamal, A., Srinivasulu, V., Seshadri, B.N. et al. (2012). *Green Chem.* 14 (9): 2513–2522.
29 Dai, Y., Gao, X., Chu, X. et al. (2018). *J. Catal.* 364: 192–203.
30 Anderson, J.A., McKenna, F.M., Linares-Solano, A., and Wells, R.P. (2007). *Catal. Lett.* 119 (1): 16–20.
31 Bereta, T., Mieczyńska, E., Ronka, S. et al. (2021). *New J. Chem.* 45 (11): 5023–5028.
32 Alsalahi, W. and Trzeciak, A.M. (2018). *Chem. Select* 3: 1727–1736.
33 Pascanu, V., Miera Gonzalez, G., Inge Ken, A., and Matute-Martin, B. (2019). *J. Am. Chem. Soc.* 141: 7223–7234.
34 Geng, K., He, T., Liu, R. et al. (2020). *Chem. Rev.* 120: 8814–8933.
35 Freund, R., Zaremba, O., Arnauts, G. et al. (2021). *Angew. Chem. Int. Ed.* 60: 23975–24001.
36 Wang, C., Liu, X., Demir, N.K. et al. (2016). *Chem. Soc. Rev.* 45: 5107–5134.
37 Yuan, B., Pan, Y., Li, Y. et al. (2010). *Angew. Chem. Int. Ed.* 49: 4054–4058.
38 Carson, F., Pascanu, V., Bermejo Gómez, A. et al. (2015). *Chem. Eur. J.* 21: 10896–10902.
39 Kardanpour, R., Tangestaninejad, S., Mirkhani, V. et al. (2014). *J. Organomet. Chem.* 761: 127–133.
40 Singha Roya, A., Mondal, J., Banerjee, B. et al. (2014). *Appl. Catal. A: Gen.* 469: 320–327.
41 Niknam, E., Panahi, F., and Khalafi-Nezhad, A. (2020). *Appl. Organomet. Chem.* 34: 5470.
42 Augustyniak, A.W., Zawartka, W., Navarro, J.A.R., and Trzeciak, A.M. (2016). *Dalton Trans.* 45: 13525–13531.
43 Huang, J., Wang, W., and Li, H. (2013). *ACS Catal.* 3: 1526–1536.
44 Alamgholiloo, H., Rostamnia, S., Hassankhani, A. et al. (2018). *Appl. Organomet. Chem.* 32: 4539.
45 Shang, N., Gao, S., Zhou, X. et al. (2014). *RSC Adv.* 4: 54487–54493.
46 Chen, L., Rangan, S., Li, J. et al. (2014). *Green Chem.* 16: 3978–3985.
47 Yuan, N., Pascanu, V., Huang, Z. et al. (2018). *J. Am. Chem. Soc.* 140 (26): 8206–8217.
48 Lima, C.F.R.A.C., Rodrigues, A.S.M.C., Silva, V.L.M. et al. (2014). *ChemCatChem* 6: 1291–1302.
49 Goncalves, R.S.B., de Oliveira, A.B.V., Sindra, H.C. et al. (2016). *ChemCatChem* 8: 743–750.
50 Wang, J.-C., Liu, C.-X., Kan, X. et al. (2020). *Green Chem.* 22: 1150–1155.
51 Liu, H.L., Li, Y.W., Luque, R., and Jiang, H.F.A. (2011). *Adv. Synth. Catal.* 353: 3107–3113.
52 Augustyniak, A.W. and Trzeciak, A.M. (2022). *Inorganica Chim. Acta* 538: 120977/1–120977/6.
53 Gutterød, E.S., Øien-Ødegaard, S., Bossers, K. et al. (2017). *Ind. Eng. Chem. Res.* 56 (45): 13206–13218.
54 Tonga, L., Songa, X., Jiang, Y. et al. (2022). *Int. J. Hydrog. Energy.* 47: 15753–15763.

55 Cirujano, F.G., López-Maya, E., Navarro, J.A.R., and De Vos, D.E. (2018). *Top. Catal.* 61: 1414–1423.
56 Kuznetsov, M.L. and Pombeiro, A.J. (2009). *Inorg. Chem.* 48 (1): 307–318.
57 Goldsmith, B.R., Hwang, T., Seritan, S. et al. (2015). *J. Am. Chem. Soc.* 137 (30): 9604–9616.
58 Pavlova, A. and Meijer, E.J. (2012). *ChemPhysChem.* 13 (15): 3492–3496.
59 Polo, V., Schrock, R.R., and Oro, L.A. (2016). *Chem. Comm.* 52 (96): 13881–13884.
60 Ohnishi, Y.Y., Nakao, Y., Sato, H., and Sakaki, S. (2006). *Organomet.* 25 (14): 3352–3363.
61 Baumgarten, S., Lesage, D., Gandon, V. et al. (2009). *ChemCatChem.* 1 (1): 138–143.
62 Lorenzo-Luis, P., Romerosa, A., and Serrano-Ruiz, M. (2012). *ACS Catal.* 2 (6): 1079–1086.
63 Scalambra, F., Lorenzo-Luis, P., de los Rios, I., and Romerosa, A. (2019). *Coord. Chem. Rev.* 393: 118–148.
64 Varela-Álvarez, A., Sordo, J.A., Piedra, E. et al. (2011). *Chem. Eur. J.* 17 (38): 10583–10599.
65 Bellarosa, L., Díez, J., Gimeno, J. et al. (2012). Highly efficient redox isomerisation of allylic alcohols catalysed by pyrazole-based ruthenium (IV) complexes in water: mechanisms of bifunctional catalysis in water. *Chem. Eur. J.* 18 (25): 7749–7765.
66 Scalambra, F., Holzmann, N., Bernasconi, L. et al. (2018). Water participation in catalysis: an atomistic approach to solvent effects in the catalytic isomerization of allylic alcohols. *ACS Catal.* 8 (5): 3812–3819.
67 Neeve, E.G. and Geier, S. (2016). Diboron(4) compounds: from structural curiosity to synthetic workhorse. *Chem. Rev.* 116: 9091–9161.
68 Rao, S., Joy, M.N., and Prabhu, K.R. (2018). Employing water as the hydride source in synthesis: a case study of diboron mediated alkyne hydroarylation. *J. Org. Chem.* 83 (22): 13707–13715.
69 Kocięcka, P. and Trzeciak, A.M. (2020). Efficient hydroarylation of terminal alkynes with sodium tetraphenylborate performed in water under mild conditions. *Appl. Catal. A: General* 589: 117243.
70 Alsalahi, W., Grzybek, R., and Trzeciak, A.M. (2017). N-Pyrrolylphosphines as ligands for highly regioselective rhodium-catalyzed 1-butene hydroformylation: effect of water on the reaction selectivity. *Catal. Sci. Technol.* 7: 3097–3103.

26

Using Speciation to Gain Insight into Sustainable Coupling Reactions and Their Catalysts

Skyler Markham[1], Debbie C. Crans[1,2], and Bruce Atwater[2]

[1] Dept. Chemistry, Colorado State University, Fort Collins, Colorado
[2] Fort Hays State University, 600 Park Street, Hays, Kansas

26.1 Introduction

One of the most important reactions in organic chemistry to date is the cross-coupling reaction. Its importance can be demonstrated in the many products that are prepared using these reactions (including discodermolide [1], polmacoxib (Acelax) [2], and sonidegib [2]) and in the fact that it won the Nobel Prize in 2010 [3]. Of particular utility to industry are the Suzuki-Miyaura, Sonogoshira, and Buchwald-Hartwig cross couplings [4]. With these three reactions occurring in 40% of the industrial publications in 2014, it is apparent industry heavily relies on these cross-coupling reactions [4]. There have been over 45,000 papers compiled in Web of Science concerning Suzuki-Miyaura cross-coupling reactions that have primarily focused on ligands, metal catalyst, reaction type, and organoboron species. The importance of the palladium (Pd) based catalysts is one of the common factors across these reactions, as illustrated in the Suzuki-Miyaura cross-coupling cycle (Figure 26.1) [5]. However, the toxicity of these catalysts and the limited availability of these catalysts is often ignored [6]. In this review, we examine present literature focused on more sustainable catalysts along with exploring the utility of speciation for future catalyst design for use in Suzuki-Miyaura cross-coupling reactions.

26.2 The First Cross-coupling Reaction

The first coupling between an alkyl halide and organoboron reagent was reported by Suzuki and Miyaura in 1981, Figure 26.2 [7]. While using metal catalysts to form carbon-carbon bonds was well-known, none had demonstrated a reaction suitable for frequent use in industry [8]. The use of relatively non-toxic boron compounds complemented by their simple and efficient synthesis allowed for extensive use in both industrial and academic applications [8].

Catalysis for a Sustainable Environment: Reactions, Processes and Applied Technologies Volume 2, First Edition. Edited by Armando J. L. Pombeiro, Manas Sutradhar, and Elisabete C. B. A. Alegria.
© 2024 John Wiley & Sons Ltd. Published 2024 by John Wiley & Sons Ltd.

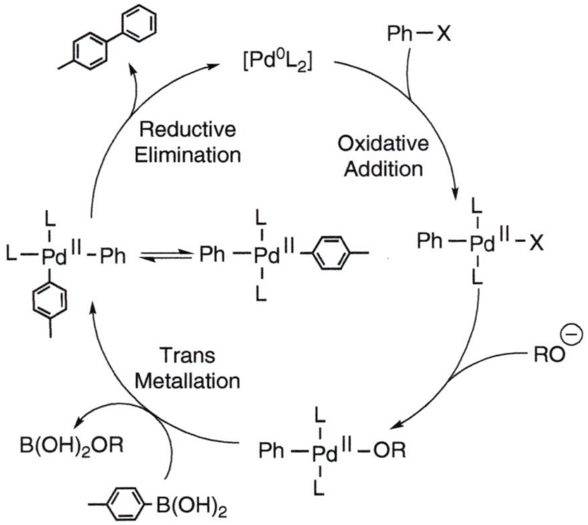

Figure 26.1 General Suzuki-Miyaura Cross-Coupling Catalytic Cycle [5] / American Chemical Society.

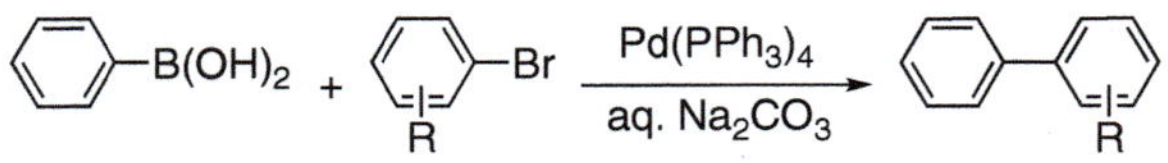

Figure 26.2 First Suzuki-type cross-coupling reported by Suzuki and Miyaura [7] / Taylor & Francis.

26.3 Phosphine Ligands for Catalysts of Cross-Coupling Reactions

Since the initial report by Suzuki and Miyaura, there have been major advances to the reaction. One major area is in improving the ligands surrounding the catalyst leading to two major ligand classes: phosphines and *N*-heterocyclic carbenes (NHC), Figure 26.3. Phosphine based ligands have a long and rich history for Pd-catalyzed cross-coupling reactions [7, 9]. Both aromatic and aliphatic phosphine ligands have been used in cross-coupling reactions which demonstrates the ligands require both electronic and steric properties of such substituents to be successful, Figure 26.4 [10–15]. The success of these ligands comes as no surprise, as they are electron rich which increases the metal center's reactivity for oxidative addition [16]. The ligands also readily dissociate and reassociate with Pd(II) which can allow a coordination site to open or close to facilitate oxidative insertion or reductive elimination at the metal center [17–19]. Phosphine ligands have been diversified over time to enhance reactivity, including developing both sterically hindered aliphatic groups as well as heteroatom based groups. The first study demonstrating a wide substrate scope using [1,1'-bis(diphenylphosphino)ferrocene]dichloro-palladium (II) (Pd(dppf)Cl$_2$) was in 1988 by Thompson et al. [20]. Some of the more recent developments into phosphine ligands belong to the Phos class (Figure 26.4) which have been further modified by multiple groups including Buchwald et al. and Chan et al. [21, 22]. This class has been rapidly growing to include several ligands derivatives, all of which have demonstrated their utility across multiple cross-coupling reactions, Figure 26.4.

The Phos ligand design is based on the Phos scaffold (Figure 26.4). The R groups bound to the phosphorus will increase electron density at the phosphorus, allowing an increased rate of oxidative addition. Reductive elimination will also be increased with large R groups (R = t-Bu and

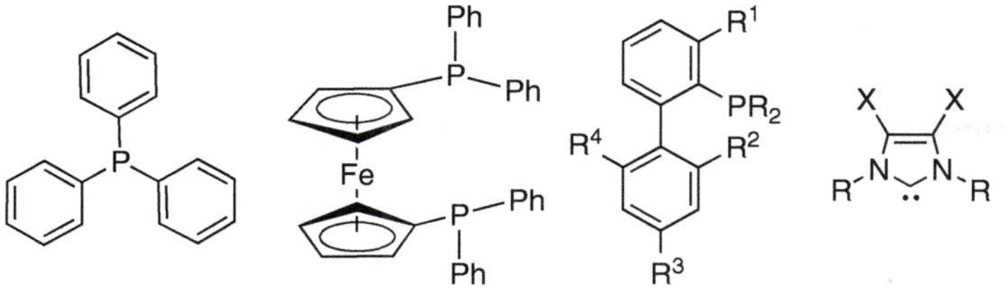

Figure 26.3 Examples of two major classes of ligands, phosphines and carbenes. From left to right: three examples of phosphine class, triphenylphosphine, ferrocenylphosphine ligand (dppf), and Phos ligand; and from the carbene class, N-heterocyclic carbene.

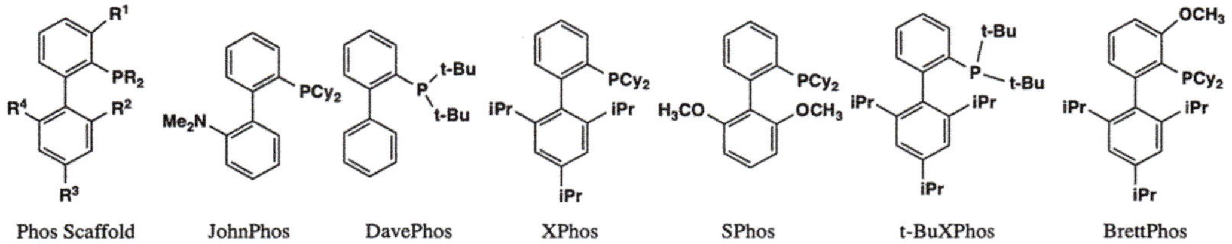

Figure 26.4 Various phoshine (Phos) ligands commonly used in cross-coupling reactions beginning with the general Phos scaffold (where R can be alkyl, aryl, H, or a heteroatom, $R^3 \neq H$ typically for ease of synthesis). Then we follow by showing ligands in a historical order: JohnPhos [10], DavePhos [11], Xphos [12], Sphos [13], t-BuXPhos [14] and BretPhos [15].

Cy, shown in Figure 26.4). The steric strain caused by R^1 next to the phosphorus will force a conformational change that pushed PR_2 over the lower aryl ring and improves the rate of reductive elimination. The lower aryl ring increases the ligands size, slowing down the oxidation of the ligand by O_2. This ring also allows stabilization of the Pd complex through Pd-arene interactions. In addition, R^2 and R^4 increase stability by preventing cyclometallation [23].

Overall, the main goal of catalyst development was working toward a universal catalyst system. For a universal catalyst to be achieved, several criteria must be met: i) possess a broad substrate scope; ii) operate at low catalyst loadings; and iii) operate at or near ambient temperature [13]. These goals have led Buchwald and others to further refine their catalysts to develop new and more advanced catalysts [24, 25]. Although a true universal catalyst has not yet been reported, there have been several advances in catalyst design that have given rise to multiple catalysts which meet most of the goals outlined previously [10–15]. The large number of ligands that have been synthesized has allowed for a systematic understanding of the effects of substituents on the ligand and coordination of the ligand class to the Pd in the catalyst [15]. This high degree of understanding of the ligand has allowed Buchwald and others to further their journey in search of a truly universal ligand for the (Pd)II catalyst for cross-coupling reactions.

Through the work of Buchwald et al.to find a universal catalyst, the XPhos ligand was developed (Figure 26.4) [12]. The XPhos catalyst system allowed for heteroaromatic couplings in high yields (80% or larger) at mild temperatures (up to 40 °C) and short reaction times (30 minutes to two hours) for several different ligands. This is a step in the direction of developing the new general catalyst design for heteroaromatic coupling reactions and the variety of ligands and their conditions and yields summarized in Figure 26.5 [26].

A major advantage of the phosphine ligands over other ligand classes is their ability to be readily interchanged with other phosphine based ligands. The interchangeable nature of the ligands on

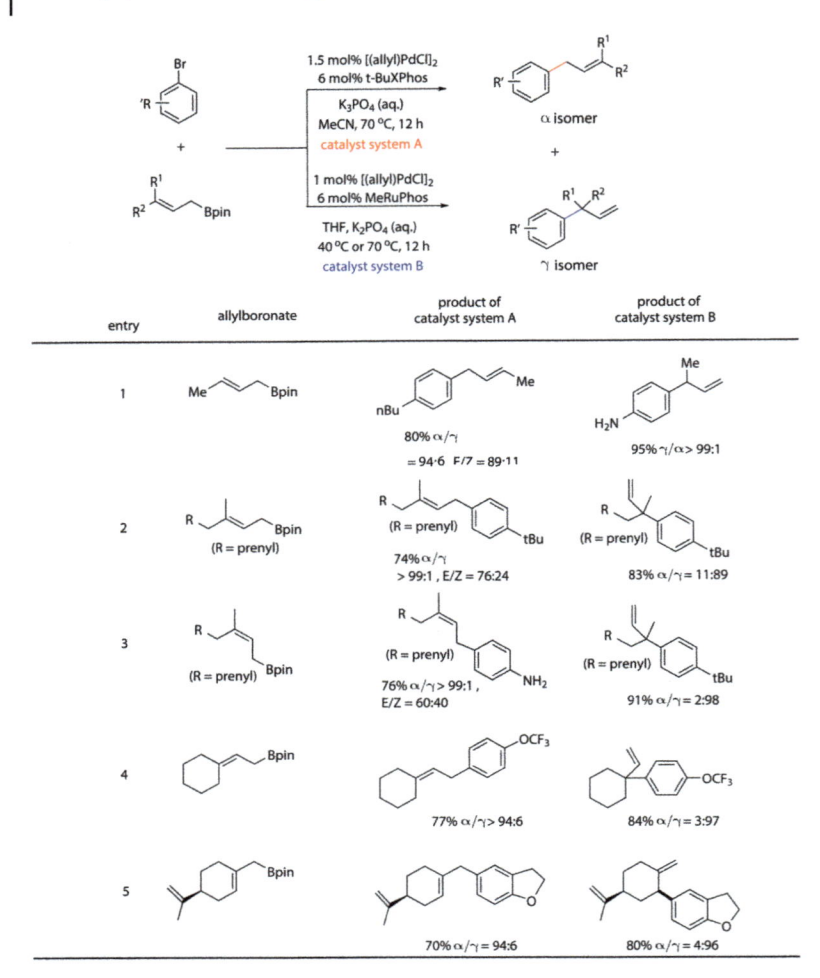

Figure 26.5 Regioselectivity in product distribution based on phosphine (Phos) structures [26] / American Chemical Society. Two catalytic systems are shown (A and B).

the Pd(II) allows a researcher to use similar reaction conditions thereby allowing catalyst flexibility in similar applications, Figure 26.5. The ability to vary the ligand while keeping the conditions identical makes these catalyst systems highly attractive for high-throughput screenings in industrial and academic settings [26]. These variable features of the phosphine-based ligand systems make them prominent ligands in effective catalysts in cross-coupling reactions.

The second major class of ligand family for Pd-catalyzed Suzuki-Miyaura cross-coupling reactions is the NHC) ligand, shown in Figure 26.6 [27, 28]. This ligand family generates two general subclasses of Pd coordination complexes: the Organ PEPPSI class of complexes (left) [28, 29] and Nolan's class of complexes (right) [29], Figure 26.6. NHCs have a history as effective ligands for organometallic transformations [30–32] and organocatalytic transformations [33], as well as for cross-coupling reactions [29, 34]. These two classes of complexes differ by the ternary ligand system attached to the Pd center. For the PEPPSI ligands, the Pd(II) center typically has the NHC coordinated *anti* to a pyridine ligand and two halide ligands to complete the coordination sphere [28]. For the Nolan family of catalysts, the NHC ligand is accompanied by a halide and an η-three bound allyl anion species *trans* to the NHC group [35]. Both of these catalyst families have demonstrated excellent performance in various cross-coupling reactions [36–38]. Of particular note is their ability to catalyze Suzuki-Miyaura reactions for a variety of boron co-catalyst species [28, 39–43].

The various positions in the coordination complex of these catalysts are highly tunable and alters the reactivity of the catalysts. The groups on the backbone of the NHC ring typically force the aromatic groups on the nitrogens closer to the Pd(II) center [28]. This impacts the catalysts in two major ways. First, it accelerates reductive elimination of the transmetallated species through the steric crowding of the metal center. The second major advantage is that it forces the alkyl groups on the aromatic rings into a position over the Pd(II) center [44, 45]. This positioning of the alkyl groups over the Pd(II) center thereby further accelerates reductive elimination due to steric crowding of the metal center. It also has the added benefit of limiting the formation of agostic interactions with β-hydrogens on the transmetallated species which limits β-hydride elimination allowing for some of the catalysts to be using in secondary $C(sp^3)$-$C(sp^2)$ couplings, Figure 26.7 [44]. The latter reaction has not so far been observed with Suzuki-Miyaura reactions [46–48].

One major advantage of NHC catalysts relative to the phosphorus-based ligands is the high degree of stability demonstrated by the pre-catalyst NHC species. Greater catalyst stability allows for convenient storage of the catalysts under ambient benchtop conditions [49]. The PEPPSI

Figure 26.6 Two classes of N-Heterocyclic carbene (NHC) ligands (NHC) catalysts: pyridine enhanced precatalyst preparation stabilization and initiation (PEPPSI) catalyst on the left, and Nolan Catalyst on the right [27, 28].

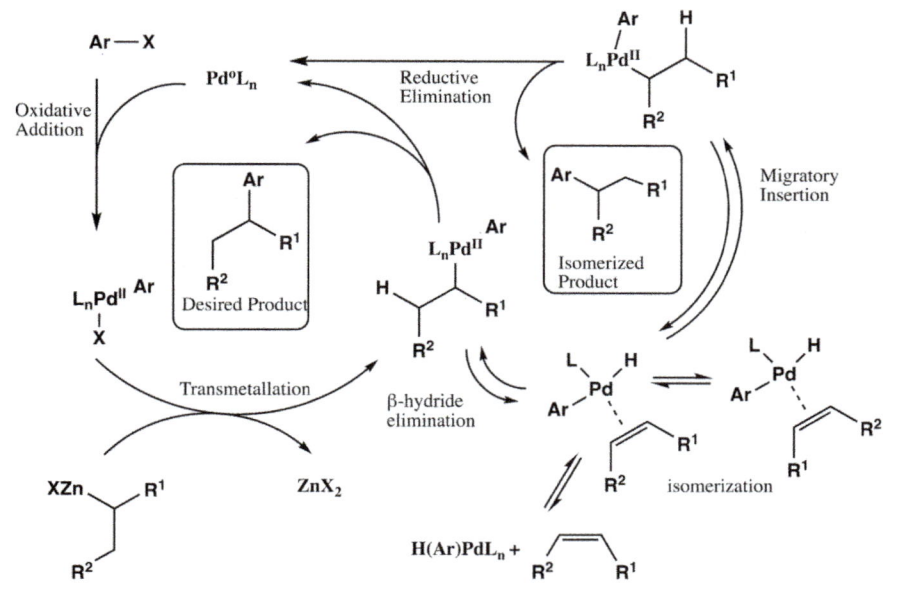

Figure 26.7 Agostic interactions due to pyridine enhanced precatalyst preparation stabilization and initiation (PEPPSI) catalysts [44] / John Wiley & Sons.

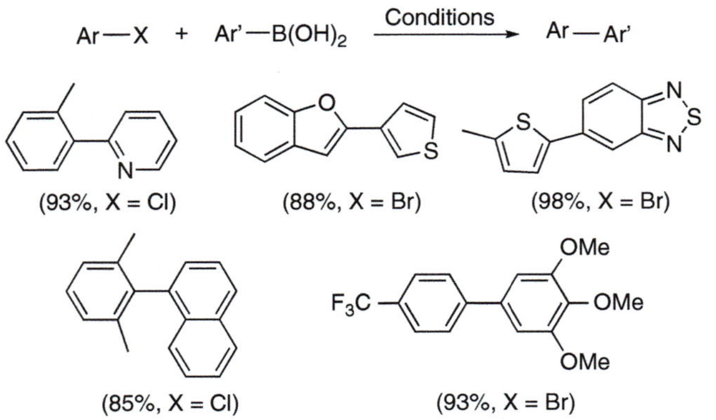

Figure 26.8 Compounds synthesized using pyridine enhanced precatalyst preparation stabilization and initiation (PEPPSI) catalysts [36] / Thieme Medical Publishing Group.

catalyst robustness has been routinely demonstrated. For example, they have been used on a solid support, and they have been recycled in flow reactions as well [50].

The PEPPSI catalyst has been used to generate a wide array of products, Figure 26.8 [36]. Unlike the phosphorus family of ligands, the NHC ligands are more on track toward the development of a true universal Pd(II) catalyst that can be utilized in all cross-coupling reactions. The main drawback for this class of coordination complex is that if the catalyst fails, a new catalyst must be prepared rather than merely altering one ligand on the coordination complex as done with the phosphorus ligands. However, this is remedied by the wide substrate scope demonstrated by the catalysts generated from the NHC ligands along with the large number of different catalysts that have been developed based on these ligands [40, 44, 45].

26.4 Speciation

Many Pd catalysts, particularly those formed from Phos ligands, are rather stable and readily support loss of ligands which allows the complexes to engage as catalysts under relevant reaction conditions. However, many Pd catalysts are coordination complexes and thus defined as a metal ion surrounded by ligands which are formed in equilibrium reactions. This is the case even if the ligand is carbon-based which will ultimately lead to C-C bond formation in the cross-coupling reaction.

At this time it appears that the properties of the palladium catalysts used in the cross-couplings may not be as well understood as initially presumed [51, 52]. For example, it has been suggested that the active catalyst may not be a homogenous Pd(II) species which carries out the reaction but rather a palladium nanoparticle [51, 52]. Mechanistic work continues to be reported which adds to the discussion of the reaction mechanism [51, 52]. Hence as the development of the system moves toward green and sustainable catalysts, more consideration of the structure and electronic properties of the catalyst and associated speciation chemistry would be required examples of which are described below.

26.5 Palladium Nanoparticle Catalysts and Speciation

Increasing interest in the nature of the specific Pd species that carries out the cross-coupling reactions suggest several alternative options for the central metal in the catalyst exist. This is of particular importance to industry as homogenous catalysts are often expensive and unrecoverable

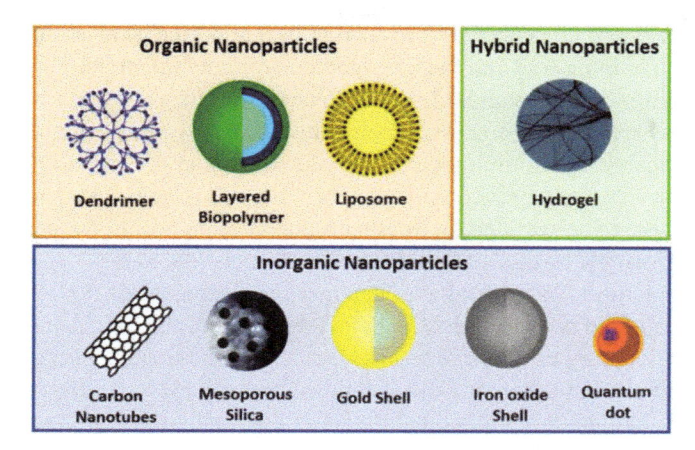

Figure 26.9 Schematic representation of different types of nanoparticles (NPs) divided into organic, hybrid, and inorganic NPs. Ref [54] / MDPI / CC BY 4.0.

[53]. Nanoparticles provide an alternative solution to these problems. For example, the Pd could be supported by any of the nanoparticle types shown in Figure 26.9 [54]. Nanoparticles can be readily be recovered through physical separations [55]. However, there many different nanoparticles (NPs) which vary in shape, size, structure, and composition and which have different catalytic properties. In Figure 26.9 [54], We show organic, hybrid, and inorganic NPs in circular, spherical, and elongated shapes. Shape, structure, size, and molecular composition will affect the properties of the materials as supports for the catalysts and varies types of NP-coatings will further provide flexibility in use of the NPs. Examples of different reactions utilizing such nanoparticles are discussed later and summarized in Figure 26.10.

One of the first nanoparticle catalysts employed for this process was palladium on carbon (Pd/C), Figure 26.10 [56]. Initial work with this catalyst system demonstrated that it can be readily employed with good to excellent yields, and it is the most common catalyst system used under Suzuki-Miyaura coupling conditions [56]. Subsequently, the Pd/C catalyst was utilized in a one-pot process to form biaryl aniline species from aryl boronic acids and halogenated nitroaromatics

Figure 26.10 Various reactions with different nanosupports [55–57].

in good to excellent yields [57]. This two-step process involved utilizing the supported catalysts to carry-out a cross-coupling between the boronic acid and the halogenated aromatic substrate. The nitro group was then reduced to an aniline under standard conditions followed by a subsequent reductive amination to give a secondary amine. This process demonstrates the utility and power of the support nanoparticle catalysts for carrying out multiple sequential reactions in one pot with the same catalyst. In addition, the catalysts for these reactions were recovered by a simple filtration [57]. This thereby allows for the catalysts to be recycled for another reaction sequence; however, this was only suggested and not demonstrated by the authors this in their papers [56, 57].

Carbon based supports have also been investigated but have been met with mixed levels of success, Figure 26.10 [55]. With Pd/C supported catalysts, leaching of a Pd species to the bulk solution is possible where observed cross-coupling activity can be attributed to the homogeneous Pd species [58]. Leaching was also reported from other supported catalyst systems [55], and these observations raise the possibility that nanomaterial based catalysts are exerting cross-coupling activity through a homogenous catalyst and/or from a catalyst located at the surface of the nanoparticle. A recent study by Zeng et al. deposited Pd catalysts onto a preformed iron oxide nanoparticle and then coated the nanoparticles with the microporous Stöber silica coating of varying pore sizes [51]. Zeng utilized filters with differing pore sizes to separate any soluble catalyst from the nanoparticle catalysts to determine whether the cross-coupling reaction was catalyzed by soluble or surface active catalysts. In addition, they varied the size of the halide and the boron species to determine whether the leached material or the Pd from the surface was the most effective cross-coupling catalyst [51]. They demonstrated that their catalyst had good reactivity with small boronic acids and halides, but the system showed no activity when tested with larger halides and a larger boronic acid than those commonly used. However, when a small aryl halide was employed with a large boronic acid they were able to achieve the same level of conversion demonstrated with the small halide and small boronic acid [51]. Their results suggest the cross-coupling reactions must be happening in bulk solution due to leached Pd catalyst. Therefore, any product that is produced must originate from cross-coupling reactions in the bulk solution rather than on the nanoparticles themselves [51]. This data lends credence to the notion that these reactions occur with homogenous catalysts and greatly contributed to the debate whether soluble or surface associated Pd species are the most active cross-coupling catalysts [52].

Numerous reports regarding the advantages of heterogeneous vs. homogeneous catalysts have been made particularly in industrial settings [51, 52, 58]. Extensive literature is available regarding the complex physical and chemical phenomena involved in such processes [51, 52]. The investigations into nanoparticle catalysts have revealed that such approaches have considerable potential [55]. However, Pd coated nanoparticle catalysts suffers from the same problems as coordination complex catalysts, such that in solution the catalyst engages in speciation equilibria, and more than one species is present in solution. In the following section we will address speciation of the catalysts and its impact on cross-coupling reactions.

26.6 Speciation of Palladium (Pd) Catalysts

Given the great variety of reaction parameters that are critical to the success of the cross-coupling reactions, we here will describe a few selected cases when formation of different Pd species affect the stereo- and regiochemistry as well as yield of product formation.

The speciation of $Pd(OAc)_2$ is particularly relevant because $Pd(OAc)_2$ is often used as a Pd catalyst precursor by forming an active Pd catalyst under some conditions of cross-coupling reactions. $Pd(OAc)_2$ is reduced in situ to form catalytically active Pd(0) species that can catalyze cross-coupling reactions at ambient temperatures [58, 59]. Pd(OAc)$_2$ is also used in ligand free ("ligandless") Suzuki-Miyaura reactions [58] however, its structure is important to the function. Of the Pd-material Trimeric [60] and polymeric [61] structures of Pd(OAc)$_2$ have been reported in the solid state; however, the solution structure of Pd(OAc)$_2$ is sensitive to its solvent environment. A trimer has been reported in benzene [62], acetic acid [63], chloroform [64] and methanol [64]. A linear dimer has been reported in chloroform-acetic acid [65], whereas a monomer has been reported in N-methyl-2-pyrrolidinone (NMP) [66]. Studies have now shown that the Pd atom on the trimer is rapidly subjected to a nucleophilic attack by H$_2$O, resulting in the loss of acetic acid and followed by a slower conversion to the monomeric species. The dissociation of the trimeric structure to the monomeric structure increases as the dipole moment of the solvent increases [58]. For example, in a solution of 40 mM Pd(OAc)$_2$ in toluene, 71% was trimer, 21% was monomer and 2% particles, whereas in DMF, there was 56% of the trimer, 42% monomer and 8% particles [58].

Pd coordination complexes, including those used as catalysts, are dependent on three factors: (i) the oxidation state of the Pd, (ii) the ligand and the ratio of Pd to ligand, and (iii) other reaction conditions such as the presence of O$_2$, water, solvent, temperature, and reaction time. Phosphine ligands are strong ligands and often remain coordinated to the complex even at low Pd: ligand ratios. Ligands such as acetate (OAc$^-$) more readily dissociate as shown in Eq. 26.1 and 26.2.

$$Pd(OAc)_2 + 3PPh_3 \rightleftharpoons \textit{trans-}[Pd(OAc)_2(PPh_3)_2] + PPh_3 \qquad (26.1)$$

$$Pd(OAc)_2 + PPh_3 \rightleftharpoons [Pd(\mu_2\text{-}OAc)(\kappa\text{-}OAc)(PPh_3)]_2 \qquad (26.2)$$

Specifically in the case of solutions of Pd(OAc)$_2$ and triphenylphosphine (PPh$_3$), the formation of the species depend on the Pd(OAc)$_2$:PPh$_3$ ratio when other reaction conditions remain the same [52]. When PPh$_3$ is in threefold or more excess of Pd(OAc)$_2$, the major species in solution is as shown in Eq (1) is the *trans*-[Pd(OAc)$_2$(PPh$_3$)$_2$]. When the Pd(OAc)$_2$ is about the same concentration as PPh$_3$, the major species in solution is the dimer [Pd(μ_2-OAc) (k-OAc)(PPh$_3$)]$_2$, as shown in Eq (2). This difference as been explored and found to result in a change in cross-coupling site selectivity in dihalogenated heteroarenes as shown in Figure 26.11 [52].

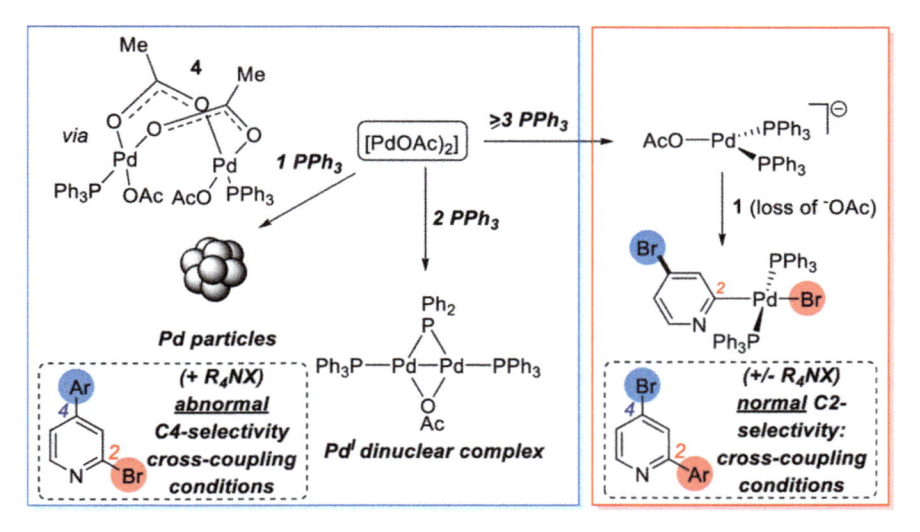

Figure 26.11 Different Pd species arising from different ratios of Pd(OAc)$_2$/nPPh$_3$ (PPh$_3$ = triphenylphosphine) result in different cross-coupling selectivities under cross-coupling conditions. Adapted from Ref [52].

Approaches toward the preparation of sp^2- and sp^3-hybridized carbon frameworks using Suzuki–Miyaura cross-coupling reactions include a strategy for the chemoselective synthesis and conversion of boronic esters. This approach provides direct access to reactive sp^2-hybridized pinacol protected boronic acid (BPin) products by controlling boron speciation during the cross-coupling reaction, as illustrated in Figure 26.12 [67]. Suzuki–Miyaura cross-coupling of an aryl BPin with a conjunctive haloaryl methyl-iminodiacetic protected boronic acid (BMIDA) generates an intermediate biaryl BMIDA [67, 68]. The byproduct from the Suzuki–Miyaura coupling reaction is HOBPin. Through appropriate control of the basicity in the biphasic reaction system these species can be hydrolyzed, and the resulting biaryl boronic acid and pinacol can be driven toward formation of the aryl BPin derivative. Accordingly, under this protocol a protected boronic acid affords reactive boronic ester products with no need for deprotection reactions. Solution speciation of boronic acids can be chemoselectively controlled to enable the formal homologation of boronic acid pinacol esters [69]. The reaction is tolerant of aryl and vinyl functionality as both the pinacol donor and acceptor, respectively, and enables streamlining of the iterative cross-coupling reaction as well as a method for controlled oligomerization.

Another reported approach involves manipulating speciation so that a different Pd intermediate within the catalytic cycle becomes the resting state through starvation of a reagent, Figure 26.13

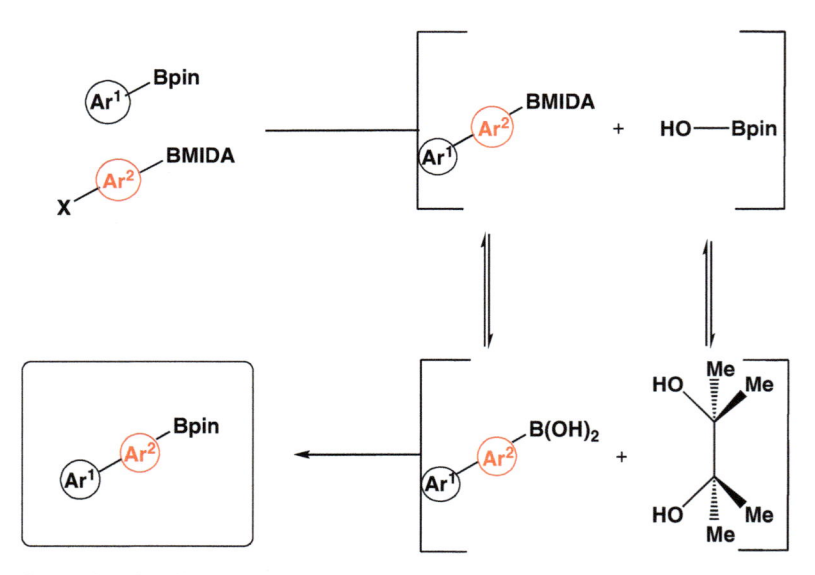

Figure 26.12 Chemoselective boronic ester synthesis by controlled speciation [67] / Thieme Medical Publishing Group.

Figure 26.13 Catalytic cycle for the cross-coupling of diazo nucleophiles with slow addition of reagent (Path A) and bulk addition of reagent (Path B) [70] / American Chemical Society.

[70]. This approach was successfully reported for the diazo cross-coupling reaction by slow addition of the diazo reagent. These conditions modified speciation within the catalytic cycle of the Pd intermediate so the oxidative addition Pd(II)ArCl intermediate becomes the resting state. By adding the diazo reagent slowly, the reaction with the diazo reagent became the resting state in the catalytic cycle. This strategy greatly expands the scope of the diazo cross-coupling reaction.

Finally, we highlight a defined heterogeneous Pd catalyst based on a phosphine-metal–organic framework (P-MOF) ligand system [71]. This system catalyzes Suzuki reactions under exceptionally mild conditions and still displayed higher selectivity than that achieved using Pd(PPh$_3$)$_4$, a standard homogeneous catalyst. The Pd-P-MOF catalyst converted a wide range of substrates and that, when formed, Pd nanoparticles promote hydrodehalogenation. The Pd reactivity is maximized by having excess phosphine ligand present in the system. The main drawbacks of these systems are the limited stability and recyclability of the MOF framework under the cross-coupling reaction conditions. We conclude that, understanding speciation is becoming an increasingly important factor and tool for improving the reaction outcome of cross-coupling reactions.

26.7 Alternative Metal Catalysts

26.7.1 Nickel

Although Pd catalysts have received the majority of the attention for Suzuki-Miyaura coupling reactions in the literature, other metals have also been explored [67]. Currently, there is an interest in the field to move toward utilizing earth abundant and "green sustainable" metals to catalyze the cross-coupling reactions. The first major alternative metal that has been employed in cross-coupling reactions is nickel (Ni) [72]. In general, Ni's main advantage is its earth abundance and significantly cheaper to produce compared to Pd.

Ni-catalyzed Suzuki-Miyaura couplings often employ phosphine or diamine ligands to carry out the reactions, Figure 26.14 [73–75] This catalyst pairing allows for several interesting transformations which are difficult to carry out with traditional Pd-catalyzed coupling reactions. Of particular interest is the ability of Ni-catalysts to form quaternary centers through the coupling of an sp^2 hybridized organoboron species with an sp^3 hybridized halide, Table 26.1 [74]. In addition, Ni can also be utilized to form chiral centers via the coupling of two sp^3 carbons with acceptable levels of stereospecificity [75, 76]. Although the stereoinduction is relatively modest, it remains significant as this kind of coupling has been difficult to accomplish with Pd catalysts [77].

Although Ni-complexes have some distinct advantages over Pd catalysts, unfortunately there are major drawbacks which limit their utility. Ni-catalysts frequently induce radical reactions during the cross-coupling reactions [78]. This is because Ni(I) is more readily accessible than Pd(I) [79]. Consequently, Ni can induce radical reactions which can interfere with the desired cross-coupling reactions [80]. However, this radical formation is the sub-reaction that supports the formation of the quaternary carbon centers (outlined previously) and thus expanding the applicability of the resulting products [75–77]. Since Ni is a first-row transition metal ion, compared to the second row Pd atom, the Ni-coordination complexes are likely to be less labile than the Pd complex, and the Ni-speciation chemistry is less complex. Therefore these cross-coupling catalysts can be a good alternative to the Pd catalysts when optimized.

Figure 26.14 Common ligands used for Ni-catalyzed Suzuki cross-couplings: phosphine ligands, bipyridine ligands, and diamine [73–75]. Phosphine [68–70].

Table 26.1 Cross-coupling of tertiary centers using Ni [72].

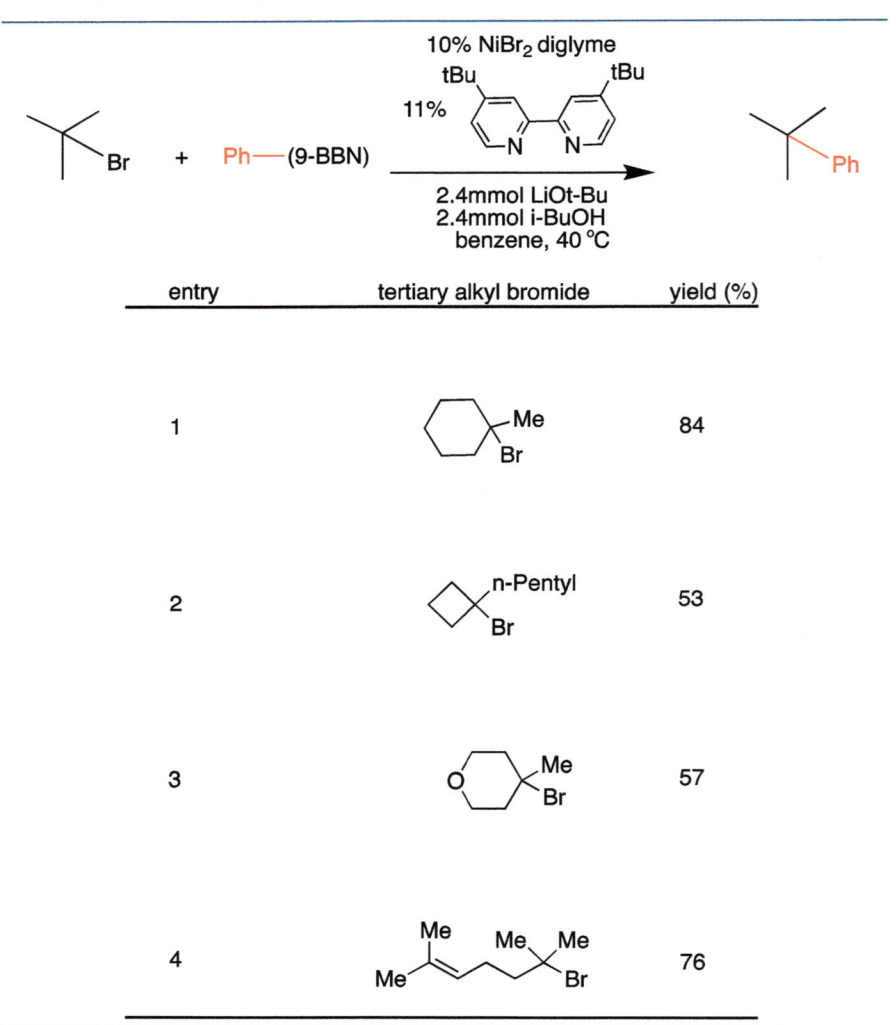

entry	tertiary alkyl bromide	yield (%)
1	1-methyl-1-bromocyclohexane	84
2	1-(n-pentyl)-1-bromocyclobutane	53
3	4-methyl-4-bromotetrahydropyran	57
4	(Me)$_2$C=CH-CH$_2$CH$_2$-C(Me)$_2$Br	76

Reaction conditions: 10% NiBr$_2$ diglyme, 11% bipyridine (tBu), 2.4 mmol LiOt-Bu, 2.4 mmol i-BuOH, benzene, 40 °C. R-Br + Ph—(9-BBN) → R-Ph

26.7.2 Cobalt

Cobalt (Co) has recently received significant attention as an alternative metal to Pd. Co, as an essential element, has significantly lower toxicity than Pd and is much more earth abundant [81]. Co was first employed in cross-coupling reactions in the early part of the 20th century [82, 83]. However, the interest in these systems were limited until recently when an emphasis on the development of more earth abundant systems became important [84, 85].

Recent work on cobalt catalysts has focused on multiple types of cross-coupling [86, 87] including Suzuki-Miyaura couplings utilizing pincer type ligands, Figure 26.15 [85, 88, 89]. These systems have allowed for the formation of $C(sp^2)$–$C(sp^3)$ bonds utilizing a diamine ligand in acceptable yields with several examples with varied functionality shown in Table 26.2[85].

Although this required a relatively high level of catalyst loading, this is not as prohibitive for Co as it is for Pd due to the lower cost and less toxic properties of many Co-complexes [81]. Other reports showed the utilization of a set of Co(II) catalysts to carry out the coupling of $B(sp^2)$ boronic acids with $C(sp^2)$ electrophiles [88]. These reactions proceeded with much lower levels of catalyst loading with good to excellent yields of the desired products [88].

The investigations into Co as a cross-coupling catalyst have opened new avenues for exploration, Table 26.2 [81]. Although parts of the mechanism of the Suzuki-Miyaura cross-coupling reaction are still under investigation [90, 91], the use of Co catalysts allows for additional methods of investigation of the cross-coupling reactions in general. Although cobalt is an NMR active nucleus, the Co-catalyst involved in the cross-coupling reaction [Co(II)], is paramagnetic and of low sensitivity. It may provide additional characterization of the inorganic portion of the reaction as well as the kinetics assessment of the organic portion of the reaction. New information of how the Suzuki-Miyaura cross-coupling reaction proceeds with earth abundant catalysts is likely to be helpful for future catalyst design and evaluation of the performance of the corresponding Pd catalysts and potentially lower the catalyst loadings.

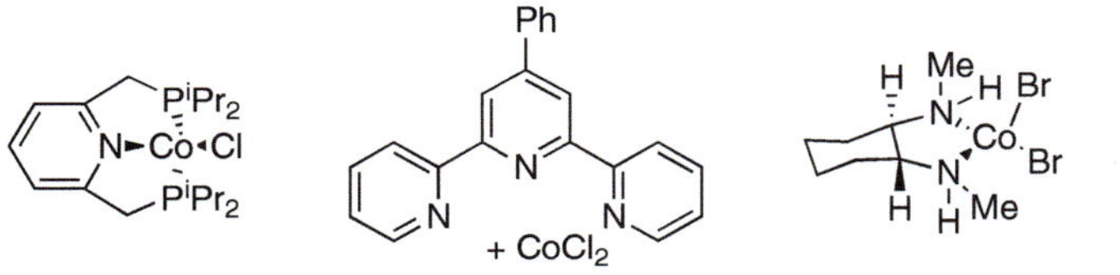

Figure 26.15 Common ligands used for Co catalyzed cross-coupling reactions [85, 88, 89].

Table 26.2 $C(sp^2)$-$C(sp^3)$ coupling catalyzed by Co [81].

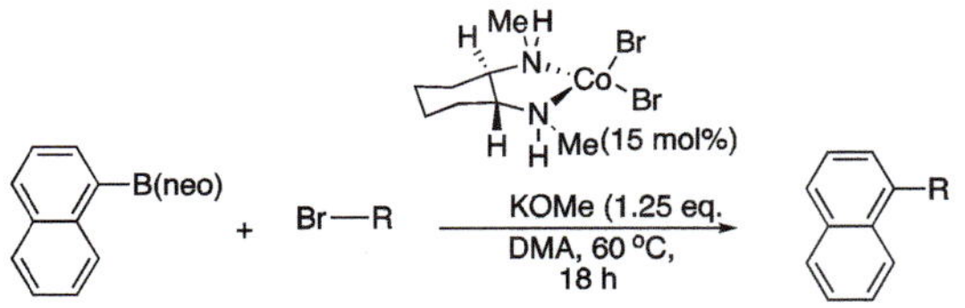

entry	alkyl bromide	yield (%)
1	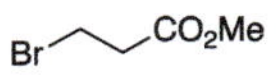 Br~CO$_2$Me	64
2	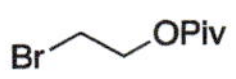 Br~OPiv	60
3	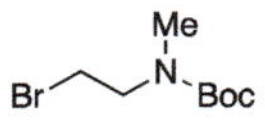 Br~N(Me)Boc	50

26.8 Speciation of Nickel and Cobalt Catalysts

Ni-pre-catalysts are a sustainable alternative to Pd pre-catalysts for the Suzuki–Miyaura cross-coupling reaction [80]. Aryl sulfamates are phenolic derivatives which can act as a directing group for the pre-functionalization of the electrophilic aromatic backbone prior to cross-coupling. Coordination complexes of Ni(0), Ni(I), and Ni(II) can form from the pre-catalyst complex, (dppf)Ni(o-tol)(Cl) (where dppf = 1,1′ bis(diphenylphosphino)-ferrocene), for Suzuki–Miyaura coupling reactions involving aryl sulfamates and boronic acids (Figure 26.16). Catalysts formed from Ni, aryl sulfamates, and boronic acids can function effectively at low catalyst loading and at milder reaction conditions than reported Pd systems.

The reactions of dppf-Ni(0) with alkyl halides proceeds through three-coordinate Ni(0) intermediates, such as [Ni(dppf)(L)] [92]. The effect of added ligand (L) on catalyst speciation and the rates of reactions of [Ni(COD)(dppf)] with alkyl halides have been investigated. Trends in reactivity with regard to steric bulk and electron donation of monodentate ligands was found to decrease the reaction rate. The halide abstraction step is generally irreversible and the subsequent recombination of a Ni(I) complex with an alkyl halide have a significant effect on the overall rate of the reaction and some ligands form very stable [Ni(dppf)(L)$_2$] species. This study established that different ligands increase the rate of the reaction between [Ni(COD)(dppf)] and a model alkyl bromide giving a 200-fold spread in rate constants.

Another study examined the cross-coupling reactions of Ni(0)(dppf) complexes with alkyl halides which proceeded through three-coordinate nickel(0) intermediates in the form [Ni(dppf)(L)], Figure 26.17 [92]. A series of monodentate ligands (L) are found to affect catalyst speciation; overall sterically bulky and electron donating ligands decrease the reaction rates of [Ni(COD)(dppf)] with alkyl halides. Some ligands form very stable [Ni(dppf)(L)$_2$] species. The yields of prototypical NI(0)(dppf)-catalyzed Kumada cross-coupling reactions of alkyl halides are significantly improved by the addition of free ligands, which demonstrates speciation provides another important variable to consider when optimizing nickel-catalyzed reactions of alkyl halides.

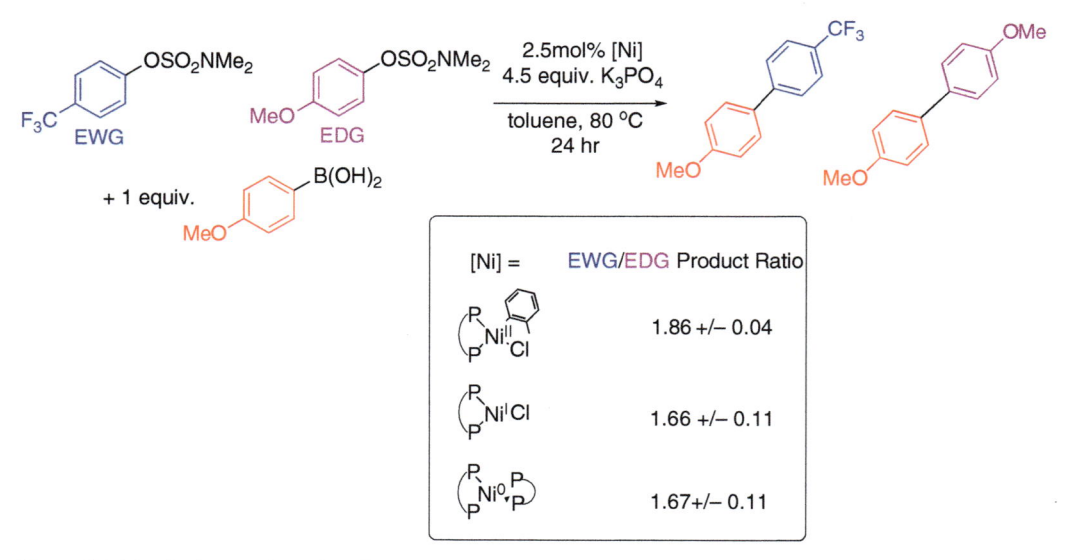

Figure 26.16 Preferential product Formation for SMC reactions catalyzed by (dppf)Ni(II)(o-tol)(Cl), (dppf)Ni(I)(Cl), (dppf)$_2$Ni(0) with aryl sulfamates of differing electronic properties and only 1 equiv. of boronic acid [80].

Figure 26.17 Excess ligand affects speciation and reaction rates from [92].

Optimizing the transmetalation reaction from a nucleophile to a metal catalyst led to the first Co-catalyzed Suzuki-Miyaura cross-coupling reaction: Cross-coupling between aryl triflate electrophiles and heteroaryl boron nucleophiles catalyzed by a Co complex, Figure 26.18 [93]. The reaction was catalyzed by a new class of high-spin, tetrahedral, bis(phosphino)pyridine Co(I) alkoxide or aryloxide complexes converting to a low-spin square planar cobalt(I) catalytic intermediate. The reactivity of methyl- or ethylalkoxide Co(I) complexes facilitated transmethylation at ambient temperature, but the concurrent β-hydride elimination reaction made these complexes unsuitable as cross-coupling catalysts. The higher stability of isopropyl or $CH(Ph)CH_3$ alkoxide Co(I) complexes facilitated transmethylation at ambient temperature and were found to be suitable as cross-coupling catalysts. The favorable speciation chemistry of the first-row cobalt complexes over state-of-the-art second row Pd complexes allowed optimizing the cross-coupling reaction conditions to result in the first Co catalyzed $C(sp^2)$-$C(sp^2)$ bond formation between neutral pinacolate boron nucleophiles and aryl triflate electrophiles [93].

Since 2016, exploration into the speciation chemistry of Co-catalysts and the applications of a range of Co-catalyst for the Suzuki-Miyaura cross-coupling reactions have dramatically expanded. At this time many more Co-catalysts including halides, NHC-ligands, peptides, and pre-catalysts, as well as a variety of nucleophiles and electrophiles, have been successfully employed.

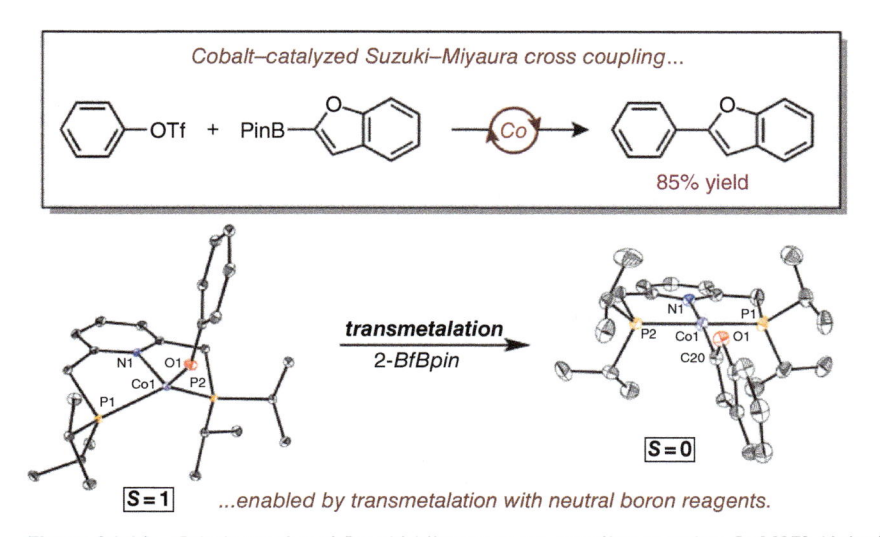

Figure 26.18 Cobalt-catalyzed Suzuki-Miyaura cross-coupling reaction. Ref [93] / John Wiley & Sons.

These recent studies have demonstrated that potential of the earth abundant metal, Co, as catalyst for the Suzuki-Miyaura cross-coupling reaction has materialized as a viable alternative for Pd-catalyzed reaction conditions in the future [88, 94–97].

26.9 Cross-coupling Reactions and Sustainability: Summary and the Future

In this review, we have described the popular cross-coupling reaction, its history, and pertinent efforts toward developing the global catalyst system as well as selected current developments toward a green catalyst system. Specifically, we have highlighted advances using Ni and Co based catalysts as representatives for the numerous earth abundant metals that are currently investigated as greener catalyst for the cross-coupling reaction. In addition to describing the suitable substrates and scope of the reactions, we show that analyzing the reactions mechanistically particularly by considering the associated speciation chemistry of the catalysts under the reaction conditions can be used to effectively improve coupling yields. Recent studies have demonstrated that simple strategies in addition to those commonly used including the addition of a small excess of catalysts ligand and continuous addition of reagents are methods that will improve yields even with earth abundant metal catalysts. Hence, we believe that speciation principles can assist the future catalyst design particularly as the systems move into more green and sustainable solvents that will be more suitable for larger scales and milder reaction conditions important for specific industrial processes.

References

1 Marshall, J.A. and Johns, B.A. (1998). *J. Org. Chem.* 63: 7885–7892.
2 Flick, A.C. et al. (2017). *J. Med. Chem.* 60 (15): 6480–6515.
3 Nobel Prizes 2010: Richard F. Heck / Ei-ichi Negishi / Akira Suzuki. *Angew. Chem. Int. Ed.* 49: 8300.
4 Brown, D.G. and Boström, J. (2016). *J. Med. Chem.* 59: 4443–4458.
5 Suzuki, A. and Miyaura, N. (1995). *Chem. Rev.* 95 (7): 2457–2483.
6 Hosseini, M.-J. et al. (2016). *Metallomics* 8 (2): 252–259.
7 Miyaura, N., Yanagi, T., and Suzuki, A. (1981). *Synth. Commun.* 11: 513.
8 Johanssen Seechurn, C.C.C., Kitching, M.O., Colacot, T.J., and Snieckus, V. (2012). *Angew. Chem. Int. Ed.* 51: 5062.
9 Dieck, H.A. and Heck, R.F. (1974). *J. Am. Chem. Soc.* 96 (4): 1133.
10 Old, D.W., Wolfe, J.P., and Buchwald, S.L. (1998). *J. Am. Chem. Soc.* 120 (37): 9722–9723.
11 Wolfe, J.P., Singer, R.A., Yang, B.H., and Buchwald, S.L. (1999). *J. Am. Chem. Soc.* 121 (41): 9550–9561.
12 Huang, X., Anderson, K., Zin, D. et al. (2003). *J. Am. Chem. Soc.* 125 (22): 6653–6655.
13 Walker, S.D., Barder, T.E., Martinelli, J.R., and Buchwald, S.L. (2004). *Angew. Chem. Int. Ed.* 43: 1871–1876.
14 Burgos, C.H., Barder, T.E., Huang, X., and Buchwald, S.L. (2006). *Angew. Chem. Int. Ed.* 45 (26): 4321–4326.
15 Fors, B.P., Watson, D.A., Biscoe, M.R., and Buchwald, S.L. (2008). *J. Am. Chem. Soc.* 130 (41): 13552–13554.

16 Pickett, T.E., Roca, F.X., and Richards, C.J. (2003). *J. Org. Chem.* 68: 2592.
17 Marion, N. and Nolan, S.P. (2008). *Acc. Chem. Res.* 41: 1440.
18 Fleckenstein, C.A. and Plenio, H. (2010). *Chem. Soc. Rev.* 39: 694.
19 Littke, A.F. and Fu, G. (2002). *Angew. Chem. Int. Ed.* 41: 4176.
20 Thompson, W.J., Jones, J.H., Lyle, P.A., and Thies, J.E. (1988). *J. Org. Chem.* 53: 2052.
21 Sather, A.C., Lee, H.G., De La Rosa, V.Y. et al. (2015). *J. Am. Chem. Soc.* 137 (41): 13433–13438.
22 Wu, J. and Chan, A.S.C. (2006). *Acc. Chem. Res.* 39: 711–720.
23 Martin, R. and Buchwald, S.L. (2008). *Acc. Chem. Res.* 41 (11): 1461–1473.
24 Dai, W. and Zhang, Y. (2005). *Tet. Lett.* 46: 1377–1381.
25 Kinzel, T., Zhang, Y., and Buchwald, S.L. (2010). *J. Am. Chem. Soc.* 132 (40):14073–14075.
26 Yang, Y. and Buchwald, S.L. (2013). *J. Am. Chem. Soc.* 135: 10642–10645.
27 Hillier, A.C. and Nolan, S.P. (2002). *Platin. Met. Rev.* 46 (2): 50–64.
28 Hadei, N., Kantchev, E.A.B., O'Brie, C., and Organ, M.G. (2005). *Org. Lett.* 7 (17): 3805–3807.
29 Froese, R.D.J., Lombardi, C., Pompeo, M. et al. (2017). *Acc. Chem. Res.* 50 (9): 2244–2253.
30 Jalaj, M., Hammouti, B., Touzani, R. et al. (2020). *Mater. Today Proc.* 31: S122–S129.
31 Herrmann, W.A. (2002). *Angew. Chem. Int. Ed.* 41 (8): 1290–1309.
32 Voloshkin, V.A., Tzourasa, N.V., and Nolan, S.P. (2021). *Daltons Trans.* 50: 12058–12068.
33 Heravi, M.M., Zadsirjan, V., Kafshdarzadeh, K., and Amiri, Z. (2020). *Asian J. Org. Chem.* 9 (12): 1999–2034.
34 Fortmana, G.C. and Nolan, S.P. (2011). *Chem. Soc. Rev.* 40: 5151–5169.
35 Hiller, A.C., Grasa, G.A., Viciu, M.S. et al. (2002). *J. Organomet. Chem.* 653: 69–82.
36 Organ, M.G., Chass, G.A., Fang, D.-C. et al. (2008). *Synthesis* 17: 2776–2797.
37 Valente, C., Pompeo, M., Sayah, M., and Organ, M.G. (2014). *Org. Process Res. Dev.* 18 (1): 180–190.
38 Zhao, Q., Meng, G., Nolan, S.P., and Szostak, M. (2020). *Chem. Rev.* 120 (4): 1981–2048.
39 Zhang, C., Huang, J., Trudell, M.L., and Nolan, S.P. (1999). *J. Org. Chem.* 64 (11): 3804–3805.
40 Farmer, J., Hunter, H.N., and Organ, M.G. (2012). *J. Am. Chem. Soc.* 134 (42): 17470–17473.
41 O'Brien, C.J., Kantchev, E.A.B., Valente, C. et al. (2006). *Chem. Eur. J.* 12: 4743–4748.
42 Marion, N., Navarro, O., Mei, J. et al. (2006). *J. Am. Chem. Soc.* 128 (12): 4101–4111.
43 Diebolt, O., Braunstein, P., Nolan, S.P., and Cazin, C.S.J. (2008). *Chem. Commun.* 3190–3192.
44 Pompeo, M., Froese, R.D.J., Hadei, N., and Organ, M.G. (2012). *Angew. Chem. Int. Ed.* 51: 11354–11357.
45 Atwater, B., Chandrasoma, N., Mitchell, D. et al. (2015). *Angew. Chem. Int. Ed.* 127: 9638–9642.
46 Yamamoto, Y., Takada, S., and Miyaura, N. (2006). *Chem. Lett.* 35: 704.
47 Yamamoto, Y., Takada, S., and Miyaura, N. (2006). *Chem. Lett.* 35: 1368.
48 Yamamoto, Y., Takada, S., and Miyaura, N. (2009). *Organometallics* 28: 152.
49 Valente, C., Belowich, M.E., Hadei, N., and Organ, M.G. (2010). *Eur. J. Org. Chem.* 2010 (26): 4343–4354.
50 Price, G.A., Hassan, A., Chandrasoma, N. et al. (2017). *Angew. Chem. Int. Ed.* 129 (43): 13532–13535,4.
51 Sun, B., Ning, L., and Zeng, H.C. (2020). *J. Am. Chem. Soc.* 142: 13823–13832.
52 Scott, N.W.J., Ford, M.J., Jeddi, N. et al. (2021). *J. Am. Chem. Soc.* 143 (25): 9682–9693.
53 Pagliaro, M., Pandarus, V., Ciriminna, R. et al. (2012). *ChemCatChem.* 4: 432–445.
54 Silva, S., Almeida, A.J., and Vale, N. (2019). *Biomolecules* 9: 22.
55 Hong, K. et al. (2020). *ACS Appl. Nano Mater.* 3: 2070–2103.
56 Marck, G., Villiger, A., and Buchecker, R. (1994). *Tet. Lett.* 35 (20): 3277–3280.
57 Pedersen, L., Mady, M.F., and Sydnes, M.O. (2013). *Tet. Lett.* 54: 4772–4775.
58 Adrio, L.A., Nguyen, B.N., Guilera, G. et al. (2012). *Catal. Sci. Technol.* 2: 316–323.

59 Deng, Y., Gong, L., Mi, A. et al. (2003). *Synthesis* 2003 (3): 337–339.
60 Skapski, A.C. and Smart, M.J.L. (1970). *J. Chem. Soc. D* 658.
61 Kirik, S.D. (2004). *Acta. Crystallogr. Sec. C. Cryst. Struct. Commun.* 60: M449–M450.
62 Stephenson, T.A., Morehouse, S.M., Powell, A.R. et al. (1965). *J. Chem. Soc.* (0): 3632–3640.
63 Pandey, R.N. et al. (1974). *Can. J. Chem.* 52: 1241–1247.
64 Bukhmutov, V.I., Berry, J.F., Cotton, F.A. et al. (2005). *Daltons Trans.* (11): 1989–1992.
65 Stoyanov, E.S. (2000). *J. Struc. Chem.* 41: 440–445.
66 Evans, J., O'Neill, L., Kambhampati, V.L. et al. (2002). *J. Chem. Soc. Dalton Trans.* (10): 2207–2212.
67 Fyfe, J.W.B. and Watson, A.J.B. (2015). *Synlett* 26: 1139–1144.
68 Fyfe, J.W.B., Valverde, E., Seath, C.P. et al. (2015). *Chem. Eur. J.* 21: 8951–8964.
69 Fyfe, J.W.B., Seath, C.P., and Watson, A.J.B. (2014). *Angew. Chem. Int. Ed.* 53: 12077–12080.
70 Sullivan, R.J., Freure, G.P.R., and Newman, S.G. (2019). *ACS Catal* 9: 5623–5630.
71 Cartagenova, D., Bachmann, S., Püntener, K. et al. (2022). *Catal. Sci. Technol.* 12: 954.
72 Tamao, K., Sumitani, K., and Kumada, M. (1972). *J. Am. Chem. Soc.* 94: 4374–4376.
73 Percec, V., Bae, J.Y., and Hill, D.H. (1995). *J. Org. Chem.* 60: 1060.
74 Zultanski, S. and Fu, G. (2013). *J. Am. Chem. Soc.* 135: 624–627.
75 Lu, Z., Wilsily, A., and Fu, G. (2011). *J. Am. Chem. Soc.* 133 (21): 8154–8157.
76 Zultanski, S. and Fu, G. (2011). *J. Am. Chem. Soc.* 133 (39): 15362–15364.
77 Cammidge, A.N. and Crépy, K.V.L. (2000). *Chem. Comm.* (18): 1723–1724.
78 Li, Y., Luo, Y., Peng, L. et al. (2020). *Nat. Commun.* 11: 417.
79 Lin, Q., Dawson, G., and Diao, T. (2021). *Synlett* 32: 1606–1620.
80 Mohadjer Beromi, M., Nova, A., Balcells, D. et al. (2017). *J. Am. Chem. Soc.* 139: 922–936.
81 Leyssens, L., Vinck, B., Van Der Straeten, C. et al. (2017). *Toxicology* 387 (15): 43–56.
82 Gilman, H. and Lichtenwalter, M.J. (1939). *J. Am. Chem. Soc.* 61: 957.
83 Kharasch, M.S. and Fuchs, C.F. (1941). *J. Am. Chem. Soc.* 63: 2316.
84 Uemara, S. and Fukuzawa, S. (1982). *Tet. Lett.* 23: 1181.
85 Ludwig, J., Simmons, E., Wisniewski, S., and Chirik, P. (2021). *Org. Lett.* 23: 625–630.
86 Gosmini, C., Bégoiun, J.-M., and Moncomble, A. (2008). *Chem. Comm.* (28): 3221–3223.
87 Hammann, J.M., Hoffmayer, M.S., Lutter, F.H. et al. (2017). *Synthesis* 49 (17): 3887–3894.
88 Asghar, S., Tailor, S.B., Elorriaga, D., and Bedford, R.B. (2017). *Angew. Chem. Int. Ed.* 56: 16367.
89 Obligacion, J.V., Semproni, S.P., Pappas, I., and Chirik, P.J. (2016). *J. Am. Chem. Soc.* 138: 10645–10653.
90 Gülak, S., Stepanek, O., Malberg, J. et al. (2013). *Chem. Sci.* 4: 776.
91 Kreyenschmidt, F., Meurer, S.E., and Koszinowski, K. (2019). *Chem. Eur. J.* 25: 5912–5921.
92 Greaves, M.E., Ronson, T.O., Maseras, F., and Nelson, D.J. (2021). *Organometallics* 40: 1997–2007.
93 Neely, J.M., Bezdek, M.J., and Chirik, P. (2016). *ACS Cent. Sci.* 2 (12): 935–942.
94 Arevalo, R. and Chirik, P.J. (2019). *J. Am. Chem. Soc.* 141 (23): 9106–9123.
95 Mills, L.R., Gygi, D., Ludwig, J.R. et al. (2022). *ACS Catal* 12: 1905–1918.
96 Tailor, S.B., Manzotti, M., Smith, G.J. et al. (2021). *ACS Catal.* 11 (7): 3856–3866.
97 Piontek, A., Ochędzan-Siodłak, W., Bisz, E., and Szostak, M. (2020). *ChemCatChem* 13 (1): 202–205.

27

Hierarchical Zeolites for Environmentally Friendly Friedel Crafts Acylation Reactions

Ana P. Carvalho[1,2], *Angela Martins*[2,3], *Filomena Martins*[1,2], *Nelson Nunes*[2,3], *and Rúben Elvas-Leitão*[2,3]

[1] *Departamento de Química e Bioquímica, Faculdade de Ciências da Universidade de Lisboa, Campo Grande, Lisboa, Portugal*
[2] *Centro de Química Estrutural, Institute of Molecular Sciences, Faculdade de Ciências, Universidade de Lisboa, Campo Grande, Lisboa, Portugal*
[3] *DEQ, Instituto Superior de Engenharia de Lisboa, IPL, R. Conselheiro Emídio Navarro, Lisboa, Portugal*

27.1 Introduction

Friedel-Crafts reactions are among the best-known types of organic reaction of aromatic molecules. Their name derives from the two chemists who first studied them c. 1877, the French Charles Friedel and the American James Mason Crafts [1, 2].

These reactions can be included in a broader classification; that is, electrophilic aromatic substitutions, in which the aromatic rings can be substituted by halogen, nitro groups, alkyl groups, and others. Because aromatic molecules are much less reactive compared with simple alkanes, electrophilic aromatic substitutions require a catalyst or specific experimental conditions to increase the electrophilic nature of the substituent. Classical examples of this strategy include the use of $FeBr_3$ in the bromination of benzene or the mixture of concentrated HNO_3 and H_2SO_4 to generate NO_2^+ used in aromatic nitration.

Starting from the original Friedel and Crafts papers, knowledge regarding these reactions has expanded over the years, leading to the publication of numerous books and papers on the subject. One of the most important works was the four-volume book *Friedel-Crafts and Related Reactions*, edited by George Olah [3], in which the author stated that different types of reactions could be considered as Friedel-Crafts type if taking place under the catalytic effect of a Lewis acid type or of protic acids [4]. Despite the strict definition of Friedel-Craft reactions, the term is still used in a broader sense and includes different processes, even biological pathways. An example is vitamin K1 biosynthesis where, in the absence of a catalyst, the carbocation electrophile is formed by the dissociation of an organodiphosphate in a Friedel-Crafts look alike reaction.

Friedel-Crafts reactions can be divided in two types: alkylations and acylations and, as the name implies, the difference is based on the type of electrophile substrate, an alkyl or an acyl group. The original methods used to perform these reactions combined $AlCl_3$ as catalyst and an alkyl halide or an acyl halide, respectively, for alkylations or acylations.

Catalysis for a Sustainable Environment: Reactions, Processes and Applied Technologies Volume 2, First Edition. Edited by Armando J. L. Pombeiro, Manas Sutradhar, and Elisabete C. B. A. Alegria.
© 2024 John Wiley & Sons Ltd. Published 2024 by John Wiley & Sons Ltd.

However, Friedel-Crafts reactions currently include those with a range of different alkylating agents, namely: activated and unactivated alkenes, alkynes, paraffins, alcohols, ethers, carbonyl, and other acylating agents such as carboxylic acids, carboxylic acid derivatives, esters, and anhydrides [3]; as well as the use of different Lewis acid catalysts such as $AlBr_3$, $FeCl_3$, $SbCl_5$, $TiCl_4$, $ZnCl_2$, $SnCl_4$, and BF_3 and of strong Brönsted acids such as HF, H_3PO_4, and H_2SO_4. Progress in developing these types of reactions lead, additionally, to the current use heterogeneous catalysts, as described further ahead in this chapter, and even of asymmetric catalysts.

The simplified mechanism for the classic acylation reaction (analogous to alkylation) is depicted in Figure 27.1 showing its main steps: carbocation formation with assistance of $AlCl_3$, attack of the aromatic ring leading to the formation of a C–C bond and a new carbocation intermediate, and, finally, proton loss with the formation of the neutral acylated substitution product. The formation of acylium ions is favored in polar solvents or when using anhydrides esters with strong proton acids.

Despite the synthetic opportunities provided by this class of reactions, namely the creation of C–C bonds, there are several intrinsic limitations to Friedel-Crafts reactions: aromatic and vinyl halides usually do not react, and the aromatic ring should not have electron-withdrawing groups. Another major challenge for these reactions is in finding ways to reduce possible carbocation rearrangements and limiting the number of substitutions in the aromatic substrate. However, these constraints are not problematic in acylation reactions, since being the acyl carbocation more stabilized rearrangements do not occur. Moreover, as the aromatic acylation products are less reactive, this reaction produces essentially monosubstituted products. In fact, the most efficient way to introduce a -CH_2- group in a benzene ring is to perform an acylation, produce a ketone, and afterwards reduce the carbonyl group. Traditional Friedel-Crafts acylation reactions have also some drawbacks compared to alkylations, as they require equimolar quantities of catalysts (e.g. $AlCl_3$) due to the formation of an acid/base complex of the acyl halide.

Both Friedel-Crafts alkylations and acylations are important synthetic tools used in industry, as they are used to prepare chemical feedstock, synthetic intermediates, and fine chemicals [5]. The large

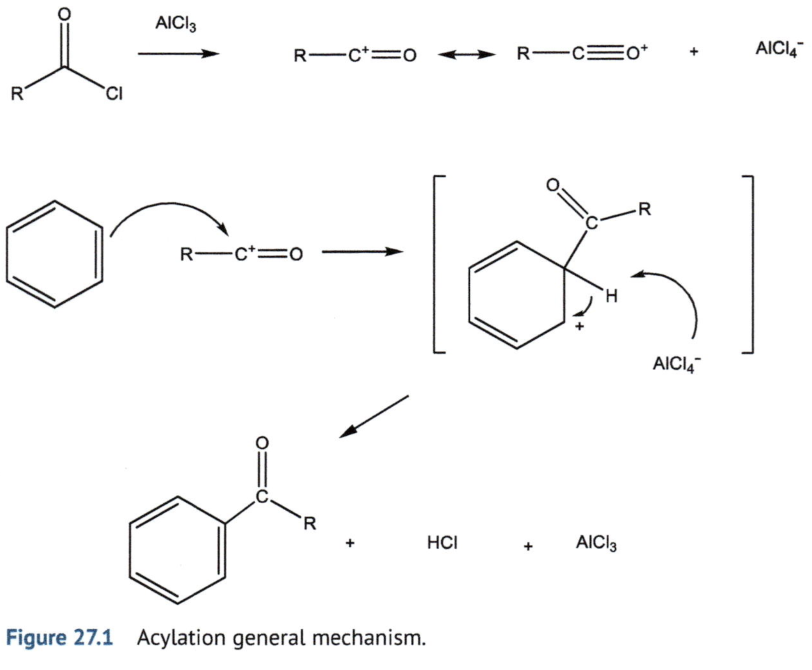

Figure 27.1 Acylation general mechanism.

number of reactions also means that different chemical, engineering, environmental factors, and economic factors can affect them. Because the focus of this chapter is on acylation reactions, only their practical challenges will be detailed next.

The most important products of Friedel-Crafts acylations are aromatic ketones for which synthetic methods can use different procedures, solvents, acylating agents, and catalysts. Table 27.1 illustrates several examples that demonstrate the diverse nature of the obtained ketones and their uses, according to the different experimental options taken.

The choice of catalyst plays a major role in Friedel-Crafts acylation reactions. The traditional catalyst is $AlCl_3$, and less active catalysts include BF_3 and $SnCl_4$. Other strong acids such as H_2SO_4, $HClO_4$, and $HO[P(OH)(O)O]_nH$ can also be used when the acylating agents are anhydrides, esters, or carboxylic acids.

One of the main problems of traditional catalysts is the large stoichiometric amounts needed, that is a threefold amount in the case of anhydrides (i.e. these catalysts act more like reagents). Separation is also a drawback, and it is even more problematic since these catalysts cannot be recycled. Additionally, they are corrosive, making the separation difficult and creating the need to neutralize and dispose of large quantities of waste.

Economic and environmental problems of homogeneous catalysts have pushed the development of heterogeneous acylation catalysts. These materials are strong solid acids and different alternatives are possible: modified clays, solid superacids, surface mounted acids, and Nafion are all examples of these materials; however, zeolites are the most important ones, and their use will be detailed further in the text. It is worth mentioning that commercial scale processes for the acylation of aromatic compounds (combined with acetic anhydride) have been already developed by Rhône-Poulenc (now Rhodia) using the Beta zeolite (BEA) structure [7].

Other variables, such as the solvents used, can also have an impact on the obtained products by influencing regioselectivity [8]. It is also possible to find in the literature examples of acylating

Table 27.1 Aromatic ketones obtained by Friedel-Crafts acylation reactions [6].

Acylating agent	Aromatic compound	Product	Uses
acetic anhydride	benzene	acetophenone	perfumes, plasticizer pharmaceutical, solvent,
acetic anhydride	toluene	4-methylacetophenone	perfumes
acetic anhydride	anisole	4-methoxyacetophenone	perfumes
acetic anhydride	isobutylbenzene	4-isobutylactophenone	pharmaceuticals
dichloroacetyl chloride	1,2-dichloro benzene	α,α,2,4-tetrachloro acetophenone	insecticides
chlorobutyroyl chloride	fluorobenzene	chloropropyl 4-fluorophenyl ketone	pharmaceuticals
tetrachloromethane	benzene	benzophenone	pharmaceuticals, insecticides, perfumes
benzoyl chloride	benzene	benzophenone	pharmaceuticals, insecticides, perfumes
phosgene	N,N-dimethyl aniline	4,4'-bis-dimethyl aminobenzophenone	dyes
phthalic anhydride	benzene	2-benzoylbenzoic acid	anthraquinone

agents acting as solvents in reactions that are considered by some authors as solvent-free acylation reactions [9, 10], or even the resort to more unusual solvents such as ionic liquids [11–13].

Experimental procedures also play a role when working with certain solvents in acyl halide acylations. Three different methodologies are commonly used [6]: reagents are all mixed and cooled and the catalyst added slowly (Elbs method); the acylating agent is added to a cooled solution/suspension of catalyst, with the substrate added afterwards (Perrier method); or the aromatic substrate is the solvent and the acylating agents is added slowly (Bouveault method).

Regarding acylating agents, as stated previously, different compounds can be used in Friedel-Crafts reactions and the selection is obviously dependent on the required substituents (R). However, other aspects should be taken into account, namely the relative reactivity of the acylating agents. Roughly, the reactivity order is the following: $[RCO]^+[BF_4]^- \approx [RCO]^+[ClO_4]^- > RCOO^- > SO_3H > RCOX > (RCO)_2O > RCO_2R' > RCONR'_2$. This order is not absolute because different R substituents influence reactivity, and different acyl halides may also show different reactivities: $RCOI > RCOBr > RCOCl > RCOF$ [6].

The great flexibility of Friedel-Crafts acylation reactions is also connected to the variety of substrates that can be used. There are numerous examples of substrates used in the production of aromatic ketones including benzene and benzene derivatives, among which toluene and aromatic amines should be considered. Also important, in this regard, is the acylation reactions of polynuclear aromatic compounds, viz. naphthalene, biphenyls, anthracene, and phenanthrene. Additionally, heteroaromatic compounds like furan, thiophene, and pyrrole, which are normally activated, can be acylated in milder conditions eventually with significantly lower environmental impact. If one uses a broader definition of Friedel-Crafts acylation, it is also possible to include substrates such as olefins.

Friedel-Crafts acylations have been part of numerous synthetic routes for multiple molecules. Some of these routes are still used at present, whereas others have been replaced by more economic or environmentally friendly ones. An example is the already mentioned acetophenone, a product that was commercially manufactured using Friedel-Crafts acylation of benzene by acetic anhydride at 30 °C, catalyzed by $AlCl_3$ in an 85% yield reaction. However, acetophenone is now produced industrially as a by-product of the oxidation of ethylbenzene or cumene. The opposite occurs with the synthesis of anthraquinone from phthalic anhydride and benzene, which uses a Friedel-Crafts reaction to produce o-benzoylbenzoic acid in the first step. This method was first described by Heller in 1906 and is still used despite the development of a newer process developed by BASF using styrene (which has also drawbacks).

The number of products currently manufactured using synthesis steps involving Friedel-Crafts acylations is massive. The synthesis of tolmetin (1-methyl-5-(4-methylbenzoyl)-1H-pyrrole-2-acetic acid) an anti-inflammatory, analgesic–antipyretic medicine, and Naproxen ((+)-6-methoxy-α-methyl-2-naphthaleneacetic acid), an anti-inflammatory agent, are only two of the many examples of different products that could be listed here.

27.2 Zeolites and Hierarchical Zeolites

Zeolites are solid materials known for more than 250 years that became important catalysts for heterogeneous processes only after the discoveries made by Milton and Barrer in the middle of the last century [14, 15]. In fact, the studies of these scientists were the first ones to establish the synthesis protocols to obtain zeolitic structures, some found in nature and others without a natural counterpart, making the use of these solids in industrial applications possible. At present, according to the Structure Commission of the International Zeolite Association, the number of synthetic zeolites is higher than 240, as over 67 natural structures have been identified [16].

Structurally, zeolites are very organized frameworks resulting from the different forms by which SiO_4 and AlO_4^- tetrahedrons are linked, sharing one or more oxygens to create cavities and channels with openings in the range of microporosity (<2 nm), as depicted in Figure 27.2 for two of the most used zeolites: zeolite Y (faujasite [FAU] structure) and zeolite A (LTA structure).

Due to the regularity of the pore structure, zeolites present what is known as shape selectivity, which is of paramount importance in several industrial processes (i.e. in petroleum refining and the petrochemical industry) [15]. Processes involving cracking, isomerization, or transalkylation reactions use zeolite-based catalysts, not only due to the shape selectivity that allows the production to be driven toward a specific and more valuable product, but also because zeolites gather other important properties, such as high mechanical and thermal stability. The latter feature allows not only the use of zeolites in high temperature processes, but also the regeneration of the poisoned catalysts by thermal treatment. An example is the zeolite Y derived catalyst used in the fluid catalytic cracking (FCC) process where, after only 2–10 seconds of residence time in the reactor, the poisoned catalyst is continuously regenerated at around 750 °C under air, and reintroduced in the reactor [17].

Ion exchange is another important property of zeolite structures which is linked to the negative charge associated with the AlO_4^- tetrahedron. This negative charge is usually compensated for by alkaline cations (e.g. Na^+), thus assuring the electroneutrality of the framework. This attribute drives the application of zeolites in the detergent industry, where they are used as water-softening agents. This same feature is also responsible for an equally important property of zeolites: a high surface acidity when the compensating ions are H^+. The presence of Brönsted and/or Lewis acid centers allows the use of zeolite-based catalysts in several industrial transformations such as, for example, long chain hydrocarbons hydroisomerization, because, according to the accepted mechanism for this process, the isomerization step occurs on the acid active sites of the metal supported catalyst while the dehydrogenation/hydrogenation reactions take place in the metal function [18].

As mentioned before, the highly organized micropore framework of zeolites is fundamental for the exceptional behavior of these solids in a large number of applications. Actually, because the dimensions of channel openings and internal diameter of cavities are of the same order as those of a large number of organic molecules, especially those involved in refining and petrochemical

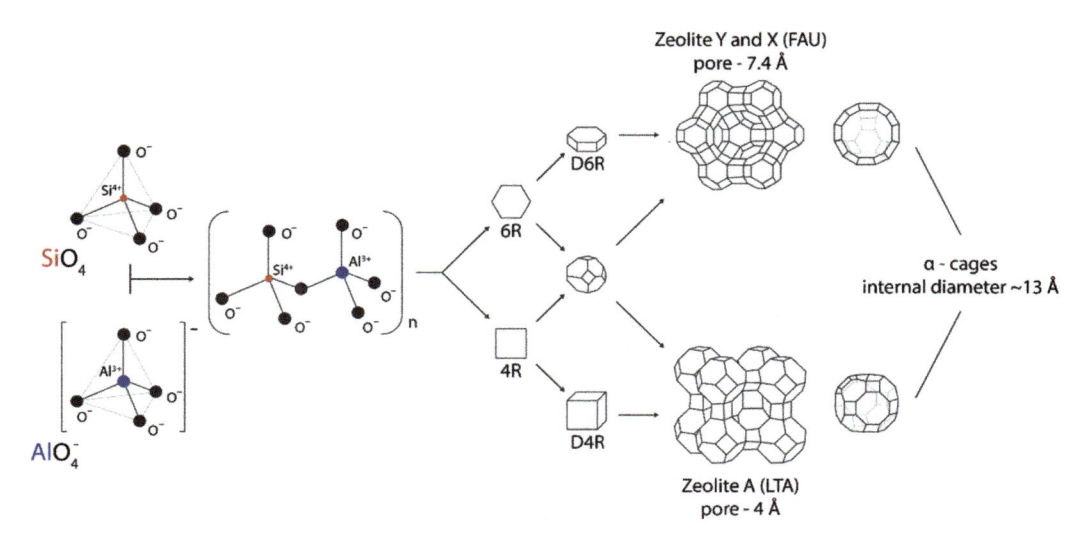

Figure 27.2 Building units and pore/cage sizes of zeolites Y and X (faujasite [FAU] structure) and zeolite A (LTA structure).

processes, and as transformations occur in the confined nanoscale porosity, zeolites are often considered as nanoreactors. Nevertheless, the strict micropore nature, that is in fact an advantage when small molecules are considered, becomes a disadvantage when the objective is to process bulky substrates for which the access to active sites may suffer from serious diffusion constraints or even be hindered. As schematized in Figure 27.3, several methodologies have been proposed to overcome this issue; that is, to prepare zeolites that along with microporosity also present a mesopore network (pores with widths between 2 and 50 nm). These various methods can be divided in two major groups, as the process involves the modification of a previously synthetized structure (top-down methods) or the mesopore structure results from a change in the synthesis protocol (bottom-up methods) [19].

The materials thus obtained are named hierarchical zeolites, highlighting the presence of, at least, two levels of porosity, even though textural modifications are not the only important changes promoted by treatments because, in many cases, acidity is also considerably affected. It must be noted that the designation of hierarchical zeolites to encompass a broad set of materials is quite recent, so it is not surprising that in studies dating from late 1990s to the beginning of the 21st century, this designation was not applied even if some mesoporosity was developed by the treatments applied to modify the zeolite structures.

Regarding the different experimental approaches that can be followed to create a secondary pore structure, it must be mentioned that, depending on the methodology, the mesoporosity developed may result, for example, from pore opening enlargement through the controlled destruction of the pristine microporous framework (intracrystalline mesoporosity). In other cases, the mesopore structure is formed by spaces created by the aggregation of small crystallites (intercrystalline mesoporosity) of the microporous zeolite.

Among the processes mentioned in Figure 27.3, dealumination treatments were the first to be explored [20] aiming not specifically to the creation of an additional mesopore structure, but to improve the hydrothermal stability of the structure. In fact, despite several experimental procedures having been proposed to remove aluminum from the zeolite framework, the most important process, from an industrial point of view, is steaming, which is used to prepare the ultrastable Y zeolite (USY) used in the FCC process [21]. Steaming is a hydrothermal treatment where the sample is heated at temperatures that can reach 850 °C under steam carried by an inert flow [22–24]. In most cases, the hydrothermal treatment is followed by an acid leaching to remove the amorphous

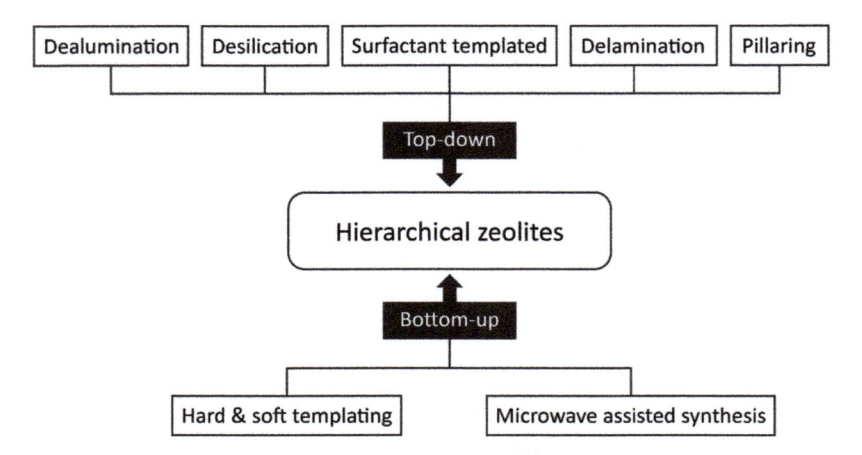

Figure 27.3 Various synthesis/modification methodologies to prepare hierarchical zeolites.

extra-framework species formed by the framework destruction, thus disclosing the mesoporous structure [25, 26]. Acid treatment (e.g. with HCl, HNO$_3$ or C$_2$H$_2$O$_4$) by itself can be applied to remove framework aluminum atoms [27–29], as well as, even though much less explored, treatments involving reaction with chelating agents [30] both in aqueous phase, or with SiCl$_4$ [21], in this case a gas phase reaction. Besides the expected textural changes, due to the bigger or smaller enlargement of the pristine microporosity [24, 26] these treatments also affect acidic properties since once the number of aluminum framework atoms decreases, the number of acid sites also, consequently, decreases [22, 23, 31].

Desilication is another top-down process that aims to promote the formation of a secondary pore structure, in this case by removing silicon atoms of the zeolite framework. Experimentally, the process consists of treating the zeolite with a basic solution being, by far, NaOH the most used base for this purpose [32–34]. This methodology is especially interesting to prepare hierarchical structures of silicon rich zeolites, as is the case of ZSM-5 (MFI structure) which is one of the most studied materials for different desilication approaches [25, 32–39].

The first studies focused on zeolite desilication reported quite simple experimental procedures where the mixture of zeolite and a basic aqueous solution (various concentrations) was stirred for different times and at different temperatures at atmospheric pressure [32, 33, 40]. In several studies a subsequent acid treatment was considered to remove debris created by the framework corrosion promoted by the treatment unblocking the porosity created by the alkaline treatment [35, 41–43]. As expected, in optimized conditions, samples with high crystallinity with a more or less broad mesopore size distribution (see Figure 27.4A) were obtained and, even though not anticipated, changes in the surface acidity were also reported [41, 42].

Despite the good results showing improvement in the diffusional properties and catalytic behaviour of samples modified by this simple procedure [32, 41], the search for strategies that would allow a more effective tailoring of the mesopore structure resulted in the preparation of templated zeolites. According to the methodology proposed by Garcia-Martinez et al. [44, 45], surfactant templated zeolites is a post-synthesis strategy that aims to introduce tailored intracrystalline mesoporosity. Experimentally, the zeolite structure is submitted to an alkaline treatment in the presence of a surfactant which, under optimized conditions, results in the development of an ordered mesopore structure (see Figure 27.4B). The process is interpreted as a local crystal rearrangement phenomenon, that is, the silica units removed from specific locations of the framework due the reaction with the alkali will rearrange around the surfactant micelles self-assembled inside the zeolite crystals. Besides the different composition of the reaction media there is another important difference compared to the simple alkaline treatment, since to prepare a surfactant templated zeolite the reaction is made under autogeneous pressure created in the autoclave where the reaction mixture is submitted to a heat treatment at, typically, 150 °C [46, 47].

The use of different surfactants will prompt the development of a mesopore structure formed by pores with widths tuned by the dimensions of the surfactant micelles, that is, depending on the length of the hydrophobic tail, the modified material will present a mesopore distribution centered at higher or lower values. Cetyltrimethylammonium bromide (CTAB) and dodecyltrimethylammonium bromide (DTAB) are the most assayed surfactants in combination with NaOH and NH$_4$OH. This top-down strategy has been applied to several zeolite structures, such as MOR, BEA, MFI, and LTA [47–50] but special reference must be made to the studies focused on zeolite Y by Garcia-Martinez et al. that not only gave an important contribution to elucidate the local crystal rearrangement mechanism previously mentioned, but is also responsible for the industrial application of a surfactant template Y structure as FCC catalyst [51].

Delamination and pillaring are other top-down methods to obtain hierarchical structures. The first delaminated zeolitic structure was reported by Corma et al. [52], who obtained ITQ-2

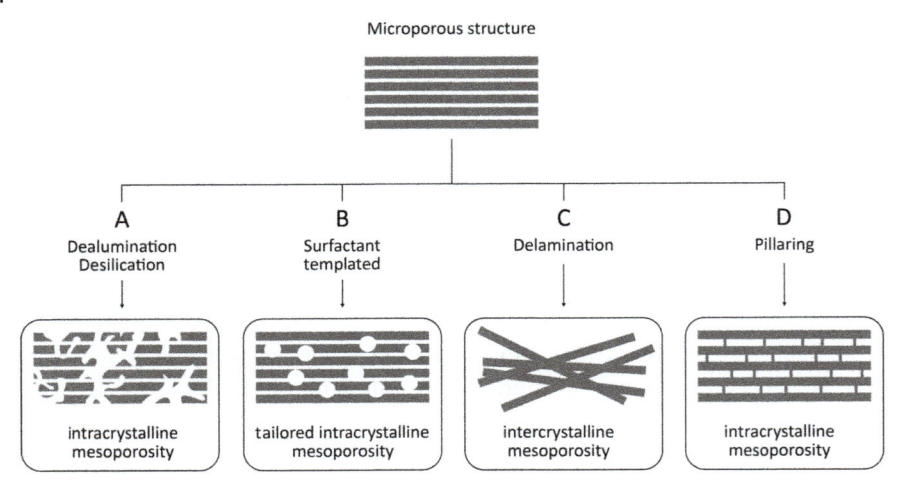

Figure 27.4 Main structural modifications of top-down methodologies.

structure through sonication of CTAB intercalated MCM-22 precursor (i.e. the MCM-22 structure before calcination). This is not a very common strategy to prepare hierarchical structures that can be considered only in the case of lamellar structures (e.g. MCM-22 or FER) and, in opposition to the methods described previously, in this case an intracrystalline mesopore structure is formed due to the random aggregation of the small crystals (see Figure 27.4C). Lamellar structures can also be modified through a pillaring process where the interlayer distance is increased by the intercalation of a metal oxide precursor between the layers, which is subsequently transformed in the corresponding oxide after calcination. The material obtained presents a 2D structure with broad intracrystalline mesopore size distribution (see Figure 27.4D). An example of this type of hierarchical structure is MCM-48 which, as the ITQ-2 mentioned previously, is also derived from the MCM-22 structure, in this case modified by the introduction of silica pillars [52].

Bottom-up methods are basically changes in the synthesis protocol to create the supplementary pore system during the synthesis process. As indicated in the scheme of Figure 27.3, the use of hard or soft templates is a possible strategy to obtain hierarchical structures that will present intra or intercrystalline mesoporosity if the crystallization of the zeolite occurs, respectively, around the template particles or in the voids created by the template aggregation. In any case, in the final step of the synthesis process the template is removed by a combustion to disclose the mesopore system.

Carbon black particles have been extensively used as hard template to obtain hierarchical materials, namely MFI structures [53, 54]. However, some studies report the use of biomass (e.g. leaves and stems [55]), inorganic materials (e.g. nanosized $CaCO_3$ [56]) and polymers (e.g. polystyrene beads [57]) to synthesise also MFI structures. Regardless of the template used, to create intercrystalline mesoporosity a considerably higher amount of template is added to a given amount of synthesis gel, than that needed to create an intracrystalline system. In this last case, besides the mesopore system, macropores can also be formed depending on the template dimensions. The synthesis conditions have also direct influence on the crystals dimension, with larger crystals been formed when less amount of template is added to the synthesis gel.

Soft templates are macromolecules, in general surfactants that in the reaction medium aggregate to form micelles, with dimension controlled by the tail length, thus tailoring the creation of an ordered secondary pore system. Experimentally, two approaches can be followed being the difference between them the moment where the template is introduced in the synthesis protocol. In the case of what is designated as primary method, the template is added to the reaction medium at the beginning of the synthesis, while in the procedure designated as secondary method, the template is introduced just before the hydrothermal treatment [58] Primary methods allow the preparation of hierarchical structures with a very developed mesopore system but have the great disadvantage of, normally, being based on non-commercially available surfactants what makes the process more time consuming and costly [19]. Syntheses applying a secondary method are a two-steps process, the first step comprising the aging of the synthesis gel at temperature around 100 °C, leading to the formation of seeds that will be assembled in the presence of the template in the second step of the process, that is during the hydrothermal treatment. Triton X-100 (polyethylene glycol tert-octylphenyl ether) is one of the surfactants that have been successfully used to obtain hierarchical zeolite structures allying meso, and even macroporosity, while maintaining a highly crystalline micropore network, as demonstrated in the study developed by Narayanan et al. [59]. These authors reported the synthesis of ZSM-5 zeolite with and without Triton showing that, as expected, both procedures resulted in identical micropore volume but a 50% increase of the mesopore volume is attained due to the presence of Triton in the reaction medium.

Microwave assisted synthesis is another possible strategy to prepare hierarchical zeolite structures. The basic idea is that, because under microwave radiation the nucleation and crystal growth rates are higher than those under conventional conditions, the process will lead to the formation of smaller crystals whose self-assembly will create an intracrystalline mesopore network in the interstices. The results of the recent study developed by Fukazawa et al. [60] clearly exemplify the potential of this methodology. The authors reported that, while a ZSM-5 sample obtained after three hours under microwave irradiation had a mesopore volume of 0.23 $cm^3\ g^{-1}$, in the case of the sample heated in an oil bath, the value was only 0.09 $cm^3\ g^{-1}$. On the other hand, the results also demonstrated that with this methodology the tunning of the pore dimension is not effective because a quite broad mesopore size distribution is obtained.

27.3 Zeolites and Hierarchical Zeolites as Catalysts for Friedel Crafts Acylation Reactions

The set of properties previously described makes zeolites unavoidable materials when one needs to assay several heterogeneous catalysts for a given transformation. Thus, it is not surprising that they have also been considered as catalysts for Friedel-Crafts acylation reactions, as the BEA structure is already used at industrial level, as discussed before.

The first studies reporting the use of zeolite structures as catalysts for these transformations are attributed to Venuto and Landis [61] dating from the late 1960s. These authors showed the feasibility of aromatic acylation and related Fries rearrangements in the presence of acid faujasites. Unfortunately, the authors also reported that the yields obtained were very low (about 1 to 5%), due to the fast deactivation of the catalyst. These poor initial results were not enough to discourage the scientific community. In fact, in subsequent studies a wide range of substrates, acylating agents, solvents, and zeolitic structures, operating under various reaction conditions, shed light on the relevant features of zeolitic catalysts. One of the most dedicated researchers to the use of zeolites in Friedel-Crafts reactions was Eric Derouane, who, with co-workers, published a series of important papers on this subject at the end of the 20th/beginning of 21st century [62–65], mainly devoted to the study of BEA zeolite structure. Clerici [66], Sartori and Maggi [67], Liang et al. [68], and, more recently Nayak et al. [69] reviewed in detail the application of several zeolite structures in Friedel-Crafts acylation reactions in the presence of a large number of substrates, with BEA and FAU being the most studied ones in all cases. In the following paragraphs, some selected examples, representative of the studies published in last 10 years regarding the application of zeolites and hierarchical zeolites, will be presented. The relevance of the zeolite structure, acidity, and texture, especially relevant in the case of hierarchical material, as well as the effect of the experimental conditions used in the reaction, will be discussed.

The effect of zeolite structure and acidity was first studied by Procházková et al. [70]. The authors presented a complete survey on the catalytic behavior of several commercial zeolite structures: BEA, MFI, FAU, FER, and MOR in the liquid phase acylation of p-xylene using hexanoyl chloride, propionic anhydride, and isobutyric anhydride as acylating agents at 130 °C, and atmospheric pressure. The authors found that the highest conversions were obtained in the presence of large pore zeolite structures FAU (specifically USY) and BEA. The influence of the acylating agent was studied for these two most promising structures: when using propionic anhydride, the highest conversion was achieved with BEA (51,1%) against FAU (44.2%), whereas in the case of hexanoyl chloride the highest conversion was obtained with FAU (66.3%) against BEA (58.2%). The effect of the zeolite acidity was also explored for BEA structure, being the ideal Si/Al ratio of 19 in the presence of isobutyric anhydride and 25 in the case of propionic anhydride. Yamazaki et al. [71] reported the acylation of 2-methoxynaphtalene (2-MN) to produce 2-methoxy-6-acetylnaphtalene (2,6-ACMN), which is a key intermediate for the production of Naproxen, as well as 1-acetyl-2-methoxynaphtalene (1,2-ACMN), as a minor product. The reaction was performed in the presence of acetic anhydride (AA) as acylating agent and acetic acid as solvent, at 150 °C and atmospheric pressure using several zeolite structures as catalysts: BEA, LTL, FER, FAU, MOR, and MFI, with high Si/Al ratios. Among the examined zeolites, the acid form of mordenite structure, HMOR ($SiO_2/Al_2O_3 = 200$) showed the higher conversion (82%) and selectivity (86%). Reusability experiments were performed and showed that, after the third reuse cycle the conversion fell for about 30%, but the selectivity to 2,6-ACMN increased to about 98% (Figure 27.5). This behavior was

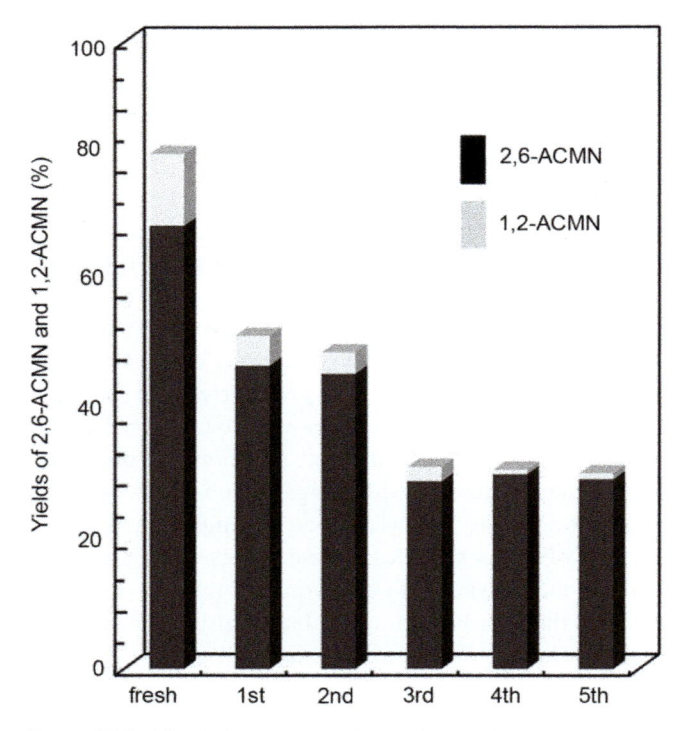

Figure 27.5 Yield of 2-methoxy-6-acetylnaphtalene (2,6-ACMN) and 1-acetyl-2-methoxynaphtalene (1,2-ACMN) for fresh and reused HMOR catalyst for the acylation of 2-methoxynaphtalene (2-MN). Reaction conditions: acetic anhydride [AA]/2-MN ratio of 10 mmol; $m_{catalyst}$ = 0.5 g; T = 150 °C and t = 48 hours. Reproduced with permission from Ref [71].

attributed to the leaching of Brönsted acid sites during the repeated catalytic cycles in the presence of acetic acid at high temperature, as well as to some collapsing of the pore structure, which was proved by N_2 adsorption data that showed some decrease in pore volume and surface area. Moreover, X-ray fluorescence (XRF) measurements also revealed a decrease in Al content.

Later, the same research group studied the effect of reaction temperature using a dealuminated MOR zeolite with $SiO_2/Al_2O_3 = 110$, obtained upon acid treatment of a commercial MOR zeolite ($SiO_2/Al_2O_3 = 25$) [72] with 8 mol L^{-1} HCl solution at 80 °C for 3 h. The catalytic performance was studied in the acylation of anisole (AN) with AA in acetic acid medium, changing the reaction temperature between 110 and 170 °C. The reaction profiles show an optimum temperature of 150 °C, with a slight decrease in catalytic activity at 170 °C and a more pronounced decrease for reaction temperatures below 130 °C, which was attributed to the boiling point of AA (140 °C). Thus, for lower temperatures, the interactions between the catalyst and the acylating agent would be less effective. Still, the authors call the attention for the fact that the more drastic reaction temperature used in this study, compared to the typical temperatures used for these reactions in the presence of heterogeneous catalysts, is related to the low content of Brönsted acid sites present on the catalysts, when compared to traditional zeolite catalysts.

The important role of zeolite acidity was also studied by Wei et al. [73], who modified Beta zeolite (BEA structure) with potassium cation (K^+), from KNO_3, and the organic base 2,4-dimethylquinoline (2,4-DMQ), and studied its effect in the acylation of AN with AA at 85 °C, 1.0 MPa and WHSV of 1.6 h^{-1}, using a ratio AN/AA of 1. It was found that the increase in K^+ loading caused a decrease in total acid strength with consequent decrease in the catalytic performance. The modification with 2,4-DMQ mainly poisoned the Brönsted acid sites and kept the Lewis sites almost intact.

For the samples treated with 2,4-DMQ (Beta-N samples), as well as those modified with different amounts of K^+ (x%K/Beta samples), the authors established linear correlations between the catalytic conversion of AN (AN conversion) and the concentration of Brönsted acid sites (C_B), as

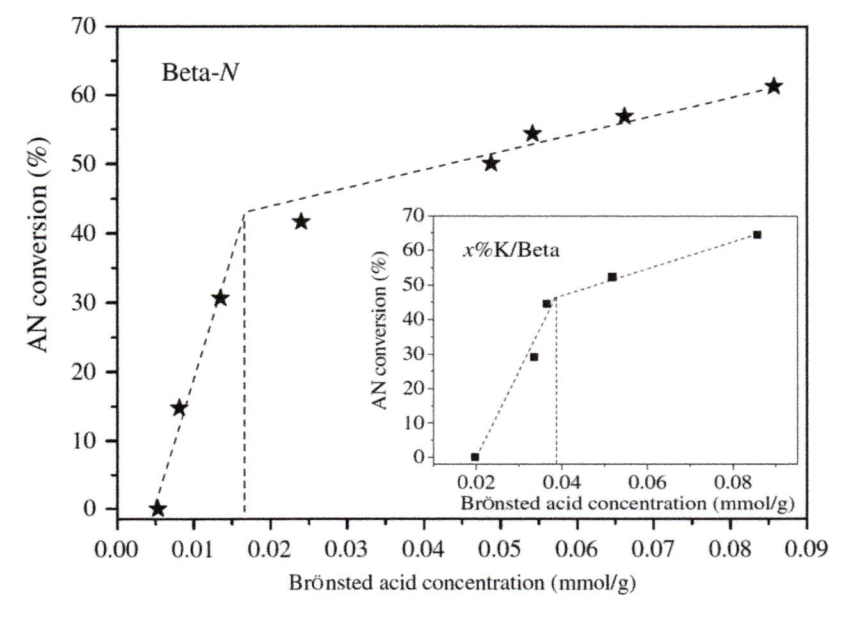

Figure 27.6 Relation between acid site concentration and catalytic performances of Beta-N samples (the insert shows the same relation for x%K/Beta samples). Reaction conditions: T = 85 °C; p = 1.0 MPa; WHSV (AA) = 1.6 h-1; n(AN):n(AA) = 1; time on stream two hours. Reproduced with permission from Ref [73] / Elsevier.

can be observed in Figure 27.6. In the case of Beta-N samples, for low C_B (0.0053–0.0167 mmol g^{-1}) the catalytic conversion increases substantially with small changes in Brönsted acidity, whereas for high C_B (0.0167–0.0857 mmol g^{-1}) the increased acidity has a small impact on the catalytic conversion. A similar behavior was observed for x%K/Beta samples (insert on Figure 27.6) though the demarcation point of the two linear relations was obtained for higher C_B value. In addition, when C_B on x%K/Beta samples was lower than 0.0197 mmol g^{-1} no catalytic activity was detected whereas for Beta-N samples the same was noted for 0.0053 mmol g^{-1}. According to the authors, these differences were attributed to the role of Lewis acidity.

The behavior of acid modified zeolites was also investigated by Chen et al. [74] on the acylation of thiophene with AA, at 80 °C and atmospheric pressure, using a trickle bed reactor. In this study the acidity of H-BEA zeolite was modified through an acid treatment with CH_3COOH, HCl, and HNO_3. The parent zeolite (Si/Al = 27) was stirred with 1 mol L^{-1} of each one of the acids at 100 °C for 12 hours, then washed, dried, and calcined at 450 °C for five hours [75]. The catalytic stability of the samples after treatment followed the order: HNO_3-BEA≈HCl-BEA>>CH_3COOH-BEA≈H-BEA. According to the authors, the reason for this trend may be that some Brönsted acid sites, which were covered by non-acidic amorphous substances on the external surface, would be exposed during the treatment with HCl and HNO_3. In the presence of CH_3COOH, a weak organic acid, only a small part of the amorphous substance was removed. Upon deactivation, mainly by deposition of carbonaceous deposits, the zeolite samples were dried and calcined at 450 °C for five hours. The reusability of BEA based catalysts was demonstrated after nine catalytic cycles. Indeed, the average lifetime of H-BEA was kept at 2911 minutes before the deactivated point (considered when the conversion of thiophene was below 90%), during the nine catalytic cycles, which was roughly identical to the capacity of CH_3COOH-BEA sample (3061 min). As expected, the HCl-BEA and HNO_3-BEA showed superior reusability performance, with 4863 and 4893 minutes, respectively.

Koehle et al. [76] investigated the catalytic behavior of Brönsted and Lewis acid zeolite catalyst in the acylation of methylfuran in the presence of AA at 110 °C at 14 bar. The reaction was performed using two acid BEA zeolites, with high and low alumina contents, (Si/Al = 23 and 138, respectively). The reaction was also carried out in the presence of various Lewis acid BEA zeolites, that is, Sn, Zr, Hf, or Ti loaded BEA. The authors showed that the highest reaction rate was obtained with the BEA (Si/Al = 23) whereas the higher turnover frequency was found for BEA (Si/Al = 138). Among Lewis acid zeolites Sn-BEA shows higher turnover frequency when compared to the other metal loaded zeolites. Electronic structure calculations showed that, the presence of both Brönsted and Lewis acid sites, follow the same mechanism (i.e. the classic addition-elimination aromatic electrophilic substitution). An unexpected finding that resulted from the authors calculations was that the most favorable catalytic pathway in Lewis rich zeolite Sn-BEA sample also involves a Brönsted acid catalysis step carried out by the silanol groups present on the material.

The operation conditions, particularly the role of the heating source was explored by Shekara et al. [77] who studied solventless liquid-phase acylation of p-cresol with different acylating agents, namely, acetic, propionic, butyric, hexanoic, octanoic, and decanoic acids, using BEA zeolite (Si/Al = 30) under conventional and microwave heating. A significant increase in activity was observed when the reaction occurred under microwave radiation. In fact, under conventional heating, the conversion with all acylating agents was less than 20%, whereas under microwave heating the conversions ranged between 50 to 80%. Moreover, the characterization of the microwave spent catalysts showed the absence of coke/coke precursors in contrast to catalysts used under conventional heating. The authors pointed out that, under conventional heating, the initial inhibition and coke formation occurred due to the preferential adsorption of the acylating agent. As p-cresol is chiefly responsible for heating up the reaction mixture, the heat-up energy released when the reaction occurs under microwave radiation might be affecting the adsorption of the acylating agent, suppressing the formation of coke/coke precursors. This study opened new perspectives to alternative heating sources in Friedel-Crafts acylation reaction, aiming to suppress one of the major drawbacks in the use of zeolitic materials (i.e. the fast deactivation due to the formation of coke/coke precursors that become trapped inside the zeolite porosity).

To overcome the disadvantages inherent to the use of purely microporous zeolites, hierarchical materials started to appear in the literature from the beginning of the 21st century, as referred before. One of the first examples on the application of hierarchical zeolites in Friedel-Crafts acylation reactions was presented by Derouane et al. in 2004 [78]. The authors prepared nano-sized H-BEA, starting from a commercial H-BEA zeolite with Si/Al = 18, through the confined space synthesis (CSS) method, where the zeolite was synthesized within the pores of a carbon black matrix, that is, a hard templated bottom-up method. Nano-sized H-BEA (n-H-BEA), with crystal sizes in the range 0.01–0.02 μm, and agglomerates of about 5 μm were obtained. The catalytic behavior of parent and modified zeolites was studied in the Friedel-Crafts acylation of AN by AA at 90 °C for five hours, where it was shown that n-H-BEA sample has a superior catalytic performance since the zeolite nanocrystals decrease the diffusional constraints that limit the egression of the voluminous product, p-methoxyacetophenone (p-MOAP) from the small crystal zeolites with very short diffusion path. At the time, the authors anticipated the future application of hierarchical zeolite materials, comprising nano-sized and micro+mesoporous materials, in many fine and specialty chemical organic synthesis reactions, when one of the products is strongly adsorbed inside the zeolite native micropores.

The optimization of Brönsted and Lewis acidity as a consequence of the generation of intracrystalline mesoporosity in HZSM-5 zeolite (MFI structure) was studied by Silva et al. [79]. The authors performed desilication treatments on commercial MFI zeolite (Si/Al = 12 and 23) with NaOH, changing the base concentration, temperature, and treatment time. The characterization of the modified samples showed a significant modification on the textural parameters, estimated from N_2 adsorption isotherms (Table 27.2). For example, sample Z23-at5, treated with a 0.2 mol L^{-1} NaOH solution for 90 minutes at 85 °C reached a mesoporous volume of 0.22 cm^3 g^{-1} and an external surface area of 160 m^2 g^{-1} against 0.02 cm^3 g^{-1} and 12 m^2 g^{-1}, respectively, for the parent Z23 (HZSM-5,

Table 27.2 Alkaline treatment conditions, Si/Al ratio, and textural properties of parent and desilicated ZSM-5 zeolite [79].

Sample	Treatment conditions			Si/Al[1]	RC (%)	External area (m^2 g^{-1})	Porous volume (cm^3 g^{-1})		
	[NaOH] (mol L^{-1})	T (°C)	t (min)				micro	meso	total[2]
Z23	—	—	—	23	100	12	0.15	0.02	0.17
Z23-at1	0.2	65	30	18	108	14	0.16	0.02	0.19
Z23-at2	0.2	65	48	17	107	21	0.16	0.04	0.19
Z23-at3	0.2	85	30	17	101	64	0.15	0.09	0.24
Z23-at4	0.2	85	48	14	87	136	0.12	0.17	0.29
Z23-at5	0.2	85	90	14	78	160	0.10	0.22	0.32
Z23-at6	0.4	85	48	9	48	172	0.07	0.36	0.53
Z23-at7	0.4	85	90	8	43	149	0.07	0.34	0.41

1) Determined by energy-dispersive X-ray spectroscopy (EDX), 2) total pore volume estimated at p/p^0=0.95.

Si/Al = 23). Nevertheless, some loss of microporosity was also noted, with 0.10 cm^3 g^{-1} for the treated sample against 0.15 cm^3 g^{-1} for the parent material.

Modifications on the profile of the acid sites also occurred, with a decrease in the number of catalytically strong active sites (*ht*-AS), and a pronounced generation of Lewis acid sites, resulting in a substantial decrease in Brönsted to Lewis acid ratio (B/L). Again, for Z23-at5 sample the B/L ratio was 3.2 against 21 for the parent Z23. The catalytic performance was studied in the acylation of AN with AA at 100 °C for two hours using a ratio of AA to AN of 2:1. Figure 27.7a presents the specific activity, calculated as the ratio of the main product, 4-methoxyacetophenone (4-MAP) per *ht*-acid site (denominated as 4-MAP/*ht*-AS). As can be observed, the development of mesoporosity caused a substantial increase in the specific activity, reaching a maximum with Z23-at5 sample. It is worth noticing that the sample with the highest mesoporous volume, Z23-at7, presents a decrease in the specific activity, which can be attributed to a high decrease in microporous volume (see Table 27.2) as well as a B/L ratio of 2.7, which is lower than that for all other samples. The most promising sample, Z23-at5, was selected for reusability assays. Thus, the sample was recovered after each cycle, and calcined at 600 °C under air atmosphere (5° C min^{-1}) for five hours. As can be observed from Figure 27.7b, sample Z23-at5 retained approximately 90% of the specific activity for 4-MAP for at least three catalytic cycles. Some loss of specific activity can be ascribed to some loss of active sites during the regeneration treatments.

To explore the influence of the textural modification performed on desilicated zeolites, Aleixo et al. [46] evaluated the catalytic behavior of MCM-22 (MWW structure), a less assayed zeolite in Friedel Crafts acylation reactions. The parent zeolite was desilicated for 45 minutes at 50 °C with 0.05 or 0.1 mol L^{-1} NaOH solutions (samples MCM-22/0.05 and MCM-22/0.1, respectively), followed by an acid treatment (AT) with 0.1 mol L^{-1} HCl at 70 °C for six hours [41], giving MCM-22/0.05/AT and MCM-22/0.1/AT samples. The characterization of the samples through low temperature N$_2$

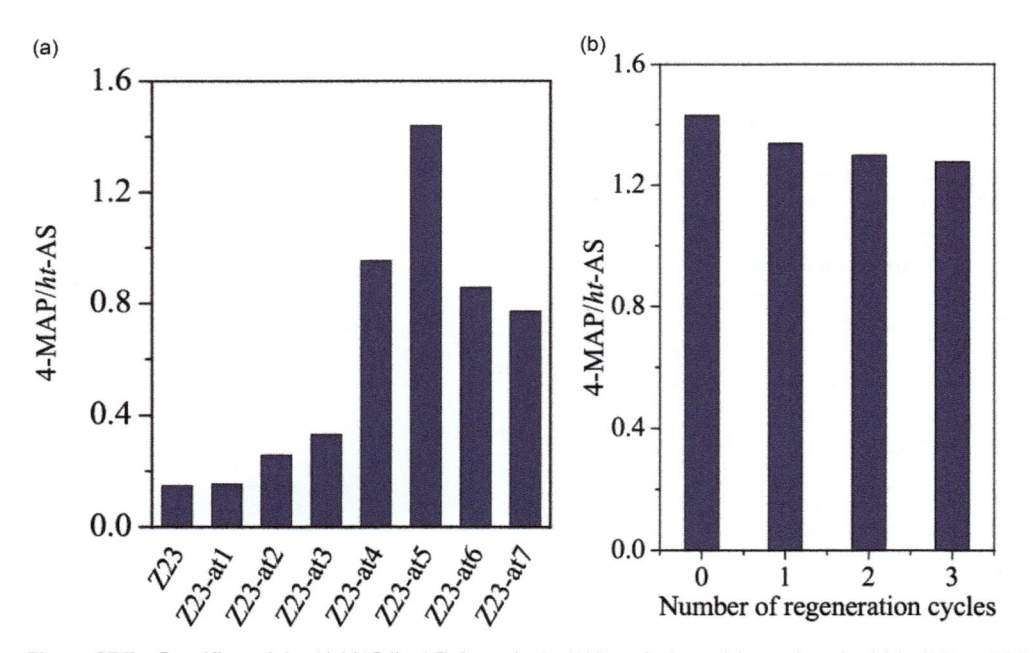

Figure 27.7 Specific activity (4-MAP/*ht*-AS) for anisole (AN) acylation with acetic anhydride (AA) at 100 °C for two hours using a ratio of AA to AN of 2:1 for fresh catalyst (a) parent (Z23) and desilicated samples and (b) regeneration cycles for desilicated Z23-at5 sample. Ref [79] / Elsevier.

adsorption isotherms showed that the development of mesoporosity occurs when 0.05 mol L^{-1} NaOH solution is used (0.22 cm^3 g^{-1} against 0.15 cm^3 g^{-1} for the parent zeolite) and is not affected by the acid treatment nor, significantly, by the increase of the base concentration. On the other hand, a decrease of the microporous volume is noted as NaOH concentration increases, becoming more pronounced upon acid treatment, in accordance with some loss of crystallinity. The acidity characterization, quantified by the integration of FTIR spectra of pyridine chemisorbed on Brönsted and Lewis acid sites, showed an important decreased on Brönsted acidity for MCM-22/0.05 sample due to the partial occlusion caused by deposition of fragments at the pore apertures, which are removed upon acid treatment, giving an even higher concentration of Brönsted acid sites when compared with the parent material. In the case of MCM-22/0.1/AT sample, the use of a more concentrated NaOH solution caused a simultaneous extraction of Si and Al from the framework that is accentuated by the acid treatment. For the catalytic assays, furan, pyrrole, and AN were used as substrates and AA was used as acylating agent. The experiments were performed for about 1 h at 60 °C using a molar ratio AA/substrate of 5 and 150 mg catalyst. According to the reaction mechanism in the case of furan and AN, the reaction is almost selective toward isomers 2-acetylfuran and p-MOAP, respectively, because the α-product is preferred because of the greater stability of the α-intermediate structure. For pyrrole substrate, two isomers were obtained, which can be attributed to a higher electrophilic susceptibility of carbon-β relative to carbon-α [65].

To quantify the catalytic behavior of parent and hierarchical catalyst, by calculating the kinetic parameters for the reaction, the authors proposed a simplified equation for the Langmuir-Hinshelwood model [42, 46]. Considering the reaction scheme, represented as

$$A_{(ads)} + S_{(ads)} \rightarrow P_{(ads)}$$

where A stands for the acylating agent, S for the substrate and P for the acylated product(s), the reactions occur in the presence of a large excess of acylating agent, to avoid a premature deactivation of the zeolite by the acylated product. In addition, the secondary product, acetic acid, is so marginally adsorbed that it can be discarded. Considering these assumptions, the reaction rate, r, can be expressed upon simplification, as

$$r \cong \frac{k[A][S]}{([A]+[S]+K_r[P])^2} \tag{27.1}$$

where k represents the rate constant of the rate determining step and K_r represents the ratio between the adsorption equilibrium constant of the product(s) and the normalized equilibrium constant of the reagents.

Figure 27.8 exemplifies the evolution of product yield as a function of reaction time for the acylation of furan by AA. As can be seen in this figure, a significant increase in product yield in the first minutes of reaction occurred, followed by a significant attenuation. The obtained results show the effect of desilication treatment where a decrease in product yield is observed as NaOH concentration increases from 0.05 to 0.1 mol L^{-1}. On the other hand, the effect of HCl treatment led, in both cases, to an increase in product yield, which is especially relevant in the first minutes of reaction (see insert in Figure 27.8).

Table 27.3 summarizes the kinetic parameters obtained from the non-linear regression treatment applied to the proposed Langmuir-Hinshelwood model equation as well as the turnover frequencies, calculated as the ratio between the kinetic constant and the concentration of accessible Brönsted acid sites [80]. From all three substrates, furan was selected as a representative example since it shows, for all samples, higher rate constant, and TOF.

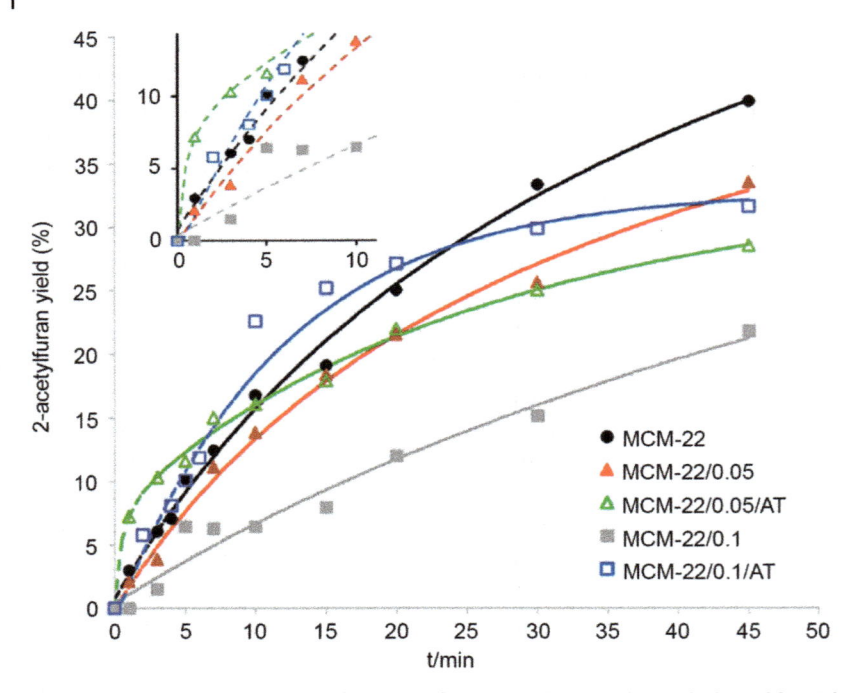

Figure 27.8 Product yield as a function of reaction time for the acylation of furan by acetic anhydride (AA). Lines represent calculated values resulting from the application of the kinetic model. Ref [46] / Elsevier.

Table 27.3 Summary of kinetics (rate constant, k, and turnover frequency, TOF), adsorption parameters, and statistical figures of merit for the catalytic reaction using furan as substrate [46].

Furan	MCM-22	MCM-22/0.05	MCM-22/0.05/AT	MCM-22/0.1	MCM-22/0.1/AT
k (mmol min^{-1} g^{-1})	14.3 ± 0.2	13.6 ± 0.3	178.8 ± 1.9	5.1 ± 0.1	23.8 ± 1.9
K_r	5.3 ± 0.3	8.8 ± 0.6	113.9 ± 4.9	5.8 ± 0.74	12.6 ± 2.5
R^2	0.992	0.989	0.999	0.955	0.924
s_{fit}	0.044	0.056	0.187	0.029	0.311
F	1269	761	11239	171	97
TOF (min^{-1})	93.5	117.2	1027.6	48.6	171.2

R^2 = determination coefficient; s_{fit}= standard deviation of fit; F = Fisher-Snedecor statistics.

As depicted, desilicated samples show lower rate constants and TOFs when compared with the parent MCM-22, which can be ascribed to the accumulation of extra-framework material at the pore mouths of the zeolite, or even some loss of active sites, especially in the case of MCM-22/0.1 sample. For the acid treated samples, a significant increase in rate constant and TOF is observed. For example, for MCM-22/0.05/AT sample, k is about 12 times and K_r is more than 20 times higher when compared with the parent zeolite. For MCM-22/0.1/AT sample the increase in the rate constant is much more moderate, as well as that in K_r, due to the loss of active sites during the desilication treatment. Interestingly, in the case of AN, a more voluminous molecule, K_r value is even lower when compared with the parent MCM-22, which was attributed to the possible interconnection of the two internal pore systems that fastens product desorption.

Aleixo et al. continued to study the effect of alkaline and alkaline+acid treatment on Friedel Crafts acylation reactions with small molecules as substrates, applying the same methodology, but now using BEA zeolite with distinct Si/Al ratios (12.5 and 32) [42]. In this case, the alkaline treatment was made with 0.1 mol L^{-1} NaOH solution at 60 °C for 30 minutes. The subsequent acid washing was performed using the experimental conditions previously mentioned [41, 46]. The catalytic experiments were performed at 60 °C, with a molar ratio AA/substrate of 5 and 150 mg catalyst, using furan, AN, and pyrrole as substrates and AA as an acylating agent.

The characterization of parent and modified samples showed that the Si/Al ratio of the parent material strongly influenced the catalytic behavior of desilicated and desilicated+acid treated samples. In the case of the BEA12.5 (SI/Al = 12.5) sample, desilication caused a decrease in products' yield and rate constant. Upon acid treatment, the unblocking of the zeolite porosity and the recovery of acid sites, that even slightly exceeds the parent material, lead to an increase in product yield and rate constant. On the other hand, for BEA32 (Si/Al = 32) a dissimilar behavior was noted since a continuous loss of acid sites occurred during the desilication and continued during the acid treatment. However, the desilicated+acid treated sample, BEA32-D-AT, presented the highest product yield in the case of furan and AN substrates, which can be assigned to the textural modifications, namely, the existence of larger mesopores that were developed in this sample.

The effect of size and chemical functionalities of substrate molecules was further explored by Elvas-Leitão et al. [81] on BEA modified zeolites through desilication and desilication+acid treatments, modified according to the procedure reported in a previous study [42]. The authors performed the catalytic assays at 60 °C, using benzofuran, thiophene, benzothiophene, and indole and compared the catalytic performance with the previously studied smaller furan, AN, and pyrrole substrates [42]. The relative size of the studied substrate molecules can be compared using the corresponding van der Waals volumes, giving the following order: benzothiophene>benzofuran>indole>furan>tiophene>pyrrole.

The kinetic curves and the parameters obtained using the simplified Langmuir-Hinshelwood model, proposed before [42, 46], show a clear relation between the size of substate molecules and their reactivity order, being in all cases significantly higher for tiophene and furan and

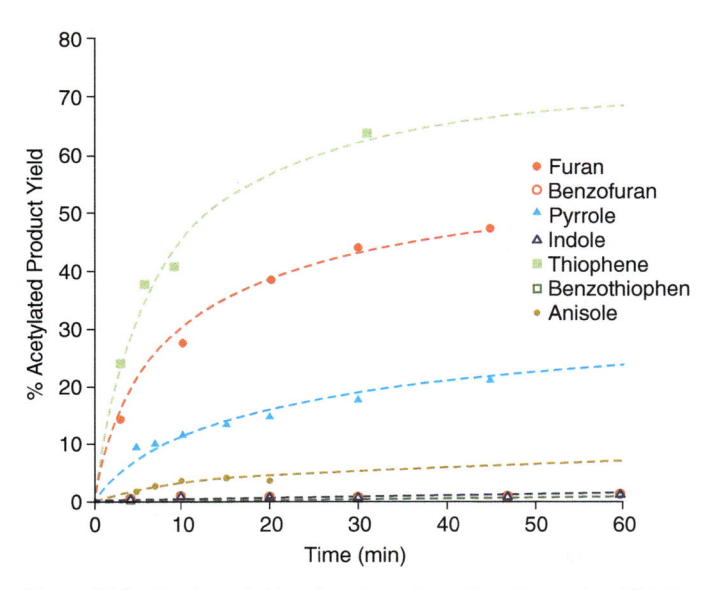

Figure 27.9 Product yield as function of reaction time using BEA-D sample catalyst [81] / MDPI / CC BY 4.0.

substantially lower for the bulkier benzothiophene and benzofurane, as can be seen for the desilicated sample, BEA-D, shown as a representative example in Figure 27.9.

When comparing the catalytic results for the three catalysts: parent, desilicated and desilicated+acid treated, the obtained results only show slight differences. In fact, the textural and chemical modifications performed on modified zeolite materials did not show in a first inspection major differences in the reactivity, neither in order or magnitude. This could only be rationalized through the mathematical analysis of the results, carried out by quantitative structure-property relationships (QSPR), as discussed later.

The use of hierarchical zeolites prepared with a surfactant templated method was reported by Linares et al. [82] where cethyltrimethylammonium bromide (CTAB) surfactant, in the presence of NH_4OH, was employed to modify USY zeolite used as heterogeneous catalyst in Friedel-Crafts alkylation of indole using alcohols as acylating agents for the synthesis of compounds of pharmaceutical interest. Recently, Martins et al. [47] modified Y zeolite using two surfactants with different sizes: CTAB and the less explored DTAB, with a slightly shorter carbon chain length, in the presence of NH_4OH. The suspensions were heated at 150 °C under autogenous pressure, changing the duration of the treatments from 6 to 48 hours. The characterization of the modified materials showed that the duration of the treatment influences the final properties of the materials. For the samples treated with DTAB (HY_D series), the six hour treatment (HY_D6) showed a significant loss of crystallinity and a decrease in the microporous volume characteristic of the zeolite structure. When the duration of the treatment was extended to 12, 24, or 48 hours, a continued enlargement of the pores occurred, ascribed to the transformation of smaller into larger micropores, in addition to the development of small diameter mesopores, along with some progressive loss of crystallinity and acidity. The size of the surfactant molecule used in the treatments also affected the zeolite properties. For samples modified for the same duration of treatments (i.e. 12 hours), the zeolite modified with the larger surfactant molecule, CTAB, showed a higher volume of larger mesopores. However, a less crystalline material was obtained, with 77% degree of crystallinity for HY_C12 against 95 % for HY_D12, as well as lower concentration of Brönsted acid sites, with 0.392 mmol g^{-1} and 0.464 mmol g^{-1} for HY_C12 and HY_D12, respectively.

The catalytic behavior of parent and surfactant templated samples was studied in the acylation of furane, using AA as acylating agent, at 60 °C, with a molar ratio AA/furane of 5. Figure 27.10 shows the product yield as a function of reaction time, considering that the reaction is almost completely selective into 2-acethylfuran.

A close inspection of Figure 27.10, discloses a sharp increase in product yield, for all samples, in the first 10 min of reaction, followed by a slope attenuation. In the case of DTAB treated samples (Figure 27.10a), in the initial stage of the reaction all modified samples present higher yield, when compared with HY material. For reaction times longer than 20 minutes, HY_D48 sample behaves distinctly, since a plateau is observed whereas a small increase is still evident for the parent and the other treated samples, attributed to the occurrence of secondary reactions that occluded the sample micro+mesoporosity. Accordingly, the best catalytic performance was obtained for HY_D24 because larger treatment time produces larger mesopores that lead to deactivation phenomena. For the samples treated with CTAB, HY_C12 sample displays a significant improvement in product since the larger pores present in this sample fasten the reaction kinetics (rate constant and TOF are two times higher when compared with HY-D12 sample). However, when the treatment time increases to 24 hours, an inversion occurs, indicating the occurrence of secondary transformations that cause diffusional limitations. The results obtained in this study clearly show that a careful optimization is needed to tune the required amount and size of mesopores needed to improve the catalytic performance for each particular reaction.

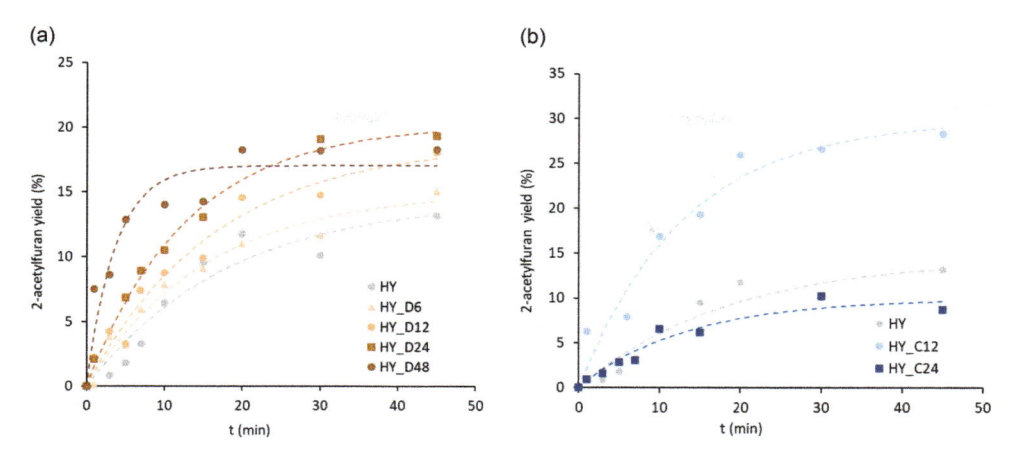

Figure 27.10 Product yield as a function of reaction time for parent HY and modified dodecyltrimethylammonium bromide (DTAB) (a) and cetyltrimethylammonium bromide (CTAB) (b) for a 6 to 48 hour treatment. Ref [47] / Elsevier.

Hierarchical zeolites where the mesoporosity is generated during the synthesis (bottom-up procedures) were also explored in Friedel-Crafts acylation reactions. Bohström and Holmberg [83], prepared hierarchical ZSM-5 (mesoporous HZSM-5) following the procedure previously described by Wang et al. [84], using poly(diallyldimethyammonium chloride) as a soft template for mesoporosity development. Mesoporous HZSM-5 showed the same Si/Al ratio as the parent material as well as the same diffraction patterns. The textural characterization revealed the maintenance of microporous volume but the mesoporosity volume almost doubled from 0.07 to 0.12 $cm^3\ g^{-1}$. The catalytic performance of micro and mesoporous HZSM-5 sample was studied in the acylation of 2-methylindole with AA, using a 2:1 molar ratio of AA:2-methylindole, at 85, 100 and 115 °C during 24 hours. The reaction over the microporous zeolite was regiospecific, giving acetylation only in the 3-position of 2-methylindole, whereas in the case of mesoporous sample a mixture of two isomers, in a 3:2 ratio, was obtained. A kinetic model, based on the Eley-Rideal reaction mechanism, allowed the determination of rate constants adsorption energy and activation energy of the reaction. In the case of microporous catalyst higher rate constant and lower activation energy was found, as compared with the mesoporous sample, but the mesoporous HZSM-5 gave lower adsorption energy and a higher yield at the highest temperature studied (89 % against 65 % for the microporous catalyst). Thus, the catalyst with higher content in mesopores has an improved resistance to deactivation and poisoning, thus overcoming what is considered as a main disadvantage in the application of zeolite materials in liquid-phase reactions.

The effects of mesopore generation in ZSM-5 zeolite, applied in liquid-phase acylation of bulky aromatic compounds was explored by Kim et al. [85]. The authors prepared MFI zeolite nanosponge, obtained through a seed-assistant hydrothermal synthesis using $C_{22}H_{45}-N^+(CH_3)_2-C_6H_{12}-N+(CH_3)_2-C_6H_{13}$. The zeolite nanosponge showed a disordered network of 2.5 nm thick zeolite layers, with very large external surface (460 $m^2\ g^{-1}$ for MFI nanosponge against 60 $m^2\ g^{-1}$ for conventional MFI zeolite), and a large volume of mesopores (0.5 $cm^3\ g^{-1}$), with a narrow distribution of diameters centered at about 4 nm, which is comparable to mesoporous MCM-41 that was used as reference material (mesopore volume of 0.6 $cm^3\ g^{-1}$). In addition, MFI nanosponge contains a high amount of strong Brönsted acid sites located at the external surface, 53 μmol g^{-1}, which largely exceeds what is usually observed for commercial bulk zeolites (6 and 7 μmol g^{-1} for bulk BEA and MFI, respectively).

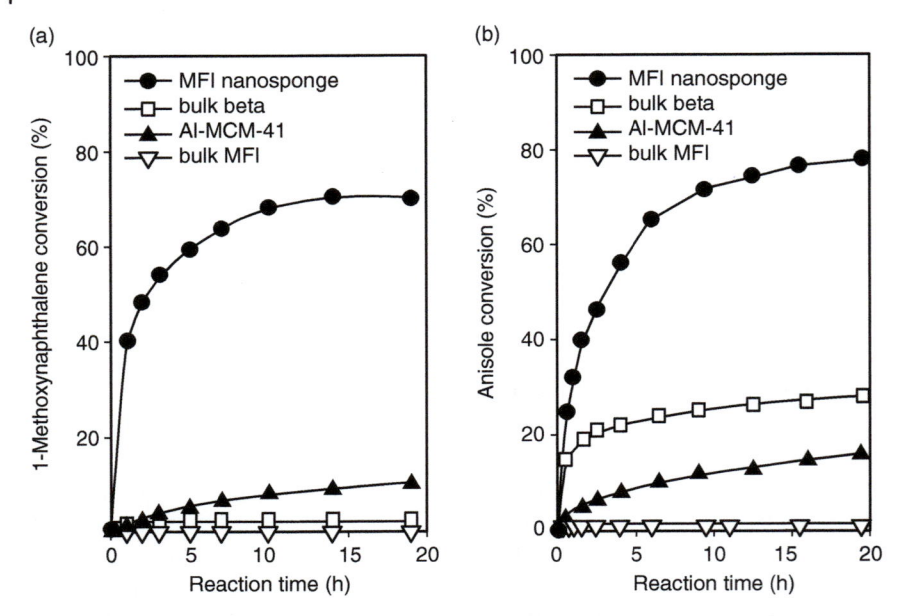

Figure 27.11 Conversion of 1-methoxinaphtalene (a) and anisole (AN) (b) as a function of reaction time. Reaction conditions for 1-methoxynapthalene (1-MM): 1 mmol of 1-methoxinaphalene, 2 mmol of acetic anhydride (AA), 4 mL of nitrobenzene, 50 mg of catalyst, T = 120 °C; for AN: 4 mmol AN, 8 mmol AA, 4 mol of nitrobenzene, 50 mg of catalyst, T = 120 °C. Ref [85] / Elsevier.

The catalytic behavior of MFI nanosponge was compared with other porous catalysts: bulk (conventional) MFI and BEA zeolites and mesoporous Al-MCM-41. The catalytic performance was studied using a series of substrate molecules, namely AN, methoxynaphthalene, naphthalene, and dimethylbenzene, in the presence of AA as the acylating agent and nitrobenzene as the solvent. The experiments were performed at temperatures ranging from 120 to 160 °C, depending on the substrate used. Figure 27.11 shows the catalytic conversion for 1-methoxynaphtalene (1-MM) and AN.

As shown, the catalytic performance of MFI nanosponge stands out, especially in the case of the bulkier substrate 1-MM. However, it must be emphasized that the high mesoporous volume cannot be the only parameter to explain the superior performance of MFI nanosponge since it is comparable with Al-MCM-41 and this catalyst reaches only a maximum conversion of about 10%. The role of the catalytically active Brönsted acid sites located at the external surface can have a determinant role since in this case the diffusional limitations are absent. In the case of a less voluminous molecule, AN, this effect is less pronounced and when using bulk BEA, a large pore zeolite with 12 membered oxygen rings (12-MR), a moderate conversion (around 10%) is observed since AN may access the active sites located inside the pores, which is not observed for medium pore (10-MR) bulk MFI. This study opens new perspectives to the development of new methodologies to produce hierarchical zeolites with high external surface and rich acidity. In fact, the active sites located at the external surface are often neglected due to shape selective effects that are highlighted in the case of zeolites but are useless when voluminous molecules are used as reagents, due to diffusional constraints. Thus, these new methods of synthesis/modifications are an elegant way to overcome this limitation often referred by Eric Derouane in liquid phase reactions using zeolites as heterogeneous catalysts [63, 64].

27.4 Understanding Friedel-Crafts Acylation Reactions through Quantitative Structure-Property Relationships

Quantitative structure-property/activity relationships (QSPR/QSAR) include, in general terms, all statistical and mathematical methods by which different types of properties can be related with the structural features of a series of compounds. This relation can be expressed as follows:

$$P = f(D_1, D_2, D_3 \ldots D_i) \tag{27.2}$$

where P (or a function of P, such as $\ln(P)$ or $1/P$) is the property of interest (e.g. biological activity, toxicity, rate or equilibrium constants, solubility, chromatographic retention coefficients, etc.) and D_i represents the molecular descriptors that encode the compounds 1D, 2D or 3D structural characteristics. The mathematical relationship between the property (P) and the descriptors (D_i), the so-called model equation, can be either linear or nonlinear. Molecular descriptors are experimentally determined quantities, either macroscopic or microscopic (in the sense of probing a given interaction at a molecular level), or, instead, theoretically derived quantities. There are thousands of molecular descriptors referenced in the literature which differ in dimensionality and nature [86]. Although the selection of descriptors is a difficult task, the following criteria should be used: relevancy, absence of redundancy, number (they should be as few as possible to avoid chance correlations), and interpretability.

QSPR/QSAR are these days one of the most important methodologies for modeling and prediction purposes in various areas of knowledge from fundamental physical chemistry to medicinal chemistry, environmental and material sciences and nanotechnologies [87–99].

QSPR/QSAR methods can be essentially divided in two large groups: regression and classification methods, being the former used to find, as referred, quantitative correlations between chemical, biological, toxicological information and structure, and the latter to visualize similarity and clustering within the compounds' space [100]. Among regression methods, multiple linear regression analysis (MLR), principal components regression (PCR), partial least squares (PLS), and artificial neural networks (ANN) have been used the most [92, 93, 100–103], with MLR occupying a prominent place in certain areas. On the side of classification methods, linear discriminant analysis (LDA), principal component analysis (PCA), decision trees (DT), random forests (RF), k-nearest neighbor (kNN), logistic regression (LR), Kohonen self-organizing maps (SOM), and support vector machines (SVM, which can be used either for classification or for regression purposes), are the most frequently cited [92, 100, 101]. All of these methodologies are considered traditional machine learning (ML) algorithms. ML is an important sub-group of artificial intelligence (AI) technology [104, 105]. Deep learning (DL) approaches (i.e. ANN with several hidden processing layers) are a more recent development in AI technology and have lately gathered special attention due to their potential ability to describe nonlinear input-output relationships and for having a more powerful generalization capability. DL techniques are considered an interesting complement to conventional ML strategies which are rather dependent on the use of human-crafted molecular descriptors [104, 105].

In the context of this chapter, we will, however, only address traditional ML algorithms, whose advantages and limitations have been thoroughly reviewed elsewhere [100–102, 104, 105].

Independently of the process and the methodology chosen, QSPR/QSAR studies always have one of two purposes (or both): either the identification of the key factors that determine a certain response (interpretative ability) or the prediction of a system's behavior when convenient structural modifications are designed and introduced in a molecule or material, prior to any synthesis, microbiological assay, or experimental measurement (predictive ability). This dichotomy between interpretability and predictability, so well-illustrated in a recent paper by Fujita and Winkler [103], should always be taken into consideration when choosing a particular methodology. There is no method ideal for all situations. The choice depends on the nature of the process being modeled (and on the quality and amount of data) and on the objective of the study: if the main drive is to understand/interpret/rationalize a certain behavior, then the researcher should select a linear method such as MLR or PLS; if, on the contrary, the researcher is mainly interested to accurately predict a given response (property), then a non-linear method like ANN appears as a more convenient alternative (Figure 27.12). In case one wants to predict a compound's class (e.g. active vs. inactive) and the activity information is scarce, classification methods are the most suitable. On the other hand, if there is enough property information (e.g. IC_{50} or k values) for a series of congeneric compounds for which it is possible to derive or experimentally obtain a set of molecular descriptors, then a regression method is the best option for quantitative prediction [100, 101].

QSPR have now been used for almost a century [106] to relate rate constants (k) of homogeneous kinetics in the liquid phase with macroscopic solvent properties such as dielectric constant, dipole moment, pH, refractive index, and other properties, or/and with microscopic solvent properties which are based in chemical probes capable of accounting for the possible role of Lewis acidity, Lewis basicity, dipolarity, polarizability, and other factors [107, 108]. Several model equations have been derived that allow a better understanding of the reaction mechanisms and/or the prediction of new k values for missing experimental values [109–111]. In the same way, equilibrium constants (K), solution enthalpies ($\Delta_{sol}H$), and many other physicochemical properties have also been correlated with different sets of parameters usually related to the same type of properties [112–114].

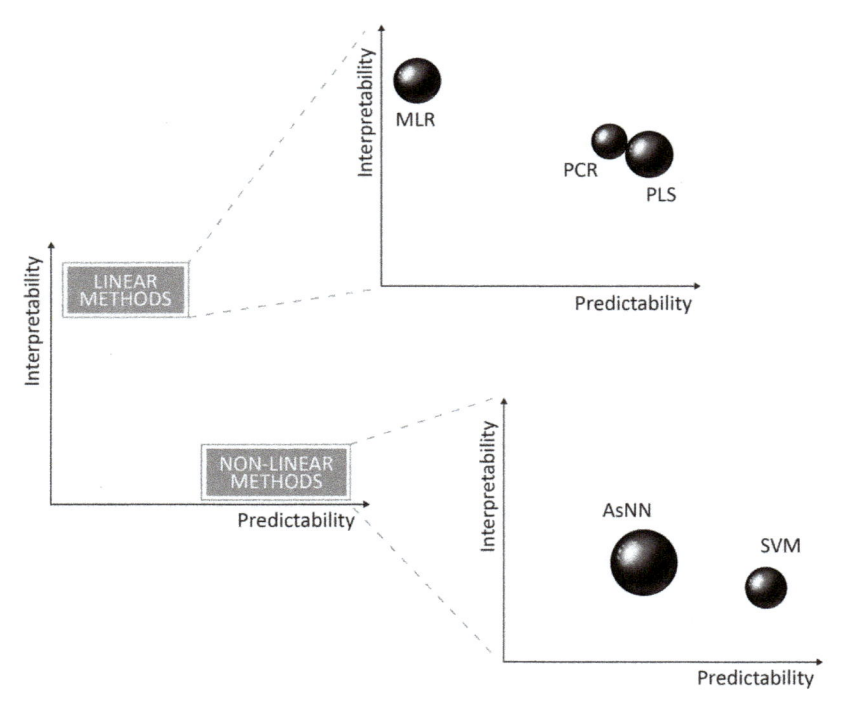

Figure 27.12 Trade-off between interpretability and predictability for the methods referred in the text. The markers' size gives an indication of how much time is consumed by each method. Adapted from Ref [100].

27.4 Understanding Friedel-Crafts Acylation Reactions through Quantitative Structure-Property Relationships

Over time, different approaches have been proposed, the most notable one being the Abraham linear free energy relationships (LFER) model, that uses solute properties as independent variables to predict several important physicochemical and biological properties [115, 116]. Some of these QSPR models have also been applied to the study of catalytic effects, although, to the best of our knowledge, the number of studies resorting to QSPR analysis in the field of catalysis is still not very expressive [117–130] and this number is even more incipient when zeolites are used as catalysts [42, 81, 131–133]. The application of a QSPR strategy is, however, of undeniable interest when the experimental assessment of catalytic behavior is expensive, time-consuming or experimentally difficult [124], or when the focus is on the assessment (and rationalization) of the simultaneous effect of various factors [128, 131, 133].

With this in mind, a couple of years ago, we started a systematic study of the catalytic behavior of hierarchical MCM-22 and BEA zeolites in Friedel-Crafts acylation reactions carried out in mild conditions using, as already mentioned, AA as an acylating agent and various heteroaromatic rings of different sizes and chemical functionalities as substrates (namely, furan, benzofuran, thiophene, benzothiophene, pyrrole, indole, and AN) [42, 46, 81]. The idea was to probe the combined effect of changes in both substrate and catalyst (zeolite) properties over rate constants, k, and relative sorption equilibrium constants, K_r, through the use of a MLR-based QSPR strategy. The use of this approach assumes that k and K_r (or better $\ln k$ and $\ln K_r$) can be expressed as linear combinations of independent and additive properties or descriptors, D, that in this case encode the structural features of both substrates and catalysts:

$$\ln k (\text{or } \ln K_r) = a_0 + \sum_{i=1}^{N} a_i D_i + x \tag{27.3}$$

In Eq. 27.3 x stands for the regression residuals, a_0 is the regression independent term and a_i the regression coefficients associated with each descriptor, D_i, obtained by minimizing the sum of the squared residuals. The order of magnitude of the regression coefficients of the normalized descriptors reflects their relative importance, whereas the corresponding signs can be interpreted in terms of positive or negative contributions to the response variable.

Experimental data (i.e. product yield vs. time) for the referred systems were first fitted to Eq. 27.1 (*vd.* 27.3) using a nonlinear regression method which allows the estimation of k and K_r values and of the corresponding coefficient parameters. Additionally, a specific software was used to calculate the parameter's uncertainties. Before performing any MLR-QSPR like those we carried out in our studies, the following general aspects are always verified: the number of parameters (descriptors) should not exceed the number of data points to avoid overfitting and thus the possibility of chance correlations (as a rule of thumb, there should always be at least 3–5 points per each parameter); intercorrelations among descriptors should be checked, so that pairs of parameters for which the determination coefficient, r^2, of the correlation matrix exceeds the cut-off value of 0.5 are excluded, thus preventing the presence of more than one parameter describing the same type of interaction in the same model equation (redundancy). If enough data points are available, the initial set should additionally be divided in two independent sub-sets, one to establish the model equation, the so-called training set, and the other, the test set or prediction set, to test the derived model, allowing therefore an external validation procedure to assess the model's predictive ability. Test and training sets should have similar sizes and should be representative of the whole chemical/structural space covered by the descriptors and also of the dependent variable's space (same variability); test set points should not be beyond training set points to avoid predictions outside the model's applicability domain. Regressions can be carried out either by a forward- or a backward-stepwise procedure, achieved by either starting with equations containing a single descriptor and adding up terms, one at a time, or the other way around, starting with equations with the "maximum" number of descriptors and then remove one by one. Descriptors are added, retained or disregarded in each step according to rigorous statistical criteria: parameters to be considered need to have a significance level (SL)>95%, and the adjusted multiple determination coefficient, R^2_{adj}, of each regression should be high and as close to 1 as possible. Also, models have to be examined for spurious points (outliers). As a rule of thumb, calculated values are kept if they are not more than 2 standard deviations (of the fit), s_{fit}, from the corresponding experimental values and also if they are not leverage points (i.e. points that tend to unbalance the regression and whose presence must therefore be prevented) [88, 134]. External validation procedures should be used whenever one seeks to evaluate the predictive capability of a model. External validation, as well as internal validation (model's robustness), must comply to stringent metrics which have been duly explained elsewhere [88, 93, 134–137]. Also, visual inspection of scatter plots of predicted vs. experimental data is strongly recommended as a complimentary indicator of good predictive power [137].

In our work, a set of descriptors frequently used to qualitatively analyze results in heterogeneous catalysis studies were used. These comprised descriptors associated with the zeolite properties, namely external area (A_{ext}), micropore volume (V_{micro}), mesopore volume (V_{meso}), Brønsted acidity ($[B]_{pyr}$), Lewis acidity ($[L]_{pyr}$), silicon/aluminium ratio (Si/Al) and crystallinity (C_{XRD}), and others linked to the substrate properties: molecular diameter (M_{Diam}), van der Waals volume (V_{vdW}), dipolarity, and Lewis acidity (as given by the normalized Dimroth and Reichardt's parameter, E_T^N).

A stepping forward regression procedure applied to ln (k or K_r) vs. descriptors in the case of the Friedel-Crafts acylations studies over hierarchical MCM-22 zeolites [46], showed that the Lewis acidity descriptor, $[L]_{pyr}$, needed not be considered in any of the final models since its statistical relevance was much lower than that of $[B]_{pyr}$ when both uniparametric equations were compared. The observed high intercorrelation between $[B]_{pyr}$ and $[L]_{pyr}$, would nevertheless eliminate the possibility of the simultaneous use of two descriptors in the same fitting equation.

In the case of K_r, the complete regression procedure led for this zeolite to four equations with three descriptors and identical R^2_{adj} and s_{fit}. From these, Eq. 27.4 was selected because its physicochemical interpretation was more straightforward and there was no real statistical justification to select one equation in detriment of any of the others.

$$\ln K_r = (-3.84 \pm 0.08) + (0.24 \pm 0.10)[B]_{pyr} + (0.89 \pm 0.11)VvdW + (1.30 \pm 0.10)E_T^N \quad (27.4)$$
$$(100\%) \qquad (100\%) \qquad (100\%) \qquad (95\%)$$
$$N = 10; R^2 = 0.972; R^2_{adj} = 0.958; s_{fit} = 0.25; F = 70$$

(in this fitting, two data points were identified as outliers and were excluded, and the model refitted).

For the rate constant k, the best-found equation was Eq. 27.5.

$$\ln k = (1.57 \pm 0.20) + (0.87 \pm 0.21)[B]_{pyr} - (2.00 \pm 0.23)VvdW - (0.47 \pm 0.23)E_T^N \quad (27.5)$$
$$(100\%) \qquad (100\%) \qquad (100\%) \qquad (93\%)$$
$$N = 12; R^2 = 0.914; R^2_{adj} = 0.882; s_{fit} = 0.69; F = 29$$

In these equations, N is the number of points, s_{fit} the standard deviation of the fit, and F the Fisher–Snedecor statistics (the numbers in brackets underneath each equation term are the significance level of each coefficient).

The mesoporous volume was removed in both cases given its low significance level. This was not surprising because diffusion is not the rate limiting step and product adsorption must be nearly insensitive to this property, in agreement with the conclusions of a Friedel-Crafts alkylation study involving Beta zeolites that showed that mesoporous volume is only relevant for large substrates [138]. A close inspection of Eq. 27.4 shows that all descriptors contribute positively for K_r, being E_T^N the most important and $[B]_{pyr}$, the less relevant. For k, Eq. 27.5 reveals that $[B]_{pyr}$ increases k while the size of the substrate and its dipolarity/Lewis acidity, as described by E_T^N, decreases it. This is again in line with Ref [138], where a good relationship between catalytic activity and $[B]_{pyr}$ was also recognized. The positive sign of the E_T^N in Eq. 27.4 shows that substrate dipolarity favors product adsorption relative to that of the reagent and, hence, active sites remain occupied by product molecules and make it more difficult for further reagent molecules to adsorb (i.e. contributing to lower k, as evidenced by the negative sign of this coefficient in Eq. 27.5). The same tendency is seen for the V_{vdW} volume, which is thermodynamically favorable (increases K_r) and kinetically unfavorable (decreases k).

The observed relative order of descriptors' importance for K_r is: E_T^N > VvdW > $[B]_{pyr}$ and for k is VvdW > $[B]_{pyr}$ > E_T^N. Notably, V_{vdW} and $[B]_{pyr}$ influence rate at least twice as much as they influence adsorption. On the contrary, the effect of the substrates' polarity coefficient, measured by E_T^N, is three times higher in K_r than in k. The interesting aspect of this QSPR analysis is that it complements and corroborates the catalytic qualitative study already mentioned.

Using a similar procedure, a second work describing results obtained for the same set of substrates but using a mesoporous BEA zeolite was published by the same authors [42]. In this case, ln k and ln K_r were correlated to a set of descriptors, in which seven related to zeolite properties (A_{ext}, V_{micro}, V_{meso}, $[B]_{pyr}$, $[L]_{pyr}$, Si/Al and C_{XRD}) and five to substrate properties (V_{vdW}; E_T^N; $f(\varepsilon)$, a function of the dielectric constant; $g(n_D)$, a function of the refractive index; and μ, the dipole moment). As in the previous study, the data set comprised 12 points limiting the maximum number of variables in the fitted equations to four. In this case, both final model equations (Eqs. 27.6 and 27.7) contained four descriptors as follows:

$$\ln k = (0.96 \pm 0.03) - (0.28 \pm 0.05) C_{XRD} + (0.72 \pm 0.08) V_{micro} - (0.53 \pm 0.05) [B]_{pyr} - (0.61 \pm 0.02) f(\varepsilon)$$
$$(100\%) \quad (99.9\%) \quad (100\%) \quad (100\%) \quad (100\%)$$
$$N = 12; R^2 = 0.992; R^2_{adj} = 0.986; s_{fit} = 0.04; F = 195 \quad (27.6)$$

and,

$$\ln K_t = (0.22 \pm .0.03) + (0.06 \pm 0.03) \text{Si/Al} - (0.24 \pm 0.03) V_{micro} + (0.83 \pm 0.03) V_{vdw} + (0.44 \pm 0.03) E_T^N$$
$$(100\%) \quad (93\%) \quad (100\%) \quad (100\%) \quad (100\%)$$
$$N = 12; R^2 = 0.990; R^2_{adj} = 0.984; s_{fit} = 0.04; F = 165 \quad (27.7)$$

Variable inflation factors (VIFs) were also computed and were all below seven, indicating that multicollinearity was not biasing the estimates of coefficients. Figure 27.13a shows a good correlation between predicted and observed ln k values. Open squares correspond to outliers, removed from the data set sequentially and the model refitted after each removal.

Comparing results with those obtained for MCM-22, there is an important difference between these two zeolites: the inclusion, for BEA, of the degree of crystallinity. C_{XRD} has a negative influence over ln k, indicating that the loss of crystallinity resulting from post-treatment modifications of BEA (selective extraction of Si or, in some cases, of Si and Al) accelerates the reaction. V_{micro} has the most significant positive contribution to ln k which reinforces the evidence from the previous work that these reactions occur mainly inside micropores and the enhanced diffusion due to mesoporosity is not relevant to the reaction. Also, $[B]_{pyr}$ exhibits a surprisingly negative coefficient probably meaning that enhancing this type of acidity as compared to the departing structure is detrimental to the catalyst. This is very clear in the case of furan reacting within zeolite BEA-32-AT, for which a decrease of over 50% of $[B]_{pyr}$ leads to a considerable increase in k, compared to parent and desilicated samples. This suggests that even for the sample with the lowest Brønsted acidity there are enough acid sites to catalyze the reaction. Therefore, a higher amount of acid sites probably promotes secondary reactions and adsorption, followed by deactivation processes. Finally, $f(\varepsilon)$, the substrates' dipolarity, also decreases the rate constant, implying very strong adsorption of product molecules leading to the occupancy of active sites and thus preventing reagent adsorption.

For K_r, Figure 27.13b displays a good correlation between predicted (by model Eq. (27.7)) and observed ln K_r values. In this case, three outliers (open squares) were detected and sequentially removed. Because VIF values were approximately unit, coefficients and their standard deviations are unbiased.

As for MCM-22, E_T^N and V_{vdW} also contribute positively to ln K_r, being V_{vdW} the most important descriptor and reflecting the effect of product bulky molecules which diffuse slowly and become more adsorbed. V_{micro} is also relevant, but contributes to the lowering of ln K_r whereas the Si/Al ratio has a very small positive contribution. The negative influence of V_{micro} relates to the formation of mesopores from the destruction/enlargement of micropores. The occurrence of secondary reactions/adsorption phenomena mostly inside mesopores promotes the formation of bulky products that become imprisoned inside the mesopores and will not be able to diffuse. The positive sign of the E_T^N coefficient, means that the combination of the substrate's HBA and dipolarity stimulates product adsorption relative to that of the substrate.

The obtained equations allowed, once again, the establishment of relative orders of importance. For K_r: $V_{vdW}(+) > E_T^N(+) > V_{micro}(-) \gg$ Si/Al$(+)$; for k: $V_{micro}(+) > f(\varepsilon)(-) > [B]_{pyr}(-) > C_{XRD}(-)$.

With BEA zeolite, the QSPR strategy was again consistent with the characterization analysis and the catalytic results. The first two QSPR analysis were performed with small substrate molecules with low interest as APIs. In order to ensure a higher variability within each descriptor, the number of substrates was increased in the last of the three works given as examples [81], by adding

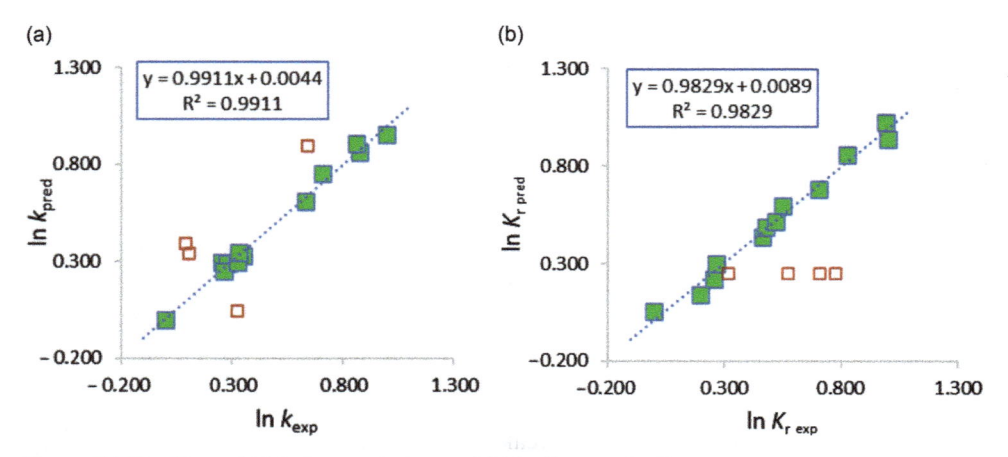

Figure 27.13 Plots of (a) ln k_{pred} vs. ln k_{exp} and (b) ln $K_{r\,pred}$ vs. ln $K_{r\,exp}$.

molecules with different chemical functionalities and larger sizes. Qualitatively, the data only showed that the rate constants of thiophene and furan were significantly higher than those for the other substrates. Also, part of the descriptors previously used referred to the substrate molecules as if they were solvents in the process. In the third of these works, only descriptors pertaining to the substrates as such were included, this time aiming to account for a higher diversity and complexity of substrate molecules and to consider them as such and not as solvents.

The developed models were based on the correlation of $\ln k$ and of $\ln K_r$, with seven descriptors related to the zeolite properties (the same ones considered previously) and seven substrate properties, namely: surface area (σ), van der Waals volume (V_{vdW}), density (ρ), molar volume (V_m), dipole moment (μ), viscosity (η), and surface tension (γ). Given the larger data set, and since some outliers were expected, fitted models went up to a maximum of five descriptors.

MLR results were filtered according to the same sequence of criteria applied earlier, plus: (i) only equations for which $R^2_{adj} \geq 0.95$ were selected; (ii) only model equations with two or less outliers were chosen; (iii) only model equations with the top 50% Fisher-Snedecor F-values were kept. This procedure led, both for $\ln k$ and $\ln K_r$, to several, virtually statistically identical models, but it was possible to determine in both cases the preferred one using also as criteria, the standard deviation of the fit, the adjusted determination coefficient and the predicted determination coefficient, R^2_{pred}, to check the predictability of each model equation. $R^2_{pred} = 1 - PRESS/SST$, where $PRESS = \sum_{i=1}^{n}(y_i - \hat{y}(i))^2$, $SST = \sum_{i=1}^{n}(y_i - \bar{y})^2$, and $\hat{y}(i)$ are the predicted values of the response of the ith observation using a model whose estimate is based on (n−1) data points excluding the ith data point. If $R^2_{adj} - R^2_{pred} < 0.2$, the model was considered to show good predictability. Solutions for which coefficients lacked physical meaning were also excluded. For $\ln k$ and $\ln K_r$, the final equations were the following:

$$\ln k = (1.06 \pm 0.07) - (0.76 \pm 0.05)V_{vdW} - (0.52 \pm 0.07)\rho + (0.14 \pm 0.04)[B]_{pyr}$$
$$(100\%) \quad\quad (100\%) \quad\quad (100\%) \quad\quad (100\%) \quad\quad (27.8)$$
$$N = 21; R^2 = 0.928; R^2_{adj} = 0.915; R^2_{pred} = 0.907; s_{fit} = 0.08; F = 73$$

and,

$$\ln K_r = (0.31 \pm 0.02) + (0.60 \pm 0.04)\sigma + (0.45 \pm 0.04)\,r - (0.06 \pm 0.02)\,V_{micro}$$
$$(100\%) \quad\quad (100\%) \quad\quad (100\%) \quad\quad (99\%) \quad\quad (27.9)$$
$$N = 16; R^2 = 0.963; R^2_{adj} = 0.953; R^2_{pred} = 0.935; s_{fit} = 0.08; F = 103$$

Because R^2_{pred} was quite close to R^2_{adj}, it was concluded that the models had good predictive ability within the covered variable's space. Also, residuals were randomly distributed around the trendline, as seen in Figure 27.14a.

The relative importance of the descriptors for k is this time $V_{vdW}(-) > \rho(-) > [B]_{pyr} (+)$. In line with previous works, $[B]_{pyr}$ is important and increases k, although it is the least contributing descriptor, possibly because a greater difference of Brønsted acidity among zeolites is needed to further reveal its importance. The most important descriptor, V_{vdW}, which accounts for the volume occupied by substrate molecules, is unfavorable to k, showing that the size of the substrate determines the ease of reagents diffusion inside zeolite pores. Furthermore, the weight of this descriptor suggests that a large electron cloud establishes weaker bonding with the catalytic surface. On the other hand, substrate density, ρ, diminishes k, since a larger density causes steric hindrance.

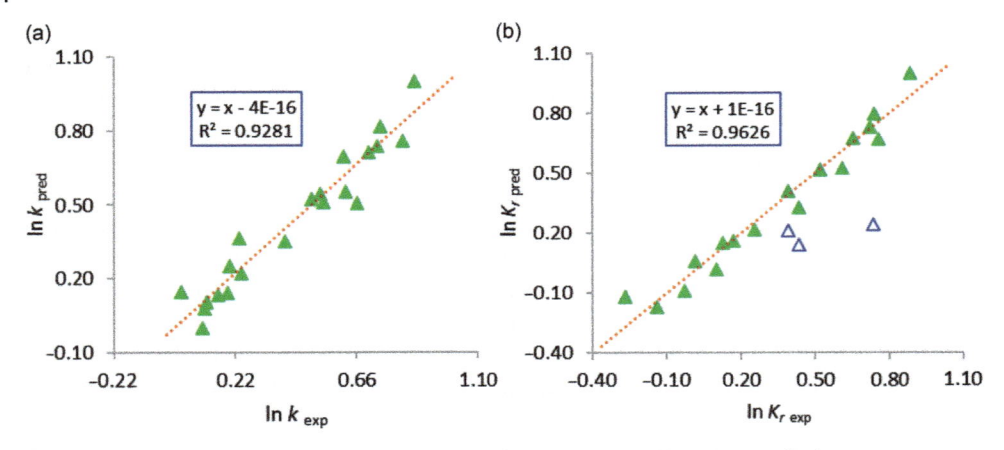

Figure 27.14 Plots of (a) $\ln k_{pred}$ vs. $\ln k_{exp}$ and (b) $\ln K_{r\,pred}$ vs. $\ln K_{r\,exp}$ (Δ = outlier).

For K_r, again only three descriptors are needed to model the adsorption/desorption behavior, and the best equation now includes σ, the substrate's surface area accessible to the zeolite, ρ, the substrate's density, and again, only one zeolite parameter, V_{micro}. The relative importance of descriptors is: $\sigma\,(+) > \rho\,(+) > V_{micro}\,(-)$ and, thus, the reagent's accessible surface area and the density increase K_r, since desorption becomes more difficult due to stronger adsorption. Larger microporous volumes facilitate desorption relative to adsorption and, so, decreases K_r. Surprisingly, Lewis acidity, $[L]_{pyr}$, has shown no significant role in the processes, although it spanned for 31% against only 14% for Brønsted acidity, $[B]_{pyr}$. Since $[B]_{pyr}$ is strongly correlated with V_{micro}, and σ is even more strongly correlated with V_{vdW}, results show a somewhat inverse influence of similar parameters in K_r and in k: the same, or similar, parameters which diminish the rate constant also increase the difficulty in product desorption (larger K_r). This can be rationalized in terms of stronger adsorption forces that at the same time increase catalytic efficiency but also make active sites less accessible to new molecules. Figure 27.14 shows the fitted values for k and K_r.

In summary, the results of these three studies show that the most relevant parameter in both k and K_r is related to the substrate molecular size. However, these parameters may be associated with different reaction steps, namely accessibility to micropores, diffusion capacity of molecules through the zeolitic atomic network and molecular polarizability, factors that were already shown to be relevant in the previous works using fewer and smaller model molecules.

It would be very interesting to better assess the predictive ability of the models. This could be achieved by estimating k and K_r for a different set of substrates not used in these correlations, for which experimental values obtained in the same conditions (i.e. same zeolite, acylating agent and temperature) would exist, but up to date there are no reports of such values in the literature.

The used multivariate QSPR approach, by taking into consideration, simultaneously, multiple changes in relevant properties of substrates and zeolites, produced in these three cases results that are consistent with data obtained from the structural characterization and from the catalytic tests. This strategy allows the identification of key properties that can be optimized through synthesis or post-synthesis treatments to design catalysts with improved catalytic performance.

27.5 Final Remarks

The present work overviews recent developments of Friedel-Crafts acylation reactions using zeolite-based catalysts. These chemical transformations play an important role in the multi-step synthesis protocol of numerous products, namely several highly consumed medicines and fine chemicals.

Although BEA zeolite structure is already being used at an industrial level as a Friedel-Crafts heterogeneous catalyst, the large majority of the published works still focus on the use for these reactions of Lewis or strong protonic acids as homogeneous catalysts, in spite of the economic and environmental drawbacks associated with them.

Zeolites have been extensively tested as acid catalysts in Friedel-Crafts acylations involving small substrates. To enlarge the applicability of zeolite-based catalysts to acylation of bulkier compounds and avoid severe deactivation phenomena *en route*, studies based on hierarchical zeolite structures but using larger substrates have been carried out with promising results. Additionally, to rationalize the combined effect of substrate and zeolite properties upon catalytic behavior, a MLR-based QSPR strategy was applied. This approach allowed the identification of key properties that could be used to improve catalytic performance. The search for better performing catalysts is indeed an important topic that continues to draw the attention of the scientific community.

In summary, the data already available in the literature shows that hierarchical zeolite derived solids can be used as catalysts in Friedel-Crafts reactions, with clear environmental advantages over traditional homogeneous catalysts. Furthermore, the application of machine learning techniques provides useful information for a better catalyst design.

Acknowledgements

This research was funded by Fundação para a Ciência e a Tecnologia (FCT) through UIDB/00100/2020. UIDP/00100/2020 and LA/P/0056/2020 and Instituto Politécnico de Lisboa (IPL) through Project IPL/2021/ZeoGemini ISEL.

References

1. Friedel, C. and Crafts, J.M. (1877). *C. R. Acad. Sci.* 84: 1392–1395.
2. Friedel, C. and Crafts, J.M. (1877). *C. R. Acad. Sci.* 84: 1450–1454.
3. Olah, G.A. (1963). *Friedel-Crafts and Related Reactions*. Geneva: Interscience Publishers.
4. Bandini, M. and Umani-Ronchi, A. (2009). *Catalytic Asymmetric Friedel-Crafts Alkylations*. Weinheim: Wiley-VCH.
5. Franck, H.-G. and Stadelhofer, J.W. (1988). *Industrial Aromatic Chemistry*, 1e. Berlin: Springer Berlin Heidelberg.
6. Röper, M., Gehrer, E., Narbeshuber, T., and Siegel, W. (2000). Acylation and alkylation. In: *Ullmann's Encyclopedia of Industrial Chemistry* (ed. C. Ley), 337–382. Hoboken: John Wiley and Sons, Ltd.
7. Cavani, F., Centi, G., Perathoner, S., and Trifiró, F. (eds.) (2009). *Sustainable Industrial Process. Principals, Tools and Industrial Examples*. Weinheim: Wiley-VCH.
8. Short, W.F., Stormberg, H., and Wiles, A.E. (1936). *J. Chem. Soc.* 319–322.
9. Li, W., Jin, H., Yang, S. et al. (2019). *Green Process. Synth.* 8: 474–479.
10. Kusumaningsih, T., Prasetyo, W.E., and Firdaus, M. (2020). *RSC Adv.* 10: 31824–31837.
11. Earle, M.J., Seddon, K.R., Adams, C.J., and Roberts, G. (1998). *Chem. Commun.* 2097–2098.
12. Boon, J.A., Levisky, J.A., Pflug, J.L., and Wilkes, J.S. (2002). *J. Org. Chem.* 51: 480–483.
13. Ross, J. and Xiao, J. (2002). *Green Chem.* 4: 129–133.
14. Barrer, R.M. (1981). *Zeolites* 1: 130–140.
15. Guisnet, M. and Ribeiro, F.R. (2006). *Les zeolithes: un nanomonde au service de la catalyse*. Les Ullis: EDP Science.

16 Structure Commission of the International Zeolite Association. www.iza-structure.org (accessed 28 June 2023).
17 Figueiredo, J.L. and Ramôa Ribeiro, F. (2007). *Catálise Heterogénea, Fundação Calouste Gulbenkian*, 2e. Lisbon.
18 Wang, W., Chang-Jun, L., and Wu, W. (2019). *Catal. Sci. Technol.* 9: 4162–4187.
19 Schwieger, W., Machoke, A.G., Weissenberger, T. et al. (2016). *Chem. Soc. Rev.* 45: 3353–3376.
20 Barrer, R.M. and Makki, M.B. (1964). *Can. J. Chem.* 42: 1481–1487.
21 Triantafillidis, C.S., Vlessidis, A.G., and Evmiridis, N.P. (2000). *Ind. Eng. Chem. Res.* 39: 307–319.
22 Xue, H., Huang, X., Zhan, E. et al. (2013). *Catal. Commun.* 37: 75–79.
23 De Baerdemaeker, T., Yilmaz, B., Müller, U. et al. (2013). *J. Catal.* 308: 73–81.
24 Ivanov, D.P., Pirutko, L.V., and Panov, G.I. (2014). *J. Catal.* 311: 424–432.
25 Kawai, T. (2014). *Colloid Polym. Sci.* 292: 533–538.
26 Janssen, A.H., Koster, A.J., and De Jong, K.P. (2002). *J. Phys. Chem. B* 106: 11905–11909.
27 You, Q., Wang, X., Wu, Y. et al. (2021). *New J. Chem.* 45: 10303–10314.
28 Xin, H., Li, X., Fang, Y. et al. (2014). *J. Catal.* 312: 204–215.
29 Kubota, Y., Inagaki, S., and Takechi, K. (2014). *Catal. Today* 226: 109–116.
30 Kim, J., Choi, M., and Ryoo, R. (2010). *J. Catal.* 269: 219–228.
31 Mériaudeau, P., Tuan, V.A., Nghiem, V.T. et al. (1999). *J. Catal.* 185: 378–385.
32 Groen, J.C., Hamminga, G.M., Moulijn, J.A., and Pérez-Ramírez, J. (2007). *Phys. Chem. Chem. Phys.* 9: 4822–4830.
33 Paixão, V., Carvalho, A.P., Rocha, J. et al. (2010). *Microporous Mesoporous Mater.* 131: 350–357.
34 Paixão, V., Monteiro, R., Andrade, M. et al. (2011). *Appl. Catal. A Gen.* 402: 59–68.
35 Verboekend, D., Mitchell, S., Milina, M. et al. (2011). *J. Phys. Chem. C* 115: 14193–14203.
36 Pérez-Ramírez, J., Christensen, C.H., Egeblad, K. et al. (2008). *Chem. Soc. Rev.* 37: 2530–2542.
37 Groen, J.C., Moulijn, J.A., and Pérez-Ramírez, J. (2007). *Ind. Eng. Chem. Res.* 46: 4193–4201.
38 Groen, J.C., Peffer, L.A.A., Moulijn, J.A., and Pérez-Ramírez, J. (2004). *Colloids Surfaces A Physicochem. Eng. Asp.* 241: 53–58.
39 Groen, J.C., Jansen, J.C., Moulijn, J.A., and Pérez-Ramírez, J. (2004). *J. Phys. Chem. B* 108: 13062–13065.
40 Ogura, M., Shinomiya, S.Y., Tateno, J. et al. (2000). *Chem. Lett.* 8: 882–883.
41 Machado, V., Rocha, J., Carvalho, A.P., and Martins, A. (2012). *Appl. Catal. A Gen.* 445–446: 329–338.
42 Aleixo, R., Elvas-Leitão, R., Martins, F. et al. (2019). *Mol. Catal.* 476: 110495.
43 Andrade, M.A. and Martins, L.M.D.R.S. (2021). *Molecules* 26: 1680.
44 Mendoza-Castro, M.J., Serrano, E., Linares, N., and García-Martínez, J. (2021). *Adv. Mater. Interfaces* 8: 1–23.
45 García-Martínez, J., Johnson, M., Valla, J. et al. (2012). *Catal. Sci. Technol.* 2: 987–994.
46 Aleixo, R., Elvas-Leitão, R., Martins, F. et al. (2017). *Mol. Catal.* 434: 175–183.
47 Martins, A., Neves, V., Moutinho, J. et al. (2021). *Microporous Mesoporous Mater.* 323: 111167.
48 Van-Dúnem, V., Carvalho, A.P., Martins, L.M.D.R.S., and Martins, A. (2018). *ChemCatChem* 10: 4058–4066.
49 Al-Ani, A., Haslam, J.J.C., Mordvinova, N.E. et al. (2019). *Nanoscale Adv.* 1: 2029–2039.
50 Ottaviani, D., Van-Dúnem, V., Carvalho, A.P. et al. (2020). *Catal. Today* 348: 37–44.
51 García-Martínez, J., Johnson, M., Valla, J. et al. (2012). *Catal. Sci. Technol.* 2: 987–994.
52 Corma, A., Fornés, V., Guil, J.M. et al. (2000). *Microporous Mesoporous Mater.* 38: 301–309.
53 Madsen, C. and Jacobsen, C.J.H. (1999). *Chem. Commun.* 673–674.
54 Jacobsen, C.J.H., Madsen, C., Houzvicka, J. et al. (2000). *J. Am. Chem. Soc.* 122: 7116–7117.
55 Valtchev, V.P., Smaihi, M., Faust, A.C., and Vidal, L. (2004). *Chem. Mater.* 16: 1350–1355.

56 Zhu, H., Liu, Z., Wang, Y. et al. (2008). *Chem. Mater.* 20: 1134–1139.
57 Holland, B.T., Abrams, L., and Stein, A. (1999). *J. Am. Chem. Soc.* 121: 4308–4309.
58 Serrano, D.P., Aguado, J., Escola, J.M. et al. (2006). *Chem. Mater.* 18: 2462–2464.
59 Narayanan, S., Vijaya, J.J., Sivasanker, S. et al. (2015). *J. Porous Mater.* 22: 907–918.
60 Fukasawa, T., Otsuka, K., Murakami, T. et al. (2021). *Colloid Interface Sci. Commun.* 42: 100430.
61 Venuto, P.B. and Landis, P.S. (1968). In: *Advances in Catalysis* (ed. D.D. Eley, H. Pines, and P.B. Weisz), vol. 18, 259–371. New York: Academic Press.
62 Derouane, E.G. (1998). *J. Mol. Catal. A Chem.* 134: 29–45.
63 Derouane, E.G., Dillon, C.J., Bethell, D., and Derouane-Abd Hamid, S.B. (1999). *J. Catal.* 187: 209–218.
64 Derouane, E.G., Crehan, G., Dillon, C.J. et al. (2000). *J. Catal.* 194: 410–423.
65 Álvaro, V.F.D., Brigas, A.F., Derouane, E.G. et al. (2009). *J. Mol. Catal. A Chem.* 305: 100–103.
66 Clerici, M.G. (2000). *Top. Catal.* 13: 373–386.
67 Sartori, G. and Maggi, R. (2009). *Advances in Friedel-crafts Acylation Reactions: Catalytic and Green Processes*. Boca raton: CRC Press.
68 Liang, J., Liang, Z., Zou, R., and Zhao, Y. (2017). *Adv. Mater.* 29: 1–21.
69 Nayak, Y.N., Nayak, S., Nadaf, Y.F. et al. (2019). *Lett. Org. Chem.* 17: 491–506.
70 Procházková, D., Kurfiřtová, L., and Pavlatová, J. (2012). *Catal. Today* 179: 78–84.
71 Yamazaki, T., Makihara, M., and Komura, K. (2017). *J. Mol. Catal. A. Chem.* 426: 170–176.
72 Makihara, M., Aoki, H., and Komura, K. (2018). *Catal. Letters* 148: 2974–2979.
73 Wei, H., Liu, K., Xie, S. et al. (2013). *J. Catal.* 307: 103–110.
74 Chen, Z., Feng, Y., Tong, T., and Zeng, A. (2014). *Appl. Catal. A, Gen.* 482: 92–98.
75 Matsukata, M., Ogura, M., Osaki, T. et al. (1999). *Top Catal.* 9: 77–92.
76 Koehle, M., Zhang, Z., Goulas, K.A. et al. (2018). *Appl. Catal. A Gen.* 564: 90–101.
77 Chandra Shekara, B.M., Jai Prakash, B.S., and Bhat, Y.S. (2012). *J. Catal.* 290: 101–107.
78 Derouane, E.G., Schmidt, I., Lachas, H., and Christensen, C.J.H. (2004). *Catal. Letters* 95: 13–17.
79 Silva, D.S.A., Castelblanco, W.N., Piva, D.H. et al. (2020). *Mol. Catal.* 492: 111026.
80 Guidotti, M., Coustard, J.M., Magnoux, P., and Guisnet, M. (2007). *Pure Appl. Chem.* 79: 1833–1838.
81 Elvas-Leitão, R., Martins, F., Borbinha, L. et al. (2020). *Molecules* 25: 5682.
82 Linares, N., Cirujano, F.G., De Vos, D.E., and García-Martínez, J. (2019). *Chem. Commun.* 55: 12869–12872.
83 Bohström, Z. and Holmberg, K. (2013). *J. Mol. Catal. A Chem.* 366: 64–73.
84 Wang, L., Zhang, Z., Yin, C. et al. (2010). *Microporous Mesoporous Mater.* 131: 58–67.
85 Kim, J.C., Cho, K., Lee, S., and Ryoo, R. (2015). *Catal. Today* 243: 103–108.
86 Todeschini, R. and Consonni, V. (2010). *Molecular Descriptors for Chemoinformatics*, Vol. 2. Hoboken: Wiley Blackwell.
87 Madzhidov, T.I., Rakhimbekova, A., Afonina, V.A. et al. (2021). *Mendeleev Commun.* 31: 769–780.
88 Martins, F., Santos, S., Ventura, C. et al. (2014). *Eur. J. Med. Chem.* 81: 119–138.
89 Winkler, D.A. (2016). *Toxicol. Appl. Pharmacol.* 299: 96–100.
90 Santana, R., Onieva, E., Zuluaga, R. et al. (2021). *Curr. Top. Med. Chem.* 21: 828–838.
91 Salahinejad, M. (2015). *Curr. Top. Med. Chem.* 15: 1868–1886.
92 Berhanu, W.M., Pillai, G.G., Oliferenko, A.A., and Katritzky, A.R. (2012). *Chempluschem* 77: 507–517.
93 Cherkasov, A., Muratov, E.N., Fourches, D. et al. (2014). *J. Med. Chem.* 57: 4977–5010.
94 Quintero, F.A., Patel, S.J., Muñoz, F., and Sam Mannan, M. (2012). *Ind. Eng. Chem. Res.* 51: 16101–16115.
95 Awfa, D., Ateia, M., Mendoza, D., and Yoshimura, C. (2021). *ACS ES&T Water* 1: 498–517.
96 Villaverde, J.J., Sevilla-Morán, B., López-Goti, C. et al. (2018). *Sci. Total Environ.* 634: 1530–1539.

97 Roy, J., Ghosh, S., Ojha, P.K., and Roy, K. (2019). *Environ. Sci. Nano* 6: 224–247.
98 Sakaguchi, K., Matsui, M., and Mizukami, F. (2005). *Appl. Microbiol. Biotechnol.* 67: 306–311.
99 Meier, M.A.R. and Barner-Kowollik, C. (2019). *Adv. Mater.* 31: 1806027.
100 Martins, F., Ventura, C., Santos, S., and Viveiros, M. (2014). *Curr. Pharm. Des.* 20: 4427–4454.
101 Yee, L.C. and Wei, Y.C. (2012). Current Modeling Methods Used in QSAR/QSPR. In: *Statistical Modelling of Molecular Descriptors in QSAR/QSPR*, Vol. 2 (eds. M. Dehmer, K. Varmuza, and D. Bonchev), 1–31. Hoboken: John Wiley and Sons, Ltd.
102 Yap, C., Li, H., Ji, Z., and Chen, Y. (2007). *Mini-Reviews Med. Chem.* 7: 1097–1107.
103 Fujita, T. and Winkler, D.A. (2016). *J. Chem. Inf. Model.* 56: 269–274.
104 Tripathi, N., Goshisht, M.K., Sahu, S.K., and Arora, C. (2021). *Mol. Divers. 2021 253* 25: 1643–1664.
105 Jiménez-Luna, J., Grisoni, F., Weskamp, N., and Schneider, G. (2021). *Expert Opin. Drug Discov.* 16: 949–959.
106 Hammett, L.P. (2002). *Chem. Rev.* 17: 125–136.
107 Gonçalves, R.M.C. and Albuquerque, L.M.P.C. (2001). *J. Phys. Org. Chem.* 14: 731–736.
108 Moreira, L., Reis, M., Elvas-Leitão, R. et al. (2019). *J. Mol. Liq.* 291: 1–8.
109 Abraham, M.H., Doherty, R.M., Kamlet, M.J. et al. (1987). *J. Chem. Soc. Perkin Trans.* 2: 913–920.
110 Laurence, C., Mansour, S., Vuluga, D., and Legros, J. (2020). *J. Phys. Org. Chem.* 33: e4067.
111 Aliaga, C., Domínguez, M., Rojas, P., and Rezende, M.C. (2020). *J. Mol. Liq.* 312: 113362.
112 Martins, F., Moreira, L., Nunes, N., and Leitão, R.E. (2010). *J. Therm. Anal. Calorim.* 100: 483–491.
113 Reis, M., Moreira, L., Nunes, N. et al. (2011). *J. Therm. Anal. Calorim.* 108: 761–767.
114 Monteiro, C., Ventura, C., and Martins, F. (2013). *J. Environ. Manage.* 122: 99–104.
115 Abraham, M.H. (1993). *Chem. Soc. Rev.* 22: 73–83.
116 Abraham, M.H., Ibrahim, A., and Zissimos, A.M. (2004). *J. Chromatogr. A* 1037: 29–47.
117 Andzelm, J.W., Alvarado-Swaisgood, A.E., Axe, F.U. et al. (1999). *Catal. Today* 50: 451–477.
118 Corma, A., Serra, J.M., Serna, P., and Moliner, M. (2005). *J. Catal.* 2: 335–341.
119 Begum, S. and Achary, P.G.R. (2018). *Int. J. Quant. Struct. Relationships* 3: 36–48.
120 Chasing, P., Maitarad, P., Wu, H. et al. (2018). *Catalysts* 8: 422.
121 Dos Santos, V.H.J.M., Pontin, D., Rambo, R.S., and Seferin, M. (2020). *J. Am. Oil Chem. Soc.* 97: 817–837.
122 Gensch, T., Dos Passos Gomes, G., Friederich, P. et al. (2022). *J. Am. Chem. Soc.* 144: 1205–1217.
123 Van Der Linden, J.B., Ras, E.J., Hooijschuur, S.M. et al. (2005). *QSAR Comb. Sci.* 24: 94–98.
124 Köhler, K., Heidenreich, R.G., Soomro, S.S., and Pröckl, S.S. (2008). *Adv. Synth. Catal.* 350: 2930–2936.
125 Hemmateenejad, B., Sanchooli, M., and Mehdipour, A. (2009). *J. Phys. Org. Chem.* 22: 613–618.
126 Aguado-Ullate, S., Guasch, L., Urbano-Cuadrado, M. et al. (2012). *Catal. Sci. Technol.* 2: 1694–1704.
127 Li, B., Dong, Y., and Ding, Z. (2013). *J. Environ. Sci.* 25: 1469–1476.
128 Rangarajan, S., Bhan, A., and Daoutidis, P. (2014). *Appl. Catal. B Environ.* 145: 149–160.
129 Shahin, A. and Ganji, S. (2016). *Curr. Drug Discov. Technol.* 13: 232–253.
130 Yang, W., Yi, J., Ma, Z., and Sun, W.H. (2017). *Catal. Commun.* 101: 40–43.
131 Prabhu, S., Murugan, G., Cary, M. et al. (2020). *Mater. Res. Express* 7: 055006.
132 Xie, P., Pu, T., Aranovich, G. et al. (2021). *Nat. Catal.* 4: 144–156.
133 Arockiaraj, M., Paul, D., Klavžar, S. et al. (2022). *J. Mol. Struct.* 1250: 131798.
134 Teixeira, C., Ventura, C., Gomes, J.R.B. et al. (2020). *Molecules* 25: 456.
135 (Q)SARs: guidance documents and reports. OCDE. https://www.oecd.org/fr/env/ess/risques/guidancedocumentsandreportsrelatedtoqsars.htm (accessed 31 March 2022).
136 Chirico, N. and Gramatica, P. (2011). *J. Chem. Inf. Model.* 51: 2320–2335.
137 Chirico, N. and Gramatica, P. (2012). *J. Chem. Inf. Model.* 52: 2044–2058.
138 Wang, Y., Sun, Y., Lancelot, C. et al. (2015). *Microporous Mesoporous Mater.* 206: 42–51.